Progress in Nonlinear Differential Equations and Their Applications
Volume 55

Evolution Equations: Applications to Physics, Industry, Life Sciences and Economics

EVEQ2000 Conference in Levico Terme (Trento, Italy), October 30–November 4, 2000

Mimmo Iannelli
Günter Lumer
Editors

Springer Basel AG

Editors' addresses:

Mimmo Iannelli
Department of Mathematics
University of Trento
via Sommarive 14
38050 Povo, Trento
Italy
e-mail: iannelli@scienze.unitn.it

Günter Lumer
Institute of Mathematics and Informatics
University of Mons-Hainaut
20, place du Parc
7000 Mons
Belgium
e-mail: dufour@umh.ac.be

2000 Mathematics Subject Classification: 34-xx, 35-xx, 47Dxx, 49Lxx, 58Jxx, 60Jxx, 76Mxx, 76Nxx, 91Bxx, 92Bxx, 92Cxx

A CIP catalogue record for this book is available from the Library of Congress, Washington D.C., USA

Bibliographic information published by Die Deutsche Bibliothek
Die Deutsche Bibliothek lists this publication in the Deutsche Nationalbibliografie; detailed bibliographic data is available in the Internet at <http://dnb.ddb.de>.

ISBN 978-3-0348-9433-3 ISBN 978-3-0348-8085-5 (eBook)
DOI 10.1007/978-3-0348-8085-5

Originally published by Birkhäuser Verlag in 2003
Softcover reprint of the hardcover 1st edition 2003

Printed on acid-free paper produced of chlorine-free pulp. TCF ∞

ISBN 978-3-0348-9433-3

9 8 7 6 5 4 3 2 1 www.birkhasuer-science

Table of Contents

Preface

The seventh International Conference on Evolution Equations and their main areas of Applications (where the emphasis evolves as time and problems change) was held October 30 to November 4 at the CIRM (Centro Internazionale per la Ricerca Matematica) in Trento, Italy.

In keeping with the basic principles and the recent tendencies governing these International Conferences, it brought together many of the world's leading experts in the fields mentioned, with particular effort on facilitating the interaction of established scientists and emerging young promising researchers, as well as the interaction of pure and applied specialists. In the latter directions, emphasis was extended here to include in addition to Physical and Life Sciences, also Industry and Economics.

Topics among the recent advances treated here concern new developments in: moving boundary problems, asymptotics in non-linear Volterra equations and other asymptotics related developments, Poincaré inequality on stratified sets, time operator and Markov processes in physics related advances, behavior of granular matter, stochastic aspects of Hamilton-Jacobi-Bellman equation, very general Paley-Wiener results applied to both classical and generalized functions, Ornstein-Uhlenbeck operators and processes, quasilinear PDEs with memory operators, semi-group approach in economics (pricing theory) and other semi-group related developments, convolution-evolution equation in aeroelasticity, new developments in the study of age-structured models, new developments in maximal regularity.

This volume, while based on, and inspired by, the material presented at the conference is, to be exact, not a mere "proceedings". The contributions submitted in 2001 and 2002, the latest up to Fall 2002, contain also much material developed after the conference by continuing research (incorporating additional new results), also incorporating significant modifications suggested and required by the referees. The volume also includes some up to date surveys of the state of the art in some of the areas considered, refereed as well. Indeed the material has been rigorously refereed (not all contributions submitted were accepted and in many cases more or less ample modifications, requested by the referees – from inside or outside the group of participants – were made).

The organizers thank the following institutions for providing support for the conference:

- CIRM (Centro Internazionale per la Ricerca Matematica, Trento)
- EU (Grant HPCFCT-2000-00049: High Level Scientific Conferences Program)
- Mathematics Department, University of Trento
- INDAM-GNAFA (Istituto Nazionale di Alta Matematica, Gruppo Nazionale Analisi Funzionale e Applicazioni).

We thank our colleagues and collaborators of the scientific committee, in particular P. Dufour, M. Miranda, A. Valli, L. Weis for assistance and advice. We thank L. Bouchez and A. Micheletti for technical and secretarial help. Finally, we thank the contributors, referees, Birkhäuser Verlag, particularly T. Hintermann and T. Hempfling for their help and cooperation in making possible and producing this volume, which we hope will in turn be valuable to the Scientific Community.

Trento, December 18, 2002

Mimmo Iannelli, Günter Lumer

Progress in Nonlinear Differential Equations
and Their Applications, Vol. 55, 1–12

Evolution Problems Associated to Linear Growth Functionals: The Dirichlet Problem

Fuensanta Andreu, Vicent Caselles, and José M. Mazón

To the Memory of Ph. Bénilan

1. Introduction and preliminaries

Let Ω be a bounded set in $\mathbb{R}^N$ with Lipschitz continuous boundary $\partial\Omega$. We are interested in the problem

$$\begin{cases} \dfrac{\partial u}{\partial t} = div\, \mathbf{a}(x, Du) & \text{in} \quad Q = (0,\infty) \times \Omega \\ u(t,x) = \varphi(x) & \text{on} \quad S = (0,\infty) \times \partial\Omega \\ u(0,x) = u_0(x) & \text{in} \quad x \in \Omega \end{cases} \tag{1.1}$$

where $\varphi \in L^1(\partial\Omega)$, $u_0 \in L^1(\Omega)$ and $\mathbf{a}(x,\xi) = \nabla_\xi f(x,\xi)$, f being a function with linear growth as $\|\xi\| \to \infty$.

A typical example of a function $f(x,\xi)$ satisfying the conditions we need is the nonparametric area integrand $f(x,\xi) = \sqrt{1 + \|\xi\|^2}$. Problem (1.1) for this particular f, that is, the time-dependent minimal surface equation, has been studied in [12] and [17]. Another examples of problems of type (1.1) included in our case, are the following: The evolution problem for plastic antiplanar shear, studied in [19], which corresponds to the plasticity functional f given by

$$f(\xi) = \begin{cases} \frac{1}{2}\|\xi\|^2 & \text{if} \quad \|\xi\| \leq 1 \\ \|\xi\| - \frac{1}{2} & \text{if} \quad \|\xi\| \geq 1 \end{cases}$$

and the evolution problems associated with the Lagrangians:

$$f(x,\xi) = \sqrt{1 + x^2 + \|\xi\|^2},$$

which was considered by S. Bernstein ([9]). On the other hand, problem (1.1) is studied in [14] for some Lagrangians f, which do not include the nonparametric area integrand, but instead include the plasticity functional and the total variation flow, that is, the case $f(\xi) = \|\xi\|$. Now, the concept of solution given in [14] is the one obtained by considering the abstract Cauchy problem in $L^2(\Omega)$ associated to the relaxed energy, but the subdifferential of the energy functional is not characterized. For the particular case of the total variation flow, we give in [3] a different approach to the Dirichlet problem. There, we studied the problem in the framework of the L^1-theory, and we characterized the subdifferential in $L^2(\Omega)$ of its relaxed energy (we refer also to [2] where we treated the L^1-theory for the Neumann problem for the total variational flow).

In general, problem (1.1) does not have a classical solution. Our aim is to introduce a concept of solution of the Dirichlet problem (1.1), for which existence and uniqueness for initial data in $L^1(\Omega)$ is obtained. To do that, first we characterize the subdifferential of the energy associated with the problem and we use the nonlinear semigroup theory to get existence and uniqueness of strong solution when the initial data are in $L^2(\Omega)$. After this, we use again the nonlinear semigroup theory to get existence of entropy solution and using the Kruzhkov's methos of doubling variables both in space and time we get the uniqueness.

In order to consider the relaxed energy we recall the definition of function of measure (see for instance, [7] or [12]). Let $g : \Omega \times \mathbb{R}^N \to \mathbb{R}$ be a Carathéodory function such that

$$|g(x,\xi)| \leq M(1 + \|\xi\|) \qquad \forall\, (x,\xi) \in \Omega \times \mathbb{R}^N, \tag{1.2}$$

for some constant $M \geq 0$. Furthermore, we assume that g possesses an asymptotic function, i.e., for almost all $x \in \Omega$ there exists the finite limit

$$\lim_{t\to 0^+} tg\left(x, \frac{\xi}{t}\right) = g^0(x,\xi). \tag{1.3}$$

It is clear that the function $g^0(x,\xi)$ is positively homogeneous of degree one in ξ, i.e.,

$$g^0(x, s\xi) = sg^0(x,\xi) \qquad \text{for all } x, \xi \text{ and } s > 0.$$

We denote by $\mathcal{M}(\Omega, \mathbb{R}^N)$ the set of all $\mathbb{R}^N$-valued bounded Radon measures on Ω. Given $\mu \in \mathcal{M}(\Omega, \mathbb{R}^N)$, we consider its Lebesgue decomposition

$$\mu = \mu^a + \mu^s,$$

where μ^a is the absolutely continuous part of μ with respect to de Lebesgue measure $\mathcal{L}^N$ of $\mathbb{R}^N$, and μ^s is singular with respect to $\mathcal{L}^N$. We denote by $\mu^a(x)$ the density of the measure μ^a with respect to $\mathcal{L}^N$ and by $(d\mu^s/d|\mu|^s)(x)$ the density of μ^s with respect to $|\mu|^s$.

Given $\mu \in \mathcal{M}(\Omega, \mathbb{R}^N)$, we define $\tilde{\mu} \in \mathcal{M}(\Omega, \mathbb{R}^{N+1})$ by

$$\tilde{\mu}(B) := \left(\mu(B), \mathcal{L}^N(B)\right),$$

for every Borel set $B \subset \mathbb{R}^N$. Then, we have

$$\tilde{\mu} = \tilde{\mu}^a + \tilde{\mu}^s = \tilde{\mu}^a(x)\mathcal{L}^N + \tilde{\mu}^s = (\mu^a(x), \chi_\Omega)\mathcal{L}^N + (\mu^s, 0).$$

Hence, we have

$$|\tilde{\mu}^s| = |\mu^s|, \qquad \frac{d\tilde{\mu}^s}{d|\tilde{\mu}^s|} = \left(\frac{d\mu^s}{d|\mu^s|}, 0\right) \quad |\mu^s| - \text{a.e.}$$

For $\mu \in \mathcal{M}(\Omega, I\!R^N)$ and g satisfying the above conditions, we define the measure $g(x,\mu)$ on Ω as

$$\int_B g(x,\mu) := \int_B g(x, \mu^a(x))\, dx + \int_B g^0\Big(x, \frac{d\mu^s}{d|\mu|^s}(x)\Big)\, d|\mu|^s \tag{1.4}$$

for all Borel set $B \subset \Omega$. In formula (1.4) we may write $(d\mu/d|\mu|)(x)$ instead of $(d\mu^s/d|\mu|^s)(x)$, because the two functions are equal $|\mu|^s$-a.e.

Another way of writing the measure $g(x,\mu)$ is the following. Let us consider the function $\tilde{g} : \Omega \times I\!R^N \times [0,+\infty[\to I\!R$ defined as

$$\tilde{g}(x,\xi,t) := \begin{cases} g(x, \frac{\xi}{t})t & \text{if} \quad t > 0 \\ g^0(x,\xi) & \text{if} \quad t = 0 \end{cases} \tag{1.5}$$

As it is proved in [7], if g is a Carathéodory function satisfying (1.2), then one has

$$\int_B g(x,\mu) = \int_B \tilde{g}\Big(x, \frac{d\mu}{d\alpha}(x), \frac{d\mathcal{L}^N}{d\alpha}(x)\Big)\, d\alpha, \tag{1.6}$$

where α is any positive Borel measure such that $|\mu| + \mathcal{L}^N \ll \alpha$.

Due to the linear growth condition on the Lagrangian, the natural energy space to study (1.1) is the space of functions of bounded variation. Let us recall several facts concerning functions of bounded variation (for further information concerning functions of bounded variation we refer to [13], [20] or [1]).

A function $u \in L^1(\Omega)$ whose partial derivatives in the sense of distributions are measures with finite total variation in Ω is called a function of bounded variation. The class of such functions will be denoted by $BV(\Omega)$. Thus $u \in BV(\Omega)$ if and only if there are Radon measures $\mu_1, \ldots, \mu_N$ defined in Ω with finite total mass in Ω and

$$\int_\Omega u D_i \varphi dx = -\int_\Omega \varphi d\mu_i \tag{1.7}$$

for all $\varphi \in C_0^\infty(\Omega)$, $i = 1, \ldots, N$. Thus the gradient of u is a vector valued measure with finite total variation

$$\| Du \| = \sup\{\int_\Omega u\, div\, \varphi\, dx : \varphi \in C_0^\infty(\Omega, I\!R^N), |\varphi(x)| \leq 1 \text{ for } x \in \Omega\}. \tag{1.8}$$

The space $BV(\Omega)$ is endowed with the norm

$$\| u \|_{BV} = \| u \|_{L^1(\Omega)} + \| Du \| . \tag{1.9}$$

For $u \in BV(\Omega)$, the gradient Du is a Radon measure that decomposes into its absolutely continuous and singular parts $Du = D^a u + D^s u$. Then $D^a u = \nabla u\, \mathcal{L}^N$ where ∇u is the Radon-Nikodym derivative of the measure Du with respect to the Lebesgue measure $\mathcal{L}^N$. There is also the polar decomposition $D^s u = \overrightarrow{D^s u} |D^s u|$ where $|D^s u|$ is the total variation measure of $D^s u$.

We shall need several results from [6] (see also [15]). Following [6], let

$$X(\Omega) = \{z \in L^\infty(\Omega, I\!R^N) : div(z) \in L^1(\Omega)\}. \tag{1.10}$$

If $z \in X(\Omega)$ and $w \in BV(\Omega) \cap L^\infty(\Omega)$ we define the functional $(z, Dw) : C_0^\infty(\Omega) \to \mathbb{R}$ by the formula

$$< (z, Dw), \varphi >= - \int_\Omega w \, \varphi \, div(z) \, dx - \int_\Omega w \, z \cdot \nabla\varphi \, dx. \tag{1.11}$$

Then (z, Dw) is a Radon measure in Ω,

$$\int_\Omega (z, Dw) = \int_\Omega z \cdot \nabla w \, dx \qquad \forall \, w \in W^{1,1}(\Omega) \cap L^\infty(\Omega) \tag{1.12}$$

and (z, Dw) is absolutely continuous with respect to $\|Dw\|$

By writing

$$z \cdot D^s u := (z, Du) - (z \cdot \nabla u) \; d\mathcal{L}^N,$$

we see that $z \cdot D^s u$ is a bounded measure. Furthermore, in [15] it is proved that $z \cdot D^s u$ is absolutely continuous with respect to $|D^s u|$ (and, thus, it is a singular measure with respect to $\mathcal{L}^N$).

In [6], a weak trace on $\partial\Omega$ of the normal component of $z \in X(\Omega)$ is defined. Concretely, it is proved that there exists a linear operator $\gamma : X(\Omega) \to L^\infty(\partial\Omega)$ such that

$$\|\gamma(z)\|_\infty \leq \|z\|_\infty$$

$$\gamma(z)(x) = z(x) \cdot \nu(x) \quad \text{ for all } x \in \partial\Omega \text{ if } z \in C^1(\overline{\Omega}, \mathbb{R}^N).$$

We shall denote $\gamma(z)(x)$ by $[z, \nu](x)$. Moreover, the following *Green's formula*, relating the function $[z, \nu]$ and the measure (z, Dw), for $z \in X(\Omega)$ and $w \in BV(\Omega) \cap L^\infty(\Omega)$, is established:

$$\int_\Omega w \; div(z) \; dx + \int_\Omega (z, Dw) = \int_{\partial\Omega} [z, \nu] w \; dH^{N-1}. \tag{1.13}$$

Let g be a function satisfying (1.2). Then for every $u \in BV(\Omega)$ we have the measure $g(x, Du)$ defined by

$$\int_B g(x, Du) = \int_B g(x, \nabla u(x)) \; dx + \int_B g^0(x, \overrightarrow{D^s u}(x)) \; d|D^s u|$$

for all Borel set $B \subset \Omega$. If we assume that Ω has a Lipschitz boundary, and that $g(x, \xi)$ is defined also for $x \in \partial\Omega$, we may consider the functional G in $BV(\Omega)$ defined by

$$G(u) := \int_\Omega g(x, Du) + \int_{\partial\Omega} g^0\left(x, \nu(x)[\varphi(x) - u(x)]\right) \; dH^{N-1}, \tag{1.14}$$

where $\varphi \in L^1(\partial\Omega)$ is a given function and ν is the outer unit normal to $\partial\Omega$. It is proved in [7] that, if $\tilde{g}(x, \xi, t)$ is continuous on $\overline{\Omega} \times \mathbb{R}^N \times [0, +\infty[$ and convex in (ξ, t) for each fixed $x \in \overline{\Omega}$, then G is the greatest functional on $BV(\Omega)$ which is lower-semicontinuous with respect to the $L^1(\Omega)$-convergence and satisfies $G(u) \leq \int_\Omega g(x, \nabla u(x)) \; dx$ for all functions $u \in C^1(\Omega) \cap W^{1,1}(\Omega)$ with $u = \varphi$ on $\partial\Omega$.

2. Strong solution for data in $L^2(\Omega)$

In this section we define the concept of solution for the Dirichlet problem (1.1) for initial data in $L^2(\Omega)$ and we state the existence and uniqueness result for this type of solutions.

Here we assume that Ω is an open bounded set in $\mathbb{R}^N$, $N \geq 2$, with boundary $\partial\Omega$ of class C^1, and the Lagrangian $f : \overline{\Omega} \times \mathbb{R}^N \to \mathbb{R}$ satisfies the following assumptions, which we shall refer collectively as (H):

(H_1) f is continuous on $\overline{\Omega} \times \mathbb{R}^N$ and is a convex diffentiable function of ξ with continuous gradient for each fixed $x \in \Omega$. Further we require f to satisfy the linear growth condition

$$C_0\|\xi\| - C_1 \leq f(x,\xi) \leq M(\|\xi\| + C_2). \tag{2.1}$$

for some positive constants C_0, C_1, C_2. Moreover, f^0 exists and $f^0(x,-\xi) = f^0(x,\xi)$ for all $\xi \in \mathbb{R}^N$ and all $x \in \overline{\Omega}$.

(H_2) $\tilde{f}(x,\xi,t)$ is continuous on $\overline{\Omega} \times \mathbb{R}^N \times [0,+\infty[$ and convex in (ξ,t) for each fixed $x \in \overline{\Omega}$.

We consider the function $\mathbf{a}(x,\xi) = \nabla_\xi f(x,\xi)$ associated to the Lagrangian f. By the convexity of f

$$\mathbf{a}(x,\xi) \cdot (\eta - \xi) \leq f(x,\eta) - f(x,\xi), \tag{2.2}$$

and the following monotonicity condition is satisfied

$$(\mathbf{a}(x,\eta) - \mathbf{a}(x,\xi)) \cdot (\eta - \xi) \geq 0. \tag{2.3}$$

Moreover, it is easy to see that

$$|\mathbf{a}(x,\xi)| \leq M \qquad \forall\ (x,\xi) \in \Omega \times \mathbb{R}^N. \tag{2.4}$$

We consider the function $h : \Omega \times \mathbb{R}^N \to \mathbb{R}$ defined by

$$h(x,\xi) := \mathbf{a}(x,\xi) \cdot \xi.$$

¿From (2.2) and (2.1), it follows that

$$C_0\|\xi\| - D_1 \leq h(x,\xi) \leq M\|\xi\| \tag{2.5}$$

for some positive constant D_1.

We assume that

(H_3) $h(x,\xi) \geq 0$ for all $x \in \Omega$ and $\xi \in \mathbb{R}^N$, h^0 exists and the function $\tilde{h}$ is continuous on $\overline{\Omega} \times \mathbb{R}^N \times [0,+\infty[$.

We need to consider the mapping $\mathbf{a}^\infty$ defined by

$$\mathbf{a}^\infty(x,\xi) := \lim_{t\to+\infty} \mathbf{a}(x,t\xi).$$

Observe that

$$h^0(x,\xi) = \mathbf{a}^\infty(x,\xi) \cdot \xi \quad \text{and} \quad C_0\|\xi\| \leq h^0(x,\xi) \leq M\|\xi\|.$$

(H_4) $\mathbf{a}^\infty(x,\xi) = \nabla_\xi f^0(x,\xi)$ for all $\xi \neq 0$ and all $x \in \overline{\Omega}$.

In particular, as a consequence of Euler's Theorem, we have

$$f^0(x,\xi) = \mathbf{a}^\infty(x,\xi)\cdot\xi = h^0(x,\xi),$$

for all $\xi \in I\!R^N$ and all $x \in \overline{\Omega}$.

(H_5) $\mathbf{a}(x,\xi)\cdot\eta \leq h^0(x,\eta)$ for all $\xi,\eta \in I\!R^N$, and all $x \in \overline{\Omega}$.

Either from (H_4) or (H_5) it follows that $\mathbf{a}^\infty(x,\xi)\cdot\eta \leq h^0(x,\eta)$ for all $\xi,\eta \in I\!R^N$, $\xi \neq 0$, and all $x \in \overline{\Omega}$. Indeed, it suffices to replace ξ by $t\xi$ in (H_5) and let $t \to +\infty$.

Definition 1. Let $\varphi \in L^1(\partial\Omega)$ and $u_0 \in L^2(\Omega)$. A measurable function $u : (0,T)\times\Omega \to I\!R$ is a *solution* of (1.1) in $Q_T = (0,T)\times\Omega$ if $u \in C([0,T], L^2(\Omega))$, $u(0) = u_0$, $u'(t) \in L^2(\Omega)$, $u(t) \in BV(\Omega)\cap L^2(\Omega)$, $\mathbf{a}(x,\nabla u(t)) \in X(\Omega)$ a.e. $t \in [0,T]$, and for almost all $t \in [0,T]$ $u(t)$ satisfies:

$$u'(t) = div(\mathbf{a}(x,\nabla u(t))) \quad \text{in } \mathcal{D}'(\Omega) \tag{2.6}$$

$$\mathbf{a}(x,\nabla u(t))\cdot D^s u(t) = f^0(x, D^s u(t)) \tag{2.7}$$

$$[\mathbf{a}(x,\nabla u(t)),\nu] \in \mathrm{sign}(\varphi - u(t)) f^0(x,\nu(x)) \quad H^{N-1} - a.e. \text{ on } \partial\Omega. \tag{2.8}$$

The main result of this section is the following:

Theorem 1. *Let $\varphi \in L^1(\partial\Omega)$ and assume we are under assumptions (H). Given $u_0 \in L^2(\Omega)$, there exists a unique solution u of (1.1) in Q_T for every $T > 0$ such that $u(0) = u_0$.*

To prove Theorem 1 we shall use the nonlinear semigroup theory ([10]). For $\varphi \in L^1(\partial\Omega)$ we define the energy functional associated with the problem (1.1) $\Phi_\varphi : L^2(\Omega) \to [0,+\infty]$ by

$$\Phi_\varphi(u) := \begin{cases} \displaystyle\int_\Omega f(x,Du) + \int_{\partial\Omega} f^0(x,\nu(x)[\varphi - u])\, dH^{N-1} & \\ & \text{if } \ u \in BV(\Omega)\cap L^2(\Omega) \\ +\infty & \text{if } \ u \in L^2(\Omega)\setminus BV(\Omega). \end{cases}$$

Note that, on the boundary, the integrand can be written in the form

$$f^0(x,\nu(x)[\varphi - u]) = |\varphi - u| f^0(x,\nu(x)).$$

Functional Φ_φ is clearly convex and has the form given in (1.14). Then, as a consequence of the Anzellotti's result ([7]) we have that Φ_φ is lower-semicontinuous. Therefore, the subdifferential $\partial\Phi_\varphi$ of Φ_φ, i.e., the operator in $L^2(\Omega)$ defined by

$$v \in \partial\Phi_\varphi(u) \iff \Phi_\varphi(w) - \Phi_\varphi(u) \geq \int_\Omega v(w-u)\, dx, \quad \forall\, w \in L^2(\Omega)$$

is a maximal monotone operator in $L^2(\Omega)$. Consequently, the existence and uniqueness of a solution of the abstract Cauchy problem

$$\begin{cases} u'(t) + \partial\Phi_\varphi(u(t)) \ni 0 & t \in]0,\infty[\\ u(0) = u_0 & u_0 \in L^2(\Omega) \end{cases} \tag{2.9}$$

follows immediately from the nonlinear semigroup theory (see [10]). Now, to get the full strength of the abstract result derived from semigroup theory we need to characterize $\partial\Phi_\varphi$. To get this characterization, we introduce the following operator $\mathcal{B}_\varphi$ in $L^2(\Omega)$.

$(u,v) \in \mathcal{B}_\varphi \iff u \in BV(\Omega) \cap L^2(\Omega), v \in L^2(\Omega)$ and $\mathbf{a}(x,\nabla u) \in X(\Omega)$ satisfies :

$$-v = div\ \mathbf{a}(x,\nabla u) \qquad \text{in } \mathcal{D}'(\Omega) \tag{2.10}$$

$$\mathbf{a}(x,\nabla u) \cdot D^s u = f^0(x, D^s u) = f^0(x, \overrightarrow{D^s u})|D^s u|, \tag{2.11}$$

$$[\mathbf{a}(x,\nabla u), \nu] \in \text{sign } (\varphi - u) f^0(x, \nu(x)) \qquad H^{N-1} - \text{a.e.} \tag{2.12}$$

Theorem 2. *Let $\varphi \in L^1(\partial\Omega)$. Assume we are under assumptions (H), then the operator $\partial\Phi_\varphi$ has dense domain in $L^2(\Omega)$ and*

$$\partial\Phi_\varphi = \mathcal{B}_\varphi.$$

Theorem 1 follows from Theorem 2 and the theory of nonlinear semigroups in $L^2(\Omega)$ genereted by subdifferentials.

3. Existence and uniqueness results for L^1-data

In this section we study the problem (1.1) for initial data in $L^1(\Omega)$. We assume that the Lagrangian $f : \overline{\Omega} \times \mathbb{R}^N \to \mathbb{R}$ satisfies the assumptions (H) given in Secction 2. Moreover, here we need to consider the following assumption:

(H_6) We assume that

$$|a(x,\xi) - a(y,\xi)| \leq \omega(\|x-y\|) \tag{3.1}$$

for all $x, y \in \Omega$, and all $\xi \in \mathbb{R}^N$, where $\omega(r)$ is a modulus of continuity. ¿From the definition of $\mathbf{a}^\infty$

$$|\mathbf{a}^\infty(x,\xi) - \mathbf{a}^\infty(y,\xi)| \leq \omega(\|x-y\|).$$

Hence

$$|h^0(x,\xi) - h^0(y,\xi)| = |\mathbf{a}^\infty(x,\xi)\cdot\xi - \mathbf{a}^\infty(y,\xi)\cdot\xi| \leq \omega(\|x-y\|)\|\xi\|. \tag{3.2}$$

Remark 1. Let us remark that assumption (H_6) is only needed to prove uniqueness. The Lipschitz continuity in x of the flux is a common assumption to prove uniqueness of Kruzkov's solutions of scalar conservation laws ([16]).

To make precise our notion of solution we need to introduce a weak trace on $\partial\Omega$ of the normal component of certain vector fields in Ω. We define the space

$$Z(\Omega) := \{(z,\xi) \in L^\infty(\Omega, I\!R^N) \times BV(\Omega)^* \ : \ div(z) = \xi \ in \ \mathcal{D}'(\Omega)\}.$$

We denote $R(\Omega) := W^{1,1}(\Omega) \cap L^\infty(\Omega) \cap C(\Omega)$. For $(z,\xi) \in Z(\Omega)$ and $w \in R(\Omega)$ we define

$$\langle (z,\xi), w\rangle_{\partial\Omega} := \langle \xi, w\rangle_{BV(\Omega)^*, BV(\Omega)} + \int_\Omega z \cdot \nabla w.$$

Then, working as in the proof of Theorem 1.1. in [6], we obtain that if $w, v \in R(\Omega)$ and $w = v$ on $\partial\Omega$ one has

$$\langle (z,\xi), w\rangle_{\partial\Omega} = \langle (z,\xi), v\rangle_{\partial\Omega} \qquad \forall\ (z,\xi) \in Z(\Omega). \tag{3.3}$$

As a consequence of (3.3), we can give the following definition: Given $u \in BV(\Omega) \cap L^\infty(\Omega)$ and $(z,\xi) \in Z(\Omega)$, we define $\langle (z,\xi), u\rangle_{\partial\Omega}$ by setting

$$\langle (z,\xi), u\rangle_{\partial\Omega} := \langle (z,\xi), w\rangle_{\partial\Omega}$$

where w is any function in $R(\Omega)$ such that $w = u$ on $\partial\Omega$. Again, working as in the proof of Theorem 1.1. in [6], we can prove that for every $(z,\xi) \in Z(\Omega)$ there exists $M_{z,\xi} > 0$ such that

$$|\langle (z,\xi), u\rangle_{\partial\Omega}| \leq M_{z,\xi} \|u\|_{L^1(\partial\Omega)} \qquad \forall\ u \in BV(\Omega) \cap L^\infty(\Omega). \tag{3.4}$$

Now, taking a fixed $(z,\xi) \in Z(\Omega)$, we consider the linear functional $F : L^\infty(\partial\Omega) \to I\!R$ defined by

$$F(v) := \langle (z,\xi), w\rangle_{\partial\Omega}$$

where $v \in L^\infty(\partial\Omega)$ and $w \in BV(\Omega) \cap L^\infty(\Omega)$ is such that $w_{|\partial\Omega} = v$. By estimate (3.4), there exists $\gamma_{z,\xi} \in L^\infty(\partial\Omega)$ such that

$$F(v) = \int_{\partial\Omega} \gamma_{z,\xi}(x) v(x) \ dH^{N-1}.$$

Consequently there exists a linear operator $\gamma : Z(\Omega) \to L^\infty(\partial\Omega)$, with $\gamma(z,\xi) := \gamma_{z,\xi}$, satisfying

$$\langle (z,\xi), w\rangle_{\partial\Omega} = \int_{\partial\Omega} \gamma_{z,\xi}(x) w(x) \ dH^{N-1} \qquad \forall\ w \in BV(\Omega) \cap L^\infty(\Omega).$$

In case $z \in C^1(\overline{\Omega}, I\!R^N)$, we have $\gamma_z(x) = z(x) \cdot \nu(x)$ for all $x \in \partial\Omega$. Hence, the function $\gamma_{z,\xi}(x)$ is the weak trace of the normal component of (z,ξ). For simplicity of the notation, we shall denote $\gamma_{z,\xi}(x)$ by $[z,\nu](x)$.

We need to consider the space $BV(\Omega)_2$, defined as $BV(\Omega) \cap L^2(\Omega)$ endowed with the norm

$$\|w\|_{BV(\Omega)_2} := \|w\|_{L^2(\Omega)} + \|Du\|.$$

It is easy to see that $L^2(\Omega) \subset BV(\Omega)_2^*$ and

$$\|w\|_{BV(\Omega)_2^*} \leq \|w\|_{L^2(\Omega)} \qquad \forall\ w \in L^2(\Omega). \tag{3.5}$$

Now, it is well known (see [18]) that the dual $\left(L^1(0,T; BV(\Omega)_2)\right)^*$ is isometric to the space $L^\infty(0,T; BV(\Omega)_2^*, BV(\Omega)_2)$ of all weakly* measurable functions f :

$[0,T] \to BV(\Omega)_2^*$, such that $v(f) \in L^\infty([0,T])$, where $v(f)$ denotes the supremum of the set $\{|\langle w, f\rangle| \ : \ \|w\|_{BV(\Omega)_2} \leq 1\}$ in the vector lattice of measurable real functions. Moreover, the dual paring of the isometric is defined by

$$\langle w, f\rangle = \int_0^T \langle w(t), f(t)\rangle \, dt,$$

for $w \in L^1(0,T;BV(\Omega)_2)$ and $f \in L^\infty(0,T;BV(\Omega)_2^*, BV(\Omega)_2)$.

To make precise our notion of solution we need the following definitions.

Definition 2. Let $\Psi \in L^1(0,T,BV(\Omega))$. We say Ψ admits a *weak derivative* in $L^1_w(0,T,BV(\Omega)) \cap L^\infty(Q_T)$ if there is a function $\Theta \in L^1_w(0,T,BV(\Omega)) \cap L^\infty(Q_T)$ such that $\Psi(t) = \displaystyle\int_0^t \Theta(s)ds$, the integral being taken as a Pettis integral.

Definition 3. Let $\xi \in \left(L^1(0,T,BV(\Omega)_2)\right)^*$. We say that ξ is *the time derivative* in the space $\left(L^1(0,T,BV(\Omega)_2)\right)^*$ of a function $u \in L^1((0,T)\times\Omega)$ if

$$\int_0^T < \xi(t), \Psi(t) > dt = -\int_0^T \int_\Omega u(t,x)\Theta(t,x)dxdt$$

for all test functions $\Psi \in L^1(0,T,BV(\Omega))$ which admit a weak derivative $\Theta \in L^1_w(0,T,BV(\Omega)) \cap L^\infty(Q_T)$ and have compact support in time.

Observe that if $w \in L^1(0,T,BV(\Omega)) \cap L^\infty(Q_T)$ and $z \in L^\infty(Q_T, \mathbb{R}^N)$ is such that there exists $\xi \in \left(L^1(0,T,BV(\Omega))\right)^*$ with $div(z) = \xi$ in $\mathcal{D}'(Q_T)$, we can define, associated to the pair (z,ξ), the distribution (z, Dw) in Q_T by

$$\langle (z,Dw), \phi\rangle := -\int_0^T \langle \xi(t), w(t)\phi(t)\rangle - \int_0^T \int_\Omega z(t,x)w(t,x)\nabla_x\phi(t,x) \qquad (3.6)$$

for all $\phi \in \mathcal{D}(Q_T)$.

Definition 4. Let $\xi \in \left(L^1(0,T,BV(\Omega)_2)\right)^*$, $z \in L^\infty(Q_T, \mathbb{R}^N)$. We say that $\xi = div(z)$ in $\left(L^1(0,T,BV(\Omega)_2)\right)^*$ if (z,Dw) is a Radon measure in Q_T with normal boundary values $[z,\nu] \in L^\infty((0,T)\times\partial\Omega)$, such that

$$\int_{Q_T} (z,Dw) + \int_0^T < \xi(t), w(t) > dt = \int_0^T \int_{\partial\Omega} [z(t,x),\nu]w(t,x)dH^{N-1}dt,$$

for all $w \in L^1(0,T,BV(\Omega)) \cap L^\infty(Q_T)$.

We consider the set

$$\mathcal{T} = \{T_k, T_k^+, T_k^- \ : \ k > 0\}.$$

where $T_k(r) = [k-(k-|r|)^+]sign_0(r)$, $k \geq 0$, $r \in \mathbb{R}$.

Our concept of solution for the Dirichlet problem (1.1) is the following.

Definition 5. A measurable function $u : (0,T) \times \Omega \to I\!R$ is an *entropy solution* of (1.1) in $Q_T = (0,T) \times \Omega$ if $u \in C([0,T]; L^1(\Omega))$, $p(u(\cdot)) \in L^1_w(0,T,BV(\Omega))$ for all $p \in \mathcal{T}$, and there exists $\xi \in \left(L^1(0,T,BV(\Omega)_2)\right)^*$ such that:

(i) $(\mathbf{a}(x,\nabla u(t)), \xi(t)) \in Z(\Omega)$ a.e. $t \in [0,T]$,

(ii) ξ is the time derivative of u in $\left(L^1(0,T,BV(\Omega)_2)\right)^*$ in the sense of Definition 3,

(iii) $\xi = div(\mathbf{a}(x,\nabla u(t)))$ in $\left(L^1(0,T,BV(\Omega))\right)^*$ a.e. $t \in [0,T]$ in the sense of Definition 4,

(iv) $[\mathbf{a}(x,\nabla u(t)),\nu] \in \operatorname{sign}\big(p(\varphi) - p(u(t))\big) f^0(x,\nu(x))$ a.e. in $t \in [0,T]$ for all $p \in \mathcal{T}$,

(v) the following inequality is satisfied

$$
\begin{aligned}
-\int_0^T \int_\Omega j(u(t)-l)\eta_t \, dxdt &+ \int_0^T \int_\Omega \eta(t) h(x, Dp(u(t)-l))\, dt \\
&+ \int_0^T \int_\Omega \mathbf{a}(x,\nabla u(t)) \cdot \nabla\eta(t) p(u(t)-l)\, dxdt \\
&\leq \int_0^T \int_{\partial\Omega} [\mathbf{a}(x,\nabla u(t)),\nu]\eta(t) p(u(t)-l)\, dH^{N-1} dt,
\end{aligned}
$$

for all $l \in I\!R$, for all $\eta \in C^\infty(\overline{Q_T})$, with $\eta \geq 0$, $\eta(t,x) = \phi(t)\psi(x)$, being $\phi \in \mathcal{D}(]0,T[)$, $\psi \in C^\infty(\overline{\Omega})$, and $p \in \mathcal{T}$, where $j(r) = \int_0^r p(s)\, ds$.

Our main result is the following existence and uniqueness theorem.

Theorem 3. *Assume we are under asumptions (H) + (H_6) and let $u_0 \in L^1(\Omega)$ and $\varphi \in L^1(\partial\Omega)$. Then there exists a unique entropy solution of (1.1) in $(0,T)\times\Omega$ for every $T > 0$ such that $u(0) = u_0$. Moreover, if $u(t), \hat{u}(t)$ are the entropy solutions corresponding to initial data $u_0, \hat{u}_0$, respectively, then*

$$
\|(u(t)-\hat{u}(t))^+\|_1 \leq \|(u_0-\hat{u}_0)^+\|_1 \quad \text{and} \quad \|u(t)-\hat{u}(t)\|_1 \leq \|u_0-\hat{u}_0\|_1 \tag{3.7}
$$

for all $t \geq 0$.

To prove the existence part of Theorem 3 we use the techniques of completely accretive operators ([8]) and the Crandall-Liggett's semigroup generation Theorem. To define the operator in $L^1(\Omega)$ associated with the problem we need to consider a more general set of truncature functions, concretely, the set $\mathcal{P}$ of all nondecreasing continuous fuctions $p : I\!R \to I\!R$, such that there exists p' except a finite set and $\operatorname{supp}(p')$ is compact. Obviously, $\mathcal{T} \subset \mathcal{P}$.

We define the operator $\mathcal{A}_\varphi$ in $L^1(\Omega)$ by:

$$
\begin{aligned}
(u,v) \in \mathcal{A}_\varphi \iff \; & u, v \in L^1(\Omega), p(u) \in BV(\Omega) \text{ for all } p \in \mathcal{P} \\
& \text{and } \mathbf{a}(x,\nabla u) \in X(\Omega) \text{ satisfies :} \\
& -v = div\, \mathbf{a}(x,\nabla u) \qquad \text{in } \mathcal{D}'(\Omega),
\end{aligned} \tag{3.8}
$$

$$\mathbf{a}(x,\nabla u)\cdot D^s p(u) = f^0(x, D^s p(u)) \qquad \forall\, p\in\mathcal{P}, \tag{3.9}$$

$$[\mathbf{a}(x,\nabla u),\nu] \in \text{sign}\,(p(\varphi)-p(u)) f^0(x,\nu(x)), \quad \forall\, p\in\mathcal{P} \quad H^{N-1}-\text{a.e.} \tag{3.10}$$

In order to prove the existence of an entropy solution of (1.1) for every initial data in $L^1(\Omega)$, we need the following result.

Theorem 4. *Let $\varphi\in L^1(\partial\Omega)$. Assume we are under assumptions (H_1)-(H_5). Then, $\mathcal{B}_\varphi\subset\mathcal{A}_\varphi$ and the operator $\mathcal{A}_\varphi$ is m-completely accretive in $L^1(\Omega)$ with dense domain. Moreover, if $(T(t))_{t\geq 0}$ is the semigroup of order preserving contractions in $L^1(\Omega)$ generated by the operator $\mathcal{A}_\varphi$, then its restriction to $L^2(\Omega)$ coincides with the semigroup generated by the operator $\mathcal{B}_\varphi$.*

Using the above result and some a priori estimates we prove that for every initial data $u_0\in L^1(\Omega)$, the mild-solution $T(t)u_0$ is an entropy solution (1.1). To prove uniqueness of entropy solutions, we follow the same technique than in [3], which was inspired by the doubling of variables method introduced by Kruzhkov [16] (see also [11]) to prove L^1-contraction for entropy solutions for scalar conservation laws.

The proofs of the L^2-results of this note can be found in [4] and the L^1-results in [5].

References

[1] L. Ambrosio, N. Fusco and D. Pallara, *Functions of Bounded Variation and Free Discontinuity Problems*, Oxford Mathematical Monographs, 2000.

[2] F. Andreu, C. Ballester, V. Caselles and J. M. Mazón, *Minimizing Total Variational Flow*, Diff. and Int. Equt. **14** (2001), 321–360.

[3] F. Andreu, C. Ballester, V. Caselles and J. M. Mazón, *The Dirichlet problem for the total variational flow*, J. Funct. Anal. **180** (2001), 347–403.

[4] F. Andreu, V. Caselles and J. M. Mazón, *A Parabolic Quasilinear Problem for Linear Growth Functionals*, To appear in Rev. Mat. Iberoamericana.

[5] F. Andreu, V. Caselles and J. M. Mazón, *Existence and Uniqueness of a Solution for a Parabolic Quasilinear Problem for Linear Growth Functionals with L^1 data*, To appear in Math. Ann.

[6] G. Anzellotti, *Pairings Between Measures and Bounded Functions and Compensated Compactness*, Ann. di Matematica Pura ed Appl. IV (135) (1983), 293–318.

[7] G. Anzellotti, *The Euler equation for functionals with linear growth*, Trans. Amer. Math. Soc. **290** (1985), 483–500.

[8] Ph. Benilan and M.G. Crandall, *Completely Accretive Operators*, in Semigroups Theory and Evolution Equations, Ph. Clement et al. editors, Marcel Dekker, 1991, pp. 41–76.

[9] S. Bernstein, *Sur les équations du calcul des variations*, Ann. Sci. Ecole Norm. Sup. **29** (1912), 431–485.

[10] H. Brezis, *Operateurs Maximaux Monotones*, North Holland, Amsterdam, 1973.

[11] J. Carrillo, *On the uniquenes of solution of the evolution dam problem*, Nonlinear Anal. **22** (1994), 573–607.

[12] F. Demengel and R. Teman, *Convex Functions of a Measure and Applications*, Indiana Univ. Math. J. **33** (1984), 673–709.

[13] L. C. Evans and R. F. Gariepy, *Measure Theory and Fine Properties of Functions*, Studies in Advanced Math., CRC Press, 1992.

[14] R. Hardt and X. Zhou, *An evolution problem for linear growth functionals*, Commun. Partial Differential Equations **19** (1994), 1879–1907.

[15] R. Kohn and R. Temam, *Dual space of stress and strains with application to Hencky plasticity*, Appl. Math. Optim. **10** (1983), 1–35.

[16] S.N. Kruzhkov, *First order quasilinear equations in several independent variables*, Math. USSR-Sb. **10** (1970), 217–243.

[17] A. Lichnewski and R. Temam, *Pseudosolutions of the Time Dependent Minimal Surface Problem*, Journal of Differential Equations **30** (1978), 340–364.

[18] L. Schwartz, *Analyse IV. Applications a la théorie de la mesure*, Hermann, 1993.

[19] X. Zhou, *An Evolution Problem for Plastic Antiplanar Shear*, Appl. Math. Optm. **25** (1992), 263–285.

[20] W. P. Ziemer, *Weakly Differentiable Functions*, GTM 120, Springer Verlag, 1989.

Fuensanta Andreu
Departamento de Analisis Matematico
Universitat de Valencia
E-46100 Burjassot (Valencia), Spain
E-mail address: fuensanta.andreu@uv.es

Vicent Caselles
Departament de Tecnologia
Escola Superior Politecnica
Universitat Pompeu Fabra
Passeig Circumval.lació, 8
E-08003 Barcelona, Spain
E-mail address: vicent.caselles@tecn.upf.es

José M. Mazón
Departamento de Analisis Matematico
Universitat de Valencia
E-46100 Burjassot (Valencia), Spain
E-mail address: jose.m.mazon@uv.es

Progress in Nonlinear Differential Equations
and Their Applications, Vol. 55, 13–23

Time Operators Associated to Dilations of Markov Processes

Ioannis Antoniou and Zdzislaw Suchanecki

1. Introduction

The original goal of dilation theory is to find a way to study general bounded operators through isometries or unitary operators (see [1] for an excellent treatment of the subject). This is moreover a natural framework to study the question whether a Markov process may arise as a projection of deterministic dynamical system. The latter question is closely related to the long standing problem of the relationship between deterministic laws of dynamics and probabilistic description of physical processes, known as the *problem of irreversibility.*

In contrast to the usual coarse graining approach B. Misra, I. Prigogine and M. Courbage [2, 3] formulated the problem of irreversibility taking as a fundamental physical principle the law of entropy increase. In this theory the problem of reconciling the dynamical evolution with the irreversible thermodynamical evolution is viewed in terms of establishing a non-unitary equivalence between the unitary dynamical group and probabilistic Markov processes. More explicitly, let $\{U_t\}$ be a unitary group of evolution, which is induced by point dynamics $\{S_t\}$ of the phase space, acting on the Hilbert space $\mathcal{H}$ spanned by the square integrable phase space functions. One considers now the possibility of relating the group $\{U_t\}$ with a Markovian semigroup $\{W_t\}$ through a non-unitary positivity preserving intertwining transformation Λ in such a way that

$$\Lambda U_t = W_t \Lambda\,, \quad \text{for } t \geq 0\,. \tag{1}$$

If the transformation Λ is invertible then the intertwining between the unitary dynamics and the Markov evolution involves no loss of information (non-unitary "equivalence"). If, on the other hand, Λ is an orthogonal projection on a subspace of $\mathcal{H}$, then the relation between $\{U_t\}$ and $\{W_t\}$ can be seen as a coarse-graining compatible with dynamics. The conservative systems for which invertible transformation to dissipative systems have been constructed are qualified by the existence of internal time. An internal time operator for the unitary evolution evolution $\{U_t\}$ is a self-adjoint operator T satisfying the canonical commutation relation:

$$U_{-t} T U_t = T + tI\,, \quad \text{for each } t\,. \tag{2}$$

Internal time operators for unitary dynamics were introduced [4] in the context of unstable dynamical systems of the Kolmogorov type. Misra, Prigogine and

Courbage showed that the unitary evolution $\{U_t\}$ of a Kolmogorov system can be intertwined (1) either by a similarity

$$W_t = \Lambda U_t \Lambda^{-1}$$

or by a coarse-graining projection $\Lambda = P$,

$$W_t = PU_tP\,,$$

with the Markov semigroup $\{W_t\}$.

Since 1980's the Misra, Prigogine and Courbage theory of irreversibility has been developed further in several directions: classical dynamical systems [5, 6, 7, 8, 9, 10], relativistic systems [11], or quantum systems [12], to mention just the main direction. Of course, the above list of references is far from being complete. It is therefore natural that also new problems have been raised. One of them is the problem of implementability of the resulting Markov semigroup, i.e., the question whether the semigroup $\{W_t\}$ is implementable by a (noninvertible) point transformation of the phase space. Another problem is to characterize the class of dynamical systems, both classical and quantum, which can be intertwined with dissipative systems. Yet another problem, which is in a sense converse to the latter, is to characterize the Markov semigroups that can be intertwined with unitary groups or which can arise as their projections.

The first question was raised because the formulation of invertible dynamics in terms of point transformations of the phase space is equivalent to the formulation in terms of unitary measure preserving groups. The question of implementability of $\{W_t\}$ is then about whether the irreversibility can be observed on the level of trajectories. Although all unitary and positivity preserving transformations on L^2 (isometries on L^p, $p \neq 2$) are implementable [13, 14] the semigroup $\{W_t\}$ related with $\{U_t\}$ through a similarity transformation Λ is, in general, not implementable [15, 16, 17] (see also [18] for quantum dynamics).

Concerning the second problem, it is known that K-systems, both classical and quantum, admit time operators thus can be intertwined with dissipative systems. It is also known that if a unitary dynamics admits a time operator then it has the ergodic property of mixing.

The third problem has been so far only partially solved. One of the reasons of the difficulties to resolve this problem is the above-mentioned negative answer to the problem of implementability. A unitary group of positivity preserving operators which acts on the Hilbert space spanned by phase space functions is implemented by some transformations of the phase space. But the induced Markov semigroup is in general not related to the underlying phase space and its dynamics. It should be noticed that in all known constructions of the similarity between the unitary and Markov semigroups the knowledge of the underlying dynamics plays a crucial role. It is in fact crucial from the point of view of the possibility of constructing a time operator which relates both semigroups. Therefore the third question is actually about the ability of relating a Markov semigroup with some phase space dynamics and with the existence of a time operator. We shall show how to resolve this problem by relating Markov semigroups with canonical Markov processes acting on larger spaces. Namely that the original Markov semigroup can be dilated to some unitary group which also admits a time operator. Thus the concept of time operator goes beyond the unitary dynamics where it was initially introduced (see also [19, 20, 21]). In view of the above mention answer on the first problem, that Markov semigroups are not connected with the underlying dynamics on the phase space, the dilation approach to the problem of associating time operators with Markov semigroups, appears to be not only justified but the only possible, which guarantees a general solution.

The plan of the paper is as follows. Section 2 contains the basic facts concerning Markov processes. In Section 3 we introduce rigorously the concept of time operator and show its importance in the resolution of the problem of irreversibility in physics. Section 4 is devoted to the construction of time operators associated with the dilations of Markov semigroups.

2. Markov processes and Markov semigroups

The purpose of this section is to recall the correspondence between Markov processes and semigroups. Let $(\Omega, \mathcal{F}, P)$ be a probability space and X_t an $\mathcal{X}$-valued stochastic process ($\mathcal{X} \subset \mathbb{R}^n$) on Ω. X_t is a *Markov process* if for any time instants $t_1 < \cdots < t_n < t_{n+1}$, states $x_1, \ldots, x_n, x_{n+1} \in \mathcal{X}$ and a Borel set $B \subset \mathcal{X}$ we have

$$P\{X_{t_{n+1}} \in B | X_{t_1} = x_1, \ldots, X_{t_n} = x_n\} = P\{X_{t_{n+1}} \in B | X_{t_n} = x_n\}. \qquad (3)$$

Because of (3) all finite-dimensional distributions

$$P\{X_{t_1} \in B_1, \ldots, X_{t_n} \in B_n\},$$

where $B_1, \ldots, B_n$ are Borel subset of $\mathcal{X}$, are determined by the transition probabilities $Q_{s,t}$:

$$Q_{s,t}(x, B) = P\{X_t \in B | X_s = x\},$$

for $s < t$. Assuming that the transition probabilities are *stationary*, i.e., $Q_{s,t}(x, B)$ depend only on the difference $t - s$ and that the set of indices t is the interval $[0, \infty)$, we can consider instead of $Q_{s,t}(x, B)$ the transition probabilities:

$$Q_t(x, B) = P\{X_t \in B | X_0 = x\}, \qquad (4)$$

for all $t > 0$. Recall that the transition probabilities have the following properties:

1) For each t and fixed B, $Q_t(x, B)$ is measurable as a function of x
2) For each t and fixed x, $Q_t(x, .)$ is a probability measure on Borel subsets of $\mathcal{X}$
3) For each s, t

$$Q_{s+t}(x, B) = \int_{\mathcal{X}} Q_t(y, B) Q_s(x, dy)$$

(Chapman-Kolmogorov equation).

B by a theorem Ionescu Tulcea [22] an arbitrary family of transition probabilities, i.e., any family of functions $Q_t(x, B)$ satisfying conditions 1)-3), determines a Markov process. On the other hand suppose that the set $\mathcal{X}$ is equipped with the measure structure, i.e., with a σ-algebra Σ of subsets of $\mathcal{X}$ and a measure μ. Then each transition probability $Q_t(x, B)$ determines an operator M_t on $L^2_{\mathcal{X}}$ by putting

$$M_t(\mathbb{1}_B) = \int_{\mathcal{X}} Q_t(x, B)\mu(dx)\,,$$

and then extending M_t to a bounded and positivity preserving linear operator on $L^2_{\mathcal{X}}$. The family $\{M_t\}_{t\geq 0}$, which forms a semigroup of Markov operators on $L^2_{\mathcal{X}}$, is called a *Markov semigroup.*

By the Markov operator we mean a linear operator M on the space $L^2_{\mathcal{X}}$ (or on any of the spaces $L^p_{\mathcal{X}}$, for $p \geq 1$) with the following properties:

(a) M is a contraction

$$\|Mf\| \leq \|f\|$$

(b) M preserves positivity

$$Mf \geq 0 \quad \text{if} \quad f \geq 0$$

(c) M preserves probability normalization

$$\int_{\mathcal{X}} Mf d\mu = \int_{\mathcal{X}} f d\mu\,.$$

If, in addition, the measure μ on $(\mathcal{X}, \Sigma)$ is normalized then

(d) M preserve constants

$$M1 = 1\,.$$

The properties (a)–(d) characterize Markov process. Namely (see [22]), each operator M_t with properties (a)–(d) determines the transition probability

$$Q_t(x, B) \stackrel{\text{df}}{=} M_t^* \mathbb{1}_B(x) \tag{5}$$

where M_t^* is the M_t adjoint on the space $L^2_{\mathcal{X}}$, and the transition probabilities define in turns a stochastic process, called the *canonical Markov process.*

3. Time operator and the problem of irreversibility

We begin with the formal definition of time operator in a general setting. Let $\mathcal{H}$ be a separable Hilbert space and $\{V_t\}$ a semigroup of bounded operators, called an *evolution semigroup*, acting on $\mathcal{H}$ (time t can be discrete or continuous). The operator T acting on $\mathcal{H}$ is called a *(internal) time operator* with respect to $\{V_t\}$ if the domain of T is dense in $\mathcal{H}$ and, for each t, the following equality holds:

$$TV_t = V_tT + tV_t\,. \tag{6}$$

The physical meaning of the time operator T is that it allows to define the average age $\langle \rho, T\rho \rangle$ of states (normalized elements) $\rho \in \mathcal{H}$. If the evolution operators V_t

preserve the scalar product in $\mathcal{H}$ (the evolution of a *conservative system*) then the relation (6) can be written

$$\langle V_t h, T V_t h\rangle = \langle h, Th\rangle + t\|h\|^2 . \tag{7}$$

Following [23] we can say that "the average age of the evolved state keeps step with the external clock time t". Typical examples of conservative systems are the evolution of phase densities of Hamiltonian systems and the evolution of wave functions or density operators of quantum systems.

All known constructions of the operator Λ have been done so far for dynamical systems which are K-flows. By a dynamical system we mean a measure space $(\mathcal{X}, \Sigma, \mu)$, where $\mathcal{X}$ is a phase space, μ is a finite measure on the σ-algebra Σ and $\{S_t\}$ is a one parameter evolution semigroup of measurable transformations of $\mathcal{X}$. For reversible systems $\{S_t\}$ is a group of automorphisms of $\mathcal{X}$ and the measure μ represents in the case $\mu(\mathcal{X}) = 1$ an equilibrium distribution. The phase functions f evolve according to the Koopman operators

$$V_t f(x) = f(S_t x) \ , \quad x \in \mathcal{X} . \tag{8}$$

The Koopman operators are isometries on the Banach space $L^p_{\mathcal{X}} = L^p(\mathcal{X}, \Sigma, \mu)$, $p \geq 1$, of p-integrable functions provided that $\{S_t\}$ are measure preserving transformations. If S_t are automorphisms the Koopman operators restricted to the Hilbert space L^2 are unitary.

Let us recall that a dynamical system is a K-flow [24] if there exists a sub-σ-algebra Σ_0 of Σ such that for $\Sigma_t = S_t(\Sigma_0)$ we have

(i) $\Sigma_s \subset \Sigma_t$, for $s < t$
(ii) $\sigma\big(\bigcup_{t\in\mathbf{R}} \Sigma_t\big) = \Sigma$
(iii) $\bigcap_{t\in\mathbf{R}} \Sigma_t = \Sigma_{-\infty}$ – the trivial σ-algebra, i.e., the algebra of sets of measure 0 or 1

where $\sigma\big(\bigcup_{t\in\mathbf{R}} \Sigma_t\big)$ stands for σ-algebra generated by Σ_t, $t \in \mathbf{R}$.

In the Misra-Prigogine-Courbage theory of irreversibility it is considered an K-system and the associated unitary group $\{U_t\}$ of Frobenius-Perron operators, which act on the Hilbert space L^2 of square integrable functions on the phase space

$$U_t \rho(x) = V_t^* \rho(x) = \rho(S_{-t} x) . \tag{9}$$

The group $\{U_t\}$ defines, in particular, the evolution of densities in the Liouville space.

With any K-flow we can associate a family of conditional expectations $\{E_t\}$ with respect to the σ-algebras $\{\Sigma_t\}$ (projectors if we confine ourselves to the Hilbert space L^2). These projectors determine the time operator T:

$$T = \int_{-\infty}^{+\infty} t dE_t . \tag{10}$$

K-systems are related through non-unitary transformations to a class of dissipative systems whose evolution is described by a Markov semigroup $\{W_t\}$ on L^2

that approaches monotonically the equilibrium. The non-unitary transformation Λ relating both systems is defined, up to constants, as a function of the operator T:

$$\Lambda = f(T) + E_{-\infty}\,, \tag{11}$$

where $E_{-\infty}$ is the expectation (projection on constants). The function f is assumed to be positive, non increasing, $f(-\infty) = 1$, $f(+\infty) = 0$ and such that lnf is concave on $\mathbf{R}$. The Markov operators W_t are of the form

$$W_t = \left(\int_{-\infty}^{\infty} \frac{f(s)}{f(s-t)}\, dE_s + E_{-\infty} \right) U_t\,. \tag{12}$$

4. Time operators associated with Markov processes

The problem of the association of time operators with Markov semigroups can be seen a part of a more general problem of the association of time operators with stochastic processes. The most natural way to associate a time operator with a stochastic process $\{X_t\}$ would be to associate it with the filtration, i.e., the increasing family of σ-algebras $\{\mathcal{F}_t\}$, where $\mathcal{F}_t = \sigma\{X_s : s \leq t\}$, through (10) (in such case E_t would be the conditional expectations $E(\cdot|\mathcal{F}_t)$). Such correspondence can be established for any stochastic process. However, the time operator must be, by the definition, related to the time evolution of the stochastic processes expressed in terms of a semigroup of operators acting on a Hilbert space. Such dynamical semigroups (groups) can be specified for some classes of stochastic processes. If, for example, a stochastic process is stationary in the wide sense then it defines a dynamical group of unitary operators (shifts of the time) acting on the Hilbert space generated by the realizations of the stochastic process. As we shall see below Markov semigroups can also be extended (dilated) to unitary groups associated with stationary processes and, consequently, a time operator can be associated with such stationary extension.

Recall that a semigroup of isometries $\{\tilde{M}_t\}$ on a Banach space $\tilde{X}$ is a dilation of the contractive semigroup $\{M_t\}$ on the Banach space X if there exist two linear operators $I : X \longrightarrow \tilde{X}$ and $E : \tilde{X} \longrightarrow X$ such that

$$M_t = E\tilde{M}_t I\,, \quad \text{for each } t\,. \tag{13}$$

In other words the following diagram commutes

$$\begin{array}{ccc} X & \xrightarrow{M_t} & X \\ I \downarrow & & \uparrow E \\ \tilde{X} & \xrightarrow{\tilde{M}_t} & \tilde{X} \end{array}$$

Usually, a dilation is realized in such a way that X is isomorphic to a subspace of $\tilde{X}$, I is the canonical injection of X into $\tilde{X}$, and E is the projection of $\tilde{X}$ onto X. If X is a Hilbert space and $\tilde{M}_t$ are unitary operator we may speak about unitary dilations of contractions.

Let us assume for the sake of the clarity of presentation that the time is discrete, $t = 0, 1, 2, \ldots$. This assumption allows to consider a single Markov operator M instead of the whole Markov semigroup because $M_t = M^t$. The case of continuous time can be however treated in the same way. Our aim is to construct a positive dilation $\tilde{M}$ of M which is implemented by a measure preserving transformation S of some probability space $(\Omega, \mathcal{F}, P)$.

Dilations of contractive semigroups such as the constructed in Ref. [1] cannot be applied in our case. We are interested in more specific dilations of the Markov semigroup $\{M^n\}$, where M satisfies (a)–(d), acting on the Hilbert space $X = L^2_{\mathcal{X}}$ to a group acting on the larger space $\tilde{X} = L^2_{\Omega}$ which corresponds to a stationary process. In particular we are looking for such dilation that both $\tilde{M}^n$, $n = 0, \pm 1, \pm 2, \ldots$ and E are positivity preserving.

Faced with the difficulty to obtain meaningful positive dilations of Markov semigroups using the Sz-Nagy dilation theory, we constructed [25] a positive dilation based on the natural extensions of dynamical systems through canonical Markov processes. There the positivity is clearly related to that of the inducing semigroup. It is now our task to apply the dilation approach to an arbitrary Markov Semigroup in order to associate it with a time operator. Let us consider first the case of a single Markov operator M on $L^2_{\mathcal{X}}$ that satisfies (a)–(d).

Suppose then it is given a Markov operator M on the space $L^2_{\mathcal{X}}$. Recall that M corresponds to a transition probability $Q(x, A)$ through (5). Define the product space

$$\Omega = \prod_{n=-\infty}^{\infty} \mathcal{X}_n \,, \quad \text{where } \mathcal{X}_n = \mathcal{X} \,,$$

and the σ-algebra $\mathcal{F}$ generated by the cylindrical sets

$$C_{A_1, \ldots, A_k} = \cdots \times \Omega \times A_1 \times A_2 \times \cdots \times A_k \times \Omega \times \cdots \,, \tag{14}$$

where $k \in \mathbf{N}$, $A_i \in \mathcal{A}$.

On $(\mathcal{X}, \mathcal{F})$ we define the cylindrical measure P

$$P(C_{A_1, \ldots, A_k}) = \int_{A_1} \int_{A_2} \cdots \int_{A_k} Q(x_{k-1}, dx_k) \ldots Q(x_1, dx_2) \, \mu(dx_1) \,, \tag{15}$$

and show [25] that P is correctly defined and normalized. Thus using the Kolmogorov theorem [22] we extend P to a probability measure on the whole σ-algebra $\mathcal{F}$.

Denote by S the left shift on Ω, i.e.,

$$S\omega = S(x_n)_{-\infty}^{\infty} \stackrel{\text{def}}{=} (x_{n+1})_{-\infty}^{\infty} \,, \quad \omega \in \Omega \,. \tag{16}$$

It can be shown [25] that P is invariant with respect to S, which amounts to showing that to check that if either $A_0 = \Omega$ or $A_{k+1} = \Omega$ then

$$P(C_{A_0, \ldots, A_k}) = P(C_{A_1, \ldots, A_{k+1}}) = P(C_{A_1, \ldots, A_k}) \,. \tag{17}$$

Now let $\tilde{M}$ be the operator on L^2_Ω associated with S as follows:

$$\tilde{M}\tilde{f}(\omega) = \tilde{f}(S^{-1}\omega), \quad \tilde{f} \in L^2_\Omega . \tag{18}$$

The operator $\tilde{M}$ is a dilation of M. Precisely, we show that

$$M^n = E\tilde{M}^n I , \quad n \in \mathbf{N}, \tag{19}$$

where I is the canonical injection $L^2_\mathcal{X}$ into L^2_Ω:

$$L^2_\mathcal{X} \ni f \longmapsto If = \tilde{f} \in L^2_\Omega$$

defined as

$$(If)(\ldots, x_{-1}, x_0, x_1, \ldots) \stackrel{\text{def}}{=} f(x_0) \tag{20}$$

and E is the conditional expectation with respect to the σ-algebra generated by the cylinders

$$C^0_A = \{\omega = (x_n)^\infty_{-\infty} \ : \ x_0 \in A\} , \quad A \in \mathcal{A}. \tag{21}$$

Consequently, for any $f \in L^2_\mathcal{X}$, the function $E\tilde{M}^n If$ depends only on the 0–coordinate and can be uniquely identified with a function on Ω. The complete proof of the above constructed dilation the reader can find in Ref. [25].

In order to construct the time operator $\tilde{T}$ associated with the dilation $\{\tilde{M}^n\}$ we can proceed as in the case of K-flows. We distinguish the σ-algebra $\mathcal{F}_0$ generated by those cylinders $C^{0,\ldots,k}_{A_0,\ldots,A_k}$ $(k = 0, 1, \ldots \ , A_0, \ldots, A_k \in \Sigma)$ for which the set A_0 is placed on the 0th coordinate and note that the σ-algebra $S^{-n}\mathcal{A}_0 \ , \quad n \in \mathbf{Z}$, is the σ-algebra generated by $C^{n,\ldots,n+k}_{A_0,\ldots,A_k}$ with A_0 placed on the nth coordinate. Then we see that

(i) $S\mathcal{F}_0 \supset \mathcal{F}_0$

(ii) $\sigma(\bigcup_{n=-\infty}^{\infty} S^n\mathcal{F}_0) = \mathcal{F}$.

However, the the third condition of K-flows:

(iii) $\bigcap_{n=-\infty}^{\infty} S^n\mathcal{F}_0$ is the trivial σ-algebra

is, in general not satisfied. The condition (iii) is satisfied, i.e., the dilated evolution is a K-flow, provided the Markov operator has additionally the following property [25]:

(e) M strongly converges to equilibrium

$$\|M^n f - 1\|_{L^p} \longrightarrow 0, \text{ as } n \to \infty \text{ for each probability density } f \in L^p .$$

The dynamical system with the Frobenius-Perron operator satisfying (e) is called the *exact system.*

The above consideration leads to the conclusion that Markov operators satisfying conditions (a)–(e) can be dilated to K-flows. The corresponding time operator can be then constructed by the means of the conditional expectations

$$E_n = E(\cdot|\mathcal{F}_n)$$

by putting

$$T = \sum_{n=-\infty}^{\infty} n(E_n - E_{n-1}) .$$

The case of continuous Markov semigroup $\{M_t\}_{t\geq 0}$ is treated analogously. As before we assume that each M_t corresponds through (4) to the transition probability $Q_t(x, A)$. Now the dilation space Ω will be the product

$$\Omega = \prod_{t\in \mathbf{R}} \mathcal{X}_t , \quad \text{where } \mathcal{X}_t = \Omega .$$

The σ-algebra $\mathcal{F}$ is generated by the cylinders

$$C^{t_0,\ldots,t_k}_{A_0,\ldots,A_k} , \quad \text{where } t_0, \ldots, t_k \in \mathbf{R} , \quad A_0, \ldots, A_k \in \Sigma .$$

The probability measure P on $(\Omega, \mathcal{F})$ and the distinguish σ-algebra $\mathcal{F}_0$ are also defined analogously (see [25] for details). Then we may define the group of shifts transformations of Ω

$$S_t\omega = x(\cdot + t) , \quad \text{for each function } \omega = x(\cdot) \in \Omega .$$

Finally, we define the operators

$$\tilde{M}_t f(\omega) = f(S_{-t}\omega) , \quad f \in L^2_\Omega ,$$

and check in essentially the same way as before that $\tilde{M}_t$ are the dilations of M_t. The time operator T is now of the form (10) (or stochastic integral)

$$T = \int_{-\infty}^{\infty} t dE_t$$

where E_t are the conditional expectations $E_t(\cdot|S_t\mathcal{F}_0)$. In this way we have defined the time operator T on the Hilbert space L^2_Ω, but it can be defined as linear operators on any space $L^p(\Omega, \mathcal{F}, P)$, where $1 \leq p \leq \infty$.

It can be checked directly that T satisfies the relation

$$T\tilde{M}_t = \tilde{M}_t T + t\tilde{M}_t . \tag{22}$$

However, since T is the time operator associated with a K-system $(\Omega, \mathcal{F}, P; S)$ (or a K-flow) the relation (22) also follows the previous results [2, 15, 9].

We may now ask the question whether the above constructed time operator can be projected on the space $L^2_{\mathcal{X}}$ giving rise to the time operator associated with the Markov semigroup $\{M_t\}$, i.e., whether the operator ETI satisfies (6) with respect to M_t. The answer to this question is however negative. To see this it is enough to notice that for any $f \in L^2_{\mathcal{X}}$ the injection If is the function depending only on the "zero-coordinate" but T places zero for the projection on this coordinate thus $ETIf = 0$. Therefore the time operator of the dilated semigroup is, like the resulting K-flow, generically associated with the extended space.

References

[1] B. Sz-Nagy, C Foias, *Harmonic Analysis of Operators in Hilbert Space*, North-Holland, Amsterdam, 1970.

[2] B. Misra, I. Prigogine, M. Courbage, *From deterministic dynamics to probabilistic descriptions.* Physica **98A** (1979), 1–26.

[3] I. Prigogine, *From being to becoming.* Freeman, San Francisco, 1980.

[4] B. Misra, *Nonequilibrium entropy, Lyapunov variables, and ergodic properties of classical systems.* Proc. Natl. Acad. USA **75** (1978), 1627–1631.

[5] M. Courbage, *Intrinsic irreversibility of Kolmogorov dynamical systems.* Physica **A 122** (1983), 459–482.

[6] Y. Elskens and I. Prigogine, *From instability to irreversibility.* Proc. Natl. Acad. Sci. **83** (1986), 5756–5760.

[7] R. de la Llave, *Rates of convergence to equilibrium in the Prigogine-Misra-Courbage theory of irreversibility.* J. Stat. Physics **29** (1982), 17–31.

[8] B. Misra and I. Prigogine, *Time Probability and Dynamics.* In Long Time Predictions in Dynamical Systems, ed. C. Horton, L. Reichl, V. Szebehely, Wiley, N.Y., (1983), 21–43.

[9] Z. Suchanecki and A. Weron, *Applications of an operator stochastic integral in Prigogine's theory of irreversible dynamical systems.* Expo. Math. **8** (1990), 67–79.

[10] Z. Suchanecki, *On lambda and internal time operators.* Physica **A 187** (1992), 249–266.

[11] I. Antoniou and B. Misra, *The relativistic internal time algebra.* Nuclear Physics, Proceed. Suppl. Sect. **6** (1989), 240–242.

[12] C.M. Lockhart and B. Misra, *Irreversibility and measurement in quantum mechanics.* Physica **136A** (1986), 47–76.

[13] K. Goodrich, K. Gustafson and B. Misra, *On converse to Koopman's lemma.* Physica **102A** (1980), 379–388.

[14] J. Lamperti, *On the isometries of certain function spaces.* Pacific J. Math. **8** (1958), 459–466.

[15] S. Goldstein, B. Misra and M. Courbage, *On intrinsic randomness of dynamical systems.* J. Stat. Phys. **25** (1981), 111–126.

[16] I. Antoniou and K. Gustafson, *From probabilistic description to deterministic dynamics.* Physica **A 197** (1993), 153–166.

[17] Z. Suchanecki, I. Antoniou and S. Tasaki, *Nonlocality of the Misra-Prigogine-Courbage Semigroup.* J. Stat. Physics **75** (1994), 919–928.

[18] I. Antoniou W.A. Majewski and Z. Suchanecki, *Implementability of Liouville evolution, Koopman and Banach-Lamperti theorems in classical and quantum dynamics.* Physica **A** (submitted).

[19] I. Antoniou and Z. Suchanecki, *Non-uniform time operator, chaos and wavelets on the interval.* Chaos Solitons and Fractals **11** (2000), 423–435.

[20] I. Antoniou, V.A. Sadovnichii and S.A. Shkarin, *Time operators and shift representation of dynamical systems.* Physica **A 269** (1999), 299–313.

[21] I. Antoniou, I. Prigogine, V.A. Sadovnichii and S.A. Shkarin, *Time operator for diffusion.* Chaos, Solitons and Fractals **11** (2000), 465–477.

[22] J. Neveu, *Mathematical Foundations of the Calculus of Probability,* Holden–Day Inc., 1965.

[23] I. Antoniou and B. Misra, *Non-unitary transformation of conservative to dissipative evolutions.* J. Phys. A: Math. Gen. **24** (1991), 2723–2729.

[24] I. Cornfeld, S. Fomin and Ya. Sinai *Ergodic Theory,* Springer-Verlag, Berlin, 1982.

[25] I. Antoniou, K. Gustafson and Z. Suchanecki, *On the inverse problem of statistical physics: from irreversible semigroups to chaotic dynamics.* Physica A **252** (1998), 345–361.

Ioannis Antoniou
International Solvay Institute for Physics and Chemistry
CP 231, ULB Campus Plaine
Bd. du Triomphe
B-1050 Brussels, Belgium
and Theoretische Natuurkunde
Free University of Brussels
E-mail address: iantonio@vub.ac.be

Zdzislaw Suchanecki
International Solvay Institute for Physics and Chemistry
CP 231, ULB Campus Plaine
Bd. du Triomphe
B-1050 Brussels, Belgium
Theoretische Natuurkunde
Free University of Brussels
and Institute of Mathematics
University of Opole

Progress in Nonlinear Differential Equations
and Their Applications, Vol. 55, 25–60

Boundary-Value Problems for Systems of Hamilton-Jacobi-Bellman Inclusions with Constraints

Jean-Pierre Aubin

Introduction

It is well known that value functions of optimal control problems are solutions to Hamilton-Jacobi partial differential equation of the form

$$-\frac{\partial}{\partial t}v(t,x) + \inf_{u\in P(x,v(t,x))}\left(\frac{\partial}{\partial x}v(t,x)f(x,v(t,x),u) - g(x,v(t,x),u)\right) = 0$$

with adequate boundary conditions.

Observe nevertheless that in this equation, the infimum hides two inequalities:

1. there exist $u \in P(x,v(t,x))$ such that
$$-\frac{\partial}{\partial t}v(t,x) + \frac{\partial}{\partial x}v(t,x)f(x,v(t,x),u) - g(x,v(t,x),u) \leq 0$$
2. for all $u \in P(x,v(t,x))$,
$$\frac{\partial}{\partial t}v(t,x) - \frac{\partial}{\partial x}v(t,x)f(x,v(t,x),u) + g(x,v(t,x),u) \leq 0$$

However, several other problems of control theory lead to the study of controlled systems of first-order partial differential equations (or systems of first-order partial differential inclusions): Let $P : X \times Y \rightsquigarrow \mathcal{U}$ be a set-valued map associating with any pair (x,y) a feasible set $P(x,y)$ of controls and f and g be single-valued maps from $X \times Y \times \mathcal{U}$ to finite-dimensional vector spaces X and Y respectively.

The problem is to find a set-valued map $V : \mathbf{R}_+ \times X \rightsquigarrow Y$ satisfying

1. there exists $u \in P(x,V(t,x))$ such that
$$0 \in \text{“}-\frac{\partial}{\partial t}V(t,x) + \frac{\partial}{\partial x}V(t,x)f(x,V(t,x),u)\text{''} - g(x,V(t,x),u) \qquad (1)$$
2. for all $u \in P(x,v(t,x))$,
$$0 \in \text{“}\frac{\partial}{\partial t}V(t,x) - \frac{\partial}{\partial x}V(t,x)f(x,V(t,x),u)\text{''} + g(x,V(t,x),u) \qquad (2)$$

where we shall give a meaning to the derivative

$$\text{“}\frac{\partial}{\partial t}V(t,x) - \frac{\partial}{\partial x}V(t,x)f(x,V(t,x),u)\text{''}$$

in Theorem 3.1 below. Indeed, even in the absence of controls, it is well known that such solutions may have shocks – i.e., can be set-valued – and even, when they happen to be single-valued, are not necessarily differentiable in the usual sense.

The definition of solution shall be taken in a generalized sense – Frankowska solution[1] that we shall define later in the paper.

In order to obtain uniqueness, we have to impose boundary conditions. Furthermore, problems arising in economics, finance and other fields lead to introduce constraints bearing both on the state and on the solution. We shall describe these boundary conditions and constraints by introducing two set-valued maps $\Phi : \mathbf{R}_+ \times X \rightsquigarrow Y$ and $\Psi : \mathbf{R}_+ \times X \rightsquigarrow Y$ such that $\Phi \subset \Psi$, the first one encompassing initial and/or boundary-value conditions, or other conditions as we shall see, the second one viability constraints both on the state variables x – that must remain in the domain of Ψ – and on the solution $V(t,x)$.

We shall prove that **there exists a unique** “solution” $(t,x) \rightsquigarrow V(t,x)$ to this general problem (1,2) satisfying the conditions

$$\forall\, (t,x) \in \mathbf{R}_+ \times X,\ \ \Phi(t,x) \ \subset\ V(t,x) \ \subset\ \Psi(t,x)$$

in the class of closed set-valued maps (i.e., set-valued maps with closed graph), that depends continuously of the data Φ (in the “graphical sense”, mapping graphical limits to graphical limits, as it is explained later).

Even more, we shall provide an explicit formula when $f(x,u)$ and $P(x)$ do not depend on the variable y and when

$$g(x,y,u) \ :=\ -M(x,u)y - L(x,u)$$

is affine with respect to y where

1. M is a continuous matrix-valued function

$$M : (x,u) \in X \times \mathcal{U} \mapsto M(x,u) \in \mathcal{L}(X,Y)$$

[1] Hélène Frankowska proved that the epigraph of the value function of an optimal control problem – assumed to be only lower semicontinuous – is semi-permeable, (i.e., invariant and backward viable) under a (natural) auxiliary system. Furthermore, when it is continuous, she proved that its epigraph is viable and its hypograph invariant ([43, 44, 46, Frankowska]). By duality, she proved that the latter property is equivalent to the fact that the value function is a viscosity solution of the associated Hamilton-Jacobi equation in the sense of M. Crandall and P.-L. Lions. See also [32, Barron & Jensen] and [31, Bardi & Capuzzo-Dolcetta] for more details. We refer also to [39, 40, Carja & Ursescu] for the study of these equations through the characteristics method using the contingent derivative (and not epi and hypo derivatives). Such concepts have been extended to solutions of systems of first-order partial differential equation without boundary conditions by Hélène Frankowska and the author (see [19, 20, 21, 22, 24, Aubin & Frankowska] and chapter 8 of [2, Aubin]). See also [16, 17, Aubin & Da Prato]. This point of view is used here in the case of boundary value problems.

2. L is a continuous "vector-Lagrangian"

$$L : (x,u) \in X \times \mathcal{U} \mapsto L(x,u) \in Y$$

Let us denote by $\mathcal{C} : x \in X \rightsquigarrow \mathcal{C}(x) \subset \mathcal{C}(0,\infty;X) \times L^1(0,\infty;\mathcal{U})$ the set-valued map associating with $x \in X$ the set $\mathcal{C}(x)$ of the pairs $(x(\cdot),u(\cdot))$ solutions to the control system

$$\begin{cases} \text{i)} & x'(t) = f(x(t),u(t)) \\ \text{ii)} & u(t) \in P(x(t)) \end{cases}$$

starting at x at $t = 0$.

In the absence of constraints ($\Psi(t,x) := Y$), we shall prove that setting

$$\begin{cases} J_{\Phi}(t;(x(\cdot),u(\cdot)))(T,x) \\ := e^{\int_0^t M(x(s),u(s))ds}\Phi(T-t,x(t)) + \int_0^t e^{\int_0^\tau M(x(s),u(s))ds} L(x(\tau),u(\tau))d\tau \end{cases}$$

then the set-valued solution V is defined by

$$V(\Phi)(T,x) := \bigcup_{(x(\cdot),u(\cdot))\in\mathcal{C}(x)} \bigcup_{t\in[0,T]} J_{\Phi}(t;(x(\cdot),u(\cdot)))(T,x)$$

With an adequate choice of the set-valued map Ψ associated with the set-valued map Φ, we find as solution the set-valued map defined by

$$W(\Phi)(T,x) := \bigcup_{(x(\cdot),u(\cdot))\in\mathcal{C}(x)} \bigcap_{t\in[0,T]} J_{\Phi}(t;(x(\cdot),u(\cdot)))(T,x)$$

We shall find as many formulas as pairs (Ψ,Φ) of set-valued maps (see formula (24) of Theorem 5.1 below).

We can read this type of results the other way around: For instance, the set-valued map $V(\Phi)$ defined by

$$\begin{cases} V(\Phi)(T,x) := \bigcup_{(x(\cdot),u(\cdot))\in\mathcal{C}(x)} \bigcup_{t\in[0,T]} \\ \left(e^{\int_0^t M(x(s),u(s))ds}\Phi(T-t,x(t)) + \int_0^t e^{\int_0^\tau M(x(s),u(s))ds} L(x(\tau),u(\tau))d\tau \right) \end{cases}$$

is the unique "solution" to the Hamilton-Jacobi partial differential inclusion (1,2) satisfying the initial condition

$$V(0,x) = \Phi(0,x)$$

and

$$\forall\, t \geq 0,\ x \in X,\ \ \Phi(t,x) \subset V(t,x)$$

They define set-valued analogue of optimal control problems, where the "$\cup$ operation" replaces the "inf operation" and the "$\cap$ operation" the "sup operation".

Actually, when $Y := \mathbf{R}$ and when we associate with two extended functions $\mathbf{c} : \mathbf{R}_+ \times X \rightsquigarrow \mathbf{R} \cup \{+\infty\}$ and $\mathbf{b} : \mathbf{R}_+ \times X \rightsquigarrow \mathbf{R} \cup \{+\infty\}$ the set-valued maps

$$\begin{cases} \text{i)} & \Phi(t,x) := \mathbf{c}(t,x) + \mathbf{R}_+ \\ \text{ii)} & \Psi(t,x) := \mathbf{b}(t,x) + \mathbf{R}_+ \end{cases}$$

we find problems of dynamic valuation and management of portfolios in mathematical finance, used in particular for valuating options, as in [55, Pujal], [30, Aubin, Pujal & Saint-Pierre]) and [56, Pujal & Saint-Pierre]. For instance, we deduce that

$$\begin{cases} \inf_{y \in V(\Phi)(T,x)} y = \inf_{(x(\cdot),u(\cdot)) \in \mathcal{C}(x)} \inf_{t \in [0,T]} \\ \left(e^{\int_0^t M(x(s),u(s))ds} \mathbf{c}(T-t, x(t)) + \int_0^t e^{\int_0^\tau M(x(s),u(s))ds} L(x(\tau), u(\tau)) d\tau \right) \end{cases}$$

is the **valuation function** of a stopping time problem (see Section 5 below).

These two explicit formulas are given by the **caliber** $V : \mathbf{R}_+ \times X \rightsquigarrow Y$ defined in the following way: y belongs to $V(T,x)$ if there exists a control $t \in [0,T] \mapsto u(t)$ and a time $T^* \in [0,T]$ such that the solution $(x(\cdot), u(\cdot), y(\cdot))$ to the control system

$$\begin{cases} \text{i)} & x'(t) = f(x(t), y(t), u(t)) \\ \text{ii)} & y'(t) = g(x(t), y(t), u(t)) \\ \text{iii)} & u(t) \in P(x(t), y(t)) \end{cases}$$

starting at $x(0) = x$, $y(0) = y$ satisfies

$$\begin{cases} \text{i)} & \forall\, t \in [0, T^*], \;\; y(t) \in \Psi(T-t, x(t)) \\ \text{ii)} & y(T^*) \in \Phi(T - T^*, x(T^*)) \end{cases}$$

Observe that taking $\Phi(t,x) = \emptyset$ whenever $t > 0$ guarantees that $T^* = T$.

We shall prove that this set-valued map is the unique solution to our problem (1,2). Actually, this is a reformulation dictated by problems arising in dynamic economic theory, finance mathematics and control theory of the celebrated "method of characteristics".

We shall revisit this method using the tools of set-valued analysis and viability theory which go back to the early 80's[2]. They find here an unexpected relevance to assert the existence and the uniqueness of the solution to this problem, since such solutions may have shocks – i.e., can be set-valued – and even, when they happen to be single-valued, they are not differentiable in the usual sense. The tools forged by set-valued analysis and viability theory happen to allow us to prove existence

[2]See [18, Aubin & Frankowska], [2, Aubin], [60, Rockafellar & Wets], etc. for instance.

and uniqueness in the class of set-valued maps with closed graph only, instead of classes of vector-distributions[3].

The basic concept useful in our framework is the concept of **viable-capture basin** of a "target" $C \subset K$ viable in a constrained subset $K \subset X$ under a differential inclusion $x' \in F(x)$: It is the subset $\mathrm{Capt}_F^K(C)$ of initial states $x_0 \in K$ such that C is reached in finite time before possibly leaving K by at least one solution $x(\cdot) \in \mathcal{S}(x_0)$, where $\mathcal{S}(x_0)$ denotes[4] the set of solutions to differential inclusion $x' \in F(x)$ starting at x_0.

---

[3]The strong requirement of pointwise convergence of differential quotients can be weakened in (at least) two ways, each way sacrificing different groups of properties of the usual derivatives:

- **Distributional Derivatives**: Fix the direction v and take the limit of the function $x \mapsto \nabla_h f(x)(v)$ in the weaker sense of distributions. The limit $D_v f$ may then be a distribution, and no longer a single-value map. However, it coincides with the usual limit when f is Gâteaux differentiable. Moreover, one can define difference quotients of distributions, take their limit, and thus, differentiate distributions.
  Distributions are no longer functions or maps defined on $\mathbf{R}^n$, so they loose the pointwise character of functions and maps, but retain the linearity of the operator $f \mapsto D_v f$, mandatory for using the theory of linear operator for solving partial differential equations.
- **Graphical Derivatives**: Fix the direction x and take the limit of the function $v \mapsto \nabla_h f(x)(v)$ in the weaker sense of "graphical convergence" (the graph of the graphical limit is the Painlevé-Kuratowski upper limit of the graphs). The limit $Df(x)$ may then be a set-valued map, and no longer a single-value map. However, it coincides with the usual limit when f is Gâteaux differentiable. Moreover, one can define difference quotients of set-valued maps, take their limit, and thus, differentiate set-valued maps. These graphical derivatives keep the pointwise character of functions and maps, mandatory for implementing the Fermat Rule, proving inverse function theorems under constraints or using Lyapunov functions, for instance, but loose the linearity of the map $f \mapsto Df(x)$.

In both cases, the approaches are similar: They use (different) convergences *weaker than the pointwise convergence* for increasing the possibility for the difference-quotients to converge. But the price to pay is the loss of some properties by passing to these weaker limits (the pointwise character for distributional derivatives, the linearity of the differential operator for graphical derivatives).

[4]When we are studying the viable-capture basins of targets under differential inclusions, we observe that they are not specific to differential inclusions. They involve only few properties of the solution map $\mathcal{S}$ associating with any initial state x the set $\mathcal{S}(x)$ of solutions $t \mapsto x(t)$ that are solutions to the above differential inclusion starting at x at initial time 0. These properties (translation and concatenation properties, as well as continuity properties) of the solution map $x \rightsquigarrow \mathcal{S}(x)$ are common to other control problems, such as

1. control problems with memory (see the contributions of [50, 51, 52, Haddad], some of them being presented in [2, Aubin]) – before known under the name of functional control problems, the new fashion calling them as "path dependent control systems",
2. parabolic (diffusion-reaction) type partial differential inclusions (see the contributions of [64, 65, 67, Shi Shuzhong], some of them being presented in [2, Aubin]) – also known as distributed control systems
3. "mutational equations" governing the evolution in metric spaces, including "morphological equations" governing the evolution of sets (see [4, Aubin] for instance).

Then we shall prove that *the graph of the solution* $(t, x) \rightsquigarrow V(t, x)$ *to the above boundary value problem is the viable-capture basin of the graph of the set-valued map* Φ *viable in the graph of the set-valued map* Ψ under the auxiliary differential inclusion

$$\begin{cases} \text{i)} & \tau'(t) = -1 \\ \text{ii)} & x'(t) = f(x(t), y(t), u(t)) \\ \text{iii)} & y'(t) = g(x(t), y(t), u(t)) \\ \text{iv)} & u(t) \in P(x(t), y(t)) \end{cases}$$

and that this solution is **unique** among the solutions with closed graph to this boundary value problem.

In some instances, this viable-capture basin can be computed analytically, and we obtain in this case an explicit formula of the solution to the above boundary value problem.

In all cases, the "Viability/Capturability Algorithm" designed by Patrick Saint-Pierre provides numerically the viable-capture basins, and thus, the solutions to systems of Hamilton-Jacobi-Bellman equations, bypassing finite-difference methods (see [61, Saint-Pierre], [38, Cardaliaguet, Quincampoix & Saint-Pierre], [55, Pujal] and [56, Pujal & Saint-Pierre]) for solutions of Hamilton-Jacobi-Bellman equations derived from mathematical finance.)

This existence and uniqueness result follows from the three following steps:

1. from a characterization proved in [15, Aubin & Catté] stating the viable-capture basin $\mathrm{Capt}^K(C)$ is the **unique** closed subset D between C and K satisfying

$$\begin{cases} \text{i)} & \mathrm{Capt}^D(C) = D \\ \text{ii)} & \mathrm{Capt}^K(D) = D \end{cases} \tag{3}$$

2. from a characterization stated below of the viable-capture basin of a target C viable in a closed subset K under a differential inclusion $x' \in F(x)$ proved in [12, Aubin]: Let us recall that
   (a) $K \backslash C$ is a repeller means that all solutions $x(\cdot) \in \mathcal{S}(x)$ starting from $x \in K \backslash C$ reach C or leave K in finite time,
   (b) The subset $D \backslash C$ is said **locally viable** under $\mathcal{S}$ if from any initial state $x \in D \backslash C$ starts at least one solution viable in $D \backslash C$ on a nonempty interval,
   (c) A subset $D \subset K$ is **locally backward invariant relatively to** K under $\mathcal{S}$ if for every $x \in D$, all backward solutions starting from x and viable in K on an interval $[0, T]$ are viable in D on $[0, T]$.

Theorem 0.1. *Let us assume that F is Marchaud*[5] *and that a closed subset $C \subset K$ satisfies property*

$$K \backslash C \text{ is a repeller under } F \tag{4}$$

Then the viable-capture basin $\text{Capt}^K(C)$ *is the* **unique** *closed subset D satisfying $C \subset D \subset K$ and*

$$\begin{cases} i) & D \backslash C \text{ is locally viable under } \mathcal{S} \\ ii) & D \text{ is is locally backward invariant relatively to } K \text{ under } \mathcal{S} \end{cases} \tag{5}$$

3. from the Viability and Invariance Theorems that translate the necessary and sufficient conditions (5) in terms of tangential conditions. We recall that the contingent cone to a subset K at a point $x \in K$, introduced in the early thirties independently by Bouligand and Severi, adapts to any subset the concept of tangent space to manifolds: A direction $v \in X$ belongs to $T_K(x)$ if there exist sequences $h_n > 0$ and $v_n \in X$ converging to 0 and v respectively such that

$$\forall n \geq 0, \quad x + h_n v_n \in K$$

This means that the contingent cone is the Painlevé-Kuratowski upper limit of the subsets $\frac{K-x}{h}$ when h converges to 0 (see for instance [18, Aubin & Frankowska] and [60, Rockafellar & Wets] for more details).

We shall use the following statements of [12, Aubin]:

Theorem 0.2. *Let us assume that F is Marchaud, that $C \subset K$ and K are closed and that $K \backslash C$ is a repeller. Then the viable-capture basin* $\text{Capt}_F^K(C)$ *is*

(a) *the* **largest** *closed subset D satisfying $C \subset D \subset K$ and*

$$\forall x \in D \backslash C, \quad F(x) \cap T_D(x) \neq \emptyset \tag{6}$$

(b) *if furthermore F is Lipschitz, the* **unique** *closed subset D satisfying $C \subset D \subset K$ and the* Frankowska properties*:*

$$\begin{cases} \text{i)} & \forall x \in D \backslash C, \quad F(x) \cap T_D(x) \neq \emptyset \\ \text{ii)} & \forall x \in D \cap \text{Int}(K), \quad -F(x) \subset T_D(x) \\ \text{iii)} & \forall x \in D \cap \partial K, \quad -F(x) \subset T_D(x) \cup T_{X \backslash K}(x) \end{cases} \tag{7}$$

4. from the property

$$T_{\text{Graph}(V)}(t, x, y) = \text{Graph}(DV(t, x, y))$$

[5] F is a Marchaud map if

$$\begin{cases} \text{i)} & \text{the graph and the domain of } F \text{ are nonempty and closed} \\ \text{ii)} & \text{the values } F(x) \text{ of } F \text{ are convex} \\ \text{iii)} & \text{the growth of F is linear:} \\ & \exists\, c > 0 \mid \forall x \in X, \quad \|F(x)\| := \sup_{v \in F(x)} \|v\| \leq c(\|x\| + 1) \end{cases}$$

of the contingent derivative $DV(t,x,y)$ of the set-valued map $V : (t,x) \rightsquigarrow V(t,x)$ at the point (t,x,y) of its graph introduced in [1, Aubin]: The graph of the set-valued map $DV(t,x,y)$ from $\mathbf{R} \times X$ to Y is equal to the contingent cone to the graph of V at (t,x,y) (see [1, Aubin]). This is how Fermat defined in 1637 the derivative of a function as the slope of the tangent to its graph. Leibniz and Newton provided the characterization in terms of limits of difference quotients. Here, too, *the contingent derivative* $DV(t,x,y)$ *is the upper graphical limit of the difference quotients*, the graph of which being by definition the upper limit of the graphs of the difference quotients $\nabla_h V(t,x,y)$ of V at $(t,x,y) \in \text{Graph}(V)$ defined by

$$(\lambda, f) \mapsto \nabla_h V(t,x,y)(\lambda, f) := \frac{V(t+\lambda h, x+hf) - y}{h}$$

Indeed, we observe that

$$\text{Graph}(\nabla_h V(t,x,y)) = \frac{\text{Graph}(V) - (t,x,y)}{h}$$

so that the contingent cone to the graph of V being the upper limit of the graphs of the difference quotients, is equal by definition the graph of the upper graphical limit of the difference quotients.

Consequently, to say that $g \in Y$ belongs to the contingent derivative $DV(t,x,y)(\pm 1, f)$ of V at (t,x,y) in the direction $(\pm 1, f) \in \mathbf{R} \times X$ means that

$$\liminf_{h \to 0+,\ f' \to f} \left\| \frac{V(t \pm h, x+hf') - y}{h} - g \right\| = 0$$

Since the contingent cone is a closed subset, the graph of a contingent derivative is always closed and positively homogenous (this is what remains of the required linearity of the derivative in classical analysis, but, fortunately, we can survive pretty well without linearity).

When $u : \mathbf{R} \times X \mapsto Y$ is single-valued, we set

$$Du(t,x) := Du(t,x,u(t,x)).$$

We see at once that

$$Du(t,x)(\pm 1, f) = \pm \frac{\partial u(t,x)}{\partial t} + \frac{\partial u(t,x)}{\partial x} \cdot f$$

whenever u is differentiable at (t,x). When u is Lipschitz on a neighborhood of (t,x) and when the dimension of X is finite, the domain of $Du(t,x)$ is not empty. Furthermore, the Rademacher Theorem stating that a locally Lipschitz single-valued map is almost everywhere differentiable implies that $x \rightsquigarrow Du(t,x)$ is almost everywhere single-valued.

However, in this case, equality $Du(t,x)(-1,-f) = -Du(t,x)(1,f)$ is not true in general. We refer to [18, Aubin & Frankowska], [60, Rockafellar & Wets] for more details.

The above results – which are interesting by themselves for other mathematical models of evolutionary economics, population dynamics, epidemiology – can be applied to many other problems. Dealing with subsets, they can be applied to graphs of single-valued maps as well as set-valued maps, to epigraphs and hypographs of (extended) real-valued functions for solving Hamilton-Jacobi-Bellman equation, to graph of "impulse" maps (which take empty values except in a discrete sets, useful in the study of hybrid systems or inventory management), etc, (see for instance [5, 8, 6, 9, Aubin], [25, 26, 27, 28, Aubin & Haddad], [29, Aubin, Lygeros, Quincampoix, Sastry & Seube], [34, Bensoussan & Menaldi], [62, Saint-Pierre] and their bibliographical references).

Outline. We begin in Section 1 with two non standard motivations arising in macroeconomic problems faced by central banks (filtering informations on the economy from past informations and future expectations) and in the study of impulse and hybrid systems, leading to systems of Hamilton-Jacobi-Bellman inclusions. This section (stating the problem without solving them, this task being done in specific articles) can be skipped by true believers in mathematics ... We define in Section 2 the "caliber" of a pair of set-valued map (Φ, Ψ) under a control system, that appears naturally in some control problems and in economic and financial mathematics. We next prove in Section 3 that the caliber is the unique solution to the system of Hamilton-Jacobi-Bellman inclusions satisfying the imposed conditions. In Section 4, we provide a useful stability result, stating roughly that the caliber of graphical limits is the graphical limit of calibers. Section 5 deals with the explicit formula of the caliber when the control system is structured and the exosystem is affine with respect to the second variable. We also prove that in this case, the caliber is the unique solution to a system of "fixed-set equations" and provide interesting "barrier properties" of the boundary of the caliber. We derive in Section 6 the usual characterization theorems of the valuation functions of a large class of control and stopping time problems as Frankowska episolutions to the scalar Hamilton-Jacobi-Bellman equation that justify the usefulness of the results they are derived from, and extend this scalar situation in Section 7 to the case of "dynamical vector optimization", where we look for intertemporal Pareto minima. □

1. Motivations

We shall provide two motivations coming from recent issues arising

1. in macroeconomics (filtering informations on the economy from past informations and future expectations),
2. in hybrid systems and impulse control

leading to systems of Hamilton-Jacobi-Bellman partial differential inclusions.

We refer to specific articles for more details, since they use the basic theorems of this present paper to solve the Hamilton-Jacobi-Bellman partial differential inclusions that pop-up in those articles.

Further applications to the value functions of optimal control and stopping time problems are given in Section 6 and to dynamic vector optimization in Section 7.

1.1. Selector through Past Informations and Future Expectations

As a first motivation, we present a problem originating in a research programme under current investigation by Noël Bonneuil, Halim Doss, Georges Haddad, Henri Pages, Dominique Pujal, Patrick Saint-Pierre and the author on macro-economic problems faced by central banks.

We quote excerpts of the introduction of [68, Svensson & Woodford] *" It is a truism that monetary policy operates under considerable uncertainty about the state of the economy and the size and nature of the disturbances that hit the economy... But in a more realistic case where important variables are forward-looking [and not only backward-looking] variables, the problem of efficient signal-extraction is inherently more complicated ... In the real world, many important indicator variables are forward-looking variables [routinely watched by central banks] ... "*

We suggest to take up this issue by using nonlinear continuous evolutionary models controlled by instruments such as interest rates, keeping the problem of extracting the "real evolution" knowing only backward-looking measurements and forward-looking expectations, that we shall describe by "expectation tubes".

The use of the Kalman filter for extracting information is replaced by the recent concept of detector introduced in [13, Aubin, Bicchi & Pancanti] and [27, Aubin & Haddad] in the case of "inpulse and hybrid control systems". We adapt this concept of detector in the case of both backward-looking and forward-looking informations and expectations.

For that purpose, we introduce three time variables for describing the evolution of the system: the current (or present) time T, the past time $t \in [0, T]$, and a prediction time or forward-looking time $s \geq T$ where $a := s - T \in \mathbf{R}_+$ is the prediction horizon, used to take into account anticipations, expectations (or make predictions) in the future.

At each past date, the state is measured, or informations on the state are gathered: This is mathematically described by a "detectability tube" (as in [13, Aubin, Bicchi & Pancanti])

$$t \in \mathbf{R}_+ \rightsquigarrow I(t) \subset X := \mathbf{R}^n$$

that provides the limited amount of information about the states at time t. We take $I(t) := X$ when no information is recorded at time t. Hence "discrete" measurements are obtained when $I(t) \neq X$ only for a discrete number of instants t_n.

An example of detectability tubes is given by $I(t) := h^{-1}(y(t))$ where $h : X \mapsto Y$ is an observation map (or measurement map) and where $t \mapsto y(t)$ is the evolution of the observed output:

$$\forall\, t \in [0, T], \;\; y(t) \;=\; h(x(t))$$

The same framework houses also the case when the observation map is set-valued: We set $I(t) := H^{-1}(y(t))$ and the above viability condition reads

$$\forall\, t \in [0,T],\;\; y(t) \;\in\; H(x(t))$$

so that "tychastic" uncertainties (by opposition to stochastic uncertainties) on the measurements can be incorporated in this framework.

In order to take into account expectations made at each instant t for future dates $s := t + a$, $a \geq 0$, we describe them mathematically by an expectation tube $(t,a) \in \mathbf{R}^2_+ \mapsto I(t,a) \subset X$ where we set $I(t,0) := I(t)$ for obtaining the detectability tube.

We may assume that if $a_1 \leq a_2$, then $I(t,a_2) \subset I(t,a_1) \subset I(t,0) =: I(t)$, since the predictions made at time t up to time $s_2 := t + a_2$ are valid up to time $s_1 := t + a_1 \leq s_2$.

Therefore, we associate with any current time T, any horizon $s \geq t$ and any backward-looking time $t \in [0,T]$ the set $I(t, s-t)$ of states measured at time t and that depend upon the duration $a := s - t$ of the expectation interval between t and the horizon s.

Hence, the information/expectation constraint can be summarized by

$$\forall\, T \geq 0,\; \forall\, s \geq T \;\&\; \forall\, t \in [0,T],\; x(t) \;\in\; I(t, s-t) \tag{8}$$

Let $\mathcal{U}$ be a space of controls, regulees, prices, interest-rate instruments, etc. The dynamics of the state are described by a map $f : (t,a,x,u) \in \mathbf{R}^2_+ \times X \times \mathcal{U} \mapsto X$ and by a set-valued map $P : \mathbf{R}_+ \times X \rightsquigarrow \mathcal{U}$ depicting the state-dependent constraints on the controls $u \in P(a,x)$.

For any current time T and horizon $s \geq T$, we assume that the evolution of the state of the system is governed by the control system

$$\forall\, t \in [0,T], \begin{cases} \text{i)} & x'(t) \;=\; f(t, s-t, x(t), u(t)) \\ \text{ii)} & u(t) \;\in\; P(s-t, x(t)) \end{cases} \tag{9}$$

In other words, at each time $t \in [0,T]$, the velocity $x'(t)$ depends upon time t, the time $s - t$ left to the horizon s and a control $u(t)$ subjected to constraints depending both upon the expected time $s-t$ left to the horizon and the state $x(t)$ at tine t.

We also introduce a set-valued map $a \in \mathbf{R}_+ \rightsquigarrow C(a)$ specifying another constraint on the subsets $C(a)$ of initial states that may depends upon the term $a \geq 0$ satisfying

$$\forall\, a \geq 0,\;\; C(a) \;\subset\; I(0,a)$$

Definition 1.1. *Let us consider a control system (f,P), an expectation tube $I : \mathbf{R}^2_+ \rightsquigarrow X$ and a tube $C : \mathbf{R}_+ \rightsquigarrow X$ satisfying $\forall\, a \geq 0,\;\; C(a) \subset I(0,a)$.*

The selector *is the set-valued map $\mathbf{S}_{(I,C)} : \mathbf{R}^2_+ \rightsquigarrow X$ that associates with any current time T and any expectation $a := s - T$ the (possibly empty) subset*

$\mathbf{S}_{(I,C)}(t,a)$ *of states* $x \in I(t,a)$ *such that there exists a solution* $x(\cdot) \in \mathcal{S}(t,a,x)$ *to the system*

$$\begin{cases} \forall\, t \in [0,T],\ \forall\, s := T+a \geq 0, \\ \text{i)} \quad x'(t) \ = \ f(t, s-t, x(t), u(t)) \\ \text{ii)} \quad u(t) \ \in \ P(s-t, x(t)) \end{cases} \tag{10}$$

such that

$$\begin{cases} \text{i)} \quad x(0) \ \in \ C(s) \\ \text{ii)} \quad \forall\, t \in [0,T], \ \ x(t) \ \in \ I(t, s-t) \\ \text{iii)} \quad x(T) \ = \ x \end{cases}$$

In other words, both the dynamics and the constraints depend upon horizon $s \geq T$ and take into account the informations gathered at any preceding time $t \in [0,T]$ and expectations at time $s-t$ left to the horizon s. The selector is thus a tube associating with any horizon $s \geq T$ the set of states x such that there exists a control $u(\cdot)$ governing the evolution $x(\cdot)$ through control system (9):

$$\forall\, t \in [0,T], \begin{cases} \text{i)} \quad x'(t) \ = \ f(t, s-t, x(t), u(t)) \\ \text{ii)} \quad u(t) \ \in \ P(s-t, x(t)) \end{cases}$$

that satisfies for all anterior time $t \in [0,T]$ the expected constraints made at that time t for the future time $s-t$.

We can check easily that

Lemma 1.2. *The graph of the selector* $\mathbf{S}_{(I,C)}$ *is the capture basin of* $\{0\} \times \mathrm{Graph}(C)$ *viable in the graph of the tube* I *under the auxiliary system*

$$\begin{cases} \text{i)} \quad \tau'(t) \ = \ -1 \\ \text{ii)} \quad \alpha'(t) \ = \ +1 \\ \text{iii)} \quad x'(t) \ = \ -f(\tau(t), \alpha(t), x(t)) \end{cases}$$

Proof. Indeed, to say that (T,a,x) belongs to the viable-capture basin of $\{0\} \times \mathrm{Graph}(C)$ viable in $\mathrm{Graph}(I)$ means that there exist an evolution $\widehat{x}(\cdot)$ to $\widehat{x}'(t) \in -f(T-t, a+t, \widehat{x}(t), \widetilde{u}(t))$ starting at $\widehat{x}(0) := x$ and a time $t^\star \geq 0$ such that

$$\begin{cases} \text{i)} \quad \forall\, t \in [0,t^\star], \quad (T-t, a+t, \widehat{x}(t)) \ \in \ \mathrm{Graph}(I) \\ \text{ii)} \quad\quad (T-t^\star, a+t^\star, \widehat{x}(t^\star)) \ \in \ \{0\} \times \mathrm{Graph}(C) \end{cases}$$

The second condition means that $t^\star = T$ and that $\widehat{x}(T)$ belongs to $C(a+T)$. The first one means that for every $t \in [0,T]$, $\widehat{x}(t) \in I(T-t, a+t)$. This amounts to saying that the evolution $x(\cdot) := \widehat{x}(T-\cdot)$ is a solution to the control system

$$x'(t) \ = \ f(t, a+T-t, x(t), u(t))$$

starting at $x(0) := \widehat{x}(T) \in C(T+a)$, satisfying $x(T) = x$ and

$$\forall\, t \in [0,T],\ \forall\, a \geq 0, \ \ x(t) \in I(t, a+T-t)$$

This means that $x \in \mathbf{S}_{(I,C)}(T,a)$. □

We shall therefore characterize the selector as a solution to a system of Hamilton-Jacobi-Bellman partial differential inclusions

$$\forall\, x \in V(t,a),\ \exists\, u \in P(a,x)$$

such that

$$-\frac{\partial V(t,a,x)}{\partial t} + \frac{\partial V(t,a,x)}{\partial a} + f(t,a,x,u) \;=\; 0$$

satisfying the initial condition

$$\forall\, a \geq 0,\ \ V(0,a) \;=\; C(a)$$

and the viability constraints

$$\forall\, (t,a) \in \mathbf{R}^2_+,\ \ V(t,a) \;\subset\; I(t,a)$$

We deduce from the knowledge of the derivatives of the selector the **regulation map** $R : \mathbf{R}^2_+ \times X \rightsquigarrow \mathcal{U}$ providing the controls (or regulees, prices, interest-rate instruments) that at each time t, for any future date a and any state x answer the detection/prediction problems. The regulation map associates with any triple (t,a,x) the set $R(t,a,x)$ of controls $u \in P(a,x)$ such that the solutions to the new control system

$$\left\{\begin{array}{l} \forall\, t \in [0,T],\ \forall\, s \geq T, \\ \text{i)}\ \ x'(t) \;=\; f(t,s-t,x(t),u(t)) \\ \text{ii)}\ \ u(t) \;\in\; R(t,s-t,x(t)) \end{array}\right. \tag{11}$$

satisfy the constraints (1.1)

$$\left\{\begin{array}{l} \text{i)}\ \ x(0) \;\in\; C(s) \\ \text{ii)}\ \ \forall\, t \in [0,T],\ \ x(t) \;\in\; I(t,s-t) \\ \text{iii)}\ \ x(T) \;=\; x \end{array}\right.$$

Finally, the Capture Basin Algorithm allows us to compute the selector and the regulation map.

1.2. The Substratum of an Impulse Differential Inclusion

Impulse differential inclusions are described by two set-valued maps F – the right-hand side of the differential inclusion $x' \in F(x)$ governing the continuous evolution of a hybrid system – and Φ, describing the **reset map** reinitializing the system when required and a constrained set K inside which the evolution of the impulse differential equation must remain. We denote by $\mathcal{S}(x)$ the set of solutions $x(\cdot)$ to the differential inclusion starting at x.

Let us set $x(^-t) := \lim_{\tau \mapsto t-} x(\tau)$ when $x(\cdot)$ is defined on some interval $[t-\eta, t[$ where $\eta > 0$, and, for consistency purposes, $x(s) = x(^-t)$ if $s = t$.

An evolution of the impulse differential inclusion, called a "run" or an "execution" in the hybrid system community – is a finite or infinite sequence $x(\cdot) := \{(\tau_n, x_n, x_n(\cdot))\}_{n \geq 0}$ made of triples

1. of nonnegative **cadences** $\tau_n \in [0, +\infty[$,
2. a sequence of **reinitialized states** x_n,
3. a sequence of **motives** $x_n(\cdot) \in \mathcal{S}(x_n)$ satisfying the end-point condition $x_n(\tau_n) \in \Phi^{-1}(x_{n+1})$

defining the sequence of impulse times $t_{n+1} := t_n + \tau_n$ and, on each interval $[t_n, t_{n+1}[$,

$$\forall\, t \in [t_n, t_{n+1}[,\;\; x(t) \;:=\; x_n(t - t_n) \tag{12}$$

If the sequence is finite and stops at τ_N, we set $\tau_{N+1} := +\infty$ and take $x_N(\cdot) \in \mathcal{S}(x_N)$ (this definition is taken from [25, 26, Aubin & Haddad]).

We say that a run $x(\cdot)$ is viable in K if for any $t \geq 0$, $x(t) \in K$.

At this stage, a run $x(\cdot)$ can just be a (discrete) sequence of states $x_{n+1} \in \Phi(x_n)$ at a fixed time, or just a (continuous) solution $x(\cdot)$ to the differential inclusion $x' \in F(x)$, or an hybrid of these two modes, the discrete and the continuous.

We just define the concept of **substratum** of an impulse differential inclusion introduced in [9, Aubin], that summarizes the salient features of a run by considering only its sequences of cadences τ_n and of reinitialized states x_n. Knowing them, we can reconstruct the motives of the run by taking solutions $x_n(\cdot) \in \mathcal{S}(x_n)$ satisfying the end-point condition $x_n(\tau_n) \in \Phi^{-1}(x_{n+1})$. The question that arises is to provide an algorithm that provides these sequences of cadences τ_n and of reinitialized states x_n without solving the impulse differential inclusion, but through an algorithm. The substratum just does that:

Definition 1.3. *We associate with the dynamics (F, Φ) of the impulse differential inclusion its* substratum $\Gamma : \mathbf{R}_+ \times K \rightsquigarrow K$*, that is the set-valued map associating with any $(T, x) \in \mathbf{R}_+ \times K$ the subset*

$$\Gamma(T, x) \;:=\; \bigcup_{x(\cdot) \in \mathcal{S}_F^K(x)} \Phi(x(T))$$

of the elements $y \in \Phi(x(T))$ where $x(\cdot) \in \mathcal{S}_F^K(x)$ is a solution to the differential inclusion $x' \in F(x)$ starting at x and viable in K until it reaches $x(T) \in C := K \cap \Phi^{-1}(K)$ at time T.

We associate with the substratum Γ

1. *the* cadence map

$$\mathbf{C}(x) \;:=\; \{t \geq 0 \;\; \textit{such that}\; \Gamma(t, x) \;\neq\; \emptyset\}$$

2. *and the* initialization map $\mathbf{I} : K \rightsquigarrow X$

$$\mathbf{I}(x) \;=\; \bigcup_{t \in \mathbf{C}(x)} \Gamma(t, x)$$

Knowing the substratum Γ of (K, F, Φ), and thus the cadence map $\mathbf{C}$ and the initialization map $\mathbf{I}$, we can **reconstruct** a viable run of the impulse differential inclusion (F, Φ) through the following algorithm: Given the cadence τ_n and the initial state x_n, we take

$$\left\{\begin{array}{ll} \text{i)} & \text{the next cadence } \tau_{n+1} \in \mathbf{C}_{(F,\Phi)}(x_n), \\ \text{ii)} & \text{the next reinitialized state } x_{n+1} \in \Gamma_{(F,\Phi)}(\tau_{n+1}, x_n) \subset \mathbf{I}(x_n), \\ \text{iii)} & \text{the next motive } x_n(\cdot) := x(\cdot + t_n), \text{ a solution to } x' \in F(x) \text{ satisfying} \\ & x_n(0) = x_n \;\&\; x_n(\tau_{n+1}) \in \Phi^{-1}(x_{n+1}) \end{array}\right. \tag{13}$$

Assume for a while that the impulse differential inclusion is actually an impulse differential equation (f, φ) where the maps f and φ are single-valued and that the substratum is single-valued and differentiable. We define

$$\Phi(t,x) := \left\{\begin{array}{lll} \varphi(x) & \text{if} & x \in C := K \cap \varphi^{-1}(K) \\ \emptyset & \text{if} & x \notin C \end{array}\right.$$

and

$$\Psi(t,x) := \left\{\begin{array}{lll} K & \text{if} & x \in K \\ \emptyset & \text{if} & x \notin K \end{array}\right.$$

Then we shall prove that the substratum is a "solution" $v(t,x)$ to the system of first-order partial differential inclusions

$$\forall\, i = 1, \ldots n, \quad -\frac{\partial v_i(t,x)}{\partial t} + \sum_{j=1}^{n} \frac{\partial v_i(t,x)}{\partial x_j} f_j(x) = 0$$

satisfying the "condition"

$$\forall\, x \in C := K \cap \varphi^{-1}(K), \;\; v(0,x) = \varphi(x)$$

that is either single-valued or takes empty values – and thus, is a set-valued map.

Actually, we shall extend this result to general impulse differential inclusions by characterizing the substratum as a generalized (set-valued) solution – a Frankowska solution – to the system of first-order partial differential inclusions

$$\left\{\begin{array}{ll} \text{i)} & -\frac{\partial}{\partial t} V(t,x) + \frac{\partial V(t,x)}{\partial x} \cdot u = 0 \\ \text{ii)} & u \in F(x) \end{array}\right.$$

satisfying the "condition"

$$\forall\, x \in C := K \cap \Phi^{-1}(K), \;\; V(0,x) = \Phi(x)$$

and the constraints $V(t,x) \subset K$.

Indeed, the substratum is a particular case of a caliber with $f(x,y,u) := u$, $g(x,y,u) := 0$, $P(x,y) := F(x)$, $\Phi(0,x) := \Phi(x)$ if $x \in K$, $\Phi(0,x) := \emptyset$ if $x \notin K$, $\Phi(t,x) := \emptyset$ if $t > 0$, $\Psi(t,x) := K$ if $x \in K$ and $\Psi(t,x) := \emptyset$ otherwise.

2. The Caliber of Dynamical Constraints and Objective

The purpose of this section is to show how "viability techniques" may be efficient for solving systems of first-order partial differential inclusions arising in different fields of control theory and hybrid systems.

We denote by $L^1(0,\infty;\mathcal{U})$ the set of measurable functions from $[0,+\infty[$ to a vector space $\mathcal{U}$, the **control space**.

We consider a control system of the form

$$\begin{cases} \text{i)} & x'(t) = f(x(t),y(t),u(t)) \\ \text{ii)} & y'(t) = g(x(t),y(t),u(t)) \\ \text{iii)} & u(t) \in P(x(t),y(t)) \end{cases}$$

We denote by $\mathcal{B}(x,y)$ the set of solutions $(x(\cdot),y(\cdot),u(\cdot)) \in \mathcal{C}(0,\infty;X\times Y)\times L^1(0,\infty;\mathcal{U})$ to the above system starting at (x,y) at time 0.

Definition 2.1. *We say that the* control system *is*

1. Marchaud *if the set-valued map $P : X\times Y \rightsquigarrow \mathcal{U}$ is Marchaud, if $f : X\times Y\times \mathcal{U} \mapsto X$ and $g : X\times Y\times\mathcal{U}\mapsto Y$ are continuous and affine*[6]*,*
2. Lipschitz *if the set-valued map $P : X \rightsquigarrow \mathcal{U}$ is Lipschitz and if $f : X\times Y\times\mathcal{U}\mapsto X$ and $g : X\times Y\times\mathcal{U}\mapsto Y$ are Lipschitz.*

We associate the set-valued map $G : \mathbf{R}_+\times X\times Y \rightsquigarrow \mathbf{R}_+\times X\times Y$ defined by

$$G(T,x,y) := \{\{-1\}\times\{f(x,y,u)\}\times\{g(x,y,u\}\}_{u\in P(x,y)} \tag{14}$$

and we denote by $\mathcal{R}$ the set-valued map defined the formula

$$\mathcal{R}(T,x,y) = \{(T-\cdot,x(\cdot),y(\cdot),u(\cdot))\}_{(x(\cdot),y(\cdot),u(\cdot))\in\mathcal{B}(x,y)}$$

We infer that *the set-valued map G is a Marchaud (resp. Lipschitz) map whenever the control system is Marchaud (resp. Lipschitz).*

We introduce now dynamical constraints and objectives defined by

1. a set-valued map $\Phi : \mathbf{R}_+\times X \rightsquigarrow Y$ defining an **objective**, regarded as an obstacle in problems of unilateral mechanics, for instance.
2. a set-valued map $\Psi : \mathbf{R}_+\times X \rightsquigarrow Y$ defining dynamical constraints. **State constraints** are involved in the domain

$$\mathrm{Dom}(\Psi) := \{(t,x)\in\mathbf{R}_+\times X \mid \Psi(t,x) \neq \emptyset\}$$

of the set-valued map Ψ.

We shall assume that

$$\begin{cases} \text{i)} & \forall\,(t,x)\in\mathbf{R}_+\times X, \quad \Phi(t,x) \subset \Psi(t,x) \\ \text{ii)} & \forall\, t<0,\ \forall\, x\in X, \quad \Psi(t,x) := \emptyset \end{cases}$$

[6]We actually need only that the values $\{(f(x,y,u),g(x,y,u))\}_{u\in P(x,y)}$ are closed and convex.

Definition 2.2. *The two constraint and objective set-valued maps being given, the* caliber $(T,x) \rightsquigarrow V_\Psi(\Phi)(T,x)$ of the pair (Ψ,Φ) under the controlled system *is the set-valued map associating with the pair (T,x) made of the horizon T and the initial state x the set of initial observations y such that there exist a control $t \in [0,T] \mapsto u(t)$ and a time $T^\star \in [0,T]$ such that a solution $(x(\cdot),u(\cdot),y(\cdot)) \in \mathcal{B}(x,y)$ starting at $x(0)=x$, $y(0)=y$ satisfies*

$$\begin{cases} \text{i)} & \forall\, t \in [0,T^\star], \;\; y(t) \in \Psi(T-t,x(t)) \\ \text{ii)} & y(T^\star) \in \Phi(T-T^\star,x(T^\star)) \end{cases} \tag{15}$$

We observe at once the following property: *The caliber satisfies the initial condition*

$$\forall\, x \in X, \;\; V_\Psi(\Phi)(0,x) \;=\; \Phi(0,x)$$

Indeed, condition (15)ii) with $T=0$ means that $y \in V_\Psi(\Phi)(0,x)$, implying that $T^\star=0$ and $(0,x,y) \in \text{Graph}(\Phi)$, i.e., $y \in \Phi(0,x)$. Hence $V_\Psi(\Phi)(0,x) \subset \Phi(0,x) \subset V_\Psi(\Phi)(0,x)$. □

What is the connection between this problem and the basic viability theorems? The answer is simple: The graph of the caliber is the capture basin of the graph of the set-valued map Φ viable in the graph of Ψ under the auxiliary control system G.

Proposition 2.3. *The graph of the caliber $V_\Psi(\Phi)$ is equal to the viable-capture basin of* Graph(Φ) *viable in* Graph(Ψ) *under the auxiliary system $\mathcal{R}$:*

$$\text{Graph}(V_\Psi(\Phi))) \;=\; \text{Capt}_{\mathcal{R}}^{\text{Graph}(\Psi)}(\text{Graph}(\Phi))$$

Proof. It is enough to translate conditions (15) in the form

$$\begin{cases} \text{i)} & \forall\, t \in [0,T^\star], \;\; (T-t,x(t),y(t)) \in \text{Graph}(\Psi) \\ \text{ii)} & (T-T^\star,x(T^\star),y(T^\star)) \in \text{Graph}(\Phi) \end{cases} \tag{16}$$

to recognize that

$$\text{Graph}(V_\Psi(\Phi)) \;=\; \text{Capt}_{\mathcal{R}}^{\text{Graph}(\Psi)}(\text{Graph}(\Phi)) \qquad \square$$

This being checked, it will be sufficient to translate the properties of capture basins in terms of caliber.

3. Set-Valued Solutions to Systems of Hamilton-Jacobi-Bellman Inclusions

In the case of controlled systems, we shall relate the caliber with the set-valued solution to the controlled Hamilton-Jacobi-Bellman partial differential equations

1. there exists $u \in P(x,V(t,x))$ such that

$$0 \in \text{“}-\frac{\partial}{\partial t}V(t,x)+\frac{\partial}{\partial x}V(t,x)f(x,V(t,x),u)\text{''} - g(x,V(t,x),u)$$

2. for all $u \in P(x, v(t,x))$,

$$0 \in \text{“}\frac{\partial}{\partial t}V(t,x) - \frac{\partial}{\partial x}V(t,x)f(x,V(t,x),u)\text{''} + g(x,V(t,x),u)$$

satisfying the initial condition

$$V(0,x) = \Phi(0,x)$$

and the constraints

$$\forall\, t \geq 0,\; x \in X,\;\; \Phi(t,x) \subset V(t,x) \subset \Psi(t,x)$$

in a sense that we make precise in the following Theorem.

Theorem 3.1. *Let us assume that the system is Marchaud.*

1. *Then the caliber $V_\Psi(\Phi)$ of (Ψ,Φ) is the* **largest** *closed set-valued map $V : \mathbf{R}_+ \times X \rightsquigarrow Y$ satisfying*

$$\forall\,(t,x) \in \mathbf{R}_+ \times X,\;\; \Phi(t,x) \subset V(t,x) \subset \Psi(t,x) \qquad (17)$$

and

$$\left\{ \begin{array}{l} \forall\, y \in V(t,x)\backslash\Phi(t,x),\; \exists\, u \in P(x,y) \\ \text{such that} \\ 0 \in DV(-1, f(x,y,u)) - g(x,y,u) \end{array} \right. \qquad (18)$$

2. *Let us set*

$$\mathbf{R}(t,x,y) := \{u \in P(x,y) \mid 0 \in DV_\Psi(\Phi)(t,x,y)(-1, f(x,y,u)) - g(x,y,u)\}$$

Knowing the caliber, any solution satisfying the constraints (15)i) and reaching the objective (15)ii) in finite time is obtained in the following way: Starting from (x_0, y_0) such that $y_0 \in V_\Psi(\Phi)(T,x_0)\backslash\Phi(T,x_0)$, any solution $(x(\cdot), y(\cdot), u(\cdot))$ to the control system: for almost all $t \in [0,T]$,

$$\left\{ \begin{array}{lrcl} \text{i)} & x'(t) & = & f(x(t), y(t), u(t)) \\ \text{ii)} & y'(t) & = & g(x(t), y(t), u(t)) \\ \text{iii)} & u(t) & \in & \mathbf{R}(T-t, x(t), y(t)) \end{array} \right. \qquad (19)$$

starting at (x,y) is a solution satisfying

$$y(t) \in V_\Psi(\Phi)(T-t, x(t))$$

until the first time $t^\star \in]0,T]$ when

$$y(t^\star) \in \Phi(T-t^\star, x(t^\star))$$

3. *If we assume furthermore that the system is Lipschitz, then the caliber $V_\Psi(\Phi)$ of (Ψ,Φ) is the* **unique** *closed set-valued map $V : \mathbf{R}_+ \times X \rightsquigarrow Y$ satisfying (17), (18) and*

$$\left\{ \begin{array}{l} \forall\, y \in V(t,x) \cap \Psi^\circ(t,x),\; \forall\; u \in P(x,y) \\ 0 \in DV(1, -f(x,y,u)) + g(x,y,u) \\ \& \\ \forall\, y \in V(t,x) \cap \Psi^\partial(t,x),\; \forall\; u \in P(x,y) \\ 0 \in DV(1, -f(x,y,u)) \cup D\Psi^c(+1, -f(x,y,u)) + g(x,y,u) \end{array} \right. \qquad (20)$$

where

$$\mathrm{Graph}(\Psi^\circ) := \mathrm{Int}(\mathrm{Graph}(\Psi)), \quad \mathrm{Graph}(\Psi^\partial) := \partial\mathrm{Graph}(\Psi)$$

and

$$\mathrm{Graph}(\Psi^c) := X\backslash\mathrm{Graph}(\Psi).$$

Proof.

1. The graph of the caliber $V_\Psi(\Phi)$ being equal to the capture basin

$$\mathrm{Capt}_G^{\mathrm{Graph}(\Psi)}(\mathrm{Graph}(\Phi))$$

by Proposition 2.3, we can apply the first part of Theorem 0.1 since

$$\mathrm{Graph}(\Psi) \subset \mathbf{R}_+ \times X \times Y$$

is a repeller under G. Hence it is the largest graph of a set-valued map $V : X \rightsquigarrow Y$ between $\mathrm{Graph}(\Phi)$ and $\mathrm{Graph}(\Psi)$ such that, for any $(t,x,y) \in \mathrm{Graph}(V)\backslash\mathrm{Graph}(\Phi))$ – i.e., whenever $y \in V(t,x)\backslash\Phi(t,x)$ – there exists a control $u \in P(x)$ such that

$$(-1, f(x,y,u), -g(x,y,u)) \in T_{\mathrm{Graph}(V)}(t,x,y) \tag{21}$$

In other words, it is the graph of the largest closed set-valued map V satisfying

$$\Phi(t,x) \subset V(t,x) \subset \Psi(t,x)$$

and, whenever $y \in V(t,x)\backslash\Phi(t,x)$, there exists $u \in P(x,y)$ such that

$$0 \in DV(-1, f(x,y,u)) - g(x,y,u)$$

2. The solutions $(T - \cdot, x(\cdot), y(\cdot))$ viable in the graph of Ψ until they reach the graph of Ψ satisfy: for almost all $t \geq 0$.

$$\left\{\begin{array}{l} u(t) \in P(x(t),y(t)) \\ \& \\ (-1, f(x(t),y(t),u(t)), g(x(t),y(t),u(t))) \in T_{\mathrm{Graph}(V)}(T-t, x(t), y(t)) \end{array}\right.$$

This condition can be rewritten

$$\text{for almost all } t \geq 0, \quad u(t) \in R(T-t, x(t), y(t))$$

3. Under the Lipschitz conditions, Theorem 0.2 states that the graph of $V_\Psi(\Phi)$ is the unique closed subset $\mathrm{Graph}(V)$ satisfying the Frankowska properties:

$$\left\{\begin{array}{ll} \text{i)} & \forall\, (t,x,y) \in \mathrm{Graph}(V)\backslash\mathrm{Graph}(\Phi), \\ & \exists\, u \in P(x,y) \mid (-1, f(x,y,u), g(x,y,u)) \in T_{\mathrm{Graph}(V)}(t,x,y) \\ \text{ii)} & \forall\, (t,x,y) \in \mathrm{Graph}(V) \cap \mathrm{Int}(\mathrm{Graph}(\Psi)), \forall\, u \in P(x,y) \\ & (1, -f(x,y,u), -g(x,y,u)) \in T_{\mathrm{Graph}(V)}(t,x,y) \\ \text{iii)} & \forall\, (t,x,y) \in \mathrm{Graph}(V) \cap \partial(\mathrm{Graph}(\Psi)), \forall\, u \in P(x,y) \\ & (1, -f(x,y,u), -g(x,y,u)) \in T_{\mathrm{Graph}(V)}(t,x,y) \cup T_{X\backslash\mathrm{Graph}(\Psi)}(t,x,y) \end{array}\right.$$

Using the fact that $\mathrm{Graph}(DV)(t,x,y) = T_{\mathrm{Graph}(V)}(t,x,y)$, we infer the third part of the Theorem. □

4. Stability Properties

We state the following Theorem of [12, Aubin]:

Theorem 4.1. *Let us consider a sequence of closed subsets C_n satisfying* $\mathrm{Viab}(K) \subset C_n \subset K$ *and*

$$\mathrm{Lim}_{n\to+\infty} C_n := \mathrm{Limsup}_{n\to+\infty} C_n = \mathrm{Liminf}_{n\to+\infty} C_n$$

If the set-valued map F is Marchaud and Lipschitz and if K is closed and backward invariant under F, then

$$\mathrm{Lim}_{n\to+\infty}\mathrm{Capt}^K(C_n) = \mathrm{Capt}^K(\mathrm{Lim}_{n\to+\infty} C_n) \tag{22}$$

Theorem 4.1 implies "continuity properties" of the caliber regarded as a map $\Phi \mapsto V_\Psi(\Phi)$.

For that purpose, let us recall the following definitions of graphical convergence (see [18, Aubin & Frankowska] and/or [60, Rockafellar & Wets] for instance):

1. the upper limit of the graphs

$$\mathrm{Limsup}_{n\to+\infty}\mathrm{Graph}(\Phi_n) =: \mathrm{Graph}(\mathrm{Lim}^\sharp{}_{n\to+\infty}\Phi_n)$$

is the graph of the **graphical upper limit** $\mathrm{Lim}^\sharp{}_{n\to+\infty}\Phi_n$ defined by

$$(\mathrm{Lim}^\sharp{}_{n\to+\infty}\Phi_n)(T,x) = \liminf_{n\to+\infty,\ s\to t,\ y\to x} V(\Phi_n)(s,y)$$

2. the lower limit of the graphs

$$\mathrm{Liminf}_{n\to+\infty}\mathrm{Graph}(\Phi_n) =: \mathrm{Graph}(\mathrm{Lim}^\flat{}_{n\to+\infty}\Phi_n)$$

is the graph of the **graphical lower limit**.

Then we derive the following "continuity" properties of the calibers:

Theorem 4.2. *Let us consider a sequence of nontrivial set-valued maps $\Psi_n : \mathbf{R}_+ \times X \rightsquigarrow Y \cup \{+\infty\}$ and $\Phi_n : \mathbf{R}_+ \times X \rightsquigarrow Y \cup \{+\infty\}$ such that*

$$\forall\,(T,x) \in \mathbf{R}_+ \times X,\ \ \Phi_n(T,x) \subset \Psi_n(T,x)$$

The calibers satisfy also the following properties.

1. *Let us assume that the auxiliary control system F is Marchaud. Then*

$$\lim{}^\sharp{}_{n\to+\infty} V_{\Psi_n}(\Phi_n)(T,x) \subset V_{\lim^\sharp{}_{n\to+\infty}\Psi_n}\left(\lim{}^\sharp{}_{n\to+\infty}\Phi_n\right)(T,x)$$

2. *Let us assume that the auxiliary control system F is Marchaud and Lipschitz. Then*

$$\begin{cases} V\left(\lim{}^\flat{}_{n\to+\infty}\Phi_n\right)(T,x) \subset \lim{}^\flat{}_{n\to+\infty} V(\Phi_n)(T,x) \\ \subset \lim{}^\sharp{}_{n\to+\infty} V(\Phi_n)(T,x) \subset V\left(\lim{}^\sharp{}_{n\to+\infty}\Phi_n\right)(T,x) \end{cases}$$

Therefore, if the sequence of set-valued maps Φ_n converges graphically to Φ, the caliber of the graphical limit is the graphical limit of the calibers.

Proof. If the system is Marchaud, Theorem 4.1 implies that

$$\begin{aligned}
&\mathrm{Limsup}_{n\to+\infty}\mathrm{Capt}_{\mathcal{R}}^{\mathrm{Graph}(\Psi_n)}(\mathrm{Graph}(\Phi_n)) \\
&\quad\subset \mathrm{Capt}_{\mathcal{R}}^{\mathrm{Limsup}_{n\to+\infty}\mathrm{Graph}(\Psi_n)}(\mathrm{Limsup}_{n\to+\infty}\mathrm{Graph}(\Phi_n))
\end{aligned}$$

Hence we deduce from Proposition 2.3 characterizing the graph of the caliber as a viable-capture basin and from the definitions of graphical limits that

$$\mathrm{lim}^{\sharp}{}_{n\to+\infty}V_{\Psi_n}(\Phi_n)(T,x) \subset V_{\mathrm{lim}^{\sharp}{}_{n\to+\infty}\Psi_n}\left(\mathrm{lim}^{\sharp}{}_{n\to+\infty}\Phi_n\right)(T,x)$$

Under Lipschitz conditions of G, Theorem 4.1 implies that

$$\mathrm{Capt}_{\mathcal{R}}(\mathrm{Liminf}_{n\to+\infty}\mathrm{Graph}(\Phi_n)) \subset \mathrm{Liminf}_{n\to+\infty}\mathrm{Capt}_{\mathcal{R}}(\mathrm{Graph}(\Phi_n))$$

Hence we deduce that

$$V\left(\mathrm{lim}^{\flat}{}_{n\to+\infty}\Phi_n\right)(T,x) \subset \mathrm{lim}^{\flat}{}_{n\to+\infty}V(\Phi_n)(T,x)$$

This completes the proof. □

5. Caliber of Structured Problems with Linear Exosystems

Of special interest is the particular case when the first differential equation does not depend upon the variable y and when the set-valued map $P : X \rightsquigarrow \mathcal{U}$ does not depend on the observation variable y: We thus obtain a **structured system** – as **age structured systems** in demography, when x plays the role of the age variable – of the form

$$\begin{cases}
\text{i)} & x'(t) = f(x(t),u(t)) \\
\text{ii)} & y'(t) = g(x(t),y(t),u(t)) \\
\text{iii)} & u(t) \in P(x(t))
\end{cases}$$

where $y(\cdot)$ is often regarded as an **observation** of the state (see for instance [14, Aubin, Bonneuil & Maurin]). In control theory, the second controlled equation is called the **exosystem**.

We denote by $\mathcal{C} : x \in X \rightsquigarrow \mathcal{C}(x) \in \mathcal{C}(0,\infty;X) \times L^1(0,\infty;\mathcal{U})$ the set-valued map associating with $x \in X$ the set $\mathcal{C}(x)$ of the pairs $(x(\cdot),u(\cdot))$ solutions to the control system

$$\begin{cases}
\text{i)} & x'(t) = f(x(t),u(t)) \\
\text{ii)} & u(t) \in P(x(t))
\end{cases}$$

starting at x at $t = 0$.

We shall also set $g(x,y,u) := -M(x,u)y - L(x,u)$ where

1. M is a bounded continuous matrix-valued function

$$M : (x,u) \in X \times \mathcal{U} \mapsto M(x,u) \in \mathcal{L}(X,Y)$$

2. L is a continuous[7] "vector-Lagrangian"

$$L : (x,u) \in X \times \mathcal{U} \mapsto L(x,u) \in Y$$

with linear growth.

defining the map $(y;(x(\cdot),u(\cdot))) \rightsquigarrow \mathcal{S}(y;(x(\cdot),u(\cdot)))$ associating the subset of evolutions

$$y(t) = e^{-\int_0^t M(x(s),u(s))ds}\left(y - \int_0^t e^{\int_0^\tau M(x(s),u(s))ds} L(x(\tau),u(\tau))d\tau\right) \qquad (23)$$

to the linear dynamical system

$$y'(t) = -M(x(t),u(t))y(t) - L(x(t),u(t))$$

starting at $y \in Y$.

We already know that the caliber is the unique Frankowska solution to

1. there exists $u \in P(x, V(t,x))$ such that

$$0 \in \text{“} -\frac{\partial}{\partial t}V(t,x) + \frac{\partial}{\partial x}V(t,x)f(x,V(t,x),u)\text{”} - M(x,u)V(t,x) - L(x,u)$$

2. for all $u \in P(x, v(t,x))$,

$$0 \in \text{“}\frac{\partial}{\partial t}V(t,x) - \frac{\partial}{\partial x}V(t,x)f(x,V(t,x),u)\text{”} + M(x,u)V(t,x) + L(x,u)$$

satisfying the viability constraints

$$\forall\, t \geq 0,\ x \in X,\ \ \Phi(t,x) \subset V(t,x) \subset \Psi(t,x)$$

and the initial condition

$$V(0,x) = \Phi(0,x)$$

In summary, we now deal with a **structured problem** where the exosystem is linear with respect to the observations. In this case, we shall be able to provide an explicit formula of the caliber V.

For that purpose, we introduce the subset

$$\begin{cases} J_\Phi(t;(x(\cdot),u(\cdot)))(T,x) \\ := e^{\int_0^t M(x(s),u(s))ds}\Phi(T-t,x(t)) + \int_0^t e^{\int_0^\tau M(x(s),u(s))ds} L(x(\tau),u(\tau))d\tau \end{cases}$$

(where t ranges over $[0,T]$). The controls – most often prices or other regulees in economics, portfolio in finance – appear **both** in the matrix M and in the Lagrangian L.

We associate with Ψ the set-valued map J_Ψ defined by

$$\begin{cases} J_\Psi(t;(x(\cdot),u(\cdot)))(T,x) \\ := e^{\int_0^t M(x(s),u(s))ds}\Psi(T-t,x(t)) + \int_0^t e^{\int_0^\tau M(x(s),u(s))ds} L(x(\tau),u(\tau))d\tau \end{cases}$$

[7] We could take L to be set-valued map, but we restrict to the single-valued case for simplicity.

and

$$K_\Psi(t,x;(x(\cdot),u(\cdot))) \ := \ \bigcap_{s\in[0,t]} J_\Psi(s,x;(x(\cdot),u(\cdot)))$$

We next introduce

$$L_\Psi^\Phi(t;(x(\cdot),u(\cdot)))(T,x) \ := \ K_\Psi(t,x;(x(\cdot),u(\cdot))) \cap J_\Phi(t;(x(\cdot),u(\cdot)))(T,x)$$

5.1. Explicit Formula of the Caliber

We shall prove now that the caliber V of the pair (Ψ,Φ) under $\mathcal{B}$ is equal to the set-valued map $V_\Psi(\Phi)$ defined by

$$V_\Psi(\Phi)(T,x) \ := \ \bigcup_{(x(\cdot),u(\cdot))\in\mathcal{C}(x)} \ \bigcup_{t\in[0,T]} L_\Psi^\Phi(t;(x(\cdot),u(\cdot)))(T,x) \tag{24}$$

Note that if $\Psi_1 \subset \Psi_2$ and $\Phi_1 \subset \Phi_2$, then $V_{\Psi_1}(\Phi_1) \subset V_{\Psi_2}(\Phi_2)$ and that

$$\forall\, t \geq 0, \forall\, x \in X, \ \ \Phi(t,x) \ \subset \ V_\Psi(\Phi)(t,x) \ \subset \ \Psi(t,x)$$

We shall use the fact that its graph is the viable-capture basin of the graph of the cost function Φ viable under the graph of Ψ under the auxiliary system $\mathcal{R}$.

Theorem 5.1. *Let us assume that the set-valued maps Ψ and Φ are nontrivial. Then the caliber of the pair (Ψ,Φ) is equal to the set-valued map $V_\Psi(\Phi)$ defined by (24).*

Furthermore, the caliber is the **unique solution** *V to the two following "fixed set equations" :*

$$\left\{\begin{array}{lll} & V_V(\Phi)(T,x) & := \bigcup_{(x(\cdot),u(\cdot))\in\mathcal{C}(x)} \bigcup_{t\in[0,T]} L_V^\Phi(t,x;(x(\cdot),u(\cdot))) \\ = & V(T,x) & \\ = & V_\Psi(V)(T,x) & := \bigcup_{(x(\cdot),u(\cdot))\in\mathcal{C}(x)} \bigcup_{t\in[0,T]} L_\Psi^V(t,x;(x(\cdot),u(\cdot))) \end{array}\right. \tag{25}$$

Moreover, when F is Marchaud and the set-valued maps Φ and Ψ are closed, the graph of the caliber $V_\Psi(\Phi)$ is closed.

5.2. Examples

1. We see that the set-valued map defined by

$$V(\Phi)(T,x) \ := \ \bigcup_{(x(\cdot),u(\cdot))\in\mathcal{C}(x)} \ \bigcup_{t\in[0,T]} J_\Phi(t;(x(\cdot),u(\cdot)))(T,x) \ = \ V_{\mathbf{Y}}(\Phi)$$

is the caliber of (Ψ,Φ) where the set-valued map $\Psi = \mathbf{Y}$ is defined by

$$\mathbf{Y}(t,x) \ := \left\{\begin{array}{ll} Y & \text{if} \ \ t \geq 0 \\ \emptyset & \text{if} \ \ t < 0 \end{array}\right.$$

Indeed, we observe that taking $\Psi = \mathbf{Y}$, then

$$K_{\mathbf{Y}}(t,x;(x(\cdot),u(\cdot))) \ = \ J_{\mathbf{Y}}(t,x;(x(\cdot),u(\cdot))) \ = \ Y$$

so that, for any set-valued map Φ, we have $L_{\mathbf{Y}}^\Phi = J_\Phi$ and thus, $V(\Phi) = V_{\mathbf{Y}}(\Phi)$.

2. We also observe that the set-valued map

$$W(\Psi)(T,x) := \bigcup_{(x(\cdot),u(\cdot))\in\mathcal{C}(x)} \bigcap_{t\in[0,T]} J_\Psi(t;(x(\cdot),u(\cdot)))(T,x) = V_\Psi(\Psi_\emptyset)$$

is the caliber of (Ψ,Φ) where we take $\Phi := \Psi_\emptyset : \mathbf{R}_+ \times X \rightsquigarrow Y$ defined by

$$\Psi_\emptyset(t,x) := \begin{cases} \Psi(0,x) & \text{if } t = 0 \\ \emptyset & \text{if } t \neq 0 \end{cases}$$

Indeed, we see that $J_{\Psi_\emptyset}(t,x;(x(\cdot),u(\cdot))) = \emptyset$ if $t < T$ and

$$J_{\Psi_\emptyset}(T,x;(x(\cdot),u(\cdot))) = J_\Psi(T,x;(x(\cdot),u(\cdot))).$$

Therefore

$$L_\Psi^{\Psi_\emptyset}(t,x;(x(\cdot),u(\cdot))) := \begin{cases} K_\Psi(T,x;(x(\cdot),u(\cdot))) & \text{if } t = T \\ \emptyset & \text{if } t < T \end{cases}$$

and thus, $W(\Psi) = V_\Psi(\Psi_\emptyset)$.

3. Let us introduce a time-independent set-valued map $U : X \rightsquigarrow X$. We shall associate with three pairs (Ψ,Φ) of set-valued maps associated with U the three following calibers:
   (a) Taking $\Phi := U_\emptyset$ and $\Psi = \mathbf{Y}$, we obtain

$$\bigcup_{(x(\cdot),u(\cdot))\in\mathcal{C}(x)} \left(e^{\int_0^T M(x(s),u(s))ds} U(x(T)) + \int_0^T e^{\int_0^\tau M(x(s),u(s))ds} L(x(\tau),u(\tau))d\tau \right)$$

   When $M = 0$, the above problem boils down to the set-valued equivalent of the "Bolza map"

$$\bigcup_{(x(\cdot),u(\cdot))\in\mathcal{C}(x)} \left(U(x(T)) + \int_0^T L(x(\tau),u(\tau))d\tau \right)$$

   and the "Mayer map"

$$\bigcup_{(x(\cdot),u(\cdot))\in\mathcal{C}(x)} U(x(T))$$

   when furthermore $L = 0$. This is the case of the substratum of an impulse differential inclusion defined above (see also [9, Aubin]).
   (b) Taking $\Phi := U_\emptyset$ and $\Psi := U$, we obtain

$$\bigcup_{(x(\cdot),u(\cdot))\in\mathcal{C}(x)} \bigcap_{t\in[0,T]} \left(e^{\int_0^t M(x(s),u(s))ds} U(x(t)) + \int_0^t e^{\int_0^\tau M(x(s),u(s))ds} L(x(\tau),u(\tau))d\tau \right)$$

(c) Taking $\Phi := U$ and $\Psi := \mathbf{Y}$, we obtain

$$\bigcup_{(x(\cdot),u(\cdot))\in\mathcal{C}(x)}\ \bigcup_{t\in[0,T]}\left(e^{\int_0^t M(x(s),u(s))ds}U(x(t)) + \int_0^t e^{\int_0^\tau M(x(s),u(s))ds}L(x(\tau),u(\tau))d\tau\right)$$

When $M = 0$, we find

$$V(U)(T,x) := \bigcup_{(x(\cdot),u(\cdot))\in\mathcal{C}(x)}\ \bigcup_{t\in[0,T]}\left(U(x(t)) + \int_0^t L(x(\tau),u(\tau))d\tau\right)$$

This map – the set-valued analogue of the valuation function of "obstacle problems" – involves "max-plus" operations and is the set-valued equivalent of the mathematical fear – or faith with respect to a Maslov measure introduced in dynamical optimization by Pierre Bernhard (see [35, 35, 37, Bernhard]).

Regarding the caliber as a transform $\Phi \mapsto V(\Phi)$ mapping closed set-valued maps to closed set-valued map, we observe that the caliber satisfies

$$V\left(\bigcup_{i=1,\ldots,n}\Phi_i\right) = \bigcup_{i=1,\ldots,n}V(\Phi_i)$$

the extensivity property: $\Phi \subset V(\Phi)$ and the monotonicity property: If $\Phi_1 \subset \Phi_2$, then $V(\Phi_1) \subset V(\Phi_2)$.

We recall that Theorem 4.2 implies that the value transform is also "upper continuous" in the sense that

$$\mathrm{Limsup}_{n\to+\infty,\ s\to t,\ y\to x}V(\Phi_n)(s,y) \subset V\left(\mathrm{Limsup}_{n\to+\infty,\ s\to t,\ y\to x}\Phi_n(s,y)\right)(t,x)$$

and actually, under stronger assumptions, "continuous" in the sense that the caliber of a graphical limit is the graphical limit of calibers.

5.3. Proof of the Explicit Formula

First, to say that a pair (T,x,y) belongs to the viable-capture basin

$$\mathrm{Graph}(V) := \mathrm{Capt}_{\mathcal{R}}^{\mathrm{Graph}(\Psi)}(\mathrm{Graph}(\Phi) \subset \mathrm{Graph}(\Psi)$$

means that there exist a solution $(T-\cdot,\widetilde{x}(\cdot),\widetilde{u}(\cdot),y(\cdot)) \in \mathcal{R}(T,x,y)$ to the auxiliary control problem and some $t^* \geq 0$ such that

$$\begin{cases} \text{i)} & \forall\, t\in[0,t^*],\ (T-t,\widetilde{x}(t),y(t)) \in \mathrm{Graph}(\Psi) \\ \text{ii)} & (T-t^*,\widetilde{x}(t^*),y(t^*)) \in \mathrm{Graph}(\Phi) \end{cases}$$

or, equivalently, such that

$$\begin{cases} \text{i)} & \forall\, t\in[0,t^*],\ y(t) \in \Psi(T-t,\widetilde{x}(t)) \\ \text{ii)} & y(t^*) \in \Psi(T-t^*,\widetilde{x}(t^*)) \end{cases}$$

By the very definition of $\mathcal{R}(T,x,y)$ and by the definition (23) of the auxiliary system that its component $y(\cdot)$ satisfies

$$y(t) \; = \; e^{-\int_0^t M(\widetilde{x}(s),\widetilde{u}(s))ds}\left(y - \int_0^t e^{\int_0^\tau M(\widetilde{x}(s),\widetilde{u}(s))ds} L(\widetilde{x}(\tau),\widetilde{u}(\tau))d\tau\right)$$

this implies that (T,x,y) belongs to $\mathrm{Capt}_{\mathcal{R}}^{\mathrm{Graph}(\Psi)}(\mathrm{Graph}(\Phi)$ if and only if there exists a solution $(\widetilde{x}(\cdot),\widetilde{u}(\cdot)) \in \mathcal{C}(x)$ satisfying for almost all $t \in [0,\overline{t}]$,

$$\begin{cases} \text{i)} \quad y \in e^{\int_0^t M(\widetilde{x}(s),\widetilde{u}(s))ds}\Psi(T-t,\widetilde{x}(t),\widetilde{u}(t)) \\ \qquad\qquad + \int_0^t e^{\int_0^\tau M(\widetilde{x}(s),\widetilde{u}(s))ds} L(\widetilde{x}(\tau),\widetilde{u}(\tau))d\tau \\ \text{ii)} \quad y \in e^{\int_0^{t^*} M(\widetilde{x}(s),\widetilde{u}(s))ds}\Phi(T-t^*,\widetilde{x}(t^*)) \\ \qquad\qquad + \int_0^{t^*} e^{\int_0^\tau M(\widetilde{x}(s),\widetilde{u}(s))ds} L(\widetilde{x}(\tau),\widetilde{u}(\tau))d\tau \end{cases} \tag{26}$$

Since we set

$$K_\Psi(t,x;(\widetilde{x}(\cdot),\widetilde{u}(\cdot))) \; := \; \bigcap_{s\in[0,t]} J_\Psi(s,x;(\widetilde{x}(\cdot),\widetilde{u}(\cdot)))$$

and

$$L_\Psi^\Phi(t,x;(\widetilde{x}(\cdot),\widetilde{u}(\cdot))) \; := \; K_\Psi(t,x;(\widetilde{x}(\cdot),\widetilde{u}(\cdot))) \cap J_\Phi(t,x;(\widetilde{x}(\cdot),\widetilde{u}(\cdot))$$

this is equivalent to state that y belongs to $V_\Psi(\Phi)(T,x)$.

Formula (3) of [15, Aubin & Catté] and the first part of this theorem implies that the caliber is the unique solution to the system (25).

Finally, the closedness of the graph of the caliber is an immediate consequence of Theorem 0.1 because $\mathrm{Graph}(\Psi) \subset \mathbf{R}_+ \times X \times Y$ being a repeller under G, which is Marchaud, the viable-capture basin $\mathrm{Graph}(V_\Psi)(\Phi)) := \mathrm{Capt}_{\mathcal{R}}^{\mathrm{Graph}(\Psi)}(\mathrm{Graph}(\Phi))$ is closed. □

5.4. The Barrier Property of the Caliber

We begin with a first form of the barrier property:

Proposition 5.2. *Let us consider $y_T \in \partial_{\Psi(T,x)} V_\Psi(\Phi)(T,x)$. Then, any solution $(x(\cdot),u(\cdot)) \in \mathcal{C}(x)$ starting from x satisfying inclusion: for every $t \in [0,t^*]$*

$$y_T \; \in \; L_\Psi^\Phi(t;(x(\cdot),u(\cdot)))(T,x) \tag{27}$$

until the first time t^* when

$$y_T \in \Phi(T-t^*,x(t^*)$$

actually satisfies: for every $t \in [0,t^*]$

$$\forall\, y_T \; \in \; \partial L_\Psi^\Phi(t;(x(\cdot),u(\cdot)))(T,x)$$

Proof. Since y_T belongs to $V_\Psi(\Phi)(T,x)$, there exists a solution $(x^*(\cdot),u^*(\cdot))$ and a time $t^* \in [0,T]$ such that

$$y_T \in L_\Psi^\Phi(t^*,x;(x^*(\cdot),u^*(\cdot)))$$

Since y_T belongs to $\partial_{\Psi(T,x)}V_\Psi(\Phi)(T,x)$, it can be approximated by elements $y_n \in \Psi(T,x) \setminus V_\Psi(\Phi)(T,x)$. We know that for every t and any solution $(x(\cdot),u(\cdot))$, either y_n does not belong to $K_\Psi(t,x;(x(\cdot),u(\cdot)))$ or else, it belongs to $K_\Psi(t,x;(x(\cdot),u(\cdot)))\backslash V(T,x)$.

We claim that

$$\begin{cases} \forall\,(x(\cdot),u(\cdot)) \in \mathcal{C}(x),\ \forall\, t \in [0,T], \\ K_\Psi(t,x;(x(\cdot),u(\cdot)))\backslash V(T,x) \ \subset\ X\backslash J_V(t,x;(x(\cdot),u(\cdot))) \end{cases} \tag{28}$$

Indeed, take y in $K_\Psi(t,x;(x(\cdot),u(\cdot)))\backslash V(T,x)$. This means that for every $s \in [0,t]$, y belongs to $J_\Psi(s,x;(x(\cdot),u(\cdot)))$, or, equivalently, that

$$\forall\, s \in [0,t],\ \ (T-s,x(s),y(s)) \in \text{Graph}(\Psi)$$

where

$$y(t) := e^{-\int_0^t M(x(s),u(s))ds}\left(y - \int_0^t e^{\int_0^\tau M(x(s),u(s))ds}L(x(\tau),u(\tau))d\tau\right)$$

Since (T,x,y) does not belong to the viable capture basin, we infer that every solution in $\mathcal{R}(T,x,y)$ starting at (T,x,y) is viable in the graph Graph(Ψ) of Ψ before hitting the graph Graph(V) of V: Therefore, if

$$\forall\, s \in [0,t],\ \ y(s) \in \Psi(T-s,x(s),u(s))$$

we have

$$y(t) \in X\backslash V(T-t,x(t)) \subset X\backslash\Phi(T-t,x(t))$$

that can be written in the form

$$y \in X\backslash J_V(t,x;(x(\cdot),u(\cdot))) \subset X\backslash J_\Phi(t,x;(x(\cdot),u(\cdot)))$$

Therefore, inclusion (27) holds true; thus, y_n does not belong to $J_V(t,x;(x(\cdot),u(\cdot)))$. Consequently, for any $t \in [0,T]$, for any $n \geq 0$,

$$y_n \in X\backslash L_\Psi^V(t,x;(x(\cdot),u(\cdot))) \subset X\backslash L_\Psi^\Phi(t,x;(x(\cdot),u(\cdot)))$$

This is in particular true for the solution $(x^*(\cdot),u^*(\cdot))$ and for any $t \in [0,t^*]$. Hence, letting y_n converge to y_T, we deduce the conclusion of the proposition. □

We can strengthen this result and prove the following "barrier property". Let us recall that $J_{V_\Psi(\Phi)}(0,x;(x(\cdot),u(\cdot))) = V_\Psi(\Phi)(T,x)$. We shall prove that the boundary condition

$$y_T \in \partial V_\Psi(\Phi)(T,x) = \partial J_{V_\Psi(\Phi)}(0,x;(x(\cdot),u(\cdot)))$$

propagates as long as y_T remains in the interior of $K_\Psi(t,x;(x(\cdot),u(\cdot)))$:

Theorem 5.3. *Let us consider* $y_T \in \partial V_\Psi(\Phi)(T,x) \cap \mathrm{Int}(\Psi(T,x))$. *Then, any solution* $(x(\cdot),u(\cdot)) \in \mathcal{C}(x)$ *starting from* x *satisfying the inclusion*

$$y_T \in J_{V_\Psi(\Phi)}(t,x;(x(\cdot),u(\cdot))) \tag{29}$$

until the first time $\bar{t}$ *when*

$$y_T \notin \mathrm{Int}(K_\Psi(\bar{t},x;(x(\cdot),u(\cdot))))$$

actually satisfies

$$\forall\, t \in [0,\bar{t}],\ \ y \in \partial J_{V_\Psi(\Phi)}(t;(x(\cdot),u(\cdot)))(T,x) \tag{30}$$

Proof. Since y_T belongs to $\partial V_\Psi(\Phi)(T,x)$, it can be approximated by elements $y_n \in \Psi(T,x)\backslash V_\Psi(\Phi)(T,x)$. By assumption, for any $t < \bar{t}$, there exist δ_t such that

$$B(y_T,\delta_t) \subset K_\Psi(t;(x(\cdot),u(\cdot)))(T,x)$$

and thus, that there exists some N_t such that, for any $n \geq N_t$, y_n belongs to $B(y^T,\delta_t)$, and thus

$$y_n \in K_\Psi(t;(x(\cdot),u(\cdot)))(T,x)\backslash V_\Psi(\Phi)(T,x)$$

But by (27), we know that in this case, y_n does not belong to

$$J_{V_\Psi(\Phi)}(t;(x(\cdot),u(\cdot)))(T,x).$$ □

6. Frankowska contingent epi-solutions to Hamilton-Jacobi-Bellman equations

When $Y := \mathbf{R}$, we can associate with two extended functions $\mathbf{c} : \mathbf{R}_+ \times X \rightsquigarrow \mathbf{R} \cup \{+\infty\}$ and $\mathbf{b} : \mathbf{R}_+ \times X \rightsquigarrow \mathbf{R} \cup \{+\infty\}$ such that

$$\forall\,(t,x) \in \mathbf{R}_+ \times X,\ \ 0 \leq \mathbf{b}(t,x) \leq \mathbf{c}(t,x)$$

the set-valued maps Φ and Ψ defined by

$$\begin{cases} \text{i)} & \Phi(t,x) := \mathbf{c}(t,x) + \mathbf{R}_+ \\ \text{ii)} & \Psi(t,x) := \mathbf{b}(t,x) + \mathbf{R}_+ \end{cases}$$

by setting $\Phi(t,x) := \emptyset$ whenever $\mathbf{c}(t,x) = +\infty$. We observe that $\mathrm{Graph}(\Phi) = \mathcal{E}p(\mathbf{c})$ and that

$$D\Phi(t,x,\mathbf{c}(t,x))(\pm 1,v) = D_\uparrow \mathbf{c}(t,x)(\pm 1,v) + \mathbf{R}_+$$

where $D_\uparrow \mathbf{c}(t,x)$ is the contingent epiderivative of $\mathbf{c}$ at (t,x) in the direction $(\pm 1,v)$, defined by

$$D_\uparrow \mathbf{c}(t,x)(\pm 1,v) = \liminf_{h\to 0+,v'\mapsto v} \frac{\mathbf{c}(t \pm h, x + hv') - \mathbf{c}(t,x)}{h}$$

We set

$$\begin{cases} J_{\mathbf{c}}(t;(x(\cdot),u(\cdot)))(T,x) \\ := e^{\int_0^t M(x(s),u(s))ds}\mathbf{c}(T-t,x(t)) + \int_0^t e^{\int_0^\tau M(x(s),u(s))ds} L(x(\tau),u(\tau))d\tau \end{cases}$$

and then,

$$K_{\mathbf{b}}(t,x;(x(\cdot),u(\cdot))) \;:=\; \sup_{s\in[0,t]} J_{\mathbf{b}}(s,x;(x(\cdot),u(\cdot)))$$

We next integrate this cumulated cost together with the former cost

$$J_{\mathbf{c}}(t,x;(x(s),u(s)))$$

by introducing the new cost function

$$L_{\mathbf{b}}^{\mathbf{c}}(t;(x(\cdot),u(\cdot)))(T,x) \;:=\; \max(K_{\mathbf{b}}(t,x;(x(\cdot),u(\cdot))), J_{\mathbf{c}}(t;(x(\cdot),u(\cdot)))(T,x)$$

We shall deduce from Theorem 5.1 the following consequence :

Theorem 6.1. *Let us assume that the extended functions* $\mathbf{b}$ *and* $\mathbf{c}$ *are nontrivial and non negative. The viable-capture basin* $\mathrm{Capt}_{\mathcal{R}}^{\mathcal{E}p(\mathbf{b})}(\mathcal{E}p(\mathbf{c}))$ *of* $\mathcal{E}p(\mathbf{c})$ *under* $\mathcal{R}$ *is the epigraph of the valuation function* $V_{\mathbf{b}}(\mathbf{c})$ *defined by*

$$V_{\mathbf{b}}(\mathbf{c})(T,x) \;:=\; \inf_{(x(\cdot),u(\cdot))\in\mathcal{C}(x)} \inf_{t\in[0,T]} L_{\mathbf{b}}^{\mathbf{c}}(t;(x(\cdot),u(\cdot)))(T,x)$$

Furthermore, any solution $(x(\cdot),u(\cdot)) \in \mathcal{C}(x)$ *starting from* $x \in X$ *satisfying the inequality: for every* $t \in [0,t^\star]$

$$\begin{cases} V_{\mathbf{b}}(\mathbf{c})(T,x) \\ \geq e^{\int_0^t M(x(s),u(s))ds} V_{\mathbf{b}}(\mathbf{c})(T-t,x(t)) + \int_0^t e^{\int_0^\tau M(x(s),u(s))ds} L(x(\tau),u(\tau))d\tau \end{cases} \tag{31}$$

until the first time $t^\star$ *when*

$$V_{\mathbf{b}}(\mathbf{c})(T-t^\star,x(t^\star)) = \mathbf{c}(T-t^\star,x(t^\star))$$

is an optimal solution for the optimal time $t^\star$.

Finally, the valuation function is a solution $\mathbf{v}$ *to the two following functional equations stating that the functions* $L_{\mathbf{b}}^{\mathbf{v}}$ *and* $L_{\mathbf{v}}^{\mathbf{c}}$ *have the same infimum than* $L_{\mathbf{b}}^{\mathbf{c}}$*:*

$$\begin{cases} \inf_{(x(\cdot),u(\cdot))\in\mathcal{C}(x)} \inf_{t\in[0,T]} L_{\mathbf{v}}^{\mathbf{c}}(t,x;(x(\cdot),u(\cdot))) \\ = \mathbf{v}(T,x) \\ = \inf_{(x(\cdot),u(\cdot))\in\mathcal{C}(x)} \inf_{t\in[0,T]} L_{\mathbf{b}}^{\mathbf{v}}(t,x;(x(\cdot),u(\cdot))) \end{cases} \tag{32}$$

Proof. It is enough – and easy – to check that

$$L_{\Psi}^{\Phi}(t;(x(\cdot),u(\cdot)))(T,x) \;=\; L_{\mathbf{b}}^{\mathbf{c}}(t;(x(\cdot),u(\cdot)))(T,x) + \mathbf{R}_+$$

and thus, that

$$\inf_{y\in V_{\Psi}^{\Phi}(t,x)} y \;=\; V_{\mathbf{b}}^{\mathbf{c}}(t,x)$$

since $\inf\left(\bigcup_{i\in I}[a_i,\infty[\right) = \inf_{i\in I} a_i$. □

Theorem 5.3 implies the following form of the optimality principle:

Theorem 6.2. *Let us assume that* $V_{\mathbf{b}}(\mathbf{c})(T,x) > \mathbf{b}(T,x))$. *Then any solution* $(x(\cdot),u(\cdot)) \in \mathcal{C}(x)$ *starting from* $x \in \mathrm{Dom}(V_{\mathbf{b}}(\mathbf{c}))$ *satisfying inequalities*

$$\left\{\begin{array}{l} V_{\mathbf{b}}(\mathbf{c})(T,x) \\ \geq e^{\int_0^t M(x(s),u(s))ds} V_{\mathbf{b}}(\mathbf{c})(T-t,x(t)) + \int_0^t e^{\int_0^\tau M(x(s),u(s))ds} L(x(\tau),u(\tau))d\tau \end{array}\right. \tag{33}$$

until the first time $t^\star \in [0,T]$ *when*

$$V_{\mathbf{b}}(\mathbf{c})(T-t^\star,x(t^\star)) = \mathbf{b}(T-t^\star,x(t^\star))$$

actually satisfies equality

$$\left\{\begin{array}{l} \forall\, t \in [0,t^\star],\ V_{\mathbf{b}}(\mathbf{c})(T,x) \\ = e^{\int_0^t M(x(s),u(s))ds} V_{\mathbf{b}}(\mathbf{c})(T-t,x(t)) + \int_0^t e^{\int_0^\tau M(x(s),u(s))ds} L(x(\tau),u(\tau))d\tau \end{array}\right. \tag{34}$$

The first statement of Theorem 3.1 implies that the valuation function is a Frankowska contingent episolution to Hamilton-Jacobi-Bellman equations:

Theorem 6.3 (Frankowska). *Let us assume that the control system* (P,f,L,M) *is Marchaud and that the functions* $\mathbf{b}$ *and* $\mathbf{c}$ *are nontrivial, non negative and lower semicontinuous.*

Then the valuation function $V_{\mathbf{b}}(\mathbf{c})$ *is characterized as the smallest of the nonnegative lower semicontinuous functions* $\mathbf{v} : \mathbf{R}_+ \times X \mapsto \mathbf{R}_+ \cup \{+\infty\}$ *satisfying for every* $(t,x) \in]0,\infty[\times X$

$$\left\{\begin{array}{ll} \text{i)} & \mathbf{b}(t,x) \ \leq\ \mathbf{v}(t,x) \ \leq\ \mathbf{c}(t,x) \\ \text{ii)} & \text{if } (t,x) \in \Omega(\mathbf{v}), \\ & \inf_{u\in P(x)} \left(D_\uparrow \mathbf{v}(t,x)(-1,f(x,u)) + L(x,u) + M(x,u)\mathbf{v}(t,x)\right) \ \leq\ 0 \end{array}\right.$$

Let us set

$$\mathbf{R}(t,x) := \{u\in P(x) \,|\, D_\uparrow V_{\mathbf{b}}(\mathbf{c})(t,x)(-1,f(x,u)) + L(x,u) + M(x,u)V_{\mathbf{b}}(\mathbf{c})(t,x) \leq 0\}$$

Knowing the valuation function, an optimal solution is obtained in the following way: Starting from x_0 *such that* $V_{\mathbf{b}}(\mathbf{c})(T,x_0) < \mathbf{c}(T,x_0)$, *any solution* $(x(\cdot),u(\cdot))$ *to the control system*

$$\left\{\begin{array}{lrcl} \text{i)} & x'(t) & = & f(x(t),u(t)) \\ \text{ii)} & u(t) & \in & \mathbf{R}(T-t,x(t)) \end{array}\right. \tag{35}$$

is an optimal solution, and the first time $t^\star \geq 0$ *when*

$$V_{\mathbf{b}}(\mathbf{c})(T-t^\star,x(t^\star)) \ = \ \mathbf{c}(T-t^\star,x(t^\star))$$

is the optimal time.

The second part of Theorem 3.1 implies the following existence and uniqueness result:

Theorem 6.4 (Frankowska). *Let us assume that the control system (P, f, L, M) is Marchaud and Lipschitz and that* **b** *and* **c** *are nontrivial, non negative and lower semicontinuous.*

Then the valuation function $V_{\mathbf{b}}(\mathbf{c})$ is the unique lower semicontinuous episolution **v** *to the system of differential inequalities: for every $(t, x) \in \mathrm{Dom}(\mathbf{v})$,*

$$\left\{\begin{array}{ll} \text{i)} & \mathbf{b}(t,x) \leq \mathbf{v}(t,x) \leq \mathbf{c}(t,x) \\ \text{ii)} & \text{if } \mathbf{v}(t,x) < \mathbf{c}(t,x), \\ & \inf_{u\in P(x)} (D_{\uparrow}\mathbf{v}(t,x)(-1, f(x,u)) + L(x,u) + M(x,u)\mathbf{v}(t,x)) \leq 0 \\ \text{iii)} & \text{if } \mathbf{v}(t,x) > \mathbf{b}(t,x), \\ & \sup_{u\in P(x)} (D_{\uparrow}\mathbf{v}(t,x)(1, -f(x,u)) - L(x,u) - M(x,u)\mathbf{v}(t,x)) \leq 0 \\ \text{iv)} & \text{if } \mathbf{v}(t,x) = \mathbf{b}(t,x), \\ & \sup_{u\in P(x)} [\min(D_{\uparrow}\mathbf{v}(t,x)(1, -f(x,u)), D_{\downarrow}\mathbf{b}(t,x)(1, -f(x,u))) \\ & -L(x,u)) - M(x,u)\mathbf{v}(t,x)))] \leq 0 \end{array}\right. \tag{36}$$

Remark. Condition (36)iv) is automatically satisfied whenever

$$\sup_{u\in P(x)} (D_{\downarrow}\mathbf{b}(t,x)(1, -f(x,u)) - L(x,u)) - M(x,u)\mathbf{v}(t,x)) \leq 0$$

We refer to the papers [43, 44, 46, Frankowska] for other differential properties of the value function obtained using the tools of the epigraphical approach, and in particular, by duality, the links with viscosity solutions and lower semicontinuous bilateral solutions introduced also in [32, Barron & Jensen] by PDE's methods.

7. Vector Optimal Control Problems

When $Y := \mathbf{R}^n$ is supplied with the natural order relation $\leq$ associated with the positive orthant $\mathbf{R}^n_+$, we can associate with two maps $\vec{\mathbf{c}} : \mathbf{R}_+ \times X \rightsquigarrow \mathbf{R}^n_+$ and $\vec{\mathbf{b}} : \mathbf{R}_+ \times X \rightsquigarrow \mathbf{R}^n_+$ such that

$$\forall\, (t,x) \in \mathbf{R}_+ \times X, \ \ 0 \leq \vec{\mathbf{b}}(t,x) \leq \vec{\mathbf{c}}(t,x)$$

the set-valued maps Φ and Ψ defined by

$$\left\{\begin{array}{lll} \text{i)} & \Phi(t,x) & := \vec{\mathbf{c}}(t,x) + \mathbf{R}^n_+ \\ \text{ii)} & \Psi(t,x) & := \vec{\mathbf{b}}(t,x) + \mathbf{R}^n_+ \end{array}\right.$$

In the following formulas, the supremums and the infimums are taken component by component. We set

$$\left\{\begin{array}{l} J_{\vec{\mathbf{c}}}(t; (x(\cdot), u(\cdot)))(T, x) \\ := e^{\int_0^t M(x(s),u(s))ds}\vec{\mathbf{c}}(T-t, x(t)) + \displaystyle\int_0^t e^{\int_0^\tau M(x(s),u(s))ds} L(x(\tau), u(\tau))d\tau \end{array}\right.$$

and then,

$$K_{\vec{\mathbf{b}}}(t,x;(x(\cdot),u(\cdot))) \;:=\; \sup_{s\in[0,t]} J_{\vec{\mathbf{b}}}(s,x;(x(\cdot),u(\cdot)))$$

We next integrate this cumulated cost together with the former cost

$$J_{\vec{\mathbf{c}}}(t,x;(x(s),u(s)))$$

by introducing the new cost function

$$L_{\vec{\mathbf{b}}}^{\vec{\mathbf{c}}}(t;(x(\cdot),u(\cdot)))(T,x) \;:=\; \max(K_{\vec{\mathbf{b}}}(t,x;(x(\cdot),u(\cdot))), J_{\vec{\mathbf{c}}}(t;(x(\cdot),u(\cdot)))(T,x)$$

and the subset

$$V_{\vec{\mathbf{b}}}(\vec{\mathbf{c}})(T,x) \;:=\; \bigcup_{(x(\cdot),u(\cdot))\in\mathcal{C}(x)} \bigcup_{t\in[0,T]} L_{\vec{\mathbf{b}}}^{\vec{\mathbf{c}}}(t;(x(\cdot),u(\cdot)))(T,x)$$

We shall deduce from Theorem 5.1 the following consequence :

Theorem 7.1. *Let us assume that for any $(x,u)\in X\times\mathcal{U}$ and for any $y\in\mathbf{R}^n_+$, whenever $y_i=0$, then $(M(x,u)y)_i=0$.*

Then

$$\forall\, T\ge 0, x\in X,\;\; V_\Psi(\Phi)(T,x) \;=\; V_{\vec{\mathbf{b}}}(\vec{\mathbf{c}})(T,x)+\mathbf{R}^n_+$$

Proof. It is enough to check that

$$L_\Psi^\Phi(t;(x(\cdot),u(\cdot)))(T,x) \;=\; L_{\vec{\mathbf{b}}}^{\vec{\mathbf{c}}}(t;(x(\cdot),u(\cdot)))(T,x)+\mathbf{R}^n_+$$

By assumption, the cone $\mathbf{R}^n_+$ is forward and backward invariant under the differential equation $y'(t)=M(x(t),u(t))y(t)$, so that

$$\forall\, y\in\mathbf{R}^n_+,\;\; e^{\pm\int_0^t M(x(\tau),u(\tau))d\tau}y \;\in\; \mathbf{R}^n_+$$

Therefore, since $\vec{\mathbf{c}}(t,x)\in\mathbf{R}^n_+$, we infer that

$$J_\Phi(t;(x(\cdot),u(\cdot)))(T,x) \;=\; J_{\vec{\mathbf{c}}}(t;(x(\cdot),u(\cdot)))(T,x)+\mathbf{R}^n_+$$

Next, we observe that if $a_i\in\mathbf{R}^n$,

$$\bigcap_{i\in I}(a_i+\mathbf{R}^n_+) \;=\; \sup_{i\in I} a_i+\mathbf{R}^n_+$$

and thus, that

$$K_\Psi(t;(x(\cdot),u(\cdot)))(T,x) \;=\; L_{\vec{\mathbf{b}}}(t;(x(\cdot),u(\cdot)))(T,x)+\mathbf{R}^n_+$$

and that

$$L_\Psi^\Phi(t;(x(\cdot),u(\cdot)))(T,x) \;=\; L_{\vec{\mathbf{b}}}^{\vec{\mathbf{c}}}(t;(x(\cdot),u(\cdot)))(T,x)+\mathbf{R}^n_+$$

Finally, we note that

$$\begin{cases} V_\Psi(\Phi)(T,x) \\ = \bigcup_{(x(\cdot),u(\cdot))\in\mathcal{C}(x)}\bigcup_{t\in[0,T]}\left[L_{\vec{\mathbf{b}}}^{\vec{\mathbf{c}}}(t;(x(\cdot),u(\cdot)))(T,x)+\mathbf{R}^n_+\right] \\ = V_{\vec{\mathbf{b}}}(\vec{\mathbf{c}})(T,x)+\mathbf{R}^n_+ \end{cases}$$

□

Recall that for a closed subset $A \subset \mathbf{R}^n$ satisfying $A = A + \mathbf{R}^n_+$, the interior of A is equal to

$$\mathrm{Int}(A) \;=\; A + \overset{\circ}{\mathbf{R}^n_+}$$

and thus, that the boundary of A is equal to the set of (weak) Pareto optima of A: Indeed, $y \in \partial A$ if and only if for any $z \in A$, there exists at least $i \in \{1, \ldots, n\}$ such that $y_i \leq z_i$. We say that $z \gg y$ if for any $i \in \{1, \ldots, n\}$, $z_i > y_i$.

Hence, we deduce the following consequence of Theorem 5.3:

Theorem 7.2. *Let us consider* $y_T \gg \vec{\mathbf{b}}(T, x))$ *be a Pareto minimum of the set* $J_{V_{\vec{\mathbf{b}}}(\vec{\mathbf{c}})}(T, x)$. *Consider any solution* $(x(\cdot), u(\cdot)) \in \mathcal{C}(x)$ *starting from*

$$x \in \mathrm{Dom}(V_{\mathbf{b}}(\mathbf{c}))$$

satisfying

$$y_T \;\geq\; J_{V_{\vec{\mathbf{b}}}(\vec{\mathbf{c}})}(t, x; (x(\cdot), u(\cdot))) \tag{37}$$

until the first time t^* *when, for at least one component* $i = 1, \ldots, n$,

$$y_{T_i} \;\leq\; K_{\vec{\mathbf{b}}_i}(t^*, x; (x(\cdot), u(\cdot))))$$

Then y_T *actually remains a Pareto minimum of the sets* $J_{V_{\vec{\mathbf{b}}}(\vec{\mathbf{c}})}((t; (x(\cdot), u(\cdot)))(T, x)$ *whenever* $t \in [0, t^*]$.

References

[1] J.-P. Aubin, *Contingent derivatives of set-valued maps and existence of solutions to nonlinear inclusions and differential inclusions.* Advances in Mathematics, Supplementary studies, Ed. Nachbin L., (1981), 160–232.

[2] J.-P. Aubin, *Viability Theory.* Birkhäuser, Boston, Basel, 1991.

[3] J.-P. Aubin, *Dynamic Economic Theory: A Viability Approach.* Springer-Verlag Birkhäuser, Boston, Basel, 1997.

[4] J.-P. Aubin, *Mutational and morphological analysis: tools for shape regulation and morphogenesis.* Birkhäuser, 1999.

[5] J.-P. Aubin, *Impulse Differential Inclusions and Hybrid Systems: A Viability Approach.* Lecture Notes, Université Paris-Dauphine, 2001.

[6] J.-P. Aubin, *Optimal Impulse Control Problems and Quasi-Variational Inequalities Thirty Years Later: a Viability Approach.* In Contrôle optimal et EDP: Innovations et Applications, IOS Press, (2000).

[7] J.-P. Aubin, *Boundary-Value Problems for Systems of First-Order Partial Differential Inclusions.* NoDEA **7** (2000), 67-90.

[8] J.-P. Aubin, *Lyapunov Functions for Impulse and Hybrid Control Systems.* Proceedings of the CDC 2000 Conference, (2000).

[9] J.-P. Aubin, *The Caliber of Impulse and Hybrid Control Systems.* Proceedings of "Evolution 2000", Birkhäuser, (2001).

[10] J.-P. Aubin, *The Substratum of Impulse and Hybrid Control Systems.* In Hybrid Systems: Computation and Control, 105-118, Di Benedetto & Sangiovanni-Vincentelli Eds, Proceedings of the HSCC 2001 Conference, LNCS 2034, Springer-Verlag, (2001).

[11] J.-P. Aubin, *Dynamic Core of Fuzzy Dynamical Cooperative Games.* Annals of Dynamic Games, Ninth International Symposium on Dynamical Games and Applications, Adelaide 2000, (2002).

[12] J.-P. Aubin, *Viability Kernels and Capture Basins of Sets under Differential Inclusions.* SIAM J. Control, **40** (2001), 853–881.

[13] J.-P. Aubin, A. Bicchi, and S. Pancanti, *Detectability of Evolutions by Tubes.* In preparation.

[14] J.-P. Aubin, N. Bonneuil, and F. Maurin, *Non-linear Structured Population Dynamics with Co-Variables.* Math. Popul. Studiesm, **91** (2000), 1–31.

[15] J.-P. Aubin, and F. Catte, *Fixed-Point and Algebraic Properties of Viability Kernels and Capture Basins of Sets.* Submitted.

[16] J.-P. Aubin, and G. Da Prato, *Solutions contingentes de l'équation de la variété centrale.* Comptes-Rendus de l'Académie des Sciences, Paris, **311** (1990), 295–300.

[17] J.-P. Aubin, and G. Da Prato, *Contingent Solutions to the Center Manifold Equation.* Annales de l'Institut Henri-Poincaré, Analyse Non Linéaire **9** (1992), 13–28.

[18] J.-P. Aubin and H. Frankowska, *Set-Valued Analysis,* Birkhäuser, Boston, Basel, 1990.

[19] J.-P. Aubin, and H. Frankowska, *Inclusions aux dérivées partielles gouvernant des contrôles de rétroaction.* Comptes-Rendus de l'Académie des Sciences, Paris **311** (1990), 851–856.

[20] J.-P. Aubin, and H. Frankowska, *Systèmes hyperboliques d'inclusions aux dérivées partielles.* Comptes-Rendus de l'Académie des Sciences, Paris **312** (1991), 271–276.

[21] J.-P. Aubin, and H. Frankowska, *Hyperbolic systems of partial differential inclusions.* Annali Scuola Normale di Pisa **18** (1992), 541–562.

[22] J.-P. Aubin, and H. Frankowska, *Partial differential inclusions governing feedback controls.* J. Convex Analysis **2** (1995), 19–40.

[23] J.-P. Aubin, and H. Frankowska, *The Viability Kernel Algorithm for Computing Value Functions of Infinite Horizon Optimal Control Problems.* J.Math. Anal. Appl. **201** (1996), 555–576.

[24] J.-P. Aubin, and H. Frankowska, *Set-valued Solutions to the Cauchy Problem for Hyperbolic Systems of Partial Differential Inclusions.* NoDEA **4**, (1997), 149–168.

[25] J.-P. Aubin, and G. Haddad, *Cadenced runs of impulse and hybrid control systems.* International Journal Robust and Nonlinear Control, (2001).

[26] J.-P. Aubin, and G. Haddad, *Path-Dependent Impulse and Hybrid Systems.* In Hybrid Systems: Computation and Control, 119–132, Di Benedetto and Sangiovanni-Vincentelli Eds, Proceedings of the HSCC 2001 Conference, LNCS 2034, Springer-Verlag, 2001.

[27] J.-P. Aubin, and G. Haddad, *Detectability under impulse differential inclusions.* Proceedings of the ECC 2001 Conference, (2001).

[28] J.-P. Aubin, and G. Haddad, *Detectability through Measurements under Impulse Differential Inclusions.* Submitted.

[29] J.-P. Aubin, J. Lygeros, M. Quincampoix, S. Sastry, and N. Seube, *Impulse Differential Inclusions: A Viability Approach to Hybrid Systems.* IEEE Transactions on Automatic Control, (2001).

[30] J.-P. Aubin, D. Pujal, and P. Saint-Pierre, *Dynamic Management of Portfolios with Transaction Costs under Contingent Uncertainty.* Preprint, (2001).

[31] M. Bardi, and I. Capuzzo Dolcetta, *Optimal control and viscosity solutions to Hamilton-Jacobi-Bellman equations.* Birkhäuser, 1998.

[32] E.N. Barron, and R. Jensen, *Semicontinuous viscosity solutions for Hamilton-Jacobi equations with convex Hamiltonians.* Comm. Partial Differential Equations **15** (1990), 1713–1742.

[33] E.N. Barron, and R. Jensen, *Optimal control and semicontiuous viscosity solutions.* Proceedings of the AMS **113** (1991), 393–402.

[34] A. Bensoussan, and Menaldi, *Hybrid Control and Dynamic Programming.* Dynamics of Continuous, Discrete and Impulse Systems **3** (1997), 395–442.

[35] P. Bernhard, *Expected Values, feared values and partial information optimal control.* In New trends in dynamic games and applications, Annals of the ISDG, 3, (1995), 3–24, Birkhäuser.

[36] P. Bernhard, *On the performance index of feared value control.* ISDG symposium, Sils-Maria, Switzerland, (1997).

[37] P. Bernhard, *Max-Plus algebra and mathematical fear in dynamic optimization,* J. Set-Valued Analysis, (2000).

[38] P. Cardaliaguet, M. Quincampoix, and P. Saint-Pierre, *Set-valued numerical methods for optimal control and differential games.* In Stochastic and differential games. Theory and numerical methods, Annals of the International Society of Dynamical Games, (1999), 177–247, Birkhäuser.

[39] O. Carja, and C. Ursescu, *The characteristics method for a first-order partial differential equation.* An. Stiin. Univ. Al. I. Cuza, Sect. 1 a Mat. **39** (1993), 367–396.

[40] O. Carja, and C. Ursescu, *Viscosity solutions and partial differential inequalities.* In Evolution equations, control theory and biomathematics, (1994), 39–44, Dekker.

[41] H. Frankowska, *L'équation d'Hamilton-Jacobi contingente.* Comptes-Rendus de l'Académie des Sciences, Paris, Série 1, **304** (1987), 295–298.

[42] H. Frankowska, *Optimal trajectories associated to a solution of contingent Hamilton-Jacobi equations.* IEEE, 26th, CDC Conference, Los Angeles, December 9–11, (1987).

[43] H. Frankowska, *Optimal trajectories associated to a solution of contingent Hamilton-Jacobi equations.* Applied Mathematics and Optimization **19** (1989), 291–311.

[44] H. Frankowska, *Hamilton-Jacobi equation: viscosity solutions and generalized gradients.* J. of Math. Analysis and Appl. **141** (1989), 21–26.

[45] H. Frankowska, *Lower semicontinuous solutions to Hamilton-Jacobi-Bellman equations.* Proceedings of 30th CDC Conference, IEEE, Brighton, December 11–13, (1991).

[46] H. Frankowska, *Lower semicontinuous solutions of Hamilton-Jacobi-Bellman equation.* SIAM J. on Control and Optimization, (1993).

[47] H. Frankowska, *Control of Nonlinear Systems and Differential Inclusions.* Birkhäuser, to appear.

[48] H. Frankowska, S. Plaskacz, and T. Rzezuchowski, *Measurable Viability Theorems and Hamilton-Jacobi-Bellman Equation.* Journal of Differential Equations **116** (1995), 265–305.

[49] H. Frankowska, S. Plaskacz, and T. Rzezuchowski, *Théorèmes de Viabilité Mesurable et l'équation d'Hamilton-Jacobi-Bellman.* Comptes-Rendus de l'Académie des Sciences, Série 1 (1995), Paris, 131–134.

[50] G. Haddad, *Monotone trajectories of differential inclusions with memory.* Isr. J. Math. **39** (1981), 83–100.

[51] G. Haddad, *Monotone viable trajectories for functional differential inclusions.* J. Diff. Eq. **42** (1981), 1–24.

[52] G. Haddad, *Topological properties of the set of solutions for functional differential differential inclusions.* Nonlinear Anal. Theory, Meth. Appl. **5** (1981), 1349–1366.

[53] A.S. Matveev, and A.V. Savkin, *Qualitative Theory of Hybrid Dynamical Systems.* Birkhäuser, 2000.

[54] A.S. Matveev, and A.V. Savkin, *Hybrid dynamical systems: Controller and sensor switching problems.* Birkhäuser, 2001.

[55] D. Pujal, *Valuation et gestion dynamiques de portefeuilles*, Thèse de l'Université de Paris-Dauphine, (2000).

[56] D. Pujal, and P. Saint-Pierre, *L'algorithme du bassin de capture appliqué pour évaluer des options européennes, américaines ou exotiques.* Preprint, (2001).

[57] M. Quincampoix, *Frontières de domaines d'invariance et de viabilité pour des inclusions différentielles avec contraintes.* Comptes-Rendus de l'Académie des Sciences, Paris, **311** (1990), 411–416.

[58] M. Quincampoix, *Enveloppes d'invariance pour des inclusions différentielles Lipschitziennes : applications aux problèmes de cibles.* Comptes-Rendus de l'Académie des Sciences, Paris, **314** (1992), 343–347.

[59] M. Quincampoix, and V. Veliov, *Viability with a target: theory and applications.* In Applications of mathematics in engineering, 47–54, Heron Press, (1998).

[60] R.T. Rockafellar, and R. Wets, *Variational Analysis.* Springer-Verlag, 1997.

[61] P. Saint-Pierre, *Approximation of the viability kernel.* Applied Mathematics & Optimisation **29** (1994), 187–209.

[62] P. Saint-Pierre, *Approximation of capture basins for hybrid systems.* Proceedings of the HSCC 2001 Conference, to appear.

[63] A. (van der) Shaft, and H. Schumacher, *An introduction to hybrid dynamical systems.* Springer-Verlag, Lecture Notes in Control, 251, 1999.

[64] Shuzhong Shi, *Théorèmes de viabilité pour les inclusions aux dérivées partielles.* C. R. Acad. Sc. Paris **303** (1986), Série I, 11–14.

[65] Shuzhong Shi, *Nagumo type condition for partial differential inclusions.* Nonlinear Analysis, T.M.A. **12** (1988), 951–967.

[66] Shuzhong Shi, *Optimal control of strongly monotone variational inequalities.*, SIAM J. Control & Optimization **26** (1988), 274–290.

[67] Shuzhong Shi, *Viability theorems for a class of differential-operator inclusions.* J. of Diff. Eq. **79** (1989), 232–257.

[68] L.E.O. Svensson, and M. Woodford, *Indicator variables for optimal policy.* NBRR WP 7953, (2000).

Jean-Pierre Aubin
Réseau de Recherche Viabilité, Jeux, Contrôle
14, rue Domat
F-75005 Paris, France
E-mail address: `J.P.Aubin@wanadoo.fr`

Progress in Nonlinear Differential Equations
and Their Applications, Vol. 55, 61–82

On a Convolution-Evolution Equation in Aeroelasticity

A.V. Balakrishnan

1. Introduction

Flutter is an instability endemic to aircraft wings that occurs at high enough airspeed in subsonic flight, and sets a bound (flutter boundary) on attainable airspeed in the subsonic regime. As is to be expected we need to make many simplifying assumptions before we can obtain a mathematically tractable model. Here we limit ourselves to a long, slender, "high-aspect-ratio" wing (see Figure 1)

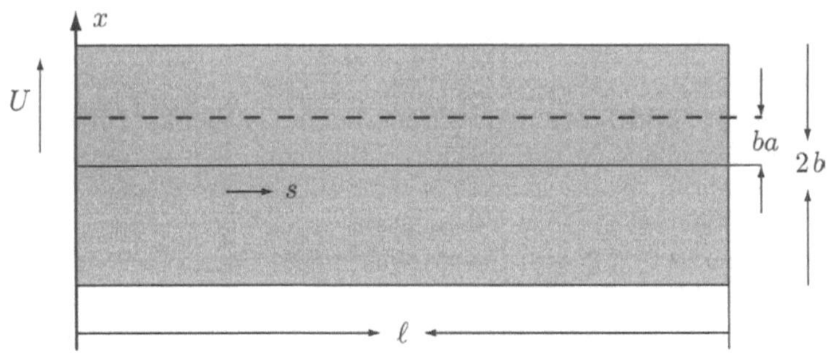

FIGURE 1. Wing: 2D Model: Long Slender High Aspect Ratio

so that we can simplify the wing model to a cantilever beam with two degrees of freedom – corresponding to plunge and pitch. The aerodynamics is assumed to be inviscid and incompressible, characterized by the small disturbance potential field equations. We simplify the problem further by invoking "typical section" theory, so that we can work with 2D aerodynamics (see Figure 2). In this way we finally arrive at a model describable in terms of a time-dependent boundary value problem for a set of partial differential (wave) equations. We show that the abstract version reduces to a convolution-semigroup equation in a Hilbert space, and the Flutter problem can then be characterized as the problem of determining system stability as a function of the airspeed, but determining the Flutter Boundary analytically is still an open problem, with room for further research. Numerical calculations on

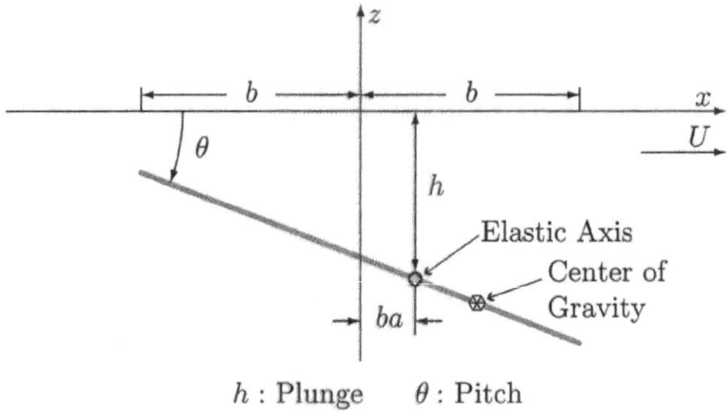

FIGURE 2. Geometry of a typical section

typical cases do however establish that the model does capture the known features of flutter.

This material has already been reported, most recently in [1]. What is new in this paper is a self-contained mathematical development of the aerodynamic theory leading to the results quoted without proof in [1] in the calculation of the lift and moment.

The structure model used derives from that of Goland [2] further elaborated in [3]. In the Aerodynamic part (which is the more difficult compared to the Structure Dynamics), key use is made of the results due to Sohngen and Tricomi [4] in their investigation of the airfoil equation.

We begin in Section 2 with the stucture model. Section 3 is devoted to the aerodynamics, and the calculation of the aerodynamic lift and moment. Section 4 assembles the total aeroelastic system and develops the abstract convolution-semigroup equation.

2. Structure Dynamics

We model each wing as a beam cantilevered to the root. With s, $0 < s < \ell < \infty$, denoting the span-wise length measured from the root and t the time, let $h(s,t)$ denote the bending (or plunge) dispacement. Let $\theta(s,t)$ denote the pitch angle and let S denote the coupling between these two motions. We use an Euler equation for the bending dynamics and a Timoshenko equation for the torsion dynamics. We have then the familiar structure equations:

$$M_S\,\ddot{x}(s,t) + D_S\,\dot{x}(s,t) + \mathcal{K}_S\,x(s,t) = \begin{vmatrix} L(s,t) \\ M(s,t) \end{vmatrix}, \quad 0 < s < \ell, \;\; 0 < t, \quad (2.1)$$

where

$$x(s,t) = \begin{vmatrix} h(s,t) \\ \theta(s,t) \end{vmatrix}.$$

The superdots represent partial derivatives with respect to time t, and

$$M_S = \begin{vmatrix} m & S \\ S & I_\theta \end{vmatrix}; \quad M_S > 0.$$

We may for our purposes take D_S to be zero. $\mathcal{K}_S$ is the differential operator

$$\mathcal{K}_s = \begin{vmatrix} EI\,\frac{d^4}{ds^4} & 0 \\ 0 & -GJ\,\frac{d^2}{ds^2} \end{vmatrix}$$

in the usual notation. The boundary conditions (clamped at 0, free at ℓ) are:

$$h(0,t) = h'(0,t) = 0 = \theta(0,t) = \theta'(\ell,t),$$

$$h'''(\ell,t) = 0 = h''(\ell,t),$$

where the prime denotes partial derivative with respect to s, $0 < s < \ell$.

The forcing functions on the right side of (2.1), $L(s,t)$, $M(s,t)$, are the aerodynamic "lift" and the aerodynamic "moment" respectively. This turns the structure problem into an "aeroelastic" problem, which we consider next.

3. Aerodynamic Loading

In this section we present a self-contained mathematical development of the relevant aerodynamic theory leading to the calculation of the aerodynamic lift and moment, while the results are known and form a substantial part of standard aeroelasticity – as in the classic text [3], our techniques are quite different.

3.1. Basic Theory

With x, y, z denoting rectangular coordinates, the velocity potential is defined as

$$Ux + \phi(x,y,z,t), \quad -\infty < x,y < \infty, \quad 0 < z < \infty, \quad 0 < t,$$

where U is the freefield air speed and has the representation

$$U = Ma_\infty,$$

where, in the usual notation, a_∞ is the speed of sound at the flight altitude, and M is the Mach number

$$0 < M < 1.$$

The "disturbance" potential $\phi(x,y,z,t)$ is assumed "small" so that it satisfies the linearized field equations: "subsonic, inviscid, small-disturbance potential equations" (see [5] for a derivation):

$$\frac{\partial^2\phi}{\partial t^2} + 2Ma_\infty \frac{\partial^2\phi}{\partial t\partial x} = a_\infty^2(1-M^2)\frac{\partial^2\phi}{\partial x^2} + a_\infty^2\left(\frac{\partial^2\phi}{\partial y^2} + \frac{\partial^2\phi}{\partial z^2}\right),$$

$$-\infty < x < \infty, \quad -\infty < y < \infty, \quad 0 < z < \infty, \quad 0 < t. \tag{3.1}$$

For our model we first invoke our "high-aspect-ratio" condition:

$$\frac{\ell}{b} \approx \infty,$$

and omit the dependence on the "spanwise" coordinate y. Then we limit ourselves to the "incompressible" case:

$$M = 0, \quad a_\infty = \infty \quad \text{but} \quad a_\infty M = U.$$

Dividing through by a_∞ in (3.1) we then have that the disturbance potential $\phi(x,z,t)$ satisfies (Laplace equation)

$$\frac{\partial^2\phi}{\partial x^2} + \frac{\partial^2\phi}{\partial z^2} = 0, \quad -\infty < x < \infty, \ 0 < z, \ 0 < t. \tag{3.2}$$

The boundary conditions.

We state the boundary conditions as stated in the aeroelasticity literature, as for example in [3, 5].

(i) Flow Tangency (or Attached Flow) Condition

$$\left.\frac{\partial\phi(x,z,t)}{\partial z}\right|_{z=0} = w_a(x,t), \quad |x| < b < \infty, \tag{3.3}$$

where $w_a(x,t)$ is the "downwash" function which we shall specify later with reference to the wing structure but for the moment we shall assume that it is of the form

$$w_a(x,t) = w_1(x)\,a_1(t) + w_2(x)\,a_2(t), \tag{3.4}$$

where $w_1(\cdot)$ and $w_2(\cdot)$ are absolutely continuous in $|x| \le b$ and $a_1(\cdot)$ and $a_2(\cdot)$ are such that the following Laplace transforms exists:

$$\int_0^\infty e^{-\sigma t}\left(|a_1(t)| + |a_2(t)|\right) dt < \infty, \quad \sigma > \sigma_a \ge 0. \tag{3.5}$$

We see that this is a Neumann Condition but only on part of the boundary. The next condition is what makes the problem nonstandard.

(ii) Kutta-Joukowski Condition.

To state this condition we first need to introduce the acceleration potential. The acceleration potential is defined as

$$\psi(x,z,t) = \frac{\partial}{\partial t}\phi(x,z,t) + U\frac{\partial}{\partial x}\phi(x,z,t). \tag{3.6}$$

The Kutta-Joukowski Condition is then that

$$\psi(x,z,t)\Big|_{z=0} = 0, \quad |x| > b, \ 0 < t \tag{3.7}$$

and "zero pressure discontinuity at trailing edge"

$$\lim_{x\to b} \psi(x,z,t)\Big|_{z=0} = 0, \quad 0 < t. \tag{3.8}$$

The latter condition is shortened in what follows as

$$\psi(b-,\,0,\,t) = 0, \quad 0 < t.$$

(iii) Finally we have the "Far Field" condition that the functions

$$\phi(x,z,t)\,, \quad \frac{\partial}{\partial x}\,\phi(x,z,t), \quad \frac{\partial}{\partial z}\,\phi(x,z,t)$$

vanish as $|x| \to \infty$, $z \to \infty$.

We note that (3.2), (3.3). (3.7), (3.8) comprise a "boundary value" problem without an "initial value" component, with time entering only through the boundary. Our aim, as usual, is to show that this problem has a solution satisfying appropriate smoothness conditions and that such solution is unique. We shall actually construct the solution – a cornerstone in Aeroelasticity theory [3,6]. Our techniques – totally mathematical – are radically different from those in [3,6]. A skeptical reader is encouraged to check this claim.

As may be expected in Neumann problems, we show that under appropriate smoothness conditions the problem can be converted to an integral equation and then show that the latter has a unique solution leading in turn to a unique solution of the field equations satisfying the assumed conditions.

We begin by detailing the conditions imposed on the solution. We require that

$$\int_{-\infty}^{\infty} |\phi(x,z,t)|^p \; dx \; < \; \infty, \quad 0 < z, \ 0 < t$$

for $1 < p < \frac{4}{3}$. Let

$$X = L^p(-\infty,\infty), \quad 1 < p < \tfrac{4}{3}\,.$$

Then

$$\phi(\cdot,z,t) \in X.$$

Let D_x denote the closed linear operator corresponding to $\frac{\partial}{\partial x}$ with domain and range in X. We assume that

$$\phi(\cdot,z,t) \in \mathcal{D}(D_x^2), \quad 0 < z,t.$$

Next we assume that the derivative

$$\frac{\partial}{\partial z}\,\phi(\cdot,z,t)$$

is absolutely continuous in z, $z > 0$, for each t (and the same fact is valid for $\phi(\cdot,z,t)$).

Next, we require that

$$\|\phi(\cdot, z, t)\| \;\to\; 0 \quad \text{as } t \to 0+$$

and that

$$\phi(\cdot, z, t), \qquad D_x^2 \phi(\cdot, z, t),$$
$$\frac{\partial^2}{\partial z^2}\, \phi(\cdot, z, t), \qquad \frac{\partial}{\partial t}\, \phi(\cdot, z, t)$$

are all Laplace transformable for

$$\operatorname{Re} \lambda > \sigma_a$$

analogous to (3.5).

We use the "superhat" notation for the Laplace transform, i.e.,

$$\hat{\phi}(\cdot, z, \lambda) \;=\; \int_0^\infty e^{-\lambda t}\, \phi(\cdot, z, t)\, dt, \quad \operatorname{Re} \lambda > \sigma_a\,.$$

Then we have

$$\int_0^\infty e^{-\lambda t}\, \frac{\partial}{\partial t}\, \phi(\cdot, z, t)\, dt \;=\; \lambda\, \hat{\phi}(\cdot, z, \lambda).$$

Defining the acceleration potential

$$\psi(\cdot, z, t) \;=\; \frac{\partial}{\partial t}\, \phi(\cdot, z, t) \;+\; U D_x\, \phi(\cdot, z, t) \tag{3.9}$$

we have that the Laplace transform

$$\begin{aligned}\hat{\psi}(\cdot, z, \lambda) \;&=\; \int_0^\infty e^{-\lambda t}\, \psi(\cdot, z, t)\, dt \\ &=\; \lambda\, \hat{\phi}(\cdot, z, \lambda) \;+\; U \int_0^\infty e^{-\lambda t}\, D_x \phi(\cdot, z, t)\, dt, \; Re\, \lambda > \sigma_a. \end{aligned} \tag{3.10}$$

Next we introduce spatial Fourier Transforms. For $f(\cdot)$ in X, we know that

$$\int_{-\infty}^{\infty} e^{-i\omega x}\, f(x)\, dx$$

defines a function denoted by $\hat{f}(i\omega)$ and such that

$$\hat{f} \,\in L^q(-\infty, \infty), \quad \tfrac{1}{p} + \tfrac{1}{q} = 1.$$

Moreoever it defines a linear bounded transformation on L^p into L^q, which we denote by $\mathcal{F}$. Then in view of our assumption, we have

$$\mathcal{F}\, \hat{\phi}(\cdot, z, \lambda) \;=\; \hat{\hat{\phi}}(\cdot, z, \lambda)$$

where

$$\hat{\hat{\phi}}(\cdot, z, \lambda) \quad \text{denotes} \quad \hat{\hat{\phi}}(i\omega, z, \lambda)$$

and

$$\mathcal{F} D_x^2\, \hat{\phi}(\cdot, z, \lambda)$$

is the function

$$-\omega^2\, \hat{\hat{\phi}}(i\omega, z, \lambda) \quad \text{in } L^q(-\infty, \infty).$$

Also

$$\mathcal{F}\, \frac{\partial^2}{\partial z^2}\, \hat{\phi}(\cdot, z, \lambda) \;=\; \frac{\partial^2}{\partial z^2}\, \hat{\hat{\phi}}(\cdot, z, \lambda).$$

Hence our first result is: Fourier + Laplace transforming the field equation (3.2):

$$-\omega^2\, \hat{\hat{\phi}}(i\omega, z, \lambda) \;+\; \frac{\partial^2}{\partial z^2}\, \hat{\hat{\phi}}(i\omega, z, \lambda) \;=\; 0, \quad 0 < z, \quad \operatorname{Re}\lambda > \sigma_n\,, \quad -\infty < \omega < \infty, \tag{3.11}$$

which we can rewrite as,

$$\frac{\partial^2}{\partial z^2}\, \hat{\hat{\phi}}(i\omega, z, \lambda) \;=\; \omega^2\, \hat{\hat{\phi}}(i\omega, z, \lambda), \quad 0 < z, \tag{3.12}$$

and regard it as a differential equation in z, $0 < z$, fixing ω and λ. Now we require that

$$\left\| \hat{\phi}(\cdot, z, \lambda) \right\|$$

is bounded in $0 < z$ for each λ, and that $\hat{\phi}(\cdot, z, \lambda)$ is continuous at $z = 0+$. Then the only possible solution of (3.12) is given by

$$\hat{\hat{\phi}}(i\omega, z, \lambda) \;=\; e^{-|\omega| z}\, \hat{\hat{\phi}}(i\omega, 0, \lambda). \tag{3.13}$$

In turn this implies that

$$\frac{\partial}{\partial z}\, \hat{\hat{\phi}}(i\omega, z, \lambda) \;=\; -|\omega|\, \hat{\hat{\phi}}(i\omega, 0, \lambda)\, e^{-|\omega| z}$$

and hence

$$\left. \frac{\partial}{\partial z}\, \hat{\hat{\phi}}(i\omega, z, \lambda) \right|_{z=0} \;=\; -|\omega|\, \hat{\hat{\phi}}(i\omega, 0, \lambda).$$

We shall use the notation

$$\hat{\hat{\nu}}(i\omega, \lambda) \;=\; \left. \frac{\partial}{\partial z}\, \hat{\hat{\phi}}(i\omega, z, \lambda) \right|_{z=0}$$

so that we have

$$\hat{\hat{\nu}}(i\omega, \lambda) \;=\; -|\omega|\, \hat{\hat{\phi}}(i\omega, 0, \lambda). \tag{3.14}$$

We now turn to the acceleration potential (3.10). We have, by taking now Fourier Transforms

$$\begin{aligned} \hat{\hat{\psi}}(i\omega, z, \lambda) \;&=\; \lambda\, \hat{\phi}(i\omega, z, \lambda) \;+\; i\omega U\, \hat{\hat{\phi}}(i\omega, z, \lambda) \\ &=\; (\lambda + i\omega U)\, \hat{\phi}(i\omega, z, \lambda). \end{aligned} \tag{3.15}$$

We rephrase the boundary conditions as follows. Condition (i) becomes

$$\hat{\nu}(x, \lambda) \;=\; \hat{w}_a(x, \lambda), \quad |x| \le b.$$

Condition (ii) becomes:

$$\psi(\cdot, z, t)$$

is continuous at $z = 0+$ for each t and

$$\psi(x, 0+, t) = 0, \quad |x| > b$$

$$\lim_{x \to b} \psi(x, 0+, t) = 0$$

which in terms of the Laplace transform bcomes

$$\hat{\psi}(x, 0, \lambda) = 0, \quad |x| > b$$

$$\lim_{x \to b} \hat{\psi}(x, 0, \lambda) = 0.$$

Following the usual notation as in [3, 6], we now define

$$A(x,t) = \tfrac{-2}{U}\, \psi(x, 0+, t), \quad |x| < b \tag{3.16}$$

and in particular then we require that

$$\lim_{x \to b} A(x,t) = 0.$$

Note that

$$A(\cdot, t) \in L^p(-b, b), \quad 1 < p < \tfrac{4}{3}\,.$$

Defining

$$\hat{A}(x, \lambda) = \int_0^\infty e^{-\lambda t}\, A(x,t)\, dt$$

and

$$\hat{\hat{A}}(i\omega, \lambda) = \int_{-b}^{b} e^{-i\omega x}\, \hat{A}(x,t)\, dx.$$

We obtain from (3.15) that

$$\begin{aligned} \hat{\hat{A}}(i\omega, \lambda) &= \tfrac{-2}{U}\, \hat{\hat{\psi}}(i\omega, 0, \lambda) \\ &= \tfrac{-2}{U}\, (\lambda + i\omega U)\, \hat{\hat{\phi}}(i\omega, 0, \lambda) \\ &= \frac{2(\lambda' + i\omega)}{|\omega|}\, \hat{\nu}(i\omega, \lambda), \end{aligned}$$

where

$$\lambda' = \frac{\lambda}{U}.$$

Thus, we have the following formula for $\hat{\nu}$:

$$\hat{\nu}(i\omega, \lambda) = \frac{1}{2}\, \frac{|\omega|}{\lambda' + i\omega}\, \hat{A}(i\omega, \lambda). \tag{3.17}$$

Here

$$\frac{1}{2}\, \frac{|\omega|}{\lambda' + i\omega}$$

is recognized as an $L^p(-\infty,\infty)$ multiplier. Actually we can say more

$$\frac{1}{2}\,\frac{|\omega|}{\lambda'+i\omega} \;=\; \frac{1}{2}\,\frac{|\omega|}{i\omega}\,\frac{i\omega}{\lambda'+i\omega} \;=\; \frac{1}{2}\,\frac{|\omega|}{i\omega}\left[1 - \frac{\lambda'}{\lambda'+i\omega}\right]$$

where

$$\frac{|\omega|}{i\omega}, \qquad -\infty < \omega < \infty$$

is recognized as the multiplier corresponding to the Hilbert Transform. Hence we see that

$$\begin{aligned}\hat{\nu}(x,\lambda) &= \frac{1}{2\pi}\int_{-b}^{b}\frac{1}{x-\xi}\,\hat{A}(\xi,\lambda)\,d\xi \\ &\quad - \frac{\lambda'}{2\pi}\int_{-b}^{\infty}\frac{d\xi}{x-\xi}\int_{-b}^{\xi}e^{-\lambda'(\xi-\sigma)}\,\hat{A}(\sigma,\lambda)\,d\sigma, \quad -b < x < \infty, \\ \hat{\nu}(x,\lambda) &= 0,\; x < -b. \end{aligned} \tag{3.18}$$

Hence we have in particular

$$\begin{aligned}\hat{w}_a(x,\lambda) &= \frac{1}{2\pi}\int_{-b}^{b}\frac{1}{x-\xi}\,\hat{A}(\xi,\lambda)\,d\xi \\ &\quad - \frac{\lambda'}{2\pi}\int_{-b}^{\infty}\frac{d\xi}{x-\xi}\int_{-b}^{\xi}e^{-\lambda'(\xi-\sigma)}\,\hat{A}(\sigma,\lambda)\,d\sigma, \quad |x| < b. \end{aligned} \tag{3.19}$$

We may regard this as an integral equation for $\hat{A}(\cdot,\lambda)$ for each λ, $\operatorname{Re}\lambda > \sigma_a$, since the left side is given. This is in fact the Possio Equation (for incompressible flow) but stated for $\lambda = i\omega$ ("oscillatory response"), involving divergent integrals. [3, 7].

Our main result is:

Theorem 3.1. *The Possio Integral Equation (3.19) has a unique solution in*

$$\bigcap_p L^p(-b,b), \quad 1 < p < \tfrac{4}{3}$$

for every λ,

$$Re\,\lambda > \sigma_a$$

such that

$$\lim_{x\to b}\hat{A}(x,\lambda) = 0.$$

Moreover, the solution can be expressed explicitly as

$$\hat{A}(x,\lambda) = r(x,\lambda) + \lambda'\int_{-b}^{x} r(s,\lambda)\,ds, \quad |x| < b, \tag{3.20}$$

$$r(x,\lambda) = u(x,\lambda) + \lambda'\,h(x,\lambda')\,\hat{c}_1(b\lambda')\int_{-b}^{b}u(x,\lambda)\,dx, \tag{3.21}$$

where

$$u(x,\lambda) = \frac{1}{\pi}\sqrt{\frac{b-x}{b+x}} \int_{-b}^{b} \sqrt{\frac{b+\xi}{b-\xi}} \frac{\hat{w}_a(\xi,\lambda)}{\xi - x} \, d\xi, \tag{3.22}$$

$$h(x,\lambda') = \frac{1}{\pi}\sqrt{\frac{b-x}{b+x}} \int_{0}^{\infty} \frac{1}{x-b-\sigma} \sqrt{\frac{2b+\sigma}{\sigma}} \, e^{-\lambda'\sigma} \, d\sigma, \tag{3.23}$$

$$\hat{c}_1(\lambda) = \frac{1}{\lambda(K_0(\lambda) + K_1(\lambda))e^{\lambda}}, \qquad Re\lambda > 0. \tag{3.24}$$

Proof. We begin with a crucial result due to Tricomi and Sohngen [4] on the "airfoil" equation.

Lemma 3.2. *On*

$$L^{\frac{4}{3}}(-b,b)$$

define the operator R by

$$g = Rf;$$

$$g(x) = \frac{2}{\pi}\sqrt{\frac{b-x}{b+x}} \int_{-b}^{b} \sqrt{\frac{b+\xi}{b-\xi}} \frac{f(\xi)}{\xi - x} \, d\xi, \quad |x| < b < \infty. \tag{3.25}$$

Then

$$g(\cdot) \in \bigcap_p L^p(-b,b), \quad 1 \le p < \tfrac{4}{3}.$$

Proof. See Tricomi [4, p. 176, *et seq.*]

Lemma 3.3. *[Tricomi-Sohngen] The Singular Integral Equation ("Airfoil Equation")*

$$f(x) = \frac{1}{2\pi} \int_{-b}^{b} \frac{g(\xi)}{x-\xi} \, d\xi, \quad |x| < b$$

has a solution $g(\cdot) \in \bigcap_p L^p[-b,b]$, $1 < p < \frac{4}{3}$ *for each*

$$f(\cdot) \quad in \quad L^{\frac{4}{3}}(-b,b)$$

given by

$$g = Rf.$$

If the solution $g(\cdot)$ *has the property that*

$$\lim_{x \to b} g(x) \quad exists \; and \; is \; finite,$$

then the solution $g(\cdot)$ *is unique.*

Proof. See Tricomi [4, p. 173 *et seq.*] where he shows that

$$g = Rf$$

is a solution and that if

$$f(x) = 0,$$

then $g(\cdot)$ must be of the form

$$g(x) = \frac{c}{\sqrt{b^2 - x^2}},$$

from which the uniqueness part follows.

Proof of Theorem 3.1. Armed with these Lemmas we now return to our integral equation (3.19), fixing λ, $\operatorname{Re}\lambda > \sigma_a$. Let

$$c(\xi, \lambda) = \frac{\lambda'}{2\pi} \int_{-b}^{\xi} e^{-\lambda'(\xi-\sigma)} \hat{A}(\sigma, \lambda)\, d\sigma, \quad |\xi| < b.$$

Then for

$$\hat{A}(\cdot, \lambda) \quad \text{in } L^p(-b, b), \quad 1 < p < \tfrac{4}{3},$$

$c(\xi, \lambda)$ is continuous in ξ in $[-b, b]$. Let

$$\begin{aligned} r_+(x) &= \frac{\lambda'}{2\pi} \int_b^{\infty} \frac{d\xi}{x-\xi} \int_{-b}^{b} e^{-\lambda'(\xi-\sigma)} \hat{A}(\sigma, \lambda)\, d\sigma, \qquad |x| < b \\ &= \frac{\lambda'}{2\pi} \int_b^{\infty} \frac{d\xi}{x-\xi}\, e^{-\lambda'\xi} \left(\int_{-b}^{b} e^{\lambda'\sigma} \hat{A}(\sigma, \lambda)\, d\sigma \right), \quad |x| < b. \end{aligned}$$

Then since

$$\int_b^{\infty} \left| e^{-\lambda'\xi} \right|^p d\xi < \infty \quad \text{for every } p > 1$$

it follows by the well-known property of the Hilbert Transform that

$$\int_{-b}^{b} |r_+(x)|^p\, dx < \infty \quad \text{for every } p > 1.$$

We rewrite (3.19) in the form

$$\hat{w}_a(x, \lambda) + r_+(x) = \frac{1}{2\pi} \int_{-b}^{b} \frac{f(\xi)}{x-\xi}\, d\xi, \quad |x| < b, \tag{3.26}$$

where

$$f(\xi) = \hat{A}(\xi, \lambda) - c(\xi).$$

Then by the Lemmas 3.1, 3.2, one solution of (3.26) is given by

$$\hat{A}(\cdot, \lambda) - c(\cdot) = R(\hat{w}_a(\cdot, \lambda) + r_+(\cdot)). \tag{3.27}$$

Let

$$g(\cdot) = R(\hat{w}_a(\cdot, \lambda) + r_+(\cdot)).$$

We shall now show that

$$\lim_{x \to b} g(x)$$

exists and is finite. First let us consider

$$h(\cdot) = Rr_+(\cdot)$$

or,

$$h(x) = \frac{2}{\pi}\sqrt{\frac{b-x}{b+x}}\int_{-b}^{b}\sqrt{\frac{b+\xi}{b-\xi}}\frac{d\xi}{\xi-x}\frac{1}{2\pi}\int_{b}^{\infty}\frac{1}{\xi-\sigma}e^{-\lambda'\sigma}\,d\sigma\int_{-b}^{b}e^{\lambda' x}\hat{A}(x,\lambda)\,dx,$$

$$|x| < b. \qquad (3.28)$$

Now

$$\frac{1}{\xi-x}\frac{1}{\xi-\sigma} = \frac{1}{x-\sigma}\left[\frac{1}{\xi-x} - \frac{1}{\xi-\sigma}\right], \qquad |x|<b,\ \ \sigma>b$$

and

$$\frac{1}{\pi}\int_{-b}^{b}\sqrt{\frac{b+\xi}{b-\xi}}\frac{1}{\xi-x}\frac{d\xi}{\xi-\sigma} = \frac{1}{x-\sigma}\sqrt{\frac{\sigma+b}{\sigma-b}}, \quad \sigma>b.$$

Hence changing the order of integration in (3.28) we have

$$h(x) = \frac{1}{\pi}\sqrt{\frac{b-x}{b+x}}\int_{b}^{\infty}\frac{1}{x-\sigma}\sqrt{\frac{\sigma+b}{\sigma-b}}\,e^{-\lambda'\sigma}\,d\sigma\left(\int_{-b}^{b}e^{-\lambda' x}\hat{A}(x,\lambda)\,dx\right)$$

$$= \frac{1}{\pi}\sqrt{\frac{b-x}{b+x}}\int_{0}^{\infty}\frac{d\sigma}{x-b-\sigma}\sqrt{\frac{2b+\sigma}{\sigma}}\,e^{-\lambda'\sigma}\left(\int_{-b}^{b}e^{-\lambda'(b-x)}\hat{A}(x,\lambda)\,dx\right).$$

Let us use the notation

$$h_+(x,\lambda) = \frac{1}{\pi}\sqrt{\frac{b-x}{b+x}}\int_{0}^{\infty}\frac{d\sigma}{x-b-\sigma}\sqrt{\frac{2b+\sigma}{\sigma}}\,e^{-\lambda'\sigma}\,d\sigma, \quad |x|<b.$$

Then expressing it as

$$h_+(x,\lambda) = \frac{-1}{\pi}\frac{2}{\sqrt{b+x}}\int_{0}^{\infty}e^{-\lambda' t^2(b-x)}\sqrt{2b + 2t^2(b-x)}\,\frac{dt}{1+t^2},$$

we see that

$$\lim_{x\to b} h_+(x,\lambda) = -1. \qquad (3.29)$$

Next let

$$g(\cdot) = R\hat{w}(\cdot,\lambda).$$

Because of our assumption (3.4), we have

$$\hat{w}_a(x,\lambda) = w_1(x)\,\hat{a}_1(\lambda) + w_2(x)\,\hat{a}_2(\lambda)$$

where $w_1(\cdot)$, $w_2(\cdot)$ are absolutely continuous in $|x| \le b$, and

$$\frac{1}{\pi}\int_{-b}^{b}\sqrt{\frac{b+\xi}{b-\xi}}\frac{w_i(\xi)}{\xi-x}\,d\xi - w_i(x) = \frac{1}{\pi}\int_{-b}^{b}\sqrt{\frac{b+\xi}{b-\xi}}\frac{(w_i(\xi)-w_i(x))}{\xi-x}\,d\xi, \quad i=1,2.$$

It follows that the integrals are bounded in $|x| \le b$ and hence it follows that

$$\lim_{x\to b} g(x) = 0. \qquad (3.30)$$

Hence it follows from Lemma 3.2 that (3.27) is the only solution. But

$$\lim_{\xi \to b} c(\xi) \;=\; \lambda' \int_{-b}^{b} e^{-\lambda'(b-\sigma)}\, \hat{A}(\sigma, \lambda)\, d\sigma$$

and hence it follows that

$$\lim_{\xi \to b} \hat{A}(\xi, \lambda) \;=\; 0.$$

Hence it follows that

$$\hat{A}(\xi, \lambda) \;-\; \lambda' \int_{-b}^{\xi} e^{-\lambda'(\xi-\sigma)}\, \hat{A}(\sigma, \lambda)\, d\sigma$$

$$= \; u(\xi, \lambda) \;+\; \lambda' h_+(x, \lambda') \int_{-b}^{b} e^{-\lambda'(b-\sigma)}\, \hat{A}(\sigma, \lambda)\, d\sigma, \quad |\xi| < b, \tag{3.31}$$

where

$$u(\cdot, \lambda) \;=\; R\,\hat{w}_a(\cdot, \lambda).$$

We shall now show that (3.31) has a unique solution in

$$\bigcap_p L^p(-b, b), \quad 1 < p < \tfrac{4}{3}.$$

This is fairly immediate because defining the operator $\mathcal{L}$ on $L^p(-b,b)$, $1 < p < \frac{4}{3}$ into itself by

$$g(\xi) \;=\; \lambda' \int_{-b}^{\xi} e^{-\lambda'(\xi-\sigma)}\, f(\sigma)\, d\sigma, \quad |\xi| < b$$

we see that $\mathcal{L}$ is Volterra. We see that (3.31) becomes

$$(I - \mathcal{L})\, \hat{A}(\cdot, \lambda) \;=\; u(\cdot, \lambda) \;+\; \lambda' h_+(\cdot, \lambda)\, L(\hat{A}(\cdot, \lambda)),$$

where $L(\cdot)$ is the continuous linear function

$$L(f) \;=\; \int_{-b}^{b} e^{-\lambda'(b-\sigma)}\, f(\sigma)\, d\sigma$$

on $L^p(b, -b)$, $1 < p < \frac{4}{3}$. Now

$$(I - \mathcal{L})^{-1} \;=\; I \;+\; \mathcal{M}$$

where

$$(\mathcal{M}f)(x) \;=\; \lambda' \int_{-b}^{x} f(\xi)\, d\xi, \quad |x| < b.$$

Hence

$$\hat{A}(\cdot, \lambda) \;=\; (I + \mathcal{M}) \Big(u(\cdot, \lambda) \;+\; \lambda' h_+(\cdot, \lambda)\, L(\hat{A}(\cdot, \lambda)) \Big)$$

and hence

$$L(\hat{A}(\cdot, \lambda)) \;=\; L((I + \mathcal{M})\, u(\cdot, \lambda)) \;+\; \lambda' L((I + \mathcal{M})\, h_+(\cdot, \lambda))\, L(\hat{A}(\cdot, \lambda)).$$

Now

$$1 - \lambda' L((I+\mathcal{M})\,h_+(\cdot,\lambda)) = \frac{1}{\hat{c}_1(\lambda' b)}$$

$$\neq 0, \qquad \operatorname{Re}\lambda > 0.$$

Hence finally we have

$$\hat{A}(\cdot,\lambda) = (I+\mathcal{M})\left(u(\cdot,\lambda) + \lambda' h_+(\cdot,\lambda)\left(\frac{L((I+\mathcal{M})\,u(\cdot,\lambda))}{1-\lambda' L((I+\mathcal{M})\,h_+(\cdot,\lambda))}\right)\right) \tag{3.32}$$

which, unravelled, using integration by parts, etc., yields (3.20).

This completes the proof of Theorem 3.1.

Let us next get back to finding the velocity potential solution $\phi(x,z,t)$. Using (3.32), we can calculate

$$\hat{\nu}(x,\lambda) \quad \text{for } |x| > b$$

from (3.18), and we obtain from (3.14) that

$$\hat{\hat{\phi}}(i\omega,0,\lambda) = \frac{-\hat{\mathcal{V}}(i\omega,\lambda)}{|\omega|}$$

and hence finally

$$\hat{\hat{\phi}}(i\omega,z,\lambda) = e^{-|\omega|z}\left(\frac{-1}{|\omega|}\right)\hat{\mathcal{V}}(i\omega,\lambda)$$

$$= i\omega\, e^{-|\omega|z}\left(\frac{i\omega}{|\omega|}\right)\hat{\mathcal{V}}(i\omega,\lambda).$$

We do not pursue this further here because in the aeroelastic application (see below) we need to work only with the boundary values.

Finally, we note that the uniqueness of solution to the field equations follows from the uniqueness of solution to the Possio Equation. Thus if there are two solutions satisfying the conditions we have imposed for given downwash function, then the difference would be a solution for zero downwash. But the Possio Equation will then have only the zero solution and by (3.32) this would make the corresponding normal derivative zero along the entire boundary. But then we can invoke uniqueness of solution to the Neumann problem, since

$$\phi(x,z,t) = 0 \quad \text{at } t=0$$

it follows that $\phi(x,z,t)$ must be zero for all $t>0$, for all x and z, $-\infty < x < \infty$, $0 < z$.

3.2. The Aerodynamic Lift and Moment Calculation

To apply this theory to calculate the lift and moment at the point s of the span variable, we need first to evaluate the corresponding normal velocity of the wing structure. This is given by

$$w_a(x,s,t) \;=\; \frac{\partial}{\partial t}\, z_a(x,s,t) \;+\; U\,\frac{\partial}{\partial x}\, z_a(x,s,t)$$

where (noting that the plunge is taken positive downward):

$$z_a(x,s,t) \;=\; -h(s,t) \;-\; (x-ba)\,\theta(s,t)$$

and hence

$$w_a(x,s,t) \;=\; -\dot{h}(s,t) \;-\; (x-ba)\,\dot{\theta}(s,t) \;-\; U\theta(s,t)$$

where the superdots indicate time derivatives. Thus $w_a(x,t)$ must from now on be replaced by $w_a(x,s,t)$ to denote the dependence on where the section is taken as well. Taking Laplace transforms where we also need to indicate the variable s, we have:

$$\hat{w}_a(x,s,\lambda) \;=\; -\lambda\,\hat{h}(s,\lambda) \;-\; (x-ba)\,\lambda\,\hat{\theta}(s,\lambda) \;-\; U\hat{\theta}(s,\lambda), \tag{3.33}$$

where, consistent with the assumption that the disturbance initial conditions are zero, we assume that the structure initial conditions are zero, i.e.,

$$h(x,s,0) \;=\; 0; \quad \theta(x,s,0) \;=\; 0.$$

We note that this simplification does not impact the system stability, which is our main concern.

In turn, we get by substitution into the solution (3.20), with $\hat{A}(x,s,\lambda)$ replacing $\hat{A}(x,\lambda)$ that:

$$\hat{A}(x,s,\lambda) = r(x,s,\lambda) \;+\; \lambda' \int_{-b}^{x} r(\xi,s,\lambda)\, d\xi,$$

$$r(x,s,\lambda) = u(x,s,\lambda') \;+\; \lambda'\, h_+(x,\lambda')\,\hat{c}_1(\lambda' b) \int_{-b}^{b} u(x,s,\lambda')\, dx,$$

where

$$u(x,s,\lambda') \;=\; \frac{1}{\pi}\sqrt{\frac{b-x}{b+x}} \int_{-b}^{b} \sqrt{\frac{b+\xi}{b-\xi}}\; \frac{\hat{w}(\xi,s,\lambda)}{\xi - x}\, d\xi.$$

We can now calculate the (differential) pressure

$$\widehat{\Delta p}(x,s,\lambda) \;=\; -2\rho\,\hat{\psi}(x,0,\lambda) \;=\; \rho U \hat{A}(x,s,\lambda)$$

ρ being the fluid density. The lift and moment, we seek, are then given by

$$\hat{L}(s,\lambda) = \int_0^\infty e^{-\lambda t}\, L(s,t)\, dt, \qquad \operatorname{Re}\lambda > \sigma_a$$
$$= \int_{-b}^{b} \widehat{\Delta p}(x,s,\lambda)\, dx$$
$$\hat{M}(s,\lambda) = \int_{-b}^{b} (x-ba)\, \widehat{\Delta p}(x,s,\lambda)\, dx$$

where ba denotes the position on the chord of the elastic axis.

We have now arrived at a stage where the final calculations reduce to routine analysis and some formulas for definite integrals. Because of page limitation we skip this tedious ordeal and go directly to the answer. We have

$$\hat{L}(s,\lambda) = \pi\rho b^2 \left[-\lambda^2 \hat{h}(s,\lambda) + \lambda^2 ba\, \hat{\theta}(s,\lambda)\right] - \pi\rho b^2 U\lambda\hat{\theta}(s,\lambda)$$
$$- bU\rho C(\lambda' b)\left(2\pi U\hat{\theta}(s,\lambda) + 2\pi\lambda\,\hat{h}(s,\lambda) + b\lambda\pi(1-2a)\,\hat{\theta}(s,\lambda)\right) \tag{3.34}$$

$$\hat{M}(s,\lambda) = \lambda^2[\pi\rho b^3]\left[a\,\hat{h}(s,\lambda) - b\left(a^2+\tfrac{1}{8}\right)\hat{\theta}(s,\lambda)\right]$$
$$+ \left[\rho b^2 U\, C(\lambda' b)\left(\tfrac{1}{2}+a\right) - \frac{\rho b^2 U}{2}\right]$$
$$\cdot \left[2\pi U\hat{\theta}(s,\lambda) + 2\pi\lambda\,\hat{h}(s,\lambda) + b\pi\lambda(1-2a)\,\hat{\theta}(s,\lambda)\right], \tag{3.35}$$

where

$$C(\lambda) = \frac{K_1(\lambda)}{K_0(\lambda) + K_1(\lambda)}, \qquad \operatorname{Re}\lambda > 0 \tag{3.36}$$

known as the Theodorsen Function. Because of its occurence in (3.34) and (3.35) it is convenient to use the notation

$$w(s,t) = 2\pi U\theta(s,t) + 2\pi\,\dot{h}(s,t) + b\pi(1-2a)\,\dot{\theta}(s,t)$$

so that its Laplace Tranoform

$$\hat{w}(s,\lambda) = 2\pi U\hat{\theta}(s,\lambda) + 2\pi\lambda\,\hat{h}(s,\lambda) + b\pi(1-2a)\lambda\,\hat{\theta}(s,\lambda).$$

Finally we note that it is known (see [3]) that

$$\frac{C(\lambda)}{\lambda} = \int_0^\infty e^{-\lambda t}\, L(t)\, dt, \quad \operatorname{Re}\lambda > 0, \quad \int_0^\infty |L(t)|\, dt < \infty, \tag{3.37}$$

where $L(\cdot)$ is the Wagner Function, for which many approximations using elementary functions have been proposed (see, e.g., [6]). It is known that

$$L(t) = \frac{1}{2} + \int_0^t \ell(s)\, ds, \quad t \geq 0 \qquad \text{and} \qquad L(t) \to 0 \quad \text{as } t \to \infty,$$

which is important as we take inverse Laplace Transforms in the next section.

4. Abstract Formulation

The abstract formulation is a "time domain" fomulation. Hence we need to invert the Laplace Transformations in (3.34) and (3.35) into the time domain. It is fortunate that they contain polynomials of degree two only which are then recognized as time derivatives. The terms containing the Theodorsen Function can be handled by using (3.37) so that

$$\rho b^2 UC(\lambda' b)\, \hat{w}(s,\lambda) \;=\; \rho b^3\, \frac{C(\lambda' b)}{\lambda' b}\, (\lambda\, \hat{w}(s,\lambda))$$

whose inverse Laplace Transform is

$$= \rho b^2 \int_0^t \ell(t-\sigma)\, \dot{w}(s,\sigma)\, d\sigma.$$

And finally:

$$\begin{vmatrix} L(s,t) \\ M(s,t) \end{vmatrix} = M_A\, \ddot{x}(s,t) + UD_A\, \dot{x}(s,t) + U^2 K_A\, x(s,t)$$
$$+ \begin{vmatrix} -\rho b^2 \int_0^t \ell(t-\sigma)\, \dot{w}(s,\sigma)\, d\sigma \\ +\rho b^3 \int_0^t \left[\left(\frac{1}{2}+a\right) \ell(t-\sigma) - \frac{1}{2}\frac{U}{b}(t-\sigma)\right] \dot{w}(s,\sigma)\, d\sigma \end{vmatrix} \tag{4.1}$$

where

$$\frac{C(\lambda' b)}{\lambda' b} = \int_0^\infty e^{-\lambda t}\, \ell(t)\, dt,$$

$$M_A = -\pi\rho b^2 \begin{vmatrix} 1 & -ab \\ -ab & (a^2+\frac{1}{8})b^2 \end{vmatrix},$$

$$D_A = (-\pi\rho b^3) \begin{vmatrix} 0 & 1 \\ -1 & 0 \end{vmatrix}; \qquad K_A = (-\pi\rho b^4) \begin{vmatrix} 0 & 0 \\ 0 & 1 \end{vmatrix}.$$

Combining the aerodynamics part (4.1) now with the structures part (2.1), it is important to note that

$$M_S - M_A = M \qquad \text{is positive definite}$$

and obtain

$$M\ddot{x}(s,t) \; - \; UD_A\,\dot{x}(s,t) \quad - \quad U^2K_A\,x(s,t) \; + \; \mathcal{K}_S\,x(s,t)$$
$$= \left| \begin{array}{c} \displaystyle\int_0^t \ell_1(t-\sigma)\,\dot{w}(s,\sigma)\,d\sigma \\ \displaystyle\int_0^t \ell_2(t-\sigma)\,\dot{w}(s,\sigma)\,d\sigma \end{array} \right| \tag{4.2}$$

where

$$\begin{aligned} \ell_1(t) &= \rho b^2 \ell(t) \\ \ell_2(t) &= \rho b^2 \left[\left(\tfrac{1}{2}+a\right) b\ell(t) \; - \; \frac{U}{2}\,t \right]. \end{aligned}$$

We have thus assembled the aeroelastic equations ready next for the abstract formulation as a convolution-semigroup equation as in [1] where the half-chord b is set equal to 1, and the model includes actuators. It should be noted that while the abstract formulation is useful conceptually, it is irrelevant for analyzing the stability problem which needs and works with only the Laplace Transform formulation.

For the abstract formulation, we start with the Hilbert spaces

$$\mathcal{H}_b \;=\; L_2[0,\ell] \times E^1, \qquad \mathcal{H}_\theta \;=\; L_2[0,\ell] \times E^1, \qquad \mathcal{H} \;=\; \mathcal{H}_b \times \mathcal{H}_\theta\,.$$

Define the linear operator

$$A_S \;=\; \left| \begin{array}{cc} A_h & 0 \\ 0 & A_\theta \end{array} \right|$$

with domain and range in $\mathcal{H}$.

$$\mathcal{D}(A_h) \;=\; \left[\left| \begin{array}{c} h(\cdot) \\ c \end{array} \right|, \quad \begin{array}{cc} \multicolumn{2}{c}{h'''' \in L_2[0,\ell]} \\ h(0) = h'(0) = 0; & h'''(\ell) = 0 \\ c = h'(\ell) & h''(\ell) = 0 \end{array} \right], \tag{4.3}$$

$$A_h \left| \begin{array}{c} h(\cdot) \\ h'(\ell) \end{array} \right| \;=\; \left| \begin{array}{c} EI\,h''''(\cdot) \\ EI\,h''(\ell) \end{array} \right|,$$

$$\mathcal{D}(A_\theta) \;=\; \left[\left| \begin{array}{c} \theta(\cdot) \\ c \end{array} \right|, \quad \begin{array}{c} \theta'' \in L_2[0,\ell] \\ \theta(0) = 0 \\ \theta'(\ell) = 0 \\ c = \theta(\ell) \end{array} \right]$$

$$A_\theta \left| \begin{array}{c} \theta(\cdot) \\ \theta(\ell) \end{array} \right| \;=\; \left| \begin{array}{c} GJ\,\theta''(\cdot) \\ GJ\,\theta'(\ell) \end{array} \right|, \tag{4.4}$$

Thus defined, A_S is self-adjoint, nonnegative definite with dense domain. Let $x \in \mathcal{D}(A_S)$. Then

$$[A_S\, x,\, x] \;=\; EI \int_0^\ell |h''(s)|^2\; ds \;+\; GJ \int_0^\ell |\theta'(s)|^2\; ds$$

where

$$x \;=\; \begin{vmatrix} h(\cdot) \\ h'(\ell) \\ \theta(\cdot) \\ \theta(\ell) \end{vmatrix}$$

and, in particular, we see that

$$A_S\, x \;=\; 0 \quad \text{implies} \;\; x \;=\; 0.$$

(There are no rigid-body modes.)

With $\sqrt{A_S}$ denoting the positive square root, we can verify that if $x \in \mathcal{D}(\sqrt{A_S}\,)$ we must have that

$$x \;=\; \begin{vmatrix} h(\cdot) \\ h'(\ell) \\ \theta(\cdot) \\ \theta(\ell) \end{vmatrix} \quad \text{is such that} \quad \begin{matrix} h''(\cdot) \in L_2[0,\ell] \\ \theta'(\cdot) \in L_2[0,\ell] \\ h(0) = h'(0) = 0 \\ \theta(0) = 0. \end{matrix}$$

Also, $\sqrt{A_S}$ has a bounded inverse.

Next we define the Hilbert space (energy space):

$$\mathcal{H}_E \;=\; \mathcal{D}(\sqrt{A_S}\,) \times L_2[0,\ell]^2$$

with inner product

$$[Y, Z]_E \;=\; \left[\sqrt{A_S}\, x_1\,,\; \sqrt{A_S}\, x_2\right] \;+\; [M z_1,\, z_2],$$

where

$$Y \;=\; \begin{vmatrix} x_1 \\ z_1 \end{vmatrix}, \qquad Y \;=\; \begin{vmatrix} x_2 \\ z_2 \end{vmatrix},$$

$$x_1 \;=\; \begin{vmatrix} h_1 \\ h_1'(\ell) \\ \theta_1 \\ \theta_1(\ell) \end{vmatrix}, \quad z_1 \;=\; \begin{vmatrix} h_2(\cdot) \\ \theta_2(\cdot) \end{vmatrix}, \quad x_2 \;=\; \begin{vmatrix} h_3 \\ h_3'(\ell) \\ \theta_3 \\ \theta_3(\ell) \end{vmatrix}, \quad z_2 \;=\; \begin{vmatrix} h_4(\cdot) \\ \theta_4(\cdot) \end{vmatrix}.$$

Define $\mathcal{A}_S$ with domain and range in $\mathcal{H}_E$ by:

$$\mathcal{D}(\mathcal{A}_S) = \left[Y = \begin{vmatrix} x_1 \\ z_1 \end{vmatrix}, \quad x_1 = \begin{vmatrix} h_1 \\ h_1'(\ell) \\ \theta_1 \\ \theta_1(\ell) \end{vmatrix} \in \mathcal{D}(A_S), \quad z_1 = \begin{vmatrix} h_2(\cdot) \\ \theta_2(\cdot) \end{vmatrix} \right.$$

$$\left. \text{and} \quad \begin{vmatrix} h_2 \\ -EI\,\frac{h_1''(\ell)}{g_h} \\ \theta_2 \\ -GJ\,\frac{\theta_1'(\ell)}{g_\theta} \end{vmatrix} \in \mathcal{D}\left(\sqrt{\mathcal{A}_S}\right) \right]$$

(the last condition implies in particular that

$$h_2'(\ell) = \frac{-EI\,h_1''(\ell)}{g_h}; \qquad \theta_2(\ell) = \frac{-GJ\,\theta_1'(\ell)}{g_\theta} \Bigg)$$

$$\mathcal{A}_S Y = \begin{vmatrix} h_2 \\ \frac{-EIh_1''(\ell)}{g_h} \\ \theta_2 \\ \frac{-GJ\theta_1'(\ell)}{g_\theta} \\ -M^{-1} \begin{vmatrix} EI\,h_1''''(\cdot) \\ -GJ\,\theta_1''(\cdot) \end{vmatrix} \end{vmatrix}.$$

Thus defined $\mathcal{A}_S$ is closed with dense domain, and compact resolvent. Moreover

$$\operatorname{Re}[\mathcal{A}_S Y, Y] = 0. \tag{4.5}$$

Next we define the linear operators D and K on $\mathcal{H}_E$ into $\mathcal{H}_E$:

$$Y = \begin{vmatrix} h_1 \\ h_1'(\ell) \\ \theta_1 \\ \theta_1(\ell) \\ h_2 \\ \theta_2 \end{vmatrix}; \qquad DY = \begin{vmatrix} 0 \\ 0 \\ 0 \\ 0 \\ M^{-1}D_A \begin{vmatrix} h_2 \\ \theta_2 \end{vmatrix} \end{vmatrix}; \qquad KY = \begin{vmatrix} 0 \\ 0 \\ 0 \\ 0 \\ M^{-1}K_A \begin{vmatrix} h_2 \\ \theta_2 \end{vmatrix} \end{vmatrix}.$$

Thus defined, D and K are bounded linear operators, their precise bounds being not of interest. Define

$$\mathcal{A} \;=\; \mathcal{A}_S \;+\; UD \;+\; U^2 K. \tag{4.6}$$

Then $\mathcal{A}$ generates a C_0-semigroup; denote it $S(t)$, $t \geq 0$. Also

$$\mathrm{Re}\,[\mathcal{A}Y,\, Y] \;=\; \mathrm{Re}\,[\mathcal{A}_S Y,\, Y] \;+\; \pi\rho U^2\,\mathrm{Re}\,[\theta_1\,,\,\theta_2]. \tag{4.7}$$

The semigroup $S(\cdot)$ is thus not necessarily a contraction for nonzero U. But the resolvent of $\mathcal{A}$, $R(\lambda, \mathcal{A})$, is compact and there are no eigenvalues in the half plane

$$\mathrm{Re}\,\lambda \;>\; \sigma_A,$$

where σ_A is the growth bound of the semigroup generated by $\mathcal{A}$. In fact

$$\mathrm{Re}\,\lambda \;\leq\; 0$$

unless

$$U^2 \;\geq\; \frac{\pi GJ}{\ell^2 \rho}, \tag{4.8}$$

which is usually larger than the range of validity of the model.

Next for each $t \geq 0$, define the linear bounded operator $\mathcal{L}(t)$ on $\mathcal{H}_E$ into $\mathcal{H}_E$ by

$$\mathcal{L}(t)\,Y \;=\; \left|\begin{array}{c} 0 \\ \vdots \\ 0 \\ M^{-1}\left|\begin{array}{c} \ell_1(t)w \\ \ell_2(t)w \end{array}\right| \end{array}\right|$$

where

$$w \;=\; 2\pi U\theta_1 \;+\; 2\pi h_2 \;+\; b\,\pi(1-2a)\theta_2\,.$$

Then (4.1) goes over into the abstract evolution-convolution equation:

$$\dot{Y}(t) \;=\; \mathcal{A}y(t) \;+\; \int_0^t \mathcal{L}(t-\sigma)\,\dot{Y}(\sigma)\;d\sigma, \quad t > 0 \tag{4.9}$$

which is then the canonical abstract version of linear subsonic incompressible aeroelasticity dynamics. Note that $\hat{\mathcal{L}}(\lambda)$ the Laplace transform:

$$\hat{\mathcal{L}}(\lambda)\,Y \;=\; \int_0^\infty e^{-\lambda t}\,\mathcal{L}(t)\,Y\;dy, \quad \mathrm{Re}\,\lambda > 0$$

is given by

$$\hat{\mathcal{L}}(\lambda)Y = (-b^2\ell)\begin{vmatrix} 0 \\ \vdots \\ 0 \\ M^{-1}\begin{vmatrix} \frac{C(k)}{k} \\ \left(\frac{1}{2k} - \left(a+\frac{1}{2}\right)\frac{C(k)}{k}\right) w \end{vmatrix} \end{vmatrix} \tag{4.10}$$

where

$$k = \lambda' b.$$

The convolution part in (4.1) is the one that gives rise to the flutter phenomenon.

Acknowledgment

I am indebted to Jensen Lin for proofing the Lift and Moment formulas.

References

[1] A.V. Balakrishnan, *Subsonic Flutter Suppression Using Self-straining Actuators.* Journal of the Franklin Institute, no. 338 (2001), 149–170.

[2] M. Goland, *The Flutter of a Uniform Cantilever Wing.* Journal of Applied Mechanics, Trans. ASME, **12**, no. 4 (1945).

[3] R.L. Bisplinghoff, H. Ashley and R.L. Halfman. *Aeroelasticity.* Addison Wesley, 1955.

[4] F.G. Tricomi, *Integral Equations.* New York: Dover Publications, 1985. Pp. 175 *et seq.*

[5] D. Nixon, *Basic Equations for Unsteady Transonic Flow.* Chapter 2 in: Unsteady Transonic Aerodynamics. Edited by David Nixon. AIAA, 1989. Pp. 57–73.

[6] Y.C. Fung, *An Introduction to the Theory of Aeroelasticity.* New York: Dover Publications, 1983.

[7] C.E. Watkins, H.L. Runyon and D.S. Wollston, *The Kernel Function of the Integral Equation Relating to the Lift and Downwash Distributions of Oscillating Finite Wings on Subsonic Flow.* NACA TN 1234, 1955.

A.V. Balakrishnan
Flight Systems Research Center
UCLA
Los Angeles
CA 90024-1594, USA
E-mail address: `bal@@ee.ucla.edu`

Progress in Nonlinear Differential Equations
and Their Applications, Vol. 55, 83–94

On the Effect of Small Delays to the Stability of Feedback Systems

András Bátkai and Bálint Farkas

1. Introduction

The problem considered in this paper is the following: Assume that some asymptotic properties are known for the solutions of the equation

$$(\mathrm{DE})_0 \qquad \begin{cases} u'(t) = (B+C)u(t), & t \geq 0, \\ + \text{initial conditions}, \end{cases}$$

where $C \in \mathcal{L}(X)$, X is a Banach space and $(B, D(B))$ is the generator of a strongly continuous semigroup in X.

The question is, whether the solutions of the equation

$$(\mathrm{DE})_\tau \qquad \begin{cases} u'(t) = Bu(t) + Cu(t-\tau), & t \geq 0, \\ + \text{initial conditions}, \end{cases}$$

have the same type of asymptotic behaviour, where $\tau > 0$ is "small".
The asymptotic property we concentrate on in this paper is uniform exponential stability.

This problem is motivated by feedback-systems and control theory. In the above equations the operators B and C could be interpreted as the system operator and the feedback respectively. It is important to know whether the stabilizing effect depends on the fact that the feedback does not act immediately but with a very small delay.

To our knowledge, the first who examined this effect on the stability was R. Datko [7, 8, 9, 10, 11]. It is known for finite-dimensional equations, i.e., for $X = \mathbb{C}^n$, that the stability can not be destroyed if the delay is sufficiently small, and there exists an extensive literature on delay dependent stability conditions, see, e.g., [13, 14]. There is a recent exposition of this problem of the small delays by J. Hale and S. Verduyn Lunel [15, 16], where many examples of functional differential and difference equations are considered. A control theoretical investigation using transfer functions was made for compact feedback in [24].

To be able to treat this question in Section 4, we collect some technical tools on partial differential equations with delay mainly from [2, 4, 5]. The main idea is a semigroup approach in L^p phase spaces. The treatment in the space of continuous functions considered in the book of Wu [26] is well developed and successfully applicable, if some regularity properties of the semigroup generated by B are known. However, if the spectral mapping theorem does not hold, we can not consider asymptotics using only characteristic equations. Whereas recent results using operator valued Fourier multipliers allow us to characterize stability, see [6, 18, 19, 20].

We recall some important results on the semigroup approach to delay equations and uniform exponential stability in Section 2.

In Section 3 we give an explanation of the problem with an example, and in Section 4 we present a rather general sufficient condition on the problem. We show that if either B generates an immediately norm continuous semigroup or C is compact, commuting with B, then the stability can not be destroyed with small delays.

2. The semigroup approach to delay equations

Let us summarize here some results from [2, 3, 4, 5, 21] on the semigroup approach to linear partial differential equations with delay and the application of it to uniform exponential stability. This will allow us to consider the problems treated in the Introduction.

Consider the equation

$$\text{(DE)} \qquad \begin{cases} u'(t) = Gu(t) + \Phi u_t, & t \geq 0, \\ u(0) = x, \\ u_0 = f, \end{cases}$$

where

- $x \in X$, X is a Banach space,
- $G : D(G) \subseteq X \longrightarrow X$ is a linear, closed and densely defined operator,
- $f \in L^p([-1,0], X)$, $p \geq 1$,
- $\Phi : W^{1,p}([-1,0], X) \longrightarrow X$ is a linear, bounded operator,
- $u : [-1, \infty) \longrightarrow X$ and $u_t : [-1,0] \longrightarrow X$ is defined by $u_t(\sigma) := u(t+\sigma)$.

Definition 2.1. *We say that a function $u : [-1,\infty) \longrightarrow X$ is a (classical) solution of (DE) if*

(i) $u \in C([-1,\infty), X) \cap C^1([0,\infty), X)$,
(ii) $u(t) \in D(G)$ *and* $u_t \in W^{1,p}([-1,0], X)$ *for all* $t \geq 0$, *and*
(iii) u *satisfies (DE) for all* $t \geq 0$.

To be able to solve (DE) by semigroup methods, we introduce the Banach space

$$\mathcal{E} := X \times L^p([-1,0], X)$$

and the operator

$$\mathcal{A}_\Phi := \mathcal{A} := \begin{pmatrix} G & \Phi \\ 0 & \frac{\mathrm{d}}{\mathrm{d}\sigma} \end{pmatrix} \tag{1}$$

with domain

$$D(\mathcal{A}) := \left\{ \binom{x}{f} \in D(G) \times W^{1,p}([-1,0],X) \ : \ f(0) = x \right\}. \tag{2}$$

Here, by $\frac{\mathrm{d}}{\mathrm{d}\sigma}$ we understand the differentiation with maximal domain in L_p.

Consider now the abstract Cauchy problem

$$\text{(ACP)} \qquad \begin{cases} v'(t) = \mathcal{A}\, v(t), & t \geq 0, \\ v(0) = v_0 \end{cases}$$

associated to the operator matrix $(\mathcal{A}, D(\mathcal{A}))$ on the Banach space $\mathcal{E}$ with initial value $v_0 := \binom{x}{f}$. There is a natural correspondence between the solutions of the two problems (see [4, Proposition 2.3 and 2.4]).

Lemma 2.2.

(i) *If u is a solution of (DE), then $t \mapsto \binom{u(t)}{u_t}$ is a solution of (ACP).*

(ii) *If $t \mapsto \binom{u(t)}{v(t)}$ is a solution of (ACP), then $v(t) = u_t$ for all $t \geq 0$ and u is a solution of (DE).*

We can now give the following definition for well-posedness.

Definition 2.3. *We say that (DE) is well-posed if*

(i) *for every $\binom{x}{f} \in D(\mathcal{A})$ there is a unique solution $u(x, f, \cdot)$, and*

(ii) *the solutions depend continuously on the initial values, i.e., if a sequence $\binom{x_n}{f_n}$ in $D(\mathcal{A})$ converges to $\binom{x}{f} \in D(\mathcal{A})$, then $u(x_n, f_n, t)$ converges to $u(x, f, t)$ uniformly for t in compact intervals.*

There is also a correspondence between the well-posedness of equation (DE) and of the abstract Cauchy problem (ACP), see also [4, Theorem 2.8].

Proposition 2.4. *Let $(\mathcal{A}, D(\mathcal{A}))$ be the operator matrix defined by (1) and (2). Then the following assertions are equivalent.*

(i) *Equation (DE) is well-posed.*

(ii) *$(\mathcal{A}, D(\mathcal{A}))$ is the generator of a strongly continuous semigroup on $\mathcal{E}$.*

As an easy consequence of Lemma 2.2 and Proposition 2.4, we have that if $(\mathcal{A}, D(\mathcal{A}))$ generates a strongly continuous semigroup $(\mathcal{T}(t))_{t \geq 0}$, then the solutions of equation (DE) are given by the first component of the function $\mathbb{R}_+ \ni t \mapsto \mathcal{T}(t)\binom{x}{f}$ for $\binom{x}{f} \in D(\mathcal{A})$.

By means of the perturbation theorem of Miyadera-Voigt (see [12, Corollary III.3.16]) one can formulate the following sufficient condition for the well-posedness of (DE) in the case where $(G, D(G))$ generates a strongly continuous semigroup $(S(t))_{t \geq 0}$, see [4, Theorem 3.3, Examples 3.4], [21].

Theorem 2.5. *Let $(S(t))_{t \geq 0}$ be a strongly continuous semigroup on X generated by $(G, D(G))$, and let $1 \leq p < \infty$. Suppose moreover that there exists a function*

$\eta : [-1,0] \to \mathcal{L}(X)$ of bounded variation such that $\Phi : C([-1,0],X) \to X$ is given by the Riemann-Stieltjes integral

$$\Phi(f) := \int_{-1}^{0} d\eta f. \tag{3}$$

Then $(\mathcal{A}, D(\mathcal{A}))$ is a generator on $\mathcal{E}$.

Important special cases are operators Φ defined by

$$\Phi(f) := \sum_{k=0}^{n} C_k f(h_k), \quad f \in W^{1,p}([-1,0],X),$$

where $C_k \in \mathcal{L}(X)$ and $h_k \in [-1,0]$ for $k = 0,\dots,n$.
We now characterize the resolvent set and the resolvent operator of $\mathcal{A}$ (see [4, Lemma 4.1]). Here $\epsilon_\lambda(t) := e^{\lambda t}$ and $\Phi_\lambda \in \mathcal{L}(X)$ is defined by

$$\Phi_\lambda x := \Phi(\epsilon_\lambda \otimes Id)x = \Phi(e^{\lambda \cdot} x)$$

for $x \in X$. Denote by $(A_0, D(A_0))$ the generator of the nilpotent left shift semigroup $(T_0(t))_{t\geq 0}$ in $L^p([-1,0],X)$.

Lemma 2.6. *Let X be a Banach space, $(G, D(G))$ be linear, closed and densely defined, and $\Phi : W^{1,p}([-1,0],X) \longrightarrow X$ be linear and bounded. Let $(\mathcal{A}, D(\mathcal{A}))$ be the operator matrix defined in (1) and (2). Then $\lambda \in \rho(\mathcal{A})$ if and only if $\lambda \in \rho(G + \Phi_\lambda)$. Moreover, for $\lambda \in \rho(\mathcal{A})$ the resolvent $R(\lambda, \mathcal{A})$ is given by*

$$\begin{pmatrix} R(\lambda, G+\Phi_\lambda) & R(\lambda, G+\Phi_\lambda)\Phi R(\lambda, A_0) \\ \epsilon_\lambda \otimes R(\lambda, G+\Phi_\lambda) & [\epsilon_\lambda \otimes R(\lambda, G+\Phi_\lambda)\Phi + Id]R(\lambda, A_0) \end{pmatrix}. \tag{4}$$

We collect the most important results from [3] concerning the Fourier multiplier method for stability of delay equations.

Definition 2.7. *We call the delay operator $\Phi \in \mathcal{L}(W^{1,p}([-1,0],X),X)$ Fourier-admissible if*

(a) *the operator $(\mathcal{A}, D(\mathcal{A}))$ is a generator for each generator $(G, D(G))$*
(b) *the function $\Phi R(\alpha + i\cdot, A_0)$ is a Fourier multiplier from $L^p(\mathbb{R}, L^p([-1,0],X))$ to $L^p(\mathbb{R}, X)$ for all $\alpha \in \mathbb{R}$.*

For the notion of operator valued Fourier multipliers we refer to [1, 17].

The following technical result allows us to apply Fourier multipliers to the stability of delay equations, see [3, Proposition 4.2].

Proposition 2.8. *Assume that the delay operator Φ satisfies the conditions of Theorem 2.5. Then Φ is Fourier admissible.*

We now formulate the main stability tool from [3].

Theorem 2.9. *Assume that Φ is Fourier-admissible, $\omega_0(G) < 0$ and let $\alpha \in (\omega_0(G), 0]$. If*

$$\sup_{\beta \geq \alpha} \|\Phi_{\beta + i\cdot} R(\beta + i\cdot, G)\|_{\mathcal{M}_p^{\mathcal{L}(X)}} < 1, \tag{5}$$

then $\omega_0(\mathcal{A}) < \alpha \leq 0$.

3. The problem

Let X be a Banach space and assume that $(B, D(B))$ generates a strongly continuous semigroup $(S(t))_{t\geq 0}$ and that $C \in \mathcal{L}(X)$. Suppose furthermore that uniform exponential stability is known for the solutions of the equation

$$(\mathrm{DE})_0 \qquad \begin{cases} u'(t) = Bu(t) + Cu(t), & t \geq 0, \\ u(0) = x, \end{cases}$$

i.e.

$$\omega_0(B + C) < 0. \tag{6}$$

We denote the semigroup generated by $(B + C, D(B))$ with $(T(t))_{t\geq 0}$.

The question is whether the same property holds for the solutions of the equation

$$(\mathrm{DE})_\tau \qquad \begin{cases} u'(t) = Bu(t) + Cu(t - \tau), & t \geq 0, \\ u(0) = x, \\ u_0 = f, \end{cases}$$

where $\tau > 0$.

It is known, see, e.g., [22, Example B-IV.3.10], that if X is a Banach lattice, C is a positive operator and $(B, D(B))$ generates a positive semigroup, then the solutions of $(\mathrm{DE})_\tau$ are exponentially stable if and only if the solutions of $(\mathrm{DE})_0$ are exponentially stable. It follows by [25] and a recent unpublished argument of Schlotterbeck that if $X = L^p(\Omega)$, C is a positive operator and $(B, D(B))$ generates a positive semigroup, then the solutions of $(\mathrm{DE})_\tau$ are uniformly exponentially stable if and only if the solutions of $(\mathrm{DE})_0$ are uniformly exponentially stable. Thus, the stability is not sensitive to small delays in this case.

It has been recently shown in [2], that if X is a Hilbert space and either $(B, D(B))$ generates a norm continuous semigroup, or C is compact commuting with B, then the stability is also not sensitive to arbitrary small delays. The question is, what can be said in the general Banach space case.

Before considering the abstract problem, we demonstrate by a simple example how the stability can be destroyed, see [2, Example 4.1] where this example first appeared.

Example 3.1. *Let $(B, D(B))$ be the (unbounded) generator of a unitary group in an infinite-dimensional Hilbert space H and let $C := d \cdot Id$ for $d < 0$. Then $(B + C, D(B))$ generates an exponentially stable semigroup. We show that there exists a sequence (τ_k), $\tau_k \to 0$, such that the solution semigroup of the equation $(DE)_{\tau_k}$ does not decay exponentially for each $k \in \mathbb{N}$.*

To construct this sequence, take $(\mu_k) \subset \mathbb{R}$, $i\mu_k \in \sigma(B)$ such that $|\mu_k| \to +\infty$ and $\mu_k \neq -d$. Defining

$$\tau_k := \begin{cases} \frac{3\pi}{2(\mu_k+d)}, & \mu_k + d > 0, \\ \frac{-\pi}{2(\mu_k+d)}, & \mu_k + d < 0, \end{cases}$$

we obtain the corresponding operators $\Phi^k = d \cdot Id\delta_{\tau_k}$ *and* $\Phi^k_\lambda = de^{-\lambda\tau_k} \cdot Id$. *Since* $\sigma(B+\Phi^k_\lambda) = \sigma(B) + de^{-\lambda\tau_k}$, *we obtain for the numbers* $\lambda_k := (\mu_k + d)i \in i\mathbb{R}$ *that* $\lambda_k \in \sigma(B + \Phi^k_{\lambda_k})$. *By the spectral characterization in Lemma 2.6 it follows that for the associated operator* $\mathcal{A}_{\tau_k} = \begin{pmatrix} B & \Phi^k \\ 0 & \frac{d}{d\sigma} \end{pmatrix}$, $\lambda_k \in \sigma(\mathcal{A}_{\tau_k})$, *and hence* $\omega_0(\mathcal{A}_{\tau_k}) \geq 0$. *This means that* $\mathcal{A}_{\tau_k}$ *can not generate a uniformly exponentially stable semigroup for all* $k \in \mathbb{N}$. *If we assume further that* $i\mu_k \in P\sigma(B)$, *which, e.g., is satisfied if* $(B, D(B))$ *has compact resolvent, then we find also classical solutions of* $(DE)_{\tau_k}$ *not decaying exponentially.*

We now recall from [2] a possible strategy to treat the question of small delays and to give sufficient conditions on B or C such that the stability can not be destroyed with small delays.

We transform the equation $(\mathrm{DE})_\tau$ into

$$u'(t) = (B+C)u(t) + C\left(u(t-\tau) - u(t)\right).$$

Now, by our assumption (6), $(B+C, D(B))$ generates a uniformly exponentially stable semigroup, thus we may be in the position to use Theorem 2.9.

We use the equality

$$u(t_2) - u(t_1) = [S(t_2 - t_1) - Id]\, u(t_1) + \int_{t_1}^{t_2} S(t_2 - s)Cu(s-\tau)\,\mathrm{d}s \tag{7}$$

for $t_2 > t_1 \geq 0$ following from $(\mathrm{DE})_\tau$.

Making the necessary calculations, we obtain that $(\mathrm{DE})_\tau$ can be written as

$$u'(t) = (B+C)\,u(t) + \Phi u_t, \tag{8}$$

where

$$\Phi f := -C\,[S(\tau) - Id]\,\delta_{-\tau} f - \int_{-\tau}^{0} CS(-s)C\delta_{s-\tau} f\,\mathrm{d}s. \tag{9}$$

To use Theorem 2.9, we have to calculate

$$\Phi_\lambda x = -C[S(\tau) - Id]e^{-\lambda\tau}x - \int_{-\tau}^{0} CS(-s)Ce^{\lambda(s-\tau)}x\,\mathrm{d}s \tag{10}$$

and estimate the Fourier multiplier norm of

$$\Phi_\lambda R(\lambda, B+C).$$

It follows from (10) that

$$\begin{aligned}\Phi_\lambda R(\lambda, B+C)x = &-C\,[S(\tau) - Id]\,e^{-\lambda\tau}R(\lambda, B+C)x \\ &- \int_{-\tau}^{0} CS(-s)Ce^{-\lambda(s-\tau)}R(\lambda, B+C)x\,\mathrm{d}s.\end{aligned} \tag{11}$$

4. The main result

We use the notations and assumptions of the previous section, i.e., we assume that X is a Banach space, $(B, D(B))$ generates a strongly continuous semigroup $(S(t))_{t\geq 0}$, $C \in \mathcal{L}(X)$, and the semigroup $(T(t))_{t\geq 0}$ generated by $(B + C, D(B))$ is uniformly exponentially stable.

Theorem 4.1. *Suppose that one of the following conditions hold.*

1. *Either* $(S(t))_{t\geq 0}$ *is an immediately norm-continuous semigroup,*
2. $(S(t))_{t\geq 0}$ *is an arbitrary strongly continuous semigroup commuting with the compact operator* C.

Assume moreover that $(T(t))_{t\geq 0}$ *is uniformly exponentially stable, and let*

$$\omega_0(B + C) < \alpha \leq 0,$$

then we have $\alpha > \omega_0(\mathcal{A}_\tau)$ *for* τ *sufficiently small.*

Here $\mathcal{A}_\tau = \begin{pmatrix} B & C\delta_\tau \\ 0 & \frac{d}{d\sigma} \end{pmatrix}$ is the corresponding delay operator to $(\mathrm{DE})_\tau$.

Proof. Before all, we fix some constants. Let $\omega \in [\omega_0(B + C), \alpha)$ and $M \geq 1$ such that $T(t) \leq Me^{\omega t}$ for all $t \geq 0$, furthermore we define $M' = \sup_{[0,1]} \|S(t)\|$ and let $\beta \geq \alpha$ arbitrary. By the Hausdorff-Young inequality we have

$$\begin{aligned} &\|\mathcal{F}^{-1}\Phi_{\beta+i\cdot}R(\beta + i\cdot, B + C)\mathcal{F}f\|_{L^p(\mathbb{R},X)} \leq \\ &\|\mathcal{F}^{-1}\Phi_{\beta+i\cdot}R(\beta + i\cdot, B + C)\|_{L^1(\mathbb{R},\mathcal{L}(X))}\|f\|_{L^p(\mathbb{R},X)}, \end{aligned} \tag{12}$$

therefore to estimate the Fourier-multiplier norm of $\Phi_{\beta+i\cdot}R(\beta+i\cdot, B+C)$ it suffices to give only L^1 estimates on the inverse Fourier transform. From (11) we have for all $x \in X$ that

$$\begin{aligned} \Phi_{\beta+i\cdot}R(\beta + i\cdot, B + C)x = &- C[S(\tau) - Id]e^{-(\beta+i\cdot)\tau}R(\beta + i\cdot, B + C)x - \\ &\int_{-\tau}^{0} CS(-s)Ce^{-(\beta+i\cdot)(s-\tau)}R(\beta + i\cdot, B + C)x \, \mathrm{d}s. \end{aligned} \tag{13}$$

Let $I_1^{\beta+i\cdot}(\tau)$ and $I_2^{\beta+i\cdot}(\tau)$ denote the first and the second part in the above expression respectively. We calculate them separately. Fix $x \in X$, then for all $t \in \mathbb{R}$

$$\begin{aligned} &- \mathcal{F}^{-1}(I_1^{\beta+i\cdot}(\tau))(t)x = -\frac{1}{2\pi i}\int_{-\infty}^{+\infty} e^{-itr}I_1^{\beta+ir}(\tau)x \, \mathrm{d}r = \\ &\frac{1}{2\pi i}\int_{-\infty}^{+\infty} e^{itr}C[S(\tau) - Id]e^{-(\beta+ir)\tau}R(\beta + ir, B + C)x \, \mathrm{d}r = \\ &\frac{1}{2\pi i}C[S(\tau) - Id]e^{-\beta\tau}\int_{-\infty}^{+\infty} e^{ir(t-\tau)}R(\beta + ir, B + C)x \, \mathrm{d}r = \\ &\frac{1}{2\pi i}C[S(\tau) - Id]e^{-\beta\tau}\int_{-\infty}^{+\infty} e^{ir(t-\tau)}R(ir, B + C - \beta)x \, \mathrm{d}r = \\ &C[S(\tau) - Id]e^{-\beta\tau}e^{-\beta(t-\tau)}\tilde{T}(t - \tau)x = C[S(\tau) - Id]e^{-\beta t}\tilde{T}(t - \tau)x, \end{aligned} \tag{14}$$

where

$$\tilde{T}(t) = \begin{cases} T(t) & \text{if } t \geq 0 \\ 0 & \text{otherwise.} \end{cases}$$

Here, we have used that the inverse Fourier transform of the resolvent $R(i\cdot, B + C - \beta)$ is $e^{-\beta\cdot}\tilde{T}(\cdot)$, since $e^{-\beta t}T(t)$ is uniformly exponentially stable. From (14), it follows immediately that

$$\mathcal{F}^{-1}(I_1^{\beta+i\cdot}(\tau))(t) = -C[S(\tau) - Id]e^{-\beta t}\tilde{T}(t-\tau), \quad \text{for all } t \in \mathbb{R}.$$

Suppose now that C is compact and commutes with the semigroup $(S(t))_{t\geq 0}$, then for any $x \in X$

$$\left\|\mathcal{F}^{-1}(I_1^{\beta+i\cdot}(\tau))(t)x\right\|_X = \left\|C[S(\tau) - Id]e^{-\beta t}\tilde{T}(t-\tau)x\right\|_X =$$
$$\left\|[S(\tau) - Id]Ce^{-\beta t}\tilde{T}(t-\tau)x\right\|_X \to 0, \text{ as } \tau \to 0,$$

and the convergence is uniform for β, x in bounded sets and t from compact intervals. Hence

$$\lim_{\tau\to 0} \left\|\mathcal{F}^{-1}(I_1^{\beta+i\cdot}(\tau))(t)\right\|_{\mathcal{L}(X)} = 0,$$

uniformly in β and t from compact intervals. Now, from Lebesgue's dominated convergence theorem we obtain

$$\lim_{\tau\to 0} \left\|\mathcal{F}^{-1}(I_1^{\beta+i\cdot}(\tau))\right\|_{L^1(\mathbb{R},\mathcal{L}(X))} =$$
$$\lim_{\tau\to 0} \int_{-\infty}^{+\infty} \left\|[S(\tau) - Id]Ce^{-\beta t}\tilde{T}(t-\tau)\right\|_{\mathcal{L}(X)} \,\mathrm{d}t =$$
$$\lim_{\tau\to 0} \int_{0}^{+\infty} \left\|[S(\tau) - Id]Ce^{-\beta(s+\tau)}T(s)\right\|_{\mathcal{L}(X)} \,\mathrm{d}s = 0,$$

uniformly in β.

Secondly, assume that $(S(t))_{t\geq 0}$ is immediately norm-continuous, therefore $(T(t))_{t\geq 0}$ being its bounded perturbation is also immediately norm-continuous, see, e.g., [23].

$$\left\|\mathcal{F}^{-1}(I_1^{\beta+i\cdot}(\tau))(t)x\right\|_X \leq \|C\| \left\|[S(\tau) - Id]e^{-\beta t}\tilde{T}(t-\tau)x\right\|_{\mathcal{L}(X)} =$$
$$\|C\| \left\|[S(\tau) - T(\tau) + T(\tau) - Id]e^{-\beta t}\tilde{T}(t-\tau)x\right\|_{\mathcal{L}(X)} \leq$$
$$\|C\| \left(\|S(\tau) - T(\tau)\|_X \cdot \left\|e^{-\beta t}\tilde{T}(t-\tau)x\right\|_X + \left\|[T(\tau) - Id]e^{-\beta t}\tilde{T}(t-\tau)x\right\|_X\right),$$

Again, the two parts will be dealt with separately. For the first term we write

$$\|S(\tau) - T(\tau)\|_X \cdot \left\|e^{-\beta t}\tilde{T}(t-\tau)x\right\|_X \leq \tau K \left\|e^{-\beta t}\tilde{T}(t-\tau)x\right\|_X \to 0,$$

as $\tau \to 0$, uniformly in β and x from bounded sets. Hence

$$\lim_{\tau \to 0} \left\| [S(\tau) - T(\tau)] e^{-\beta t} \tilde{T}(t - \tau) \right\|_{L^{(}\mathbb{R}, \mathcal{L}(X))} = 0,$$

and the convergence is uniform in β.

For the second term we have the following

$$\lim_{\tau \to 0} \left\| [T(\tau) - Id] e^{-\beta t} \tilde{T}(t - \tau) x \right\|_X = 0,$$

uniformly for β, x in bounded sets and for almost all $t \in \mathbb{R}$. Therefore, we again make use of Lebesgue's theorem, and obtain

$$\lim_{\tau \to 0} \left\| [T(\tau) - Id] e^{-\beta t} \tilde{T}(t - \tau) \right\|_{L^1(\mathbb{R}, \mathcal{L}(X))} = 0,$$

uniformly in β. Summarizing our results, we see that that under either assumptions in our theorem

$$\left\| \mathcal{F}^{-1} I_1^{\beta + i\cdot}(\tau) \right\|_{L^1(\mathbb{R}, X)} \to 0, \text{ as } \tau \to 0, \tag{15}$$

and the convergence is uniform for $\beta \in [\alpha, +\infty)$.

Now, let us try to estimate the second term $I_2^{\beta+i\cdot}(\tau)$ in (13). Let $x \in D(B)$ and $t \in \mathbb{R}$ arbitrary, then

$$\begin{aligned}
&-\mathcal{F}^{-1}(I_2^{\beta+i\cdot}(\tau))(t)x = \\
&\frac{1}{2\pi i} \int_{-\infty}^{+\infty} e^{itr} \int_{-\tau}^{0} CS(-s)Ce^{(\beta+ir)(s-\tau)} R(\beta + ir, B + C)x \, \mathrm{d}s \, \mathrm{d}r = \\
&\int_{-\tau}^{0} CS(-s)C \frac{1}{2\pi i} \int_{-\infty}^{+\infty} e^{itr} e^{(\beta+ir)(s-\tau)} R(\beta + ir, B + C)x \, \mathrm{d}r \, \mathrm{d}s = \\
&\int_{-\tau}^{0} CS(-s)Ce^{\beta(s-\tau)} \frac{1}{2\pi i} \int_{-\infty}^{+\infty} e^{ir(t+s-\tau)} R(ir, B + C - \beta)x \, \mathrm{d}r \, \mathrm{d}s = \\
&\int_{-\tau}^{0} CS(-s)Ce^{\beta(s-\tau)} e^{-\beta(t+s-\tau)} \tilde{T}(t + s - \tau)x \, \mathrm{d}s = \\
&\int_{-\tau}^{0} CS(-s)Ce^{-\beta t} \tilde{T}(t + s - \tau)x \, \mathrm{d}s.
\end{aligned}$$

The interchanging of the integrals is because of Fubini's theorem, although it is not quite obvious at the first sight that we can apply this theorem. However, we know that

$$e^{-\beta t} \tilde{T}(t)x = \lim_{r \to +\infty} \int_{-r}^{r} e^{ist} R(is, B + C - \beta)x \, \mathrm{d}r,$$

and the convergence is uniform for t in compact intervals. Hence the integration on $[-\tau, 0]$ and the limit in the above expression can be interchanged, and, after noticing that the double integral on the finite rectangle $[-\tau, 0] \times [-r, r]$ is absolutely convergent, we see that the two integrals can be also interchanged. Now, for the operator norm we obtain

$$\left\|\mathcal{F}^{-1}(I_2^{\beta+i\cdot}(\tau))(t)\right\|_{\mathcal{L}(X)} \leq \int_{-\tau}^{0} \left\|CS(-s)Ce^{-\beta t}\tilde{T}(t+s-\tau)\right\|_{\mathcal{L}(X)} \,\mathrm{d}s.$$

Therefore the L^1 estimate becomes the following. By Fubini's theorem

$$\begin{aligned}
&\|\mathcal{F}^{-1}(I_2^{\beta+i\cdot}(\tau))\|_{L^1(\mathbb{R},\mathcal{L}(X))} = \\
&\int_{-\infty}^{+\infty}\int_{-\tau}^{0} \left\|CS(-s)Ce^{-\beta t}\tilde{T}(t+s-\tau)\right\|_{\mathcal{L}(X)} \,\mathrm{d}s\,\mathrm{d}t = \\
&\int_{-\tau}^{0}\int_{\tau-s}^{+\infty} \|CS(-s)Ce^{-\beta t}T(t+\tau-s)\|_{\mathcal{L}(X)} \,\mathrm{d}t\,\mathrm{d}s \leq \\
&\|C\|^2 M' \int_{-\tau}^{0}\int_{\tau-s}^{+\infty} \|e^{-\beta t}T(t+\tau-s)\|_{\mathcal{L}(X)} \,\mathrm{d}t\,\mathrm{d}s \leq \\
&\|C\|^2 M M' \int_{-\tau}^{0}\int_{\tau-s}^{+\infty} e^{-\beta t+\omega(t+s-\tau)} \,\mathrm{d}t\,\mathrm{d}s = \\
&\frac{\|C\|^2 M M'}{\beta-\omega}\int_{-\tau}^{0} e^{-\beta(\tau-s)}\,\mathrm{d}s \leq \frac{K'}{\alpha-\omega}\tau \to 0, \text{ as } \tau \to 0,
\end{aligned} \tag{16}$$

again the convergence is uniform for β.

Combining our results in (15) and (16), we obtain that

$$\left\|\mathcal{F}^{-1}\Phi_{\beta+i\cdot}R(\beta+i\cdot, B+C)\mathcal{F}\right\|_{\mathcal{M}_p^{\mathcal{L}(X)}} < 1,$$

whenever τ is sufficiently small, which completes the proof by Theorem 2.9. □

Acknowledgments

The authors were supported by the OTKA grant Nr. F034840 and by the FKFP grant Nr. 0049/2001. A.B. was further supported by the Research Training Network HPRN-CT-2000-00116 of the European Union.

References

[1] Amann, H., *Operator-valued Fourier multipliers, vector-valued Besov spaces, and applications*, Math. Nachr. **186** (1997), 5–56.

[2] Bátkai, A., *Hyperbolicity of linear partial differential equations with delay*, to appear in Int. Eq. Oper. Th.

[3] Bátkai, A., Fašanga, E., Shvidkoy, R., *Hyperbolicity of delay equations via Fourier multipliers*, to appear in Acta. Sci. Math. (Szeged)

[4] Bátkai, A., Piazzera, S., *Semigroups and linear partial differential equations with delay*, J. Math. Anal. Appl. **264** (2001), 1–20.

[5] Bátkai, A., Piazzera, S., *Damped wave equations with delay*, Fields Institute Communications **29** (2001), 51–61.

[6] Clark, S., Latushkin, Y., Montgomery-Smith, S, Randolph, T., *Stability radius and internal versus external stability in Banach spaces: an evolution semigroup approach*, SIAM J. Control. Optim., **38** (2000), 1757–1793 (electronic).

[7] Datko, R., *Is boundary control a realistic approach to the stabilization of vibrating elastic systems?*, in: Ferreyra, Guillermo (ed.), "Evolution Equations", Marcel Dekker, 133–140 (1994).

[8] Datko, R. *Two questions concerning the boundary control of elastic systems*, J. Diff. Eq. **92** (1991), 27–44.

[9] Datko, R. *Not all feedback stabilized systems are robust with respect to small time delays in their feedback*, SIAM J. Control and Optimization **26** (1988), 697–713.

[10] Datko, R., Lagnese, J., Polis, M. P., *An example on the effect of time delays in boundary feedback of wave equations*, SIAM J. Control and Optimization **24** (1986), 152–156.

[11] Datko, R., You, Y.C., *Some second order vibrating systems cannot tolerate small time delays in their damping*, J. Optim. Th. Appl. **70** (1991), 521–537.

[12] Engel, K.-J., Nagel R., "One-parameter Semigroups for Linear Evolution Equations", Springer-Verlag, Graduate Texts in Mathematics **194**, 1999.

[13] Győri, I., Pituk, M., *Stability criteria for linear delay differential equations*, Diff. Int. Eq. **10** (1997), 841–852.

[14] Győri, I., Hartung, F., Turi, J., *Preservation of stability in delay equations under delay perturbations*, J. Math. Anal. Appl. **220** (1998), 290–313.

[15] Hale, J. K., Verduyn Lunel, S. M., *Effects of small delays on stability and control*, in: Bart, Gohberg, Ran (eds), "Operator Theory and Analysis, The M. A. Kaashoek Anniversary Volume", Operator Theory: Advances and Applications, Vol. 122, Birkhäuser, 275–301 (2001).

[16] Hale, J. K., Verduyn Lunel, S. M., *Effects of time delays on the dynamics of feedback systems*, in: Fiedler, Gröger, Sprekels (eds.), "EQUADIFF'99, International Conference on Differential Equations, Berlin 1999", World Scientific, 257–266 (2000).

[17] Hieber, M., *Operator valued Fourier multipliers*, "Topics in Nonlinear Analysis, The Herbert Amann Anniversary Volume" (J. Escher, G. Simonett, eds.), Birkhäuser, 1999, pp. 363–380.

[18] Hieber, M., *A characterization of the growth bound of a semigroup via Fourier multipliers*, "Evolution Equations and Their Applications in Physical and Life Sciences" (G. Lumer, L. Weis, eds.), Lecture Notes in Pure and Applied Mathematics Series **215**, Marcel Dekker, 2001, pp. 121–124.

[19] Latushkin, Y., Räbiger, F., *Fourier multipliers in stability and control theory*, Preprint, 2000.

[20] Latushkin, Y., Shvidkoy, R., *Hyperbolicity of semigroups and Fourier multipliers*, in: "Systems, approximation, singular integral operators, and related topics. IWOTA 2000", Birkhauser-Verlag, 2001

[21] Maniar, L., Voigt, J., *Linear delay equations in the L^p context*, Preprint, 2000.

[22] Nagel, R. (ed.), "One-parameter Semigroups of Positive Operators", Springer-Verlag, Lecture Notes Math. **1184**, 1986.

[23] Nagel, R., Piazzera, S., *On the regularity properties of perturbed semigroups*, Rend. Circ. Mat. Palermo (2) Suppl. **56** (1998), 99–110.

[24] Rebarber, R., Townly, S., *Robustness with respect to delays for exponential stability of distributed parameter systems*, SIAM J. Control and Optimization **37** (1998), 230–244.

[25] Weis, L., *The stability of positive semigroups in L^p spaces*, Proc. Amer. Math. Soc. **123** (1995), 3089–3094.

[26] Wu, J., "Theory and Applications of Partial Functional Differential Equations", Springer-Verlag, Appl. Math. Sci. **119**, 1996.

András Bátkai
ELTE TTK
Department of Applied Analysis
Pf. 120
H-1518 Budapest, Hungary

Current address:
Institut für Analysis und Technische Mathematik
TU-Wien
Wiedner Hauptstraße 8–10
A-1040 Vienna, Austria
E-mail address: batka@cs.elte.hu

Bálint Farkas
ELTE TTK
Department of Applied Analysis
Pf. 120
H-1518 Budapest, Hungary
E-mail address: fbalint@cs.elte.hu

Progress in Nonlinear Differential Equations
and Their Applications, Vol. 55, 95–100

Mean Square Continuity of Ornstein-Uhlenbeck Processes in Banach Spaces

Z. Brzeźniak, B. Goldys, and J.M.A.M. van Neerven

We consider the following stochastic abstract Cauchy problem:

$$\begin{aligned} dX(t) &= AX(t)\,dt + B\,dW_H(t), \quad t \geq 0, \\ X(0) &= x_0, \end{aligned} \tag{1.1}$$

where A is the generator of a C_0–semigroup $\mathbf{S} = \{S(t)\}_{t\geq 0}$ on a separable real Banach space E, B is a bounded linear operator from a separable real Hilbert space H into E, and $\{W_H(t)\}_{t\geq 0}$ is a cylindrical Wiener process with Cameron-Martin space H. For the precise definition of this concept we refer to [3].

It has been shown in [3] that the problem (1.1) admits a weak solution $\{X(t,x_0)\}_{t\geq 0}$ if and only if for each $t > 0$ the operator $Q_t \in \mathcal{L}(E^*, E)$ defined by

$$Q_t x^* := \int_0^t S(s)BB^*S^*(s)x^*\,ds, \qquad x^* \in E^*,$$

is the covariance of a centred Gaussian measure μ_t on E. In this case, μ_t is the distribution of the random variable $X(t,0)$, and the solution can be represented as a stochastic convolution as follows:

$$\langle X(t,x_0), x^*\rangle = \langle S(t)x_0, x^*\rangle + \int_0^t \langle S(t-s)B\,dW_H(s), x^*\rangle, \qquad x^* \in E^*.$$

We will prove that the process $\{X(t,x_0)\}_{t\geq 0}$ is mean continuous in all moments. In particular this solves the problem, left open in [3], whether $\{X(t,x_0)\}_{t\geq 0}$ is mean square continuous.

Let $C_b(E)$ denote the space of all bounded continuous real functions on E.

Lemma 1. *Let (t_n) be a sequence of nonnegative real numbers in the interval $[0,T]$ with $\lim_{n\to\infty} t_n = t$. Let $g : [0,\infty) \to \mathbb{R}$ be nondecreasing and convex with $g(\|\cdot\|) \in L^1(E,\mu_T)$. Then for all $f \in C_b(E)$ we have*

$$\lim_{n\to\infty} \int_E f(x)\,g(\|x\|)\,d\mu_{t_n}(x) = \int_E f(x)\,g(\|x\|)\,d\mu_t(x).$$

Proof. For $r > 0$ let

$$B_r = \{x \in E : \ g(\|x\|) \leq r\}.$$

This set is symmetric and convex. Symmetry is obvious, and convexity follows from

$$g(\|\alpha x + (1-\alpha)y\|) \le g(\alpha\|x\| + (1-\alpha)\|y\|) \le \alpha g(\|x\|) + (1-\alpha)g(\|y\|),$$

where $\alpha \in [0,1]$. In view of

$$\langle Q_{t_n} x^*, x^*\rangle \le \langle Q_T x^*, x^*\rangle, \qquad x^* \in E^*,$$

we may apply Anderson's inequality [2, Theorem 3.3.6] to obtain

$$\mu_{t_n}(B_r) \ge \mu_T(B_r).$$

In combination with the identity

$$\int_E |h(x)|\, d\nu(x) = \int_0^\infty \nu\{x \in E:\ |h(x)| > s\}\, ds,$$

we find, with $M = \sup_{x\in E} |f(x)|$,

$$\begin{aligned}\int_{E\setminus B_r} |f(x)|\, g(\|x\|)\, d\mu_{t_n}(x) &\le M \int_{E\setminus B_r} g(\|x\|)\, d\mu_{t_n}(x)\\ &= M \int_{g(r)}^\infty \mu_{t_n}\{x \in E:\ g(\|x\|) > s\}\, ds\\ &\le M \int_{g(r)}^\infty \mu_T\{x \in E:\ g(\|x\|) > s\}\, ds\\ &= M \int_{E\setminus B_r} g(\|x\|)\, d\mu_T(x).\end{aligned} \tag{1.2}$$

The same argument gives

$$\int_{E\setminus B_r} |f(x)|\, g(\|x\|)\, d\mu_t(x) \le M \int_{E\setminus B_r} g(\|x\|)\, d\mu_T(x). \tag{1.3}$$

It now follows easily that the family $f(X(t_n,0))g(\|X(t_n,0)\|)$ is uniformly integrable. Since $\mu_{t_n} \to \mu_t$ weakly [6], the lemma follows from [1, Theorem 5.4]. Alternatively, the weak convergence $\mu_{t_n} \to \mu_t$ implies

$$\lim_{n\to\infty} \int_E f(x)\,(g(\|x\|) \wedge g(r))\, d\mu_{t_n}(x) = \int_E f(x)\,(g(\|x\|) \wedge g(r))\, d\mu_t(x)$$

for all $r > 0$. Choosing r so large that $\int_{E\setminus B_r} g(\|x\|)\, d\mu_T(x) < \varepsilon/M$, by (1.2) and (1.3) both truncation errors are at most ε, and again the lemma follows. □

If $f : E \to \mathbb{R}$ is a bounded Borel function, then for all $t \ge 0$ we have

$$\mathbb{E}\left(f(X(t,0))\right) = \int_E f(y)\, d\mu_t(y).$$

By an easy approximation argument, for $t \ge 0$ fixed this identity extends to all functions $f \in L^1(E, \mu_t)$.

Theorem 2. *Let $g : [0,\infty) \to \mathbb{R}$ be a nondecreasing convex function with $g(0) = 0$ such that*

$$g(c\|\cdot\|) \in L^1(E, \mu_T)$$

for some $c > M + 2$, where $M = \limsup_{u\downarrow 0} \|S(u)\|$. Then for all $x \in E$ and $t \in [0, T]$ we have

$$\lim_{s\to t} \mathbb{E}\, g(\|X(t,x) - X(s,x)\|) = 0.$$

Proof. The assumption $c > M + 2$ enables us to choose $\alpha, \beta, \gamma \in (0,1)$ with $\alpha + \beta + \gamma = 1$ subject to the following two conditions:

- $\beta c > M + 1$;
- $\gamma c \geq 1$.

Step 1. First we note that for all $\tau \in [0, T]$ and $0 \leq c' \leq c$,

$$\begin{aligned}\int_E g(c'\|y\|)\, d\mu_\tau(y) &= \int_0^\infty \mu_\tau\{x \in E :\ g(c'\|x\|) > s\}\, ds \\ &\leq \int_0^\infty \mu_T\{x \in E :\ g(c'\|x\|) > s\}\, ds \\ &= \int_E g(c'\|y\|)\, d\mu_T(y) \\ &\leq \int_E g(c\|y\|)\, d\mu_T(y) < \infty.\end{aligned}$$

Hence by the condition $1/\gamma \leq c$ and the remark preceding the theorem,

$$\mathbb{E}\, g(\tfrac{1}{\gamma}\|X(\tau,0)\|) = \int_E g(\tfrac{1}{\gamma}\|y\|)\, d\mu_\tau(y),$$

and therefore by Lemma 1,

$$\lim_{\tau\downarrow 0} \mathbb{E}\, g(\tfrac{1}{\gamma}\|X(\tau,0)\|) = \lim_{\tau\downarrow 0} \int_E g(\tfrac{1}{\gamma}\|y\|)\, d\mu_\tau(y) = 0.$$

Step 2. Right continuity. Fix $t \in [0,T]$. We have, for $t \leq s \leq T$,

$$\begin{aligned}\langle X(s,x),&x^*\rangle - \langle X(t,x), x^*\rangle \\ &= \langle S(s)x, x^*\rangle + \int_0^s \langle S(s-u)B\, dW_H(u), x^*\rangle \\ &\qquad - \langle S(t)x, x^*\rangle - \int_0^t \langle S(t-u)B\, dW_H(u), x^*\rangle \\ &= \langle S(s)x - S(t)x, x^*\rangle + \langle S(s-t)X(t,0) - X(t,0), x^*\rangle \\ &\qquad + \langle Y_{s,t}, x^*\rangle,\end{aligned}$$

where

$$Y_{s,t} = \int_t^s S(s-u)B\, dW_H(u).$$

Hence,

$$X(s,x) - X(t,x) = S(s)x - S(t)x + S(s-t)X(t,0) - X(t,0) + Y_{s,t}. \tag{1.4}$$

The convexity of g implies

$$\begin{aligned} &g(\|X(s,x)-X(t,x)\|) \\ &\quad\le \alpha g\big(\tfrac{1}{\alpha}\|S(s)x-S(t)x\|\big)+\beta g\big(\tfrac{1}{\beta}\|S(s-t)X(t,0)-X(t,0)\|\big) \\ &\qquad+\gamma g\big(\tfrac{1}{\gamma}\|Y_{s,t}\|\big). \end{aligned}$$

Noting that g is continuous with $g(0)=0$, it follows that

$$\lim_{s\downarrow t}\mathbb{E}\; g\big(\tfrac{1}{\alpha}\|S(s)x-S(t)x\|\big)=\lim_{s\downarrow t} g\big(\tfrac{1}{\alpha}\|S(s)x-S(t)x\|\big)=0.$$

Arguing as in Step 1 and using the condition $(M+1)/\beta<c$ we see that for $s-t$ sufficiently small,

$$\mathbb{E}\; g\big(\tfrac{1}{\beta}\|S(s-t)X(t,0)-X(t,0)\|\big)=\int_E g\big(\tfrac{1}{\beta}\|S(s-t)y-y\|\big)\,d\mu_t(y).$$

Hence by dominated convergence,

$$\begin{aligned} &\lim_{s\downarrow t}\mathbb{E}\; g(\tfrac{1}{\beta}\|S(s-t)X(t,0)-X(t,0)\|) \\ &\qquad=\lim_{s\downarrow t}\int_E g(\tfrac{1}{\beta}\|S(s-t)y-y\|)\,d\mu_t(y)=0. \end{aligned}$$

Finally, noting that $Y_{s,t}$ and $X(s-t,0)$ have the same distribution, by Step 1 we have

$$\lim_{s\downarrow t}\mathbb{E}\; g(\tfrac{1}{\gamma}\|Y_{s,t}\|)=\lim_{s\downarrow t}\mathbb{E}\; g(\tfrac{1}{\gamma}\|X(s-t,0)\|)=0.$$

Step 3. Left continuity. Fix $t\in[0,T]$. For $0\le s\le t$ we have, using (1.4) with the rôles of s and t reversed,

$$\begin{aligned} &g(\|X(t,x)-X(s,x)\|) \\ &\quad\le \alpha g\big(\tfrac{1}{\alpha}\|S(t)x-S(s)x\|\big)+\beta g\big(\tfrac{1}{\beta}\|S(t-s)X(s,0)-X(s,0)\|\big) \\ &\qquad+\gamma g\big(\tfrac{1}{\gamma}\|Y_{t,s}\|\big). \end{aligned}$$

As in Step 2, the expectation of the first term on the right hand side tends to 0 as $s\uparrow t$ by continuity, and the expectation of the third term tends to 0 by Step 1. It remains to prove that

$$\begin{aligned} &\lim_{s\uparrow t}\;\mathbb{E}\; g(\tfrac{1}{\beta}\|S(t-s)X(s,0)-X(s,0)\|) \\ &\qquad=\lim_{s\uparrow t}\int_E g(\tfrac{1}{\beta}\|S(t-s)y-y\|)\,d\mu_s(y)=0. \end{aligned}$$

By Lemma 1, for all $s\in[0,T]$ the measure $g(c\|x\|)\,d\mu_s(x)$ is a finite Radon measure and the family

$$\big\{g(c\|x\|)\,d\mu_s(x):\; s\in[\tfrac{1}{2}t,t]\big\}$$

is tight. Fix $\varepsilon>0$ arbitrary and use Prokhorov's theorem to choose a compact set K such that

$$\int_{E\setminus K} g(c\|x\|)\,d\mu_s(x)<\varepsilon,\qquad s\in[\tfrac{1}{2}t,t].$$

Choose $0 < \tau \le \frac{1}{2}t$ so small that

$$\tfrac{1}{\beta}(\|S(u)\| + 1) \le c \quad \text{and} \quad \tfrac{1}{\beta}\|S(u)y - y\| < \varepsilon, \qquad u \in [0,\tau],\ y \in K.$$

It follows that for $s \in [t-\tau, t]$,

$$\begin{aligned}\int_E g(\tfrac{1}{\beta}\|S(t-s)y - y\|)\,d\mu_s(y)& \\ &\le \int_K g\big(\tfrac{1}{\beta}\|S(t-s)y - y\|\big)\,d\mu_s(y) \\ &\quad + \int_{E\setminus K} g\big(\tfrac{1}{\beta}\|S(t-s)y - y\|\big)\,d\mu_s(y). \\ &\le g(\varepsilon) + \varepsilon.\end{aligned}$$

Since $\lim_{\varepsilon\downarrow 0} g(\varepsilon) = 0$, this completes the proof. □

Under a slightly stronger assumption on g, we can rephrase this result in terms of Orlicz norms.

If $g : [0,\infty) \to \mathbb{R}$ is a nondecreasing convex function with $g(0) = 0$, then for a strongly measurable function $\xi : (\Omega, \mathbb{P}) \to E$ we define

$$\|\xi\|_{L_g(E)} := \inf\left\{c > 0 :\ \mathbb{E}\, g\left(\frac{\|\xi\|}{c}\right) \le 1\right\}.$$

The set $L_g(E)$ of all ξ for which $\|\xi\|_{L_g(E)}$ is finite is a Banach space; cf. [5].

Corollary 3. *Let $g : [0,\infty) \to \mathbb{R}$ be a nondecreasing convex function with $g(0) = 0$ such that $g(c\|\cdot\|) \in L^1(E, \mu_T)$ for all $c > 0$. Then for all $x \in E$ and $t \in [0,T]$ we have*

$$\lim_{s\to t} X(s,x) = X(t,x) \quad \textit{in} \quad L_g(E).$$

Proof. Let $\varepsilon > 0$ be fixed and define $g_\varepsilon(\tau) := g(\varepsilon^{-1}\tau)$. According to Theorem 2, for $|t-s|$ sufficiently small we have

$$\mathbb{E}\, g_\varepsilon\big(\|X(t,x) - X(s,x)\|\big) \le 1.$$

Hence,

$$\mathbb{E}\, g\left(\frac{\|X(t,x) - X(s,x)\|}{\varepsilon}\right) = \mathbb{E}\, g_\varepsilon\big(\|X(t,x) - X(s,x)\|\big) \le 1,$$

which means that $\|X(t,x) - X(s,x)\|_{L_g(E)} \le \varepsilon$. □

By Fernique's theorem, this result applies, e.g., to the functions

$$g(\tau) = \exp(\tau^p) - 1, \qquad 1 \le p < 2,$$

and $g(\tau) = \tau^p$, $1 \le p < \infty$. In the latter case we can apply Theorem 2 directly and obtain:

Corollary 4. *For all $x \in E$ and $t \ge 0$ we have*

$$\lim_{s\to t} \mathbb{E}\left(\|X(t,x) - X(s,x)\|^p\right) = 0, \qquad p \in [1,\infty).$$

References

[1] P. Billingsley, *Convergence of probability measures.* John Wiley and Sons, New York-London-Sydney-Toronto, 1968.

[2] V.I. Bogachev, *Gaussian Measures.* Math. Surveys and Monographs, Vol. 62, Amer. Math. Soc., Providence, R.I., 1998.

[3] Z. Brzeźniak and J.M.A.M. van Neerven, *Stochastic convolution in separable Banach spaces and the stochastic linear Cauchy problem.* Studia Math. **143** (2000), 43–74.

[4] G. Da Prato, S. Kwapień, and J. Zabczyk, *Regularity of solutions of linear stochastic equations in Hilbert spaces.* Stochastics **23** (1987), 1–23.

[5] M.A. Krasnosel'skiĭ and Ja.B. Rutickiĭ, *Convex functions and Orlicz spaces.* Noordhoff Ltd., Groningen, 1961.

[6] J.M.A.M. van Neerven, *Continuity and representation of Gaussian Mehler semigroups.* Potential Anal. **13** (2000), 199–211.

Z. Brzeźnia
Department of Mathematics
The University of Hull
Hull HU6 7RX, England
E-mail address: `Z.Brzezniak@maths.hull.ac.uk`

B. Goldys
School of Mathematics
The University of New South Wales
Sydney 2052, Australia
E-mail address: `B.Goldys@unsw.edu.au`

J.M.A.M. van Neerven
Department of Applied Mathematical Analysis
Technical University of Delft
P.O. Box 5031
2600 GA Delft, The Netherlands
E-mail address: `J.vanNeerven@its.tudelft.nl`

Progress in Nonlinear Differential Equations
and Their Applications, Vol. 55, 101–111

Some Remarks on Maximal Regularity of Parabolic Problems

Philippe Clément and Jan Prüss

1. Introduction

It has become apparent that the concept of families of *R-bounded* operators plays an important role in operator theory, in the theory of operator-valued Fourier multipliers, and also in maximal regularity of elliptic and parabolic problems. Although this theory is only of recent origin, already a number of nontrivial applications have been appeared and we expect that in the near future there will be many more.

A family $\mathcal{T} \subset \mathcal{B}(X,Y))$ of bounded linear operators on a Banach space X is called **R-bounded** if there is a constant $C > 0$ such that for all $N \in \mathbb{N}$, elements $x_j \in X$, selections $T_j \in \mathcal{T}$, and N independent symmetric $\{-1,1\}$-valued random variables ε_j on a probability space (Ω,μ) the inequality

$$||\sum_{j=1}^{N} \varepsilon_j T_j x_j||_{L_p(\Omega;Y)} \leq C||\sum_{j=1}^{N} \varepsilon_j x_j||_{L_p(\Omega;X)} \tag{1.1}$$

is valid. This can be seen as a strengthening of the concept of uniform boundedness of the family $\mathcal{T}$.

To draw the connection to maximal regularity, recall that an operator A in a Banach space X is called **positive** or **sectorial** if its domain and range are dense, $(-\infty,0)$ is contained in the resolvent set of A, and the resolvent estimate

$$|t(t+A)^{-1}|_{\mathcal{B}(X)} \leq M_0, \quad t > 0,$$

is valid. A sectorial operator A is called **$\mathcal{R}$-sectorial** if in addition the set $\{t(t+A)^{-1} : t > 0\}$ is $\mathcal{R}$-bounded. This concept was introduced in the same volume of the proceedings of this series of conferences in Clément and Prüss [5] and in Weis [16].

In that paper Weis proved, that in case X is a UMD-Banach space, $-A$ generates an exponentially stable analytic C_0-semigroup in X, then maximal L_p-regularity of the abstract Cauchy problem

$$\dot{u} + Au = f,\ t > 0, \quad u(0) = 0,$$

is equivalent to $\mathcal{R}$-boundedness of $\{i\rho(i\rho+A)^{-1} : \rho \in \mathbb{R}\}$. This can be reformulated as: A is $\mathcal{R}$-sectorial with $\mathcal{R}$-angle less than $\pi/2$; see Section 2 for more explanations.

For background and the present state of the art in this theory, we refer to the papers Clément, de Pagter, Sukochev, and Witvliet [2], Clément and Prüss [5], Kalton and Weis [12], and for surveys one should consult Denk, Hieber, and Prüss [7] and Weis [16].

It is the aim of this paper to discuss sectorial (or merely pseudo-sectorial or nonnegative) operators A in their associated real interpolation spaces $D_A(\alpha, p)$, where $\alpha \in (0,1)$ and $p \in [1,\infty]$. In Section 2 we show that the part $A_{\alpha,p}$ of A in $D_A(\alpha,p)$ is always $\mathcal{R}$-sectorial. This not surprising since $A_{\alpha,p}$ admits an $\mathcal{H}^\infty$-calculus, as was proved by Dore [8]. On the other hand, it was shown in Clément and Prüss [5] that such operators are $\mathcal{R}$-sectorial, if the space is UMD. Here we give a direct proof which is valid for nonnegative operators in arbitrary Banach spaces. Section 2 deals with the extension of the before mentioned result of Dore to the operator-valued case. Our functional calculus is strong enough to cover the theorem of Da Prato and Grisvard [6]. It is an analog in the setting of real interpolation spaces of a recent result due to Kalton and Weis [12]. Based on the counter-example of Kalton and Lancien [11] we show that our result is independent from that of Kalton and Weis. An application of Theorem 2 to a certain symbol arising in connection with Stefan problems is presented in Section 4. The corresponding theory for this symbol in B^s_{pq} and H^s_p is contained in Escher, Prüss and Simonett [9]. We conclude the paper with the extension of the famous Stein multiplier theorem for Symmetric Diffusion Semigroups on L_p-spaces to operator-valued multipliers in vector-valued setting.

2. $\mathcal{R}$-Sectoriality in Interpolation Spaces

Recall that a linear operator A in a Banach space X is called **nonnegative or pseudo-sectorial**, if $(-\infty,0) \subset \rho(A)$, the resolvent set of A, and there is a constant $M_0 > 0$ such that

$$|t(t+A)^{-1}| \leq M_0, \quad t > 0.$$

This class of operators has been studied in detail by Komatsu [13], in particular in respect of fractional powers. Note that a pseudo-sectorial operator is always closed since it has a nonempty resolvent set. If in addition, the domain $\mathcal{D}(A)$ and the range $\mathcal{R}(A)$ of A are dense in X, then A is called **positive or sectorial**. We define as usual the **spectral angle** ϕ_A as the infimum of all angles $\phi \in (0,\pi]$ such that

$$\Sigma_{\pi-\phi} := \{z \in \mathbb{C} : \; z \neq 0, |\arg(z)| < \pi - \phi\} \subset \rho(A),$$

and the resolvent estimate

$$|z(z+A)^{-1}| \leq M_{\pi-\phi}, \quad z \neq 0, |\arg(z)| < \pi - \phi,$$

is satisfied, for some constant $M_{\pi-\phi} > 0$ which is independent of z. If for a (pseudo)-sectorial operator A in addition a set $\{\lambda(\lambda+A)^{-1} : \lambda \in \Sigma_{\pi-\phi}\}$ is $\mathcal{R}$-bounded for some angle ϕ, then we call A **(pseudo)-$\mathcal{R}$-sectorial**; pseudo-$\mathcal{R}$-sectorial will also be called **$\mathcal{R}$-nonnegative**. The infimum of such angles π is called the $\mathcal{R}$-angle of A, we denote it by ϕ_A^R. Note that we always have $\phi_A \le \phi_A^R$.

It is well-known that $-A$ generates an analytic semigroup e^{-At} in X which is bounded on some sector Σ_ψ if and only if A is pseudo-sectorial with spectral angle $\phi_A < \pi/2$; e^{-At} is strongly continuous at 0 if and only if in addition $\mathcal{D}(A)$ is dense in X. A famous result due to Da Prato and Grisvard [6] states that in this case the abstract Cauchy problem

$$\dot{u} + Au = f, \quad t \in J = [0,T],\ u(0) = 0, \tag{2.2}$$

has maximal regularity in $L_p(J; D_A(\alpha,p))$, for each $\alpha \in (0,1)$, $p \in (1,\infty)$. Here, as usual, $D_A(\alpha,p)$ denotes the real interpolation space $(X, \mathcal{D}(A))_{\alpha,p}$ between X and the domain of A equipped with the graph norm of order α and exponent p. Therefore by a recent result of Weis [16], the part $A_{\alpha,p}$ of A in $D_A(\alpha,p)$ is pseudo-$\mathcal{R}$-sectorial with $\mathcal{R}$-angle $\phi^R_{A_{\alpha,p}} < \pi/2$; see also Clément and Prüss [5].

Despite its elegance, this argument has some apparent disadvantages. It only applies to the case $\phi_A < \pi/2$, it gives only the bounds $\phi^R_{A_{\alpha,p}} < \pi/2$, and it is rather indirect. The following result takes care of these shortcomings.

Theorem 1. *Let $\alpha \in (0,1)$, $p \in [1,\infty)$, and suppose that A is pseudo-sectorial in the Banach space X.*

Then the part $A_{\alpha,p}$ of A in $D_A(\alpha,p)$ is pseudo-$\mathcal{R}$-sectorial and $\phi^R_{A_{\alpha,p}} \le \phi_A$ holds.

Proof. Let A be pseudo-sectorial with spectral angle $\phi_A \in [0,\pi)$; choose $\psi > \theta > \phi_A$ and let Γ_θ denote the standard contour

$$\Gamma_\theta = (\infty, 0]e^{i\theta} \cup [0,\infty)e^{-i\theta}.$$

Then for each $\lambda \in \Sigma_{\pi-\psi}$ and $t > 0$, the Dunford calculus for A in X yields the representation

$$A(t+A)^{-1}(\lambda+A)^{-1} = \frac{1}{2\pi i}\int_{\Gamma_\theta} \frac{1}{\lambda+z} A(z-A)^{-1}\frac{dz}{z+t}.$$

Fixing finitely many $\lambda_j \in \Sigma_{\phi-\psi}$, $\varepsilon_j = \pm 1$, and $x_j \in D_A(\alpha,p)$, we obtain

$$t^\alpha A(t+A)^{-1}\sum_j \varepsilon_j\lambda_j(\lambda_j+A)^{-1}x_j = \frac{1}{2\pi i}\int_{\Gamma_\theta} z^\alpha A(z-A)^{-1}\sum_j \varepsilon_j\frac{\lambda_j}{\lambda_j+z}x_j\frac{t^\alpha dz}{(z+t)z^\alpha}.$$

Since

$$\int_{\Gamma_\theta}\frac{t^\alpha|dz|}{(|z|+t)|z|^\alpha} = \int_{\Gamma_\theta}\frac{d\zeta|}{(1+|\zeta|)|\zeta|^\alpha} < \infty,$$

Jensen's inequality yields

$$|t^\alpha A(t+A)^{-1}\sum_j \varepsilon_j\lambda_j(\lambda_j+A)^{-1}x_j|^p$$

$$\le C\int_{\Gamma_\theta}|z^\alpha A(z-A)^{-1}\sum_j \varepsilon_j\frac{\lambda_j}{\lambda_j+z}x_j|^p\frac{t^\alpha|dz|}{(|z|+t)|z|^\alpha}.$$

By means of

$$\int_0^\infty \frac{t^\alpha dt}{t|z|^\alpha(|z|+t)} = \frac{1}{|z|}\int_0^\infty \frac{s^{\alpha-1}ds}{1+s} = \frac{c_\alpha}{|z|}$$

we continue

$$\int_0^\infty |t^\alpha A(t+A)^{-1}\sum_j \varepsilon_j\lambda_j(\lambda_j+A)^{-1}x_j|^p\frac{dt}{t}$$
$$\leq c_\alpha C\int_{\Gamma_\theta}|z^\alpha||A(z-A)^{-1}\sum_j \varepsilon_j\frac{\lambda_j}{\lambda_j+z}x_j|^p\frac{|dz|}{|z|}.$$

For independent symmetric random variables ε_j on a probability space (Ω,μ), integration w.r.t. μ and application of the contraction principle then yields

$$|\sum_j \varepsilon_j\lambda_j(\lambda_j+A)^{-1}x_j|^p_{L_p(\Omega;D_A(\alpha,p))}$$
$$\leq C\int_{\Gamma_\theta}\int_\Omega|\sum_j \varepsilon_j\frac{\lambda_j}{\lambda_j+z}z^\alpha A(z-A)^{-1}x_j|^p d\mu\frac{|dz|}{|z|}$$
$$\leq 2^pC\int_{\Gamma_\theta}[\int_\Omega|\sum_j \varepsilon_j z^\alpha A(z-A)^{-1}x_j|^p d\mu]\frac{|dz|}{|z|} \leq C|\sum_j \varepsilon_j x_j|^p_{L_p(\Omega;D_A(\alpha,p))}.$$

This shows that the set of operators $\{\lambda(\lambda+A)^{-1} : \lambda\in\Sigma_{\pi-\psi}\}$ is $\mathcal{R}$-bounded in $D_A(\alpha,p)$, for each $\psi>\phi_A$, $\alpha\in(0,1)$, $p\in[1,\infty)$, which means that the part $A_{\alpha,p}$ of A in $D_A(\alpha,p)$ is pseudo-$\mathcal{R}$-sectorial and $\phi^R_{A_{\alpha,p}}\leq\phi_A$. □

3. Operator-Valued $\mathcal{H}^\infty$-calculus in Interpolation Spaces

Let the space $\mathcal{H}^\infty_0(\Sigma_\phi)$ be defined as the space of all holomorphic functions $f:\Sigma_\phi\to\mathbb{C}$ such that

$$|z^\alpha f(z)|+|z^{-\alpha}f(z)|\leq M,\quad z\in\Sigma_\phi,$$

for some $\alpha>0$. Recall that a sectorial operator A in a Banach space X is said to admit an $\mathcal{H}^\infty$**-calculus** if the usual Dunford calculus defined by

$$f(A)=\frac{1}{2\pi i}\int_{\Gamma_\theta}f(z)(z-A)^{-1}dz$$

for $f\in\mathcal{H}^\infty_0(\Sigma_\phi)$ satisfies

$$|f(A)|_{\mathcal{B}(X)}\leq C|f|_{\mathcal{H}^\infty(\Sigma_\phi)},$$

for some constant $C>0$ independent of f. The infimum of all such angles $\phi>\phi_A$ is called the $\mathcal{H}^\infty$**-angle** of A. It is well-known that in this case the Dunford calculus extends uniquely to $\mathcal{H}^\infty(\Sigma_\phi)$, for any $\phi>\phi^\infty_A$. Clément and Prüss [5] have shown that sectorial operators with $\mathcal{H}^\infty$-calculus are $\mathcal{R}$-sectorial with $\phi^R_A\leq\phi^\infty_A$, provided the underlying Banach space X has the *unconditional martingale* *difference property*, i.e., is UMD for short. The converse of this is not true, in general, since there are examples of sectorial operators in Hilbert spaces without $\mathcal{H}^\infty$-calculus; see Baillon and Clément [1] and McIntosh and Yagi [14].

This concept can easily be extended to the operator-valued case, where scalar $\mathcal{H}^\infty$-functions are replaced by functions in $\mathcal{H}_0^\infty(\Sigma_\phi;\mathcal{B}(X))$. We say that a pseudo-sectorial operator A in X admits an **operator-valued $\mathcal{H}^\infty$-calculus** if there is an angle $\phi > \phi_A$ and a constant $M > 0$ such that

$$|F(A)|_{\mathcal{B}(X)} \le C|F|_{\mathcal{H}^\infty(\Sigma_\phi;\mathcal{B}(X))}$$

is valid, for all $F \in \mathcal{H}_0^\infty(\Sigma_\phi;\mathcal{B}(X)$ commuting with A in the sense that

$$F(z)(\lambda - A)^{-1} = (\lambda - A)^{-1}F(z), \quad z \in \Sigma_\phi,\ \lambda \in \rho(A).$$

Although for many important sectorial operators, in particular partial differential operators, the $\mathcal{H}^\infty$-calculus is known to exist, it is not an easy task to prove it. However, Dore [8] has obtained the striking result that sectorial operators A always admit $\mathcal{H}^\infty$-calculi in the corresponding real interpolation spaces $D_A(\alpha,p)$, where $\alpha \in (0,1)$ and $p \in [1,\infty]$. Here we present the extension of Dore's result to the operator-valued case.

Theorem 2. *Let $\alpha \in (0,1)$, $p \in [1,\infty]$, and suppose that A is pseudo-sectorial in the Banach space X with spectral angle ϕ_A.*

Then for any $\phi > \phi_A$ there is a constant C_ϕ such that

$$|F(A)|_{\mathcal{B}(D_A(\alpha,p))} \le C_\phi|F|_{\mathcal{H}^\infty(\Sigma_\phi;\mathcal{B}(X))},$$

for each $F \in \mathcal{H}_0^\infty(\Sigma_\phi;\mathcal{B}(X))$ commuting with A.

Proof. Fix any $\phi > \phi_A$, choose $\theta \in (\phi_A,\phi)$ and let $\Gamma = \Gamma_\theta$ denote the standard path, as in the proof of Theorem 1. For $F \in \mathcal{H}_0^\infty(\Sigma_\phi;\mathcal{B}(X))$ commuting with A we have the Dunford integral

$$F(A) = \frac{1}{2\pi i}\int_\Gamma F(\lambda)(\lambda - A)^{-1}d\lambda.$$

Applying the resolvent formula as well as Cauchy's theorem this implies for $\alpha \in (0,1)$ and $t > 0$

$$t^\alpha A(t+A)^{-1}F(A) = \frac{1}{2\pi i}\int_\Gamma \lambda^\alpha F(\lambda)A(\lambda - A)^{-1}\frac{t^\alpha d\lambda}{\lambda^\alpha(\lambda+t)}.$$

Since

$$\int_\Gamma \frac{t^\alpha|dz|}{(|z|+t)|z|^\alpha} = \int_\Gamma \frac{d\zeta|}{(1+|\zeta|)|\zeta|^\alpha} < \infty,$$

Jensen's inequality yields for $1 \le p < \infty$

$$|t^\alpha A(t+A)^{-1}F(A)x|^p \le C\int_\Gamma [|\lambda^\alpha||F(\lambda)A(\lambda - A)^{-1}x|]^p\frac{t^\alpha|d\lambda|}{|\lambda^\alpha|(|\lambda|+t)}$$

$$\le C|F|_{\mathcal{H}^\infty(\Sigma_\phi;\mathcal{B}(X))}\int_\Gamma |\lambda^\alpha A(\lambda - A)^{-1}x|^p\frac{t^\alpha|d\lambda|}{|\lambda^\alpha|(|\lambda|+t)}.$$

Integrating over $t > 0$ this implies

$$\int_0^\infty |t^\alpha A(t+A)^{-1}F(A)x|^p \frac{dt}{t} \le c_\alpha C|F|_{\mathcal{H}^\infty(\Sigma_\phi;\mathcal{B}(X))} \int_\Gamma |\lambda^\alpha A(\lambda - A)^{-1}x|^p \frac{|d\lambda|}{|\lambda|},$$

since

$$\int_0^\infty \frac{t^\alpha dt}{t|z|^\alpha(|z|+t)} = \frac{c_\alpha}{|z|}.$$

This implies the inequality

$$|F(A)x|_{D_A(\alpha,p)} \le Cc_\alpha |F|_{\mathcal{H}^\infty(\Sigma_\phi;\mathcal{B}(X))}|x|_{D_A(\alpha,p)}, \quad x \in D_A(\alpha,p),$$

for each $p \in [1,\infty)$.

For the case $p = \infty$ we use the change of variables $\lambda = tz$ to obtain

$$t^\alpha A(t+A)^{-1}F(A) = \frac{1}{2\pi i}\int_\Gamma (tz)^\alpha F(tz)A(tz-A)^{-1}\frac{dz}{z^\alpha(z+1)},$$

and estimate directly to the result

$$|t^\alpha A(t+A)^{-1}F(A)x| \le C|F|_{\mathcal{H}^\infty(\Sigma_\phi;\mathcal{B}(X))}\int_\Gamma |(tz)^\alpha A(tz-A)^{-1}x|\frac{|dz|}{|z|^\alpha(|z|+1)},$$

$$\le C|F|_{\mathcal{H}^\infty(\Sigma_\phi;\mathcal{B}(X))}|x|_{D_A(\alpha,\infty)}\int_\Gamma \frac{|dz|}{|z|^\alpha(|z|+1)}.$$

This proves the theorem for all $\alpha \in (0,1)$ and $p \in [1,\infty]$. □

Having established a uniform $\mathcal{H}^\infty$-bound for the functional calculus of A, it is easy to extend it to functions $F \in \mathcal{H}^\infty(\Sigma_\phi;\mathcal{B}(X))$ which commute with A. In fact, if such F is given, set $F_\varepsilon(z) = F(z)z^\varepsilon/(1+z)^{2\varepsilon}$; then $F_\varepsilon(z) \to F(z)$ in $\mathcal{B}(X)$, locally uniformly in $z \in \Sigma_\phi$. Decomposing the path Γ into $\Gamma = \Gamma_1 \cup \Gamma_2$ where

$$\Gamma_1 = \{z \in \Gamma : |z| \le 1\}, \quad \Gamma_2 = \{z \in \Gamma : |z| \ge 1\},$$

and setting $\Gamma_0 = e^{i[-\phi,\phi]}$, by means of Cauchy's theorem we obtain the representation

$$\begin{aligned} F_\varepsilon(A)x &= \frac{1}{2\pi i}\int_{\Gamma_1} F_\varepsilon(z)z(z-A)^{-1}A^{-1}x\,dz + \frac{1}{2\pi i}\int_{\Gamma_2} F_\varepsilon(z)A(z-A)^{-1}x\frac{dz}{z} \\ &+ \left(\frac{1}{2\pi i}\int_{\Gamma_0} F_\varepsilon(z)dz\right)A^{-1}x + \left(\frac{1}{2\pi i}\int_{\Gamma_0} F_\varepsilon(z)\frac{dz}{z}\right)x, \end{aligned} \tag{3.3}$$

for each $x \in \mathcal{R}(A)$. (3.3) shows that $F(A)x := \lim_{\varepsilon\to 0} F_\varepsilon(A)x$ exists in X, for each $x \in \mathcal{R}(A) \cap D_A(\alpha,p)$. Therefore, in case $0 \in \rho(A)$, the Banach-Steinhaus theorem implies $F_\varepsilon(A)x \to F(A)x$ in X for each $x \in D_A(\alpha,p)$. On the other hand, $F_\varepsilon(A)x$ is bounded in $D_A(\alpha,p)$ by Theorem 2, hence $F(A)x \in D_A(\alpha,p)$. This shows that $F(A) \in \mathcal{B}(D_A(\alpha,p))$ is well-defined, for each $\alpha \in (0,1$ and $p \in [1,\infty]$.

The same result is valid if instead of $0 \in \rho(A)$ the function F is holomorphic in a neighbourhood of $z = 0$. This can be seen in the standard way, modyfying the path Γ near $z = 0$. Summarizing we have

Corollary 1. *Let $\alpha \in (0,1)$, $p \in [1,\infty]$, and suppose that A is pseudo-sectorial in the Banach space X with spectral angle ϕ_A, and let $\phi > \phi_A$. Assume $F \in \mathcal{H}^\infty(\Sigma_\phi; \mathcal{B}(X))$ commutes with A and either $0 \in \rho(A)$ or F is holomorphic in a neighborhood of $z = 0$.*

Then $F(A)$ is well-defined and belongs to $\mathcal{B}(D_A(\alpha,p))$.

This is a natural extension of the theorem of Da Prato and Grisvard [6]. In fact, let B be pseudo-sectorial in X commuting with A in the resolvent sense and let the *parabolicity condition*

$$\phi_A + \phi_B < \pi \tag{3.4}$$

be satisfied. Set $F_t(z) = B(tz+B)^{-1}$, $t > 0$; then because of (3.4), we have $F_t \in \mathcal{H}^\infty(\Sigma_\phi)$ for some $\phi > \phi_A$, and F_t commutes with A. Hence if either $0 \in \rho(A)$ or F_t holomorphic at $z = 0$, i.e., $0 \in \rho(B)$ then $F_t(A) = B(tA+B)^{-1}$ is well-defined and bounded in $D_A(\alpha,p)$, with bound independent of $t > 0$, i.e., AB^{-1} is sectorial in $D_A(\alpha,p)$.

It is interesting to compare Corollary 1 with a recent result due to Kalton and Weis [12], which states that whenever A is sectorial in X and admits an $\mathcal{H}^\infty$-calculus, $F \in \mathcal{H}^\infty(\Sigma_\phi; \mathcal{B}(X))$ commutes with A, $\phi > \phi_A$, and in addition $F(\Sigma_\phi) \subset \mathcal{B}(X)$ is $\mathcal{R}$-bounded, the $F(A)$ is well-defined and bounded in X. The following example shows that this result does not imply Corollary 1.

Example 1. For a Banach space Z and $\alpha \in (0,1)$, we define the so-called littele-Hölder space $\mathrm{BUC}^\alpha(R; X)$ as the space of all functions $u : \mathbb{R} \to X$ which are uniformly Hölder-continuous with exponent α and uniformly bounded on $\mathbb{R}$. Consider $L_p(G)$ for some open subset $G \subset \mathbb{R}^n$, $p \neq 2$. According to the paper of Kalton and Lancien [11] there is a sectorial operator A on $L_p(G)$ such that $-A$ generates an exponentially stable analytic C_0-semigroup in $L_p(G)$ such that problem (2.2) does not have maximal L_p-regularity. Therefore by the result of Weis mentioned in Section 2, A is not $\mathcal{R}$-sectorial resp. $\phi_A^R \geq \pi/2$. The result of Da Prato and Grisvard (or Corollary 1) shows that (2.2) has maximal regularity in $\mathrm{BUC}^\alpha(\mathbb{R}; \mathrm{L_p(G)})$ where $\alpha \in (0,1)$. Theorem 1 (or the result of Dore mentioned above) shows that $B = d/dt$ admits an $\mathcal{H}^\infty$-calculus in $\mathrm{BUC}^\alpha(\mathbb{R}; L_p(G))$, however, the result of Kalton and Weis does not apply here, since A is not $\mathcal{R}$-sectorial with $\phi_A^R < \pi/2$ in $L_p(G)$, hence also not in $\mathrm{BUC}^\alpha(\mathbb{R}; L_p(G))$.

4. Application to Symbols of Stefan Problems

In a recent paper due to Escher, Prüss and Simonett [9] it was found that the principal symbol of the localized and linearized Stefan problem with Gibbs-Thomson correction is given by

$$s(\lambda, |\xi|^2) = \lambda + |\xi|^2\sqrt{\lambda + |\xi|^2}, \quad \operatorname{Re}\lambda > 0,\ \xi \in \mathbb{R}^n.$$

Setting $D = -\Delta$, the Laplacian on $\mathbb{R}^n$, this corresponds to the abstract Volterra integro-differential problem

$$\frac{d}{dt}[u(t) + \int_0^t De^{-Ds}u(t-s)\frac{ds}{\sqrt{\pi s}}]$$
$$+ \int_0^t D^2e^{-Ds}u(t-s)\frac{ds}{\sqrt{\pi s}} = f(t), \quad t > 0, u(0) = 0. \qquad (4.5)$$

In that paper the maximal regularity properties of this problem have been studied by means of the Dore-Venni theorem in an L_p-setting, where $p \in (1, \infty)$. This result was used as the basis for the proof of well-posedness of the underlying nonlinear Stefan problem with surface tension. Here we want to apply Theorem 2 to the study of the symbol $s(\lambda, \xi)$ in interpolation spaces, in particular, in spaces of Hölder-continuous functions.

Note that $s(\lambda, |\xi|^2)$ is of first order in λ, of third order in ξ, but in addition there is the mixed order 1/2 in λ and 2 in ξ. To show maximal regularity for this symbol we therefore have to study the three symbols

$$r_1(\lambda, |\xi|^2) = \lambda s(\lambda, |\xi|^2)^{-1}, \; r_2(\lambda, |\xi|^2) = |\xi|^3 s(\lambda, |\xi|^2)^{-1},$$
$$r_3(\lambda, |\xi|^2) = \sqrt{\lambda}|\xi|^2 s(\lambda, |\xi|^2)^{-1},$$

in particular, that these symbols give rise to bounded linear operators on a certain space of functions.

We begin with $D = -\Delta$ on $\mathbb{R}^n$. It is well-known that D is sectorial in each of the Bessel-potential spaces $H_r^\beta(\mathbb{R}^n)$ and admits an $\mathcal{H}^\infty$-calculus with $\mathcal{H}^\infty$-angle $\phi_D^\infty = 0$, provided $\beta \in \mathbb{R}$ and $p \in (1, \infty)$. This is an easy consequence of, e.g., the Fourier-multiplier theorem of Mikhlin. The same result is valid in the scale of Besov-spaces $B_{r,\rho}^\beta(\mathbb{R}^n)$ for all $\beta \in \mathbb{R}$ and $r, \rho \in [1, \infty]$. For $1 < r < \infty$ the latter can be obtained by real interpolation between Bessel-potential spaces, but instead one may use Dore's result, i.e., Theorem 2 in the case of scalar-valued functions, to see this for all $r \in [1, \infty]$. Let Y^β denote any of these spaces.

Then we use the $\mathcal{H}^\infty$-calculus for D to obtain operator-valued functions $F_i(\lambda) := r_i(\lambda, D)$ in Y^β, $i = 1, 2, 3$. These functions are holomorphic except for $\lambda \in (-\infty, 0]$, they are also bounded on each sector Σ_ϕ, $\phi < \pi$.

Next fix a finite interval $J = [0, T]$ and let $X = C(J; Y^\beta)$. Define the operator A as $Au = du/dt$ with domain $\mathcal{D}(A) = \{u \in C^1(J; Y^\beta) : \; u(0) = 0\}$. It is also well-known -and in fact easy to prove- that A is pseudo-sectorial with spectral angle $\phi_A = \pi/2$ and $0 \in \rho(A)$, however $\mathcal{D}(A)$ is not dense in X. Theorem 2 yields $F_i(A) \in \mathcal{B}(D_A(\alpha, p))$, for each $\alpha \in (0, 1)$, $p \in [1, \infty]$ and $i = 1, 2, 3$. Note that for $p = \infty$ we have $D_A(\alpha, \infty) = {}_0C^\alpha(J; Y^\beta)$, where the subscript 0 means trace 0 at $t = 0$. Defining $u(t) = \int_0^t [r_1(A, D)f](s)ds$, the functional calculus implies that u belongs to the space

$$u \in {}_0C^{1+\alpha}(J; Y^\beta) \cap {}_0C^{1/2+\alpha}(J; Y^{2+\beta}),$$

and solves (4.5). These considerations yield the following maximal regularity result for (4.5).

Theorem 3. *Let Y^β, $\beta \in \mathbb{R}$, denote any of the spaces $H_r^\beta(\mathbb{R}^n)$, $r \in (1,\infty)$, or $B_{r,\rho}^\beta(\mathbb{R}^n)$, $r,\rho \in [1,\infty]$, let $\alpha \in (0,1)$ and $J = [0,T]$.*

Then for each right hand side $f \in X := {}_0C^\alpha(J;Y^\beta)$, there is a unique solution $u : J \to Y^\beta$ for (4.5) of maximal regularity type, i.e., u belongs to the class

$$u \in Z := {}_0C^{1+\alpha}(J;Y^\beta) \cap {}_0C^{1/2+\alpha}(J;Y^{2+\beta}) \cap {}_0C^\alpha(J;Y^{3+\beta}).$$

The solution map $f \mapsto u$ is bounded from X to Z.

5. The Vector-Valued Stein Multiplier Theorem

Let (Ω,μ) and (Σ,ν) be measure spaces, $p,q \in (1,\infty)$. Assume that $\mathcal{T}(t)$ is a positive C_0-semigroup of contractions in $L_p(\Omega)$ which is in addition analytic, and let $-\mathcal{A}$ denote its generator. Then $\mathcal{T}(t)$ extends to a C_0-semigroup $T(t)$ on $X = L_p(\Omega, L_q(\Sigma))$ by the definition

$$T(t)[\varphi \otimes g] = [\mathcal{T}(t)\varphi]g, \quad t \geq 0,\ \varphi \in L_p(\Omega),\ g \in L_p(\Sigma);$$

we call it the **tensor extension** of $\mathcal{T}(t)$; cf. Clément and Egberts [3]. Let A denote the negative generator of $T(t)$, the tensor extension of $\mathcal{A}$. From the papers of Clément and Prüss [4], Hieber and Prüss [10] and Clément and Prüss [5] we know that A admits an $\mathcal{H}^\infty$-calculus with $\mathcal{H}^\infty$-angle $\phi_A^\infty \leq \pi/2$ and also an operator-valued $\mathcal{H}^\infty$-calculus for functions $F \in \mathcal{H}^\infty(\Sigma_\phi;\mathcal{B}(X))$ with $F(\Sigma_\phi) \subset \mathcal{B}(X)$ $\mathcal{R}$-bounded, for each $\phi > \pi/2$. The question we address here is whether we may admit for some $\phi < \pi/2$. More precisely, we prove

Theorem 4. *Let (Ω,μ) and (Σ,ν) be measure spaces, $p,q \in (1,\infty)$. Suppose A on $X = L_p(\Omega;L_q(\Sigma))$ is the tensor extension of the negative generator $-\mathcal{A}$ of a positive analytic C_0-semigroup of contractions on $L_p(\Omega)$.*

Then there exists $\phi < \pi/2$ and $M > 0$ such that for each $F \in \mathcal{H}^\infty(\Sigma_\phi;\mathcal{B}(X))$ commuting with A and such that $F(\Sigma_\phi) \subset \mathcal{B}(X)$ is $\mathcal{R}$-bounded, $F(A) \in \mathcal{B}(X)$ is well-defined and $|F(A)|_{\mathcal{B}(X)} \leq M\mathcal{R}\{F(\Sigma_\phi)\}$, where $\mathcal{R}\{F(\Sigma_\phi)\}$ denotes the $\mathcal{R}$-bound of the set $F(\Sigma_\phi) \subset \mathcal{B}(X)$.

Proof. From Weis [16] we know $\phi_{\mathcal{A}}^\infty < \pi/2$. Therefore $\phi_A^\infty \leq \phi_{\mathcal{A}}^\infty < \pi/2$ in $L_p(\Omega, L_p(\Sigma))$. Interpolating this result with that of Hieber and Prüss we get

$$|A^{is}|_{\mathcal{B}(L_p(\Omega;L_q(\Sigma)))} \leq Me^{(\theta\varphi+(1-\theta)\pi/2)|s|}, \quad s \in \mathbb{R},$$

where

$$\varphi \in (\phi_{\mathcal{A}}^\infty, \pi/2),\ 1 < r < \infty,\ \frac{1}{q} = \frac{\theta}{p} + \frac{1-\theta}{r}.$$

Varying r from 1 to ∞ we cover all values of q, and since the exponent $\theta\varphi + (1-\theta)\pi/2$ is always strictly smaller than $\pi/2$ we see that the power angle θ_A of A in X is smaller than $\pi/2$. Clément and Prüss [5] implies $\phi_A^R \leq \theta_A$ and Kalton and Weis [12] have shown $\phi_A^\infty \leq \phi_A^R$, hence we conclude $\phi_A^\infty < \pi/2$. We may then apply the theorem of Kalton and Weis [12] mentioned in Section 3 to prove the assertion of Theorem 4. □

We obtain the following operator-valued version of the Stein Multiplier theorem [15] as a corollary.

Corollary 3. *In the situation of Theorem 4, let* $E = L_q(\Sigma)$, *and suppose that* $M : (0, \infty) \to \mathcal{B}(E)$ *is strongly measurable,* $M(t)$ *commutes with* A *for a.a.* $t > 0$, *and* $M(0, \infty) \subset \mathcal{B}(E)$ *is* $\mathcal{R}$*-bounded.*

Then with $F(z) = z\widehat{M}(z)$, *the operator defined by*

$$F(A) = A\widehat{M}(A) = \int_0^\infty M(t)Ae^{-At}dt$$

is well-defined and bounded in X.

For a proof we apply Theorem 4 to $F(z)$. Note that $F(z)$ is holomorphic in $\Sigma_{\pi/2}$, commutes with A, and is bounded in each subsector Σ_ϕ, $\phi < \pi/2$. Finiteness of the $\mathcal{R}$-bounds $\mathcal{R}\{F(\Sigma_\phi)\}$ follow from convexity of $\mathcal{R}$-bounds and the contraction principle.

It is an open question whether in Theorem 4 the space $E = L_q(\Sigma)$ can be replaced by a general UMD-space.

Acknowledgment

This work emerged while the first author was visiting the University of Halle-Wittenberg. We wish to thank the Deutsche Forschungsgemeinschaft, Bonn, for financial support.

References

[1] J.B. Baillon and Ph. Clément, *Examples of unbounded imaginary powers of operators.* J. Funct. Anal. **100** (1991), 419–434.

[2] Ph. Clément, B. de Pagter, F.A. Sukochev, and H. Witvliet, *Schauder decompositions and multiplier theorems.* Studia Math. **138** (2000), 135–163. Preprint.

[3] Ph. Clément and P. Egberts, *On the sum of maximal monotone operators.* Diff. Integral Eqns. **3** (1990), 1127–1138.

[4] Ph. Clément and J. Prüss, *Completely positive measures and Feller semigroups.* Math. Ann. **287** (1990), 73–105.

[5] Ph. Clément and J. Prüss, *An operator-valued transference principle and maximal regularity on vector-valued* L_p*-spaces.* In Evolution Equ. and Appl. Physical Life Sci., Lect. Notes Pure and Applied Math. 215, 17–26, Marcel Dekker, New York, 2001.

[6] G. Da Prato and P. Grisvard, *Sommes d'opérateurs linéaires et équations différentielles opérationelles.* J. Math. Pures Appl. **54** (1975), 305–387.

[7] W. Desch, M. Hieber, and J. Prüss, L_p*-theory of the Stokes equation in a half space.* J. Evolution Eqns. **1** (2001).

[8] G. Dore, *L^p-regularity for abstract differential equations.* In H. Komatsu, ed., Functional Analysis and Related Topics, Lecture Notes in Mathematics 1540, Springer Verlag, Berlin, 1993.

[9] J. Escher, J. Prüss, and G. Simonett, *Analytic solutions of the Stefan problem with Gibbs-Thomson correction.* To appear.

[10] M. Hieber and J. Prüss, *Functional calculi for linear operators in vector-valued L_p-spaces via the transference principle.* Adv. Diff. Eqns. **3** (1998), 847–872.

[11] N. Kalton and G. Lancien, *A solution to the problem of L^p-maximal regularity.* Math. Z. **235** (2000), 559–568.

[12] N. Kalton and L. Weis, *The H^∞-calculus and sums of closed operators.* Preprint, 2000.

[13] H. Komatsu, *Fractional powers of operators, VI. Interpolation of non-negative operators and imbedding theorems.* J. Fac. Sci. Univ. Tokyo, Sec. IA **19** (1972), 1–63.

[14] A. McIntosh and A. Yagi, *Operators of type ω without a bounded H^∞ functional calculus.* In Miniconference on Operators in Analysis, Proc. Center Math. Analysis 24, Australian Nat. Univ., Canberra, 1989.

[15] E. Stein, *Topics in Harmonic Analysis related to the Littlewood-Paley Theory.* Ann. Math. Studies 63, Princeton Univ. Press, Princeton, 1970.

[16] L. Weis, *A new approach to maximal L_p-regularity.* In Evolution Equ. and Appl. Physical Life Sci., Lect. Notes Pure and Applied Math. 215, 195–214, Marcel Dekker, New York, 2001.

Philippe Clément
Department of Mathematics
I.T.S.
Delft University of Technology
Mekelweg 4
2628 CD Delft, The Netherlands
E-mail address: ph.p.j.e.clement@@ITS.tudelft.nl

Jan Prüss
Institut für Analysis
Fachbereich Mathematik und Informatik
Martin-Luther-Universität Halle-Wittenberg
Theodor–Lieser–Str. 5
D–06120 Halle, Germany
E-mail address: anokd@@volterra.mathematik.uni-halle.de

Progress in Nonlinear Differential Equations
and Their Applications, Vol. 55, 113–126

A Semigroup Approach to No-Arbitrage Pricing Theory: Constant Elasticity Variance Model and Term Structure Models

Fabrizio Colombo, Massimiliano Giuli, and Vincenzo Vespri

1. Introduction

One of the main topics in modern mathematical finance is the pricing of contingent claims. The standard framework is built on a partial equilibrium analysis, the evolution of the underlying asset price or of the interest rate is described by a stochastic process and then the price of a contingent claim is obtained imposing the no-arbitrage condition, i.e., given the evolution of the exogenous variables (underlying asset price-interest rate) the contingent claim is priced in such a way that there are no opportunities to make money from nothing.

The price of a contingent claim can be interpreted from two different perspectives. It can be viewed as the solution of a Cauchy problem defined by the no-arbitrage Partial Differential Equation (PDE) and by the final payoff of the contingent claim as initial condition (*PDE approach*) or as the expectation of the final payoff under the risk-neutral probability measure nested in the stochastic process for the underlying asset price once the no-arbitrage condition is imposed (*Probabilistic Approach*). The two approaches are related by the well known Feynman-Kac theorem which states that the classical solution of the Cauchy problem admits a stochastic representation, i.e., the expectation of the final payoff under the probability measure associated with the no-arbitrage PDE (the risk-neutral probability measure), see [15].

The second approach in general gives in a simple way the existence of the no-arbitrage price. It is enough that the stochastic differential equation describing the price of the underlying asset or of the interest rate admits a unique weak solution and that the final payoff is integrable with respect to the risk-neutral probability measure to have a well defined no-arbitrage price for the contingent claim, but, unless we know the risk-neutral probability density, we are able to find neither the explicit no-arbitrage price nor an approximating solution. In a general setting, for a general Itô process describing the exogenous variables and for a "generic" contingent claim contract, we do not have a closed form solution and we have to resort to numerical techniques as last resource. To this end, the probabilistic approach is not of great help, many of the techniques employed are concerned with the no-arbitrage PDE. In this perspective the PDE approach seems to be more powerful than the probabilistic approach also to give estimates of the partial derivative of the no-arbitrage price with respect to the underlying asset price (to perform a delta hedging strategy) and to develop a sensitivity analysis of the solution with respect to the data of the problem. The main problem is given by the fact that the no-arbitrage PDE belongs to a class which is difficult to handle via PDE methods, mainly it is degenerate with unbounded coefficients. The existence itself is a problem via standard PDE methods, see [11, 15, 9]. In this paper we tackle the existence and regularity problem of the no-arbitrage price by means of PDE methods in constant elasticity model and term structure models. The techniques are based on the semigroup theory and in particular on contraction semigroups, which have been applied to the class of PDEs considered in this paper in [4]. This paper is intended to be a companion paper of [13] where contingent claim pricing is approached through the theory of analytic semigroups obtaining strong results about the existence and the characterization of the solution. Unfortunately in many valuation problems (and in particular in constant elasticity model and term structure models) we do not have the generation of an analytic semigroup but only of a contraction semigroup so we cannot apply the standard techniques.

Our framework can be described through the following multifactor model. The state variable x evolves as an Itô process in $\mathbf{R}^n$ solving a stochastic differential equation of the form

$$dx = \mu(x,t)dt + \sigma(x,t)dz, \tag{1.1}$$

where z is a standard Brownian motion in $\mathbf{R}^n$; $\mu : \mathbf{R}^n \times [0,\infty) \to \mathbf{R}$ and $\sigma : \mathbf{R}^n \times [0,\infty) \to \mathbf{R}^{n\times n}$, satisfy enough technical regularity conditions to guarantee that (1.1) has a unique solution for a given initial condition $x(0) = x_0 \in \mathbf{R}^n_+$. Depending on the no-arbitrage pricing problem we have that x represents the stochastic process of the underlying asset price (contingent claims), interest rates (term structure models), diffusion coefficients (stochastic volatility models), etc.

In a general setting, derivative securities are defined by two real valued functions $g : \mathbf{R}^n \to \mathbf{R}$ and $h : \mathbf{R}^n \times [0,T] \to \mathbf{R}$ denoting respectively the payoff at expiration of the contract, $t = T$, and the dividend rate at any time $t \in [0,T]$. The no-arbitrage price of the derivative is obtained by finding $f \in C^{2,1}(\mathbf{R}^n \times [0,T))$ (i.e., two times continuously differentiable in x and one in t) solving the associated Cauchy problem:

$$Df(x,t) - r(x,t)f(x,t) + h(x,t) = 0, \quad (x,t) \in \mathbf{R}^n \times [0,T) \tag{1.2}$$

with the boundary condition

$$f(x,T) = g(x,T), \quad x \in \mathbf{R}^n, \tag{1.3}$$

where

$$Df(x,t) = f_t(x,t) + f_x(x,t)\mu(x,t) + \frac{1}{2}tr[\sigma(x,t)\sigma(x,t)^\top f_{xx}(x,t)]. \tag{1.4}$$

Defining properly the functions μ, σ, g, h, r, characterizing the no-arbitrage Partial Differential Equation (PDE) (1.2), we obtain many European contingent claim pricing problems and term structure models analyzed in the literature, see [9, 21].

The existence of the solution to the Cauchy problem (1.2)–(1.3) has been demonstrated in the literature under a set of restrictive conditions on the data of the problem. Take for example the two conditions given in [9] pag. 291-297. Condition 1 essentially requires the following: μ, σ, g, h, r are continuous and satisfy a Hölder condition uniformly in t, μ, σ, r are bounded, h, g satisfy a polynomial growth condition and σ satisfies an ellipticity condition. Condition 2 for the autonomous case requires that μ, σ, g, h, r are C^2 in x and satisfy a growth condition together with their first and second derivatives with respect to x. As it is very well known, many no-arbitrage pricing problems do not satisfy these conditions.

Our framework enables us to handle some pricing problems for which an existence proof through PDE methods is not available in the literature: constant elasticity variance models and some term structure models.

The constant elasticity variance model is characterized by a stochastic process for the underlying asset price which does not give a lognormal probability distribution for the underlying asset price at the expiration date, the main feature of this process is that the volatility changes are negatively related with stock returns giving account of some experimental evidence. In this case we have a closed form solution for the no-arbitrage price only in terms of an expansion of functions.

In literature it is assumed that the interest rate r follows the stochastic process

$$dr(t) = [\alpha_1(t) + \alpha_2(t)r(t)]dt + [\beta_1(t) + \beta_2(t)r(t)]^\nu dz. \tag{1.5}$$

Changing the assumptions on the coefficients $\alpha_1(t), \alpha_2(t), \beta_1(t), \beta_2(t)$, different term structure models are obtained. In this note we consider the following ones:

1. $\beta_1 = 0, \nu = 0.5$, [7]
2. $\nu = 0.5$,[18]
3. $\alpha_1 = \alpha_2 = \beta_1 = 0, \nu = 1$, [8]
4. $\beta_1 = 0, \nu = 1$, [3]
5. $\alpha_2 = \beta_2 = 0, \nu = 1$, [17, 14]
6. $\beta_2 = 0, \nu = 1$, [20].

Applying the no-arbitrage assumption, it is possible (see, for instance [9, 21])to get the following equation for the price f of a zero coupon treasury bond.

$$\begin{cases} f_t(t,r) = \frac{1}{2}[\beta_1(t) + \beta_2(t)r]^{2\nu} f_{rr}(t,r) + & \\ +[\alpha_1(t) + \alpha_2(t)r]f_r(t,r) - rf(t,r) & t \in (0,T), \quad r \in \mathbf{R}_+ \\ f(0,r) = 1 & r \in \mathbf{R}_+ \\ f(t,0) = 1 & t \in (0,T) \end{cases}, \tag{1.6}$$

Also in this case in general we have a closed form solution only in terms of an expansion of special functions.

The aim of this note to study in a unitary way such wide class of Cauchy problems using the semigroup approach. To reach our goal we apply the preliminary results of [4]. It is our intention to prove in a forthcoming paper more smoothing properties for such class of operators. In such a way, we should be able not only to get in a simpler way the regularity results known in the existing literature (see, for instance, [6, 16]) but also to approximate numerically the solutions in a more efficient way.

Lastly the title of the paper calls for a discussion of our approach with respect to the semigroup pricing approach developed in [12, 10, 9]. Actually in [12] (see also [1]) it is observed that because of the intertemporal no-arbitrage condition, there exists a set of operators mapping future dividends into current prices and these operators satisfy the semigroup property. The author then observes that the no-arbitrage price satisfies a differential form to which the infinitesimal generator of the surrounding process is associated, if it exists. In this paper we reverse the point of view, we start from the no-arbitrage second order operator (1.4) characterizing the economy and we show that it generates, under some conditions, a Contraction Semigroup which fully describes the solution of the Cauchy problem and therefore the no-arbitrage price. This allows us to prove existence of the solution of the Cauchy problem (1.2)–(1.3) and to define estimates of the solution of the no-arbitrage price and of its derivatives.

2. Existence and Uniqueness for a Class of Parabolic equations

In [4] the following class of evolution equations have been considered:

$$\begin{cases} u_t(t,x) = A(t,x)u(t,x) + f(t,x) & t \in (0,T), \quad x \in \mathbf{R}^n \\ u(0,x) = u_0(x) & x \in \mathbf{R}^n \end{cases}, \tag{2.1}$$

where $A(t,x)$ is the linear differential operator defined by

$$\begin{aligned} Au(t,x) := & \sum_{i,j=1}^{n} D_i(\psi_i(t,x)\psi_j(t,x)q_{ij}(t,x)D_j u(t,x)) + \\ & + \sum_{i=1}^{n} \gamma_i(t,x)D_i u(t,x) + \sum_{i=1}^{n} D_i(\alpha_i(t,x))u(t,x) - \theta(t,x)u(t,x). \end{aligned}$$

Throughout the paper the following assumptions are made.

(H1)

Assumption 2.1. *$q_{ij} \in L^\infty(\mathbf{R}^n \times [0,T])$ and satisfy the ellipticity condition*

$$\sum_{i,j=1}^{n} q_{ij}(x,t)\xi_i\xi_j \geq \nu|\xi|^2, \quad a.e \ \ (x,t) \in \mathbf{R}^n \times [0,T]; \ \ \forall \xi \in \mathbf{R}^n \backslash \{0\} \tag{2.2}$$

with $\nu > 0$.

(H2)

Assumption 2.2. *$\theta(x,t)$ is bounded from below. Without loss of generality we require that*

$$\theta(x,t) \geq 1; \quad \forall (x,t) \in \mathbf{R}^n \times [0,T].$$

(H3)

Assumption 2.3. *γ_i, α_i are continuous and differentiable a.e., $i = 1, \ldots, n$.*

(H4)

Assumption 2.4. *ψ_i are continuous a.e., differentiable a.e. and nonnegative $i = 1, \ldots, n$.*

(H5)

Assumption 2.5. *$u_0 \in L^p(\mathbf{R}^n)$.*

(H6)

Assumption 2.6. *θ is continuous and differentiable a.e., $|\psi_i(x,t)\theta^{-\frac{3}{2}}(x,t)D_i\theta(x,t)|$ is bounded, $\lim_{\|x\|\to+\infty} |\psi_i(x,t)\theta^{-\frac{3}{2}}(x,t)D_i\theta(x,t)| = 0$.*

Before stating the main results let us introduce some Banach spaces that will be used in the sequel. Given a function ξ, we define for a given $p \in [1,+\infty]$ the space

$$L^p_\xi(\mathbf{R}^n) = \{f : \mathbf{R}^n \to \mathbf{R}, \quad \text{measurable}; \quad f\xi \in L^p(\mathbf{R}^n)\}$$

with the norm $\|f\|_{L^p_\xi(\mathbf{R}^n)} = \|f\xi\|_{L^p(\mathbf{R}^n)}$, and the space

$$H^{k,p}_\xi(\mathbf{R}^n) = \{f : \mathbf{R}^n \to \mathbf{R}, \quad \text{measurable}; \quad f\xi \in H^{k,p}(\mathbf{R}^n)\}$$

with the norm $\|f\|_{H^{k,p}_\xi(\mathbf{R}^n)} = \|f\xi\|_{H^{k,p}(\mathbf{R}^n)}$.

Given the functions $\psi_i : \mathbf{R}^n \to \mathbf{R}_+$ $(i = 1, \ldots, n)$ and $\gamma : \mathbf{R}^n \to \mathbf{R}_+$ we define the space

$$H^{1,p}_{\psi,\gamma}(\mathbf{R}^n) = \{f \quad \text{measurable}; \ \gamma f, \ \psi_i D_i f \in L^p(\mathbf{R}^n); \ i = 1, \ldots, n\}$$

with the norm $\|f\|_{H^{1,p}_{\psi,\gamma}(\mathbf{R}^n)} = \|\gamma f\|_{L^p(\mathbf{R}^n)} + \sum_{i=1}^n \|\psi_i D_i f\|_{L^p(\mathbf{R}^n)}$. Lastly we define $Q(t) := \mathbf{R}^n \times [0,t]$.

Some no-arbitrage PDEs and the related Cauchy problems belong to the class (2.1), expressing the final payoff as the initial datum and rewriting it with respect to the time to expiration, $\tau = T - t$ in the notation of the Introduction. Existence and uniqueness for the no-arbitrage Cauchy problem is established depending on the space to which the initial datum (final payoff of the contract g) and the continuous dividend h belong.

In what follows we report some results for the Cauchy problem (2.1) concerning existence, uniqueness and its estimates. We limit ourselves to give an outline of the proof. A detailed proof of what we state here can be found in [4].

Consider first the case where the functions g and h belong to L^p with $p \in [2,\infty)$.

Theorem 2.7. *Let $f = f_0 + \sum_{i=1}^{n} D_i f_i$ for f_0 in $L^p(Q(T))$ and f_i in $L^p_{\psi_i^{-1}}(Q(T))$ $(i = 1, \ldots, n)$. Let assumptions (H1)–(H5) be satisfied and that*

$$\theta + \frac{1}{p} \sum_{i=1}^{n} D_i \gamma_i - \frac{p-1}{p} \sum_{i=1}^{n} D_i \alpha_i \text{ is bounded from below,} \tag{2.3}$$

then equation (2.1) admits a solution in $L^p(Q(T))$. If $f_i = 0$, $i = 1, \ldots, n$, then

$$\sup_{t \in (0,T)} \frac{1}{p} \|u(t)\|^p_{L^p(\mathbf{R}^n)} + (\nu(p-1))^{\frac{1}{p}} \|\psi u^{\frac{p-2}{2}} Du\|^{\frac{2}{p}}_{L^2(Q(T))} \leq$$
$$\leq \|f_0\|_{L^p(Q(T))} + \frac{1}{p} \|u(0)\|^p_{L^p(\mathbf{R}^n)}.$$

Moreover if $f_i \neq 0$ then

$$\sup_{t \in (0,T)} \|u(t)\|^p_{L^p(\mathbf{R}^n)} + \|\psi u^{\frac{p-2}{2}} Du\|^{\frac{2}{p}}_{L^2(Q(T))} \leq$$
$$\leq C(p) \Big(\|f_0\|_{L^p(Q(T))} + \frac{1}{p} \|u(0)\|^p_{L^p(\mathbf{R}^n)} + \sum_{i=1}^{n} \|f_i\|^p_{L^p_{\psi_i^{-1}}(Q(T))} \Big), \tag{2.4}$$

where $C(p)$ is a suitable constant.

Outline of the proof. First it is considered a sequence of approximating equations with bounded, regular and no degenerate coefficients. Applying Lax-Milgram theorem it is possible to have existence, uniqueness and uniform a priori estimates for the solutions of the approximating problems. Then by passing to the limit one gets existence of a solution satisfying (2.4).

Theorem 2.8. *Let $f = f_0 + \sum_{i=1}^{n} D_i f_i$ for f_0 in $L^p(Q(T))$ and f_i in $L^p_{\sqrt{\theta}\psi_i^{-1}}(Q(T))$ $(i = 1, \ldots, n)$. Let assumptions (H1)–(H6) be satisfied. If there exists $\eta \in (0,1)$ such that*

$$(1-\eta)\theta^p + \frac{1}{p} \sum_{i=1}^{n} D_i(\gamma_i \theta^{p-1}) - \frac{p-1}{p} \sum_{i=1}^{n} D_i(\alpha_i \theta^{p-1}) \text{ is bounded from below,} \tag{2.5}$$

$$u_0 \in L^p_{\theta^{\frac{p-1}{p}}}(\mathbf{R}^n), \tag{2.6}$$

then there exists a positive constant $C(p)$ such that

$$\sup_{t \in (0,T)} \|\theta^{\frac{p-1}{p}} u(t)\|^p_{L^p(\mathbf{R}^n)} + \|\psi u^{\frac{p-2}{2}} \theta^{\frac{p-1}{2}} Du\|^2_{L^2(Q(T))} + \|\theta u\|^p_{L^p(Q(T))} \leq$$
$$\leq C(p) \Big(\|f_0\|^p_{L^p(Q(T))} + \sum_{i=1}^{n} \|f_i\|^p_{L^p_{\sqrt{\theta}\psi_i^{-1}}(Q(T))} + \|\theta^{\frac{p-1}{p}} u(0)\|^p_{L^p(\mathbf{R}^n)} \Big).$$

Outline of the proof. Again we consider the approximate problems. This time as test function we have to choose $\theta_m^{p-1} u_m |u_m|^{p-2}$ where u_m and θ_m represent respectively the approximate solutions and the approximate coefficient. Again via Lax-Milgram theorem we have the existence of solutions satisfying uniform a priori estimates. The statement is obtained passing to the limit.

Remark 2.9. In [4] the constant $C(p)$ is explicitely computed and does not explode when p goes to infinity. Therefore passing to the limit it is possible to prove that an estimate similar to (2.4) holds also in the case L^∞.

Remark 2.10. We stress that the differentiability of θ is not necessary. What we need is that $\theta = \theta_1 + \theta_2$, where θ_1 is differentiable almost everywhere, strictly greater than 1 and satisfying (H6) while θ_2 is a bounded measurable function.

Remark 2.11. It is possible to prove the previous results in the case of a suitable Banach weighted space $L^p_\chi(Q(T))$. Assume that there exists $\eta > \Xi > 0$ and a constant C such that for each $(x,t) \in Q(T)$ we have that

$$\sum_{i,j=1}^{n} \frac{4}{\nu} \|q_{ij}\|_{L^\infty(Q(T))} |\psi_i(x,t)|^2 (N+1)^2 (1+x^2)^{-1} + \sum_{i=1}^{n} |\gamma_i(x,t)|$$

$$(N+1)(1+x^2)^{-\frac{1}{2}} + \sum_{i=1}^{n} |\alpha_i(x,t)|(N+1)(1+x^2)^{-\frac{1}{2}} \leq \Xi\theta(x,t) + C. \tag{2.7}$$

In view of the applications we stress that under such an assumption the function $(1+x^2)^{-\xi}$ is an admissible weight where ξ is a positive number.

Let us state results concerning uniqueness of parabolic solutions.

Theorem 2.12. *Let* $p \geq 2$. *Assume that* (H1)–(H6) *and* (2.7) *hold. Assume moreover that in a neighborhood where* ψ_i *vanishes that*

$$|\gamma_i| + |\alpha_i| \leq C\psi_i\theta, \tag{2.8}$$

where C *is a positive constant. Then equation* (2.1) *admits a unique solution in*

$$L^p([O,T]; W^{1,p}_{\psi^{-1}\sqrt{\theta}\chi,\gamma})(\mathbf{R}^N)) \cap C([O,T]; L^p_{\theta\chi}(\mathbf{R}^N)),$$

where $\chi = (1+x^2)^{-\xi}$ *and* ξ *positive constant.*

Outline of the proof. Assume that there are two different solution v and w. Then $z = v - w$ satisfies the parabolic equation with zero data. As z is not an admissible test function we have to choose suitable approximating functions for which some a apriori estimates hold. Passing to the limit one gets that $z = 0$, i.e., the uniqueness of the solution.

Let us state the result concerning the case $1 < p < 2$. We recall that in such a context the existence and uniqueness is in Stampacchia sense (see [19], see also [4]).

Theorem 2.13. *Let* (H1)–(H6), (2.3), (2.7) *and* (2.8) *hold. Let* $f = f_0$ *with* $f_0 \in L^p(Q(T))$. *Then equation* (2.1) *admits a unique solution in Stampacchia sense in* $L^p(Q(t))$. *Moreover*

$$\sup_{t\in(0,T)} \frac{1}{p}\|u(t)\|^p_{L^p(\mathbf{R}^n)} + (\nu(p-1))^{\frac{1}{p}}\|\psi u^{\frac{p-2}{2}} Du\|^{\frac{2}{p}}_{L^2(Q(T))} \leq$$

$$\leq \|f_0\|_{L^p((Q(T))} + \frac{1}{p}\|u_0\|^p_{L^p(\mathbf{R}^n)}.$$

Assume that (H1)–(H6) *and* (2.5)–(2.6) *hold. Then there exists a positive constant* $C(p)$ *such that*

$$\sup_{t\in(0,T)} \|\theta^{\frac{p-1}{p}} u(t)\|^p_{L^p(\mathbf{R}^n)} + \|\psi u^{\frac{p-2}{2}}\theta^{\frac{p-1}{2}} Du\|^2_{L^2(Q(T))} + \|\theta u\|^p_{L^p(Q(T))} \leq$$

$$\leq C(p)\Big(\|f_0\|^p_{L^p(Q(T))} + \|\theta^{\frac{p-1}{p}} u_0\|^p_{L^p(\mathbf{R}^n)}\Big).$$

Outline of the proof. It is a consequence of duality techniques and Riesz representation theorem.

3. Constant Elasticity Variance Models

In [5, 6] the following diffusion process for the underlying asset price was proposed:

$$dS = \mu S dt + \sigma S^\rho dz.$$

with $0 \leq \rho \leq 1$. This type of process for the underlying asset price was motivated by the observation that the standard lognormal diffusion process assumed in the Black and Scholes framework ($\rho = 1$) was not supported by the data being the volatility changes negatively correlated with stock returns.

In what follows for the sake of simplicity we consider only the case $n = 1$ but all the results can be applied to multi-dimensional problems. Imposing the condition of no-arbitrage the following Cauchy problem to price an European Call Option is obtained:

$$\begin{cases} C_t(t,S) = \frac{1}{2}\sigma^2 S^\alpha C_{SS}(t,S) + r(t)SC_S(t,S) - r(t)C(t,S) & \\ & t\in(0,T), \quad S\in\mathbf{R}_+ \\ C(0,S) = (S-E)_+ & S\in\mathbf{R}_+ \\ C(t,0) = 0 & t\in(0,T) \end{cases} \tag{3.1}$$

where $\alpha = 2\rho$ and

Assumption 3.1. $r(t)$ *is a nonnegative and continuous function in* $[0,T]$.

A closed form solution is obtained in [6, 16] as an expansion in terms of the gamma density/distribution function. This approach allows to prove some regularity properties exhibiting explicitely the solution. In our approach, that we consider more general and more flexible, we deduce the regularity properties from general theorems.

Actually in what follows we establish existence and uniqueness of the solution through the results reported above.

First of all, from the Maximum Principle we have that the solution, if it exists, is nonnegative.

In order to apply the results of the previous section we extend the function $C(t,S)$ to the whole $(0,T)\times\mathbf{R}$ in the following way: $C(t,-S)=-C(t,S)$. For sake of simplicity we still denote the new function with $C(t,S)$.

Now $C(S)$ satisfies the following equation:

$$\begin{cases} C_t(t,S)=\frac{1}{2}\sigma^2|S|^\alpha C_{SS}(t,S)+r(t)SC_S(t,S)-r(t)C(t,S) & \\ & t\in(0,T),\quad S\in\mathbf{R} \\ C(0,S)=(S-E)_+-(-E-S)_+ & S\in\mathbf{R} \end{cases} \tag{3.2}$$

In what follows we need also estimates on C_S. After standard calculations one can check easily that C_S satisfies the equation

$$\begin{cases} C_{S_t}(t,S)=\frac{1}{2}\sigma^2(|S|^\alpha C_{SSS}(t,S))+ & \\ +(\frac{1}{2}\alpha\sigma^2 sign(S)|S|^{\alpha-1}+r(t)S)C_{SS}(t,S) & t\in(0,T),\quad S\in\mathbf{R}\ , \\ C_S(0,S)=H(S-E)-H(-S-E) & S\in\mathbf{R} \end{cases} \tag{3.3}$$

where H is the Heavside function. Again via the Maximum Principle we get

$$C_S(t,S)\in L^\infty((0,T)\times\mathbf{R}) \tag{3.4}$$

We remark that property (3.4) is central in the Black-Scholes analysis.

Rewrite now the equation (3.2) in variational form:

$$\begin{cases} C_t(t,S)-\frac{1}{2}\sigma^2(|S|^\alpha C_S(t,S))_S= & \\ =(r(t)S-\frac{1}{2}sign(S)\alpha\sigma^2|S|^{\alpha-1})C_S(t,S)-r(t)C(t,S) & t\in(0,T),\quad S\in\mathbf{R}\ , \\ C(0,S)=(S-E)_+-(-E-S)_+ & S\in\mathbf{R} \end{cases} \tag{3.5}$$

Using the notation of the previous Section we have:

$$\psi(S)=|S|^{\frac{\alpha}{2}},\ q=\frac{1}{2}\sigma^2,\ \gamma(t,S)=r(t)S-\frac{1}{2}sign(S)\alpha\sigma^2|S|^{\alpha-1},\ \theta(t)=r(t).$$

Consider first the case $0<\alpha\le 1$, $(0\le\rho\le\frac{1}{2})$. Assumptions (H1), (H2), (H4), (2.3) are easily verified. Condition (H6) is verified because θ is independent on S.

Condition (H3) is not totally verified because γ is differentiable a.e. but not continuous at the origin. Nevertheless the results proved in [4] can be applied because at the origin the solution vanishes (the solution is even and by (3.4) is Lipschitz continuous too) and therefore all the techniques used to prove the results of Section 2 work.

The initial datum does not belong to any $L^p(\mathbf{R})$. This difficulty can be easily overcame considering the weight function $(1+S^2)^{-\frac{1}{2}}$. In this way assumption (H5) is verified too.

Therefore applying the results of the Section 2 we get the existence of a solution. Moreover we have:

$$C(t,S)(1+S^2)^{-\frac{1}{2}} \in L^\infty((0,T)\times \mathbf{R}) \tag{3.6}$$

Combining (3.6),(3.4) and the nonnegativity of the solution of (3.1) we get that for any $t, S \geq 0$

$$0 \leq C(t,S) \leq S \tag{3.7}$$

Therefore we have that the solution is characterized by a quasilinear growth.

In this case it is not difficult to prove the convexity of the solution.

Actually after standard calculations one can check easily that C_{SS} satisfies the equation

$$\begin{cases} C_{SSt}(t,S) = \frac{1}{2}\sigma^2(S^\alpha C_{SSSS}(t,S)) + \\ +(\alpha\sigma^2 sign(S)|S|^{\alpha-1} + r(t)S)C_{SSS}(t,S) + \\ +(\alpha(\alpha-1)\sigma^2|S^{\alpha-2} + r(t))C_{SS}(t,S) \quad t\in(0,T), \;\; S\in\mathbf{R} \\ C_{SS}(0,S) = \delta(S-E) + \delta(-S-E) \quad S\in\mathbf{R} \end{cases}, \tag{3.8}$$

where δ is the Dirac function. Again via the Maximum Principle we get that C_{SS} is nonnegative, i.e., the solution is convex.

The case $1 \leq \alpha \leq 2$ $(\frac{1}{2} \leq \rho \leq 1)$ is slightly different. The main difficulty is that condition (2.3) is not verified. To overcome this difficulty we approximate this equation with equations with nondegenerate and unbounded coefficients. Let C^m be the solution of

$$\begin{cases} C^m_t(t,S) = \frac{1}{2}\sigma^2|S^m(S)|^\alpha C^m_{SS}(t,S) + \\ +r(t)S^m(S)C^m_S(t,S) - r(t)C^m(t,S) \quad t\in(0,T), \;\; S)\in\mathbf{R} \\ C^m(0,S) = (S-E)_+ - (-E-S)_+ \quad S\in\mathbf{R} \end{cases} \tag{3.9}$$

where $S^m(S)$ is equal to $-m$ if $S \leq -m$, to S if $-m \leq S \leq m$, and to m if $m \leq S$. For well known regularity results one has existence and uniqueness of such a solution C^m. After standard calculations one can check easily that C^m_S satisfies the equation

$$\begin{cases} C^m_{S\,t}(t,S) = \frac{1}{2}\sigma^2(|S^m|^\alpha C^m_{SSS}(t,S)) + \\ +(\frac{1}{2}\alpha\sigma^2(|S^m|^\alpha)_S + r(t)S^m)C^m_{SS}(t,S) \quad t\in(0,T), \;\; S\in\mathbf{R} \\ C^m_S(0,S) = H(S-E) - H(-S-E) \quad S\in\mathbf{R} \end{cases}, \tag{3.10}$$

Via the Maximum Principle we get that $C^m_S(t,S)$ are uniformly equibounded. Using this piece of information we have that the function $C^m(t,S)$ satisfies

$$\begin{cases} C^m_t(t,S) = \frac{1}{2}\sigma^2(|S^m|^\alpha C^m_S(t,S))_S + \\ +r(t)S^m C^m_S(t,S) - r(t)C^m(t,S) + g^m(t,S) & t\in(0,T), \;\; S\in\mathbf{R}_+ \\ C^m(0,S) = (S-E)_+ & S\in\mathbf{R}_+ \\ C^m(t,0) = 0 & t\in(0,T) \end{cases}, \tag{3.11}$$

where $g^m(t,S) = -\frac{1}{2}\alpha\sigma^2(|S^m|^\alpha)_S C^m_S(t,S)$. Now, g^m has a sublinear growth (we recall that C^m_S are equibounded functions) and equation (3.11) satisfies the conditions of the Theorems stated in Section 2 (obviously endowed in suitable weighted spaces). Therefore the solutions C^m satisfy uniform estimates in such weighted spaces. Passing to the limit we have existence of a solution C.

Uniqueness of the solution is more delicate because (2.8) does not hold. We get this property via the Maximum Principle. Let u and v be two week solutions of (3.1) having a linear growth at infinity. Let $w = \frac{u-v}{1+S^2}$. Then w satisfies the equation

$$\begin{cases} w_t(t,S) = \frac{1}{2}\sigma^2 S^\alpha w_{SS}(t,S)+ & \\ +(r(t)S + 2\sigma^2\frac{S^{\alpha+1}}{1+S^2})w_S(t,S)+ & \\ +(2\sigma^2\frac{S^{\alpha+2}}{(1+S^2)^3}) + \sigma^2\frac{S^\alpha}{(1+S^2)^2}) + 2r(t)\frac{S^2}{1+S^2} & -r(t))w(t,S) \\ & t\in(0,T),\ S\in\mathbf{R}_+ \\ w(0,S) = 0 & S\in\mathbf{R}_+ \\ w(t,0) = 0 & t\in(0,T) \end{cases} \tag{3.12}$$

By the Maximum Principle $w = 0$, i.e., there is a unique solution of equation (3.1).

Before concluding this section we recall that in [2] it is considered the problem of numerical approximations of equation (3.1).

4. Term Structure Models

Following [9] we can consider a single factor term structure model and the interest rate r follows the stochastic process

$$dr(t) = [\alpha_1(t) + \alpha_2(t)r(t)]dt + [\beta_1(t) + \beta_2(t)r(t)]^\nu dz. \tag{4.1}$$

where the coefficients $\alpha_1(t), \alpha_2(t), \beta_1(t), \beta_2(t)$, are the following ones:

1. $\beta_1 = 0, \nu = 0.5$, [7]
2. $\nu = 0.5$,[18]
3. $\alpha_1 = \alpha_2 = \beta_1 = 0, \nu = 1$, [8]
4. $\beta_1 = 0, \nu = 1$, [3]
5. $\alpha_2 = \beta_2 = 0, \nu = 1$, [17, 14]
6. $\beta_2 = 0, \nu = 1$, [20].

Arguing as in Section 1 is possible to prove that a zero coupon bond satisfies the equation:

$$\begin{cases} f_t(t,r) = \frac{1}{2}[\beta_1(t) + \beta_2(t)r]^{2\nu} f_{rr}(t,r)+ & \\ +[\alpha_1(t) + \alpha_2(t)r]f_r(t,r) - rf(t,r) & t\in(0,T),\quad r\in\mathbf{R}_+ \\ f(0,r) = 1 & r\in\mathbf{R}_+ \\ f(t,0) = 1 & t\in(0,T) \end{cases}, \tag{4.2}$$

By the maximum principle we get that the solution, if exists, is nonnegative and does not exceed 1.

In order to apply the results of the previous section, define a new function $u(t,r) = f(t,r) - 1$ to get

$$\begin{cases} u_t(t,r) = \frac{1}{2}[\beta_1(t) + \beta_2(t)r]^{2\nu} u_{rr}(t,r) + \\ +[\alpha_1(t) + \alpha_2(t)r]u_r(t,r) - ru(t,r) - r & t \in (0,T), \quad r \in \mathbf{R}_+ \\ u(0,r) = 0 & r \in \mathbf{R}_+ \\ u(t,0) = 0 & t \in (0,T) \end{cases}, \tag{4.3}$$

Moreover we extend the function $u(t,r))$ to the whole $(0,T) \times \mathbf{R}$ in the following way: $u(t,-r) = -u(t,r)$. For sake of simplicity we still denote the new function with $u(t,r)$. The extended function satisfies the equation

$$\begin{cases} u_t(t,r) = \frac{1}{2}(\beta_1(t) + \beta_2(t)|r|)^{2\nu} u_{rr}(t,r) + \\ +[sign(r)\alpha_1(t) + \alpha_2(t)r]u_r(t,r) - ru(t,r) - r & t \in (0,T), \quad r \in \mathbf{R} \\ u(0,r) = 0 & r \in \mathbf{R} \end{cases}, \tag{4.4}$$

Rewrite now the equation in variational form:

$$\begin{cases} u_t(t,r) + (\frac{1}{2}(\beta_1(t) + \beta_2(t)|r|)^{2\nu} u_r(t,r))_r = \\ -2\nu sign(r)\beta_2(\beta_1(t) + \beta_2(t)|r|)^{2\nu-1})u_r(t,r) + \\ +(sign(r)\alpha_1(t) + \alpha_2(t)r)u_r(t,r) - ru(t,r) - r & t \in (0,T), \quad r \in \mathbf{R} \\ u(0,r) = 0 & r \in \mathbf{R} \end{cases}, \tag{4.5}$$

Conditions (H1)–(H2) and (H4)–(H6) are easily verified. Analogously conditions (2.5) and (2.6) are satisfied. Condition (H3) is not totally verified because γ is differentiable a.e. but not continuous at the origin. Nevertheless the results proved in [4] can be applied because at the origin the solution vanishes and therefore as remarked in the previous section all the results of Section 2 work. The known function has a sublinear growth so to apply the results of the previous section we have to choose as weight function $(1 + r^2)^{-\frac{1}{2}}$. The condition on uniqueness are satisfied too. Therefore we get the existence and uniqueness of a nonnegative bounded weak solutions of equation (4.5).

Remark 4.1. In order to prove better regularity properties for the solutions of these equations arising from financial Mathematics we are extending the results of [4] in a forthcoming paper.

Acknowledgment

We wish to thank Dr. G. Cupini for useful discussions on this topic and the anonymous referee for several useful suggestions.

References

[1] E. Barucci, F. Gozzi, and V. Vespri, *On a semigroup approach to no-arbitrage pricing theory.* Proceedings of the Conference on Stochastic Analysis, Random Fields and Applications, Ascona (Switzerland),September 1996. Progress in Probability, Birkhauser-Verlag, (Basel/Switzerland), (1999) 1-14.

[2] E. Barucci , S. Polidoro, and V. Vespri, *Some results on partial differential equations and Asian options.* M3AS 11 (2001), 475–497.

[3] M. Brennan and E. Schwartz, *Analyzing convertible bonds.* J. Financial and Quantitative Analysis **17** (1982), 75–100.

[4] F. Colombo, M. Giuli, and V. Vespri, *Generation of smoothing semigroups, by elliptic operators and some applications to mathematical finance.* Com. Appl. Anal. **3** (1999), 283–299.

[5] J.C. Cox, and S. Ross, *The valuation of options for alternative stochastic process.* J. Fin. Econ. **3** (1976), 145–166.

[6] J.C. Cox, and M. Rubinstein, *Options Markets.* Prentice Hall, 1985.

[7] J.C. Cox, J. Ingersoll, and S. Ross, *A theory of the term structure of interest rate.* Econometrica **53** (1985), 385–467.

[8] M.A. Dothan, *On the term structure of interest rates.* J. Finan. Econ. **6** (1978), 59–69.

[9] D. Duffie, *Dynamic Asset Pricing Theory.* Princeton University Press, Princeton (1996).

[10] D. Duffie, and M. Garman, *Arbitraje intertemporal y valoraci ón markov de las acciones.* Cuadernos Economicos de ICE, (1991), 37–60.

[11] A. Friedman, *Stochastic Differential Equations and Applications.* Academic Press, New York, 1975.

[12] M. Garman, *Towards a semigroup pricing theory.* Journal of Finance **40** (1985), 847–862.

[13] F. Gozzi, R. Monte, and V. Vespri, *Generation of analytic semigroups for degenerate elliptic operators arising in financial mathematics.* Diff. Int Equations **15** (2002), 1085–1128.

[14] T. Ho, and S. Lee, *Term structure movements and pricing interest rate contingent claims.* J. Fin. **82** (1986), 1129–1142.

[15] I. Karatzas, and S. Shreve, *Brownian Motion and Stochastic Calculus.* Springer-Verlag, New York, 1988.

[16] L.G. Mac Millan, *Options as a strategic investment.* New York Inst. Fin. NY, 1980.

[17] R. Merton, *On the pricing of corporate debt: the risk structure of interest rate.* J. Finance **29** (1974), 449–470.

[18] N. Pearson, and T. Sun, *An empirical examination of the Cox, Ingersoll and Ross model of term structure of interest rates using the method of maximum likelihood.* J. Fin. **54** (1994), 929–959.

[19] G. Stampacchia, *Le probléme de Dirichlet pour les équations elliptiques du second ordre á coefficients discontinus.* Ann. Inst. Fourier Grenoble **15** (1965), 141–162.

[20] O.A. Vasicek, *An equilibrium characterization of the term structure.* J. Fin. Econ. **9** (1977), 177–188.

[21] P. Wilmott, J. Dewynne, and S. Howison, *Option Pricing: Mathematical Models and Computation.* Oxford Financial Press, Oxford, UK, 1993.

Fabrizio Colombo
Dipartimento di Matematica
Politecnico di Milano
Via Bonardi n. 9
I-20133 Milano, Italy
E-mail address: fabcol@@mate.polimi.it

Massimiliano Giuli
Facoltà di Economia
Università dell'Aquila
I-67100 L'Aquila, Italy
E-mail address: giuli@@univaq.it

Vincenzo Vespri
Dipartimento di Matematica
Università di Firenze
Viale Morgagni 67/a
I-50134 Firenze, Italy
E-mail address: vespri@@math.unifi.it

Progress in Nonlinear Differential Equations
and Their Applications, Vol. 55, 127–139

Perturbations of Ornstein–Uhlenbeck Operators: an Analytic Approach

Giuseppe Da Prato

1. Introduction and setting of the problem

We are given a separable Hilbert space H (with norm $|\cdot|$ and inner product $\langle\cdot,\cdot\rangle$), two linear operators $A : D(A) \subset H \to H$, $C : H \to H$ and a nonlinear mapping $F : H \to H$, such that:

Hypothesis 1.1.

(i) *A is the infinitesimal generator of a strongly continuous semigroup e^{tA} in H. There exists $\omega \in \mathbb{R}$ such that*

$$\|e^{tA}\| \le e^{\omega t}, \forall\, t \ge 0. \tag{1.1}$$

(ii) *$C \in L^+(H)$ (1) and for any $t > 0$ the linear operator Q_t, defined as*

$$Q_t x = \int_0^t e^{sA} C e^{sA^*} x ds, \quad x \in H,$$

is of trace class.

(iii) *$F \in C_b^{1,1}(H;H)$ and $F(0) = 0$.*

We shall use the following notations. If E and F are two Banach spaces we denote by $C_b^k(E;F)$, $k = 0,1,\dots$ the set of all mappings $\varphi : E \to F$ continuously differentiable and bounded together with their derivatives of order less than k and such that $\|\varphi\|_k := \sum_{h=0}^k \sup_{x\in E} |D^h\varphi(x)| < +\infty$.

Moreover $C_b^{1,1}(E;F)$ is the space of all functions $\varphi \in C_b^1(E;F)$ such that $D\varphi$ is Lipschitz continuous. We set

$$[\varphi]_{1,1} = \sup_{x\ne y} \frac{|D\varphi(x) - D\varphi(y)|}{|x-y|}, \quad \varphi \in C_b^{1,1}(E;F).$$

$C_b^{1,1}(E;F)$ is a Banach space with the norm

$$\|\varphi\|_{1,1} = \|f\|_1 + [\varphi]_{1,1}, \quad \varphi \in C_b^{1,1}(E;F).$$

We set $C_b^k(E;\mathbb{R}) = C_b^k(E)$ and $C_b^{1,1}(E;\mathbb{R}) = C_b^{1,1}(E)$.

---

[1] $L(H)$ is the Banach algebra of all linear bounded operators from H into H endowed with the sup norm $\|T\| = \sup\{|Tx| : x \in H, |x| = 1\}$, and $L^+(H) = \{T \in L(H) : \langle Tx, x\rangle \ge 0, \langle Tx, y\rangle = \langle x, Ty\rangle, \forall\, x, y \in H\}$.

Under Hypothesis 1.1–(i)–(ii) the Ornstein–Uhlenbeck semigroup

$$R_t\varphi(x) = \int_H \varphi(e^{tA}x + y)N_{Q_t}(dy),\ t > 0,\ \varphi \in C_b(H), \tag{1.2}$$

where N_{Q_t} is the gaussian measure in H with mean 0 and covariance operator Q_t, is well defined, see, e.g., [5].

We recall that R_t is not strongly continuous in $C_b(H)$. It is a weakly continuous semigroup, see [1] and a π–semigroup see [12]. The infinitesimal generator L of R_t can be defined either as in [1] through its resolvent

$$R(\lambda, L)\varphi(x) = \int_0^\infty e^{-\lambda t}R_t\varphi(x)dt,\ x \in X,\ \varphi \in C_b(H),$$

or as in [12] by setting

$$\begin{aligned} D(L) = \Big\{\varphi \in C_b(H) :\ &\lim_{h\to 0} \frac{1}{h}\,(R_h\varphi(x) - \varphi(x)) = f(x),\ \forall\, x \in H \\ &\text{and} \sup_{h\in[0,1]} \frac{1}{h}\,\|R_h\varphi - \varphi\|_0 < +\infty\Big\}, \end{aligned}$$

and

$$L\varphi(x) = \lim_{h\to 0} \frac{1}{h}\,(R_h\varphi(x) - \varphi(x)),\ x \in H,\ \ \forall\, \varphi \in D(L).$$

The main object of this paper is the following linear operator N_0 which is a perturbation of the Ornstein–Uhlenbeck generator L,

$$N_0\varphi = L\varphi + \mathcal{F}\varphi,\ \varphi \in D(L) \cap C_b^1(H),$$

where

$$\mathcal{F}\varphi = \langle F(\cdot), D\varphi\rangle,\ \varphi \in C_b^1(H).$$

Several papers have been devoted to study perturbations of Ornstein–Uhlenbeck generators, see, e.g., [5], [3], [2], [7], [10], [6], and refences therein. In all these papers some additional assumptions have been made ensuring that the term $\langle F(\cdot), D\varphi\rangle$ is "dominated" by L, (in the sense of the norms of $C_b(H)$ or of $L^2(\nu; H)$) where ν is an invariant measure for R_t). Assume for instance that

$$\varphi \in D(L) \Rightarrow \langle F(\cdot), D\varphi\rangle \in C_b(H). \tag{1.3}$$

Under this assumption, standard perturbation arguments can be used to show that the closure N of N_0 in $C_b(H)$ (or in $L^2(H,\mu)$) is an m–dissipative operator.

In this paper we consider a general Ornstein–Uhlenbeck generator L : we prove that N_0 is dissipative in $C_b(H)$ and that its closure is m–dissipative, see Theorem 3.3 below. We do not use perturbation arguments but we show, by purely analytic tools, that the range of $\lambda - N_0$ is dense in $C_b(H)$ for λ sufficiently large. This will imply the conclusion by the Lumer–Phillips theorem. Our method consists in approximating the term $\mathcal{F}\varphi = \langle F(\cdot), D\varphi\rangle$ by $\mathcal{F}_\varepsilon\varphi(x) = \frac{1}{\varepsilon}\,(\varphi(\eta(\varepsilon, x)) - \varphi(x))$, where η is the solution to the initial value problem $\eta'(t,x) = F(\eta(t,x)),\ \ \eta(0,x) =$

x. Notice that the semigroup $e^{t\mathcal{F}}$ generated by $\mathcal{F}$ is precisely

$$e^{t\mathcal{F}}(\varphi)(x) = \varphi(\eta(t,x)), \quad t \geq 0,\ x \in H,\ \varphi \in C_b(H).$$

See §2 below for details andprecise statements.

We notice that the usual probabilistic method, based on the stochastic characteristics method, seems not to work under our assumptions. This method is based on the well known representation formula for the resolvent of N :

$$R(\lambda, N)f(x) = \int_0^{+\infty} e^{-\lambda t}\mathbb{E}[f(X(t,x))]dt,\ f \in C_b(H), \tag{1.4}$$

where $X(t,x)$ is the solution of the stochastic differential equation

$$dX = (AX + F(X))dt + C^{1/2}dW(t);\ \ X(0) = x. \tag{1.5}$$

Notice that by Hypothesis 1.1–(iii), F is Lipschitz continuous, so that (1.5) has a unique solution, see, e.g., [5]. However, to prove that $\varphi := R(\lambda, N)f$ (with f regular) is a solution of the equation

$$\lambda\varphi - L\varphi - \mathcal{F}\varphi = f \tag{1.6}$$

would require to use the Itô formula, which is very delicate in the general situation (for instance if C is not nuclear).

Finally in §4 we prove that for all $f \in C_b^{1,1}(H)$ and any $\lambda > 0$, we have

$$\varphi := R(\lambda, N)f \in D(L) \cap C_b^1(H),$$

and that φ is the unique strict solution of (1.6).

In all the paper we shall use the following notation.

$$K = \sup\left\{\frac{\langle F(x) - F(y), x - y\rangle}{|x-y|^2}\ :\ x, y \in H,\ x \neq y\right\}.$$

If $K \leq 0$ then F is dissipative. Clearly $K \leq \|F\|_1$ and moreover

$$\langle F(x) - F(y), x - y\rangle \leq K|x-y|^2,\ \ \forall\, x, y \in H, \tag{1.7}$$

and

$$\langle F_x(x) \cdot h, h\rangle \leq K|h|^2,\ \ x, h \in H, \tag{1.8}$$

where F_x denotes the Fréchet derivative of F.

We conclude this section with a technical remark that will be needed in the proof of the main result of the paper, Theorem 3.3 below.

Assume that dim $H = \infty$. Then J. Kurtzweil proved in 1954 [8], that there exists a sequence $\{\varphi_\varepsilon\} \subset C_b(H) \cap C^\infty(H)$ such that $\varphi_\varepsilon(x) \to \varphi(x)$ uniformly on $x \in H$ when $\varepsilon \to 0$. However, in 1973 A. S. Nemirowski and S. M. Semenov [11] proved that $C_b^2(H)$ is not dense in $C_b(H)$, whereas $C_b^{1,1}(H)$ is. This result has been also proved through the technique of inf–sup convolution by J. M. Lasry and P. L. Lions [9].

2. Some properties of $\mathcal{F}$

It is useful to introduce a bounded approximation of $\mathcal{F}$. For any $\varepsilon > 0$ we set

$$\mathcal{F}_\varepsilon\varphi(x) = \frac{1}{\varepsilon}\,(\varphi(\eta(\varepsilon,x)) - \varphi(x)),\ \varphi \in C_b(H),\ x \in H,$$

where η is the solution to the initial value problem

$$\frac{d}{dt}\,\eta(t,x) = F(\eta(t,x)),\ \ \eta(0,x) = x,\ t \ge 0,\ x \in H. \tag{2.1}$$

Problem (2.1) has a unique solution $\eta(\cdot,x)$ since $F \in C_b^1(H;H)$.

The following lemma collects some useful properties of η.

Lemma 2.1. *The following estimates hold:*

$$|\eta(t,x)| \le e^{tK}|x|,\ t \ge 0,\ x \in H. \tag{2.2}$$

$$\|\eta_x(t,x)\| \le e^{tK},\ t \ge 0,\ x \in H. \tag{2.3}$$

$$|\eta(t,x) - x| \le t\|F\|_0,\ t \ge 0,\ x \in H. \tag{2.4}$$

$$|\eta(t,x) - x - \eta(t,\overline{x}) + \overline{x}| \le t\|F\|_1 \max\{e^{tK},1\},\ t \ge 0,\ x,\overline{x} \in H. \tag{2.5}$$

$$\left|F(x) - \frac{\eta(t,x) - x}{t}\right| \le \frac{1}{2}\,t\|F\|_0\|F\|_1,\ t \ge 0,\ x \in H. \tag{2.6}$$

$$\|\eta_x(t,x) - \eta_x(t,\overline{x})\| \le t\|F\|_{1,1} \max\{e^{2tK},1\}\ |x - \overline{x}|,\ t \ge 0,\ x,\overline{x} \in H. \tag{2.7}$$

Proof. (2.2). Multiplying the first equation in (2.1) by $\eta(t,x)$ yields

$$\frac{1}{2}\,\frac{d}{dt}\,|\eta(t,x)|^2 = \langle F(\eta(t,x)),\eta(t,x)\rangle \le K|\eta(t,x)|^2,$$

by (1.7). Now (2.2) follows from a standard comparison result.

(2.3). It is well known that $\eta(t,x)$ is differentiable in x since F is. Moreover $\eta_x(t,x)\cdot h := p^h(t,x)$, $h \in H$, fulfills the equation

$$\frac{d}{dt}\,p^h(t,x) = F_x(\eta(t,x))p^h(t,x),\ \ p^h(0,x) = h.$$

Multiplying the first identity by $p^h(t,x)$ yields, taking into account (1.8),

$$\frac{1}{2}\,\frac{d}{dt}\,|p^h(t,x)|^2 = \langle F_x(\eta(t,x))p^h(t,x),p^h(t,x)\rangle \le K|p^h(t,x)|^2.$$

Now (2.3) follows again from a standard comparison result.

(2.4). It is enough to notice that

$$\eta(t,x) - x = \int_0^t \eta_t(s,x)ds = \int_0^t F(\eta(s,x))ds \le t\|F\|_0.$$

(2.5). We have

$$\eta(t,x) - x - \eta(t,\overline{x}) + \overline{x} = \int_0^t [F(\eta(s,x)) - F(\eta(s,\overline{x}))]ds.$$

It follows, taking into account (2.3),

$$\begin{aligned} |\eta(t,x)-x-\eta(t,\overline{x})+\overline{x}| &\le \|F\|_1 \int_0^t |\eta(s,x))-\eta(s,\overline{x})|ds \\ &\le \|F\|_1 \int_0^t e^{sK} ds\, |x-\overline{x}|, \end{aligned}$$

and (2.5) follows.

(2.6). We have

$$\begin{aligned} F(x)-\frac{\eta(t,x)-x}{t} &= F(x)-\frac{1}{t}\int_0^t F(\eta(s,x))ds \\ &= \frac{1}{t}\int_0^t (F(x)-F(\eta(s,x)))ds \end{aligned}$$

Therefore, taking into account (2.4), we find

$$\left|F(x)-\frac{\eta(t,x)-x}{t}\right| \le \frac{1}{t}\,\|F\|_1 \int_0^t |\eta(s,x)-x|ds \le \frac{1}{2}\, t\|F\|_0\|F\|_1.$$

(2.7). Let $x,\overline{x},h \in H$, and set

$$r^h(t) = \eta_x(t,x)\cdot h - \eta_x(t,\overline{x})\cdot h = p^h(t,x)-p^h(t,\overline{x}),\quad t\ge 0.$$

Then $r_h(t)$ fulfills the equation

$$\begin{cases} \dfrac{d}{dt}\, r^h(t) = F_x(\eta(t,x))r^h(t) + [F_x(\eta(t,x))-F_x(\eta(t,\overline{x}))]p^h(t,x), \\ r^h(0)=0. \end{cases}$$

Since $\langle F_x(\eta(t,x))r^h(t), r^h(t)\rangle \le K|r^h(t)|^2$, it follows, by a standard argument, that

$$|r^h(t)| \le \int_0^t e^{K(t-s)}|F_x(\eta(s,x))-F_x(\eta(s,\overline{x})||p^h(s,x)|ds.$$

Now, taking into account (2.3), we find

$$\begin{aligned} |r^h(t)| &\le \|F\|_{1,1}\int_0^t e^{Kt}|\eta(s,x)-\eta(s,\overline{x})|ds\,|h| \\ &\le \|F\|_{1,1}\int_0^t e^{K(t+s)}ds\,|h|\,|x-\overline{x}|, \end{aligned}$$

that yields (2.7). □

Proposition 2.2. *For all* $\varphi \in C_b^1(H)$ *we have*

$$\lim_{\varepsilon\to 0}\mathcal{F}_\varepsilon\varphi = \mathcal{F}\varphi,\ \text{in } C_b(H), \tag{2.8}$$

$$\|F_\varepsilon\varphi\|_0 \le \|\varphi\|_1\|F\|_0, \tag{2.9}$$

and

$$|F_\varepsilon\varphi(x)-F_\varepsilon\varphi(\overline{x})| \le \frac{1}{2}\,\|\varphi\|_{1,1}(1+e^{\varepsilon K}) + \|\varphi\|_1\|F\|_1\max\{e^{\varepsilon K},1\}, \tag{2.10}$$

Moreover for all $\varphi \in C_b^{1,1}(H)$ *we have*

$$\|\mathcal{F}\varphi - \mathcal{F}_\varepsilon\varphi\|_0 \le \frac{\varepsilon}{2}\left(\|F\|_0^2\|\varphi\|_{1,1} + \|F\|_0\|F\|_1\|\varphi\|_1\right). \tag{2.11}$$

Proof. (2.8). We have for all $x \in H$,

$$\begin{aligned}
&\mathcal{F}_\varepsilon\varphi(x) - \mathcal{F}\varphi(x)\\
&= \int_0^1 \left\langle D\varphi(\xi x + (1-\xi)\eta(\varepsilon,x)), \frac{1}{\varepsilon}\,(\eta(\varepsilon,x) - x)\right\rangle d\xi - \langle F(x), D\varphi(x)\rangle\\
&= \int_0^1 \left\langle [D\varphi(\xi x + (1-\xi)\eta(\varepsilon,x)) - D\varphi(x)], \frac{1}{\varepsilon}\,(\eta(\varepsilon,x) - x)\right\rangle\, d\xi\\
&+ \left\langle \frac{1}{\varepsilon}\,(\eta(\varepsilon,x) - x) - F(x), D\varphi(x)\right\rangle.
\end{aligned} \tag{2.12}$$

Taking into account (2.6) and (2.4), we find

$$\begin{aligned}
&|\mathcal{F}_\varepsilon\varphi(x) - \mathcal{F}\varphi(x)| \le \|F\|_0 \int_0^1 \omega_{D\varphi}((1-\xi)|\eta(\varepsilon,x) - x|)d\xi + \frac{1}{2}\,\varepsilon\|F\|_0\|F\|_1\\
&\le \|F\|_0 \int_0^1 \omega_{D\varphi}((1-\xi)|\varepsilon\|F\|_0|)d\xi + \frac{1}{2}\,\varepsilon\|F\|_0\|F\|_1,
\end{aligned}$$

where $\omega_{D\varphi}$ is the uniform continuity modulus of $D\varphi$. Therefore (2.8) follows.

(2.9). We have

$$\mathcal{F}_\varepsilon\varphi(x) = \int_0^1 \left\langle D\varphi(\xi x + (1-\xi)\eta(\varepsilon,x)), \frac{1}{\varepsilon}\,(\eta(\varepsilon,x) - x)\right\rangle d\xi,$$

and so

$$|\mathcal{F}_\varepsilon\varphi(x)| \le \frac{1}{\varepsilon}\,(\eta(\varepsilon,x) - x)\|\varphi\|_1.$$

Now the conclusion follows from (2.4).

(2.10). We have

$$\begin{aligned}
&\mathcal{F}_\varepsilon\varphi(x) - \mathcal{F}_\varepsilon\varphi(\overline{x})\\
&= \int_0^1 \left\langle D\varphi(\xi x + (1-\xi)\eta(\varepsilon,x)), \frac{1}{\varepsilon}\,(\eta(\varepsilon,x) - x)\right\rangle d\xi\\
&- \int_0^1 \left\langle D\varphi(\xi \overline{x} + (1-\xi)\eta(\varepsilon,\overline{x})), \frac{1}{\varepsilon}\,(\eta(\varepsilon,\overline{x}) - \overline{x})\right\rangle d\xi\\
&= \int_0^1 \Big\langle D\varphi(\xi x + (1-\xi)\eta(\varepsilon,x)) - D\varphi(\xi\overline{x} + (1-\xi)\eta(\varepsilon,\overline{x})))\\
&\qquad\qquad , \frac{1}{\varepsilon}\,(\eta(\varepsilon,x) - x)\Big\rangle d\xi\\
&+ \int_0^1 \left\langle D\varphi(\xi \overline{x} + (1-\xi)\eta(\varepsilon,\overline{x})), \frac{1}{\varepsilon}\,(\eta(\varepsilon,x) - x - \eta(\varepsilon,\overline{x}) + \overline{x})\right\rangle d\xi.
\end{aligned}$$

It follows by (2.5)

$$
\begin{aligned}
&|\mathcal{F}_\varepsilon\varphi(x) - \mathcal{F}_\varepsilon\varphi(\overline{x})| \\
&\le \|\varphi\|_{1,1}\|F\|_0 \int_0^1 [\xi|x-\overline{x}| + (1-\xi)|\eta(\varepsilon,x)-\eta(\varepsilon,\overline{x})|]\, d\xi \\
&+\|\varphi\|_1\|F\|_1 \max\{e^{\varepsilon K},1\}|x-\overline{x}|.
\end{aligned}
$$

Now the conclusion follows.

It remains to prove (2.11). By (2.12) and (2.6) we have

$$
|\mathcal{F}\varphi(x) - \mathcal{F}_\varepsilon\varphi(x)| \le \frac{1}{\varepsilon}\,\|\varphi\|_{1,1}\int_0^1 (1-\xi)d\xi|\eta(\varepsilon,x)-x|^2 + \frac{1}{2}\,\varepsilon\|\varphi\|_1\|F\|_0\|F\|_1.
$$

Finally, taking into account (2.4), we find

$$
|\mathcal{F}\varphi(x) - \mathcal{F}_\varepsilon\varphi(x)| \le \frac{1}{2}\,\varepsilon\left[\|F\|_0^2\|\varphi\|_{1,1} + \|F\|_0\|F\|_1\|\varphi\|_1\right], \tag{2.13}
$$

and the conclusion follows. □

3. m–dissipativity of N.

Given $\varepsilon > 0$ we introduce the following approximating operator

$$
N_\varepsilon\varphi = L\varphi + \mathcal{F}_\varepsilon\varphi,\ \varphi \in D(L).
$$

The following proposition collects some basic properties of N_ε.

Proposition 3.1. *N_ε is an m–dissipative operator in $C_b(H)$ for any $\varepsilon > 0$. Moreover for any $f \in C_b^1(H)$ and any $\lambda > K+\omega$ we have $R(\lambda, N_\varepsilon)f \in C_b^1(H)$ and*

$$
\|R(\lambda,N_\varepsilon)f\|_1 \le \frac{1}{\lambda-\omega+\frac{1}{\varepsilon}\,(1-e^{K\varepsilon})}\,\|f\|_1. \tag{3.1}
$$

Finally for any $f \in C_b^{1,1}(H)$ and any $\lambda > 2(K+\omega)$ we have $R(\lambda,N_\varepsilon)f \in C_b^{1,1}(H)$ and

$$
\begin{aligned}
\|R(\lambda,N_\varepsilon)f\|_{1,1} &\le \frac{1}{\lambda-2\omega+\frac{1}{\varepsilon}\,(1-e^{2K\varepsilon})} \\
&\times\left[\|f\|_{1,1} + \max\{e^{2K\varepsilon},1\}\|F\|_{1,1}\|R(\lambda,N_\varepsilon)f\|_1\right].
\end{aligned} \tag{3.2}
$$

Proof. Let $\varepsilon > 0,\ \lambda > 0,\ f \in C_b(H)$. Consider the equation

$$
\lambda\varphi_\varepsilon - L\varphi_\varepsilon - \mathcal{F}_\varepsilon(\varphi_\varepsilon) = f, \tag{3.3}
$$

which is equivalent to

$$
\left(\lambda+\frac{1}{\varepsilon}\right)\varphi_\varepsilon - L\varphi_\varepsilon = f + \frac{1}{\varepsilon}\,\varphi_\varepsilon(\eta(\varepsilon,x)),
$$

and also to

$$
\varphi_\varepsilon = R\left(\lambda+\frac{1}{\varepsilon},L\right)f + T_\lambda(\varphi_\varepsilon), \tag{3.4}
$$

where $T_\lambda : C_b(H) \to C_b(H)$ is defined by

$$T_\lambda(\psi)(x) = R\left(\lambda + \frac{1}{\varepsilon}, L\right)\left[\frac{1}{\varepsilon}\,\psi(\eta(\varepsilon,x))\right],\ x \in H,\ \psi \in C_b(H).$$

Since, as easily seen,

$$\|T_\lambda\psi\|_0 \le \frac{1}{1+\lambda\varepsilon}\,\|\psi\|_0,\ \forall\,\psi \in C_b(H),$$

T_λ is a contraction and so equation (3.4) has a unique solution $\varphi_\varepsilon \in C_b(H)$. Moreover

$$\|\varphi_\varepsilon\|_0 \le \frac{1}{\lambda + \frac{1}{\varepsilon}}\left[\|f\|_0 + \frac{1}{\varepsilon}\,\|\varphi_\varepsilon\|_0\right],$$

that yields

$$\|\varphi_\varepsilon\|_0 \le \frac{1}{\lambda}\,\|f\|_0.$$

This shows that N_ε is m–dissipative in $C_b(H)$.

Let us prove (3.1). For any $\psi \in C_b^1(H)$ and any $t \ge 0$ we have

$$DR_t\psi(x) = \int_H e^{tA^*}(D\psi)(e^{tA}x + y)N_{Q_t}(dy).$$

It follows, for any $\lambda > \omega$

$$DR(\lambda, L)\psi(x) = \int_H\int_0^{+\infty} e^{-\lambda t}e^{tA^*}(D\psi)(e^{tA}x+y)N_{Q_t}(dy)dt, \tag{3.5}$$

and so

$$\|DR(\lambda, L)\psi\|_0 \le \frac{1}{\lambda - \omega}\,\|D\psi\|_0,\ \forall\,\psi \in C_b^1(H). \tag{3.6}$$

Now differentiating (3.4) gives

$$D\varphi_\varepsilon = D[R(\lambda + \frac{1}{\varepsilon}, L)g_\varepsilon], \tag{3.7}$$

where

$$g_\varepsilon = f + \frac{1}{\varepsilon}\,\varphi_\varepsilon(\eta(\varepsilon,x)).$$

Therefore

$$Dg_\varepsilon = Df + \frac{1}{\varepsilon}\,\eta_x^*(\varepsilon,x)D\varphi_\varepsilon(\eta(\varepsilon,x)), \tag{3.8}$$

where $\eta_x^*(\varepsilon,x)$ is the adjoint of $\eta_x(\varepsilon,x)$. Now, taking into account (2.3), (3.6) and (3.7), it follows that

$$\|D\varphi_\varepsilon\|_0 \le \frac{1}{\lambda - \omega + \frac{1}{\varepsilon}}\left(\|Df\|_0 + \frac{1}{\varepsilon}\,e^{K\varepsilon}\|D\varphi_\varepsilon\|_0\right),$$

and so, if $\lambda > 2(\omega + K)$ (3.1) follows.

Let us prove finally (3.2). Let $\psi \in C_b^{1,1}(H)$, $t \geq 0$, $x, \overline{x} \in H$. Then if $\lambda > 2\omega$ we have by (3.5)

$$|DR(\lambda, L)\psi(x) - DR(\lambda, L)\psi(\overline{x})|$$

$$\leq \int_H \int_0^{+\infty} e^{-\lambda t} e^{2\omega t} \|\psi\|_{1,1} \, |x - \overline{x}| N_{Q_t}(dy)dt \leq \frac{1}{\lambda - 2\omega} \, \|\psi\|_{1,1} \, |x - \overline{x}|.$$

Now by (3.7) it follows

$$\begin{aligned} |D\varphi_\varepsilon(x) - D\varphi_\varepsilon(\overline{x})| &= |DR(\lambda + \tfrac{1}{\varepsilon})g_\varepsilon(x) - DR(\lambda + \tfrac{1}{\varepsilon})g_\varepsilon(\overline{x})| \\ &\leq \frac{1}{\lambda - 2\omega + \frac{1}{\varepsilon}} \, \|g_\varepsilon\|_{1,1} \, |x - \overline{x}|. \end{aligned} \tag{3.9}$$

Moreover by (3.8) we have

$$\begin{aligned} &|Dg_\varepsilon(x) - Dg_\varepsilon(\overline{x})| \leq |Df(x) - Df(\overline{x})| \\ &+ \frac{1}{\varepsilon} \, |\eta_x^*(\varepsilon, x) D\varphi_\varepsilon(\eta(\varepsilon, x)) - \eta_x^*(\varepsilon, \overline{x}) D\varphi_\varepsilon(\eta(\varepsilon, \overline{x}))| \\ &\leq \|f\|_{1,1}|x - \overline{x}| + \frac{1}{\varepsilon} \, |\eta_x^*(\varepsilon, x) - \eta_x^*(\varepsilon, \overline{x})| \|\varphi\|_1 \\ &+ \frac{1}{\varepsilon} \, |\eta_x^*(\varepsilon, \overline{x})| \, |D\varphi_\varepsilon(\eta(\varepsilon, x)) - D\varphi_\varepsilon(\eta(\varepsilon, \overline{x}))|. \end{aligned}$$

Now by (2.7), taking into account (2.4) it follows that

$$\begin{aligned} |Dg_\varepsilon(x) - Dg_\varepsilon(\overline{x})| &\leq \|f\|_{1,1}|x - \overline{x}| + \|F\|_{1,1} \max\{e^{\varepsilon 2K}, 1\} |x - \overline{x}| \|\varphi_\varepsilon\|_1 \\ &+ \frac{1}{\varepsilon} \, e^{\varepsilon K} \|\varphi_\varepsilon\|_{1,1} |\eta(\varepsilon, x) - \eta(\varepsilon, \overline{x})| e^{\varepsilon K} |x - \overline{x}|. \end{aligned}$$

Then we have

$$\|g_\varepsilon\|_{1,1} \leq \|f\|_{1,1} + \|F\|_{1,1} \max\{e^{2\varepsilon K}, 1\} \|\varphi_\varepsilon\|_1 + \frac{1}{\varepsilon} \, e^{2\varepsilon K} \|\varphi_\varepsilon\|_{1,1}.$$

Finally by (3.9)

$$\begin{aligned} \|\varphi_\varepsilon\|_{1,1} &\leq \frac{1}{\lambda - 2\omega + \frac{1}{\varepsilon}} \Big(\|f\|_{1,1} + \frac{1}{\varepsilon} \, e^{2\varepsilon K} \|\varphi_\varepsilon\|_{1,1} \\ &+ \|F\|_{1,1} \max\{e^{2\varepsilon K}, 1\} \|\varphi_\varepsilon\|_1 \Big), \end{aligned}$$

and (3.2) follows. □

Lemma 3.2. *N_0 is dissipative in $C_b(H)$.*

Proof. Let $\varphi \in D(L) \cap C_b^1(H)$, and $\lambda > 0$. Set

$$\lambda\varphi - L\varphi - \mathcal{F}\varphi = f.$$

Then we have

$$\lambda\varphi - N_\varepsilon\varphi = f + \mathcal{F}\varphi - \mathcal{F}_\varepsilon\varphi.$$

It follows

$$\varphi = R(\lambda, N_\varepsilon)(f + \mathcal{F}\varphi - \mathcal{F}_\varepsilon\varphi),$$

so that

$$\|\varphi\|_0 \le \frac{1}{\lambda}\left(\|f\|_0 + \|\mathcal{F}\varphi - \mathcal{F}_\varepsilon\varphi\|_0\right).$$

Letting $\varepsilon \to 0$ yields, recalling Proposition 2.2,

$$\|\varphi\|_0 \le \frac{1}{\lambda}\,\|f\|_0,$$

that shows that N_ε is dissipative in $C_b(H)$. □

We shall denote by N the closure of N_0. It is well known, see, e.g., [4], that N is a dissipative (possibly multivalued) operator in $C_b(H)$.

Theorem 3.3. *N is m–dissipative in $C_b(H)$.*

Proof. Let $f \in C_b^{1,1}(H)$, $\lambda > 2(\omega + K)$, and let φ_ε be the solution of (3.3). Then we have

$$\lambda\varphi_\varepsilon - N_0\varphi_\varepsilon = f + \mathcal{F}_\varepsilon(\varphi_\varepsilon) - \mathcal{F}(\varphi_\varepsilon). \tag{3.10}$$

We claim that

$$\lim_{\varepsilon\to 0}[\mathcal{F}_\varepsilon(\varphi_\varepsilon) - \mathcal{F}(\varphi_\varepsilon)] = 0 \quad \text{in } C_b(H). \tag{3.11}$$

In fact, by Proposition 2.2 we have

$$\|\mathcal{F}\varphi_\varepsilon - \mathcal{F}_\varepsilon\varphi_\varepsilon\|_0 \le \frac{\varepsilon}{2}\left(\|F\|_0^2\|\varphi_\varepsilon\|_{1,1} + \|F\|_0\|F\|_1\|\varphi_\varepsilon\|_1\right),$$

and by Proposition 3.1 there exists a constant $c_1 > 0$, independent of ε, such that

$$\|\varphi_\varepsilon\|_{1,1} \le c_1.$$

Therefore (3.11) follows.

Now from (3.10) we see that

$$\lim_{\varepsilon\to 0}[\lambda\varphi_\varepsilon - N_0\varphi_\varepsilon] = f \quad \text{in } C_b(H).$$

Therefore the closure of the range of $\lambda - N_0$ includes $C_b^{1,1}(H)$ which is dense in $C_b(H)$, and by the Lumer–Phillips the closure N of N_0 is m–dissipative. □

4. Regularity results

Proposition 4.1. *Let $f \in C_b(H)$ and let $\lambda > 2(\omega + K)$. Then*

$$\lim_{\varepsilon\to 0} R(\lambda, N_\varepsilon)f = R(\lambda, N)f, \text{ in } C_b(H). \tag{4.1}$$

Proof. Let first $f \in C_b^{1,1}(H)$. Given $\varepsilon > 0$ and $\lambda > 2(\omega + K)$, let $\varphi_\varepsilon \in C_b^{1,1}(H)$ be the solution to the equation

$$\lambda\varphi_\varepsilon - L\varphi_\varepsilon - \mathcal{F}_\varepsilon(\varphi_\varepsilon) = f. \tag{4.2}$$

Then we have

$$\lambda\varphi_\varepsilon - N_0\varphi_\varepsilon = f + \mathcal{F}_\varepsilon(\varphi_\varepsilon) - \mathcal{F}(\varphi_\varepsilon).$$

It follows

$$\varphi_\varepsilon = R(\lambda, N)f + R(\lambda, N)(\mathcal{F}_\varepsilon(\varphi_\varepsilon) - \mathcal{F}(\varphi_\varepsilon)).$$

By (3.11) it follows that

$$\lim_{\varepsilon \to 0} \varphi_\varepsilon = R(\lambda, N)f,$$

as required.

If $f \in C_b(H)$ the conclusion follows from the density of $C_b^{1,1}(H)$ in $C_b(H)$. □

To prove next result we need the following proposition, see [9].

Proposition 4.2. *Assume that $\varphi \in C_b(H)$ is Lipschitz continuous, semiconcave and semiconvex. Let $K > 0$ be such that*

$$x \to \varphi(x) + \frac{K}{2}|x|^2 \text{ is convex and } x \to \varphi(x) - \frac{K}{2}|x|^2 \text{ is concave.}$$

Then $\varphi \in C_b^{1,1}(H)$ and $[\varphi]_{1,1} \le K$.

Theorem 4.3. *Let $f \in C_b^{1,1}(H)$ and $\lambda > 2(\omega + K)$. Then the equation*

$$\lambda\varphi - L\varphi - \mathcal{F}(\varphi) = f, \tag{4.3}$$

has a unique solution $\varphi \in D(L) \cap C_b^{1,1}(H)$, given by $\varphi = R(\lambda, N)f$.

Proof. Given $\varepsilon > 0$ let φ_ε be the solution of (4.2). Then by Theorem 3.3 there exists $c_1 > 0$, independent of ε, such that

$$\|\varphi_\varepsilon\|_{1,1} \le c_1. \tag{4.4}$$

Existence. Let us proceed in three steps.

Step 1. $\varphi \in C_b^{1,1}(H)$.

By (4.4) φ_ε is semiconcave (semiconvex) uniformly in ε. Consequently, since $\varphi_\varepsilon \to \varphi = R(\lambda, N)f$ in $C_b(H)$, φ is semiconcave (semiconvex) too and so it belongs to $C_b^{1,1}(H)$ by Proposition 4.2.

Step 2. $\lim_{\varepsilon \to 0} \mathcal{F}_\varepsilon(\varphi_\varepsilon)(x) = \mathcal{F}(\varphi)(x),\ x \in H.$

In view of (3.11), it is enough to show that

$$\lim_{\varepsilon \to 0} \mathcal{F}(\varphi_\varepsilon)(x) = \mathcal{F}(\varphi)(x),\ x \in H, \tag{4.5}$$

or, equivalently,

$$\lim_{\varepsilon \to 0} \langle F(x), D\varphi_\varepsilon(x) - D\varphi(x)\rangle = 0,\ \forall\, x \in H. \tag{4.6}$$

Fix $x, y \in H$ and set

$$s_\varepsilon(t) = \varphi_\varepsilon(x + ty),\ \ t \in [0, 1].$$

Then

$$s'_\varepsilon(t) = \langle D\varphi_\varepsilon(x + ty), y\rangle\ \ t \in [0, 1].$$

By (4.4) it follows that the family of functions $\{s'_\varepsilon\}$ is equicontinuous, and consequently there exists a sequence $\varepsilon_k \downarrow 0$ such that

$$s_{\varepsilon_k}(t) \to \varphi(x + ty),\ \ s'_{\varepsilon_k}(t) \to \langle D\varphi(x + ty), y\rangle,\ t \in [0, 1],$$

uniformly in t. Therefore

$$\lim_{\varepsilon\to 0}\langle D\varphi_\varepsilon(x+ty),y\rangle = \langle D\varphi(x+ty),y\rangle,\ \forall\, t\in[0,1].$$

It follows

$$\lim_{\varepsilon\to 0}\langle D\varphi_\varepsilon(x),y\rangle = \langle D\varphi(x),y\rangle,\ \forall\, x\in H,$$

and (4.6) follows setting $y = F(x)$.

Step 3. Conclusion.

Since $L\varphi_\varepsilon(x) = \lambda\varphi_\varepsilon(x) + \mathcal{F}_\varepsilon(\varphi_\varepsilon(x))$, we have

$$\lim_{\varepsilon\to 0} L\varphi_\varepsilon(x) = \lambda\varphi(x) + \mathcal{F}(\varphi(x)),\ \forall\, x\in H.$$

Moreover by (4.4), there exists $c_3 > 0$, independent of ε, such that

$$\|L\varphi_\varepsilon\|_0 \le c_3.$$

Set

$$h_\varepsilon = \lambda\varphi_\varepsilon - L\varphi_\varepsilon,\quad h = \lambda\varphi - L\varphi.$$

Then by Step 2 we have

$$\lim_{\varepsilon\to 0} h_\varepsilon(x) = f(x) + \mathcal{F}(\varphi)(x),\ \forall\, x\in H.$$

Then, since $\varphi_\varepsilon = R(\lambda,L)h_\varepsilon$, we have

$$\|h_\varepsilon\|_0 \le \frac{1}{\lambda}\,\|f\|_0 + c_3.$$

Consequently, letting $\varepsilon\to 0$ and using the dominated convergence theorem, we find

$$\varphi = R(\lambda,L)(f+\mathcal{F}(\varphi)).$$

Thus $\varphi\in D(L)$ and $\lambda\varphi - L\varphi - \mathcal{F}(\varphi) = f$.

Uniqueness. Let $\lambda > 2(\omega+K)$ and $\varphi\in D(L)\cap C^1_b(H)$ such that

$$\lambda\varphi - L\varphi - \mathcal{F}(\varphi) = f.$$

Then for any $\varepsilon > 0$ we have

$$\lambda\varphi - N_\varepsilon\varphi = f + \mathcal{F}(\varphi) - \mathcal{F}_\varepsilon(\varphi),$$

and so

$$\varphi = R(\lambda,N_\varepsilon)(f+\mathcal{F}(\varphi)-\mathcal{F}_\varepsilon(\varphi))$$

Now, letting $\varepsilon\to 0$, by Proposition 4.1, we have that $\varphi = R(\lambda,N)f$, as required. □

References

[1] S. Cerrai, *A Hille–Yosida theorem for weakly continuous semigroups.* Semigroup Forum, **49** (1994), 349–367.

[2] A. Chojnowska-Michalik, *Invariant densities for stochastic semilinear evolution equations and related properties of transition semigroups.* In: Stochastic Partial Differential Equations and Applications, G. Da Prato and L. Tubaro eds., Lecture Notes in Pure and Applied Mathematics, vol. 227, Marcel Dekker, 2002.

[3] A. Chojnowska-Michalik and B. Goldys, *Existence, uniqueness and invariant measures for stochastic semilinear equations.* Probab. Th. Relat. Fields, **102** (1995), 331–356.

[4] G. Da Prato, *Applications croissantes et équations d'évolutions dans les espaces de Banach.* Academic Press, 1976.

[5] G. Da Prato and J. Zabczyk, *Ergodicity for Infinite Dimensional Systems.* London Mathematical Society Lecture Notes, n. 229, Cambridge University Press, 1996.

[6] A. Eberle, *Uniqueness and non–uniqueness of singular diffusion operators.* Lecture Notes in Mathematics 1718, Berlin, Springer–Verlag, 1999.

[7] M. Fuhrman, *Regularity properties of transition probabilities in infinite dimensions.* Stochastics Stochastics Rep. 69, no. 1-2, 31–65, 2000.

[8] J. Kurtzweil, *On approximation in real Banach spaces.* Studia Math. **14** (1954), 213–231.

[9] J. M. Lasry and P. L .Lions, *A remark on regularization in Hilbert spaces.* Israel J. Math. **55**, 3 (1986), 257–266.

[10] H. Long and I. Simão, *On the essential self-adjointness of perturbed Ornstein-Uhlenbeck operators on Hilbert spaces.* Commun. Appl. Anal. **5**, no. 3 (2001), 371–382.

[11] A.S. Nemirovski and S. M. Semenov, *The polynomial approximation of functions in Hilbert spaces.* Mat. Sb. (N.S), **92**, 134, (1973), 257-281.

[12] E. Priola, *On a class of Markov type semigroups in spaces of uniformly continuous and bounded functions.* Studia Mathematica **136** (3), 271–295.

Giuseppe Da Prato
Scuola Normale Superiore di Pisa
piazza dei Cavalieri 7
I-56126, Pisa, Italy
E-mail address: `daprato@sns.it`

Progress in Nonlinear Differential Equations
and Their Applications, Vol. 55, 141–147

Towards an L^1-Theory for Vector-Valued Elliptic Boundary Value Problems

Robert Denk, Matthias Hieber, and Jan Prüss

1. Introduction

Vector-valued elliptic and parabolic boundary value problems subject to general boundary conditions have been investigated recently in [DHP01] in the L_p-context for $1 < p < \infty$. One of the main goals of this paper was to deduce a maximal L_p-regularity result for the solution of the parabolic initial boundary value problem. A classical reference in the elliptic context are the celebrated papers of Agmon, Douglis and Nirenberg [ADN59]. For further references and information on the scalar and vector-valued case we refer to the [Ama01] and the list of references given in [DHP01].

Vector-valued elliptic and parabolic problems on all of $\mathbb{R}^n$ were considered first by Amann [Ama01] on a large scale of function spaces, including $L_1(\mathbb{R}^n; E)$. Here E denotes an arbitrary Banach space. He proved in particular that the L_1-realization of such problems generates an analytic C_0-semigroup provided the top-order coefficients of the underlying operators are uniformly bounded and Hölder continuous.

In this note, we consider vector-valued boundary value problems with constant coefficients in the L_1-setting for a half space. Following the approach described in [DHP01], we assume the Lopatinskii-Shapiro condition to be true; we then obtain a representation of the solution u of the elliptic problem by integral operators which allows to deduce a-priori estimates for u in the $L_1(\mathbb{R}^{n+1}_+; E)$-norm. Here E denotes again an arbitrary Banach space. These estimates imply in particular that the $L_1(\mathbb{R}^{n+1}_+; E)$-realization of an elliptic boundary value problem with constant coefficients in the half space $\mathbb{R}^{n+1}_+$ generates an analytic C_0-semigroups on $L_1(\mathbb{R}^{n+1}_+; E)$. For different approaches and results with variable coefficients in the scalar-valued case we refer to Amann [Ama83], Di Blasio [DiB91], Guidetti [Gui93], Amann and Escher [AE96] and Tanabe [Tan97], Section 5.4.

2. Elliptic Problems on $L^1(\mathbb{R}^n; E)$

Throughout this section, let E be a Banach space. Following the notion of [Ama01] or Section 5 of [DHP01], we call a homogeneous $\mathcal{B}(E)$-valued polynomial $\mathcal{A}(\cdot)$ of degree $m \in \mathbb{N}$ *parameter-elliptic* if there is an angle $\phi \in [0,\pi)$ such that the spectrum $\sigma(\mathcal{A}(\xi))$ satisfies

$$\sigma(\mathcal{A}(\xi)) \subset \Sigma_\phi \quad \text{for all } \xi \in \mathbb{R}^n,\ |\xi| = 1. \tag{2.1}$$

Here $\Sigma_\theta = \{\lambda \in \mathbb{C} \setminus \{0\} : |\arg \lambda| < \theta\}$. We then call

$$\phi_{\mathcal{A}} := \inf\{\phi : \ (2.1) \text{ holds}\} = \sup_{|\xi|=1} |\arg \sigma(\mathcal{A}(\xi))|$$

the *angle of ellipticity* of $\mathcal{A}$. For $D = -i(\partial_1, \dots, \partial_n)$ we call $\mathcal{A}(D) = \sum_{|\alpha|=m} a_\alpha D^\alpha$ parameter elliptic, if its symbol $\mathcal{A}(\xi)$ is parameter-elliptic.

Assume now that $\mathcal{A}(D)$ is a parameter-elliptic operator with angle of ellipticity $\phi_{\mathcal{A}}$. It was proved in Theorem 5.2 and Corollary 5.3 of [DHP01] that for $\phi > \phi_{\mathcal{A}}$ and $k \in \mathbb{N}$ there are constants $c_{\phi,k}$, $C_{\phi,k}$ such that the solution γ_λ of

$$\lambda u + \mathcal{A}(D)u = \delta_0$$

satisfies the estimate

$$|D^\beta \gamma_\lambda(x)| \le C_{\phi,k} |\lambda|^{\frac{n+k}{m}-1} p^n_{m,k}(c_{\phi,k}|x|), \quad x \in \mathbb{R}^n, |\arg\lambda| \le \pi - \phi, |\beta| = k, \tag{2.2}$$

where $p^n_{m,k}$ is given by

$$p^n_{m,k}(r) = \int_0^\infty \frac{s^{n-2}}{(1+s)^{m-k-1}} e^{-r(1+s)} ds.$$

We define the $L_1(\mathbb{R}^n; E)$-realization A of $\mathcal{A}(D)$ by means of $A = \overline{A}_0$, where

$$[A_0 u](x) = \mathcal{A}(D)u(x), \quad x \in \mathbb{R}^n,\ u \in D(A_0) = W^m_1(\mathbb{R}^n; E).$$

The first assertion of the following proposition is a special case of Theorem 5.10 of [Ama01]. The proof given below is based on estimates on the fundamental solution. Via estimate (2.2) we also obtain some information on the domain of A, which will be important in the following.

Proposition 2.1. *Let $n, m \in \mathbb{N}$, E be a Banach space, $a_\alpha \in \mathcal{B}(E)$ and suppose that $\mathcal{A}(D)$ is parameter elliptic with angle of ellipticity $\phi_{\mathcal{A}} < \frac{\pi}{2}$. Then $-A$ generates an analytic semigroup on $L_1(\mathbb{R}^n; E)$ of angle $\frac{\pi}{2} - \phi_{\mathcal{A}}$ and we have*

$$W^m_1(\mathbb{R}^n; E) \subset D(A) \subset W^{m-1}_1(\mathbb{R}^n; E).$$

Proof. Obviously, A has dense domain. If $f \in L_1(\mathbb{R}^n; E)$, choose a sequence $f_n \in C_0^\infty(\mathbb{R}^n; E)$ such that $f_n \to f$ in $L_1(\mathbb{R}^n; E)$. For $\lambda \in \Sigma_{\pi-\phi}$, $\phi > \phi_{\mathcal{A}}$, we have $u_n := \gamma_\lambda * f_n \in W^m_1(\mathbb{R}^n; E)$ as well as $\lambda u_n + \mathcal{A}(D)u_n = f_n$. Since $u_n \to u = \gamma_\lambda * f$ in $L_1(\mathbb{R}^n; E)$ as $n \to \infty$, we see that $u \in D(A)$ and $\lambda u + Au = f$. This shows that $\lambda + A$ is invertible for each $\lambda \in \Sigma_{\pi-\phi}$ with $(\lambda + A)^{-1} f = \gamma_\lambda * f$. Thus by (2.2) and Young's inequality we obtain

$$W^{m-1}_1(\mathbb{R}^n; E) \supset D(A) \supset D(A_0) = W^m_1(\mathbb{R}^n; E).$$

Furthermore, estimate (2.2) yields $-\Sigma_{\pi-\phi} \subset \rho(A)$ and

$$|\lambda(\lambda + A)^{-1}|_{\mathcal{B}(L_1(\mathbb{R}^n;E))} \le M_{\pi-\phi},$$

for each $\phi > \phi_{\mathcal{A}}$. □

3. Elliptic Problems in a Half Space

In this section we consider boundary value problems of the form

$$\begin{aligned} \lambda u + \mathcal{A}(D)u &= f, && \text{in } \mathbb{R}^{n+1}_+ \\ \mathcal{B}_j(D)u &= g_j, && \text{on } \partial\mathbb{R}^{n+1}_+, j = 1, \dots, m. \end{aligned}$$

on a half space. Here $\mathcal{A}(D)$ as well as $\mathcal{B}_j(D)$ for $j \in \{1, \dots, m\}$ are differential operators with operator-valued coefficients. We also assume that $\mathcal{A}(D)$ and $\mathcal{B}_j(D)$ consist only of the principal parts, i.e.,

$$\begin{aligned} \mathcal{A}(D) &= \sum_{|\alpha|=2m} a_\alpha D^\alpha \\ \mathcal{B}_j(D) &= \sum_{|\beta|=m_j} b_{j\beta} D^\beta, \end{aligned}$$

where $m_j \in \{0, \dots, 2m-1\}$, $a_\alpha \in \mathcal{B}(E)$ and $b_{j\beta} \in \mathcal{B}(E)$ for $j \in \{1, \dots, m\}$. In the following we assume that $\mathcal{A}(D)$ is parameter-elliptic with angle of ellipticity $\phi_\mathcal{A} \in [0, \pi)$, i.e., there exists $\phi \in [0, \pi)$ such that

$$\sigma(\mathcal{A}(\xi)) \subset \Sigma_\phi, \quad \xi \in \mathbb{R}^{n+1}, \quad |\xi| = 1 \tag{3.1}$$

and $\phi_\mathcal{A}$ is defined as the infimum of all ϕ satisfying (3.1). Here $\mathcal{A}(\cdot)$ is the symbol of $\mathcal{A}(D)$ defined by

$$\mathcal{A}(\xi) = \sum_{|\alpha|=2m} a_\alpha \xi^\alpha, \quad \xi \in \mathbb{R}^{n+1}, a_\alpha \in \mathcal{B}(E). \tag{3.2}$$

We suppose that the following Lopatinskii-Shapiro Condition holds true:

Lopatinskii-Shapiro Condition:
For each $\xi' \in \mathbb{R}^n$ and $\lambda \in \overline{\Sigma}_{\pi-\phi}$ with $|\xi'| + |\lambda| \neq 0$, the problem

$$\begin{aligned} \lambda v(y) + \mathcal{A}(\xi', D_y)v(y) &= 0, \quad y > 0, \\ \mathcal{B}_j(\xi', D_y)v(0) &= g_j, \quad j = 1, \dots, m \end{aligned}$$

admits a unique solution $u \in C_0(\mathbb{R}_+; E)$ for each $(g_1, \dots, g_m)^T \in E^m$.

In the following we are interested in the L_1-theory of the above problem; more specifically, for $\phi > \phi_\mathcal{A}$ we consider the following problem:
Given $\lambda \in \Sigma_{\pi-\phi}$, $f \in L_1(\mathbb{R}^{n+1}_+; E)$ and $g_j \in W_1^{2m-m_j}(\mathbb{R}^{n+1}_+; E)$ for $j \in \{1, \dots, m\}$, find $u \in W_1^{2m}(\mathbb{R}^{n+1}_+; E)$ which satisfies

$$\lambda u + \mathcal{A}(D)u = f, \quad \text{in } \mathbb{R}^{n+1}_+ \tag{3.3}$$

$$\mathcal{B}_j(D)u = g_j, \quad \text{on } \partial\mathbb{R}^{n+1}_+, j = 1, \dots, m. \tag{3.4}$$

To this end, for $\mathcal{A}(D)$ and $\mathcal{B}_j(D)$ defined as above, we define an operator A in $L_1(\mathbb{R}^{n+1}_+; E)$ associated to the boundary value problem (3.3) and (3.4) with $g_j =$

0 for all $j \in \{1,\dots,m\}$ by means of $A = \overline{A_{min}}$, where $A_{min} : D(A_{min}) \to L_1(\mathbb{R}^{n+1}_+;E) \times \Pi_{j=1}^m W_1^{2m-m_j}(\mathbb{R}^{n+1}_+;E)$ is defined by

$$\begin{aligned} D(A_{min}) &:= W_1^{2m}(\mathbb{R}^{n+1}_+;E) \\ A_{min}u &:= \begin{pmatrix} \mathcal{A}(D) \\ \mathcal{B}_1(D) \\ \vdots \\ \mathcal{B}_m(D) \end{pmatrix} u. \end{aligned}$$

Moreover, we set $g := (g_1,\dots,g_m)^T$. For $\lambda \in \Sigma_{\pi-\phi}$, $f \in L_1(\mathbb{R}^{n+1}_+;E)$ and $g_j \in W_1^{2m-m_j}(\mathbb{R}^{n+1}_+;E)$ for $j = 1,\dots,m$, our boundary value problem can be rewritten as

$$\lambda Jv + Av = \begin{pmatrix} f \\ g \end{pmatrix}, \tag{3.5}$$

where $Jv = (v,0,\dots,0)^T$. We then have the following theorem.

Theorem 3.1. *Let $\mathcal{A}(D)$ be a parameter-elliptic operator of order $2m$ and angle of ellipticity $\phi_{\mathcal{A}}$. Let $\phi > \phi_{\mathcal{A}}$. For $j \in \{1,\dots,m\}$ let $\mathcal{B}_j(D)$ be boundary operators of order $m_j < 2m$. Assume that the Lopatinskii-Shapiro condition holds. Let E be a Banach space, $f \in L_1(\mathbb{R}^{n+1}_+;E)$ and $g_j \in W_1^{2m-m_j}(\mathbb{R}^{n+1}_+;E)$ for $j = 1,\dots,m$. Let $\lambda \in \Sigma_{\pi-\phi}$ and let A be defined as above. Then there exists a unique function $u \in W_1^{2m-1}(\mathbb{R}^{n+1}_+;E) \cap D(A)$ satisfying*

$$\lambda Ju + Au = \begin{pmatrix} f \\ g \end{pmatrix}. \tag{3.6}$$

Moreover, u is given by

$$u = P(\lambda + A_{\mathbb{R}^{n+1}})^{-1}E_0 f + \sum_{j=1}^m R_\lambda^j f + \sum_{j=1}^m S_\lambda^j g_j, \tag{3.7}$$

where R_λ^j and S_λ^j are kernel operators as defined in Propositions 6.8 and 6.9 of [DHP01]. *Furthermore, there exists a constant $C > 0$ such that for $0 \le |\alpha| \le 2m-1$ and $\lambda \in \Sigma_{\pi-\phi}$ we have*

$$\begin{aligned} |\lambda^{1-\frac{|\alpha|}{2m}} D^\alpha u|_{L_1(\mathbb{R}^{n+1}_+;E)} \le C[|f|_{L_1(\mathbb{R}^{n+1}_+;E)} &+ \sum_{j=1}^m |(-\Delta + |\lambda|^{\frac{1}{m}})^{\frac{2m-m_j}{2}} g_j|_{L_1(\mathbb{R}^{n+1}_+;E)} \\ &+ \sum_{j=1}^m |(-\Delta + |\lambda|^{\frac{1}{m}})^{\frac{2m-m_j-1}{2}} D_y g_j|_{L_1(\mathbb{R}^{n+1}_+;E)}]. \end{aligned}$$

Here E_0 denotes the extension of a function defined on $\mathbb{R}^{n+1}_+$ to all of $\mathbb{R}^{n+1}$ by 0 and P is the restriction of a function defined on $\mathbb{R}^{n+1}$ to $\mathbb{R}^{n+1}_+$.

Proof. The following proof is a modification of the arguments given in the proof of Theorem 6.10 of [DHP01] to the L_1-situation. Indeed, note that under the given assumptions the problem (3.6) has a unique solution $u \in W_1^{2m-1}(\mathbb{R}_+^{n+1}; E) \cap D(A)$ if and only if u defined as in (3.7) belongs to $W_1^{2m-1}(\mathbb{R}_+^{n+1}; E) \cap D(A)$.

Given $f \in L_1(\mathbb{R}_+^{n+1}; E)$, it follows from Proposition 2.1 (and its proof in case $\phi > \pi/2$) that the first term on the right hand side of (3.7) belongs to the required regularity class.

In order to treat the second term, notice that by Proposition 6.6 of [DHP01] the kernel $k_\lambda^{R,j}$ of R_λ^j satisfies for $0 \le |\alpha| \le 2m-1$ and $\lambda \in \Sigma_{\pi-\phi}$ an estimate of the form

$$|D^\alpha k_\lambda^{R,j}(\cdot, y, y')|_{L^1(\mathbb{R}^n; \mathcal{B}(E))} \le C|\lambda|^{\frac{-2m+|\alpha|+1}{2m}} p_{2m+n,|\alpha|}^{n+1}(c|\lambda|^{\frac{1}{2m}}(y+y')), \quad y, y' > 0. \tag{3.8}$$

The above kernel estimates allow us to derive L^1-estimates for the second term on the right hand side of (3.7) via the following simple lemma on L^1-continuity of integral operators acting in half spaces.

Lemma 3.2. *Let T be an integral operator in $L_1(\mathbb{R}_+^{n+1}; E)$ of the form*

$$(Tf)(x,y) = \int_0^\infty \int_{\mathbb{R}^n} k(x-x', y, y') f(x', y') dx' dy', \quad x \in \mathbb{R}^n, y > 0,$$

where $k : \mathbb{R}^n \times \mathbb{R}_+ \times \mathbb{R}_+ \to \mathcal{B}(E)$ is a measurable function. If

$$\sup_{y'>0} \int_0^\infty |k(\cdot, y, y')|_1 dy =: M < \infty,$$

then $T \in \mathcal{B}(L_1(\mathbb{R}_+^{n+1}; E))$ and $|T|_{\mathcal{B}(L_1(\mathbb{R}_+^{n+1}; E))} \le M$.

The proof of Lemma 3.2 consists only of an application of Young's and Hölder's inequality. Combining estimate (3.8) with Lemma 3.2 it follows that

$$|\lambda^{1-\frac{|\alpha|}{2m}} D^\alpha R_\lambda^j f|_{L_1(\mathbb{R}_+^{n+1}; E)} \le C|f|_{L_1(\mathbb{R}_+^{n+1}; E)}.$$

Similarly, by Proposition 6.8 of [DHP01] the kernel $k_\lambda^{S,j}$ of S_λ^j satisfies for $0 \le |\alpha| \le 2m-1$ and $\lambda \in \Sigma_{\pi-\phi}$ an estimate of the form

$$|D^\alpha K_\lambda^{S,j}(\cdot, y')|_{L^1(\mathbb{R}^n; \mathcal{B}(E))} \le C|\lambda|^{\frac{-2m+|\alpha|+1}{2m}} p_{2m+n,|\alpha|}^{n+1}(c|\lambda|^{\frac{1}{2m}} y'), \quad y' > 0.$$

Again, together with Lemma 3.2, this estimate implies that

$$\begin{aligned} |\lambda^{1-\frac{|\alpha|}{2m}} D^\alpha S_\lambda^j g_j|_{L_1(\mathbb{R}_+^{n+1}; E)} \le \; C \; & \Big(\sum_{j=1}^m |(-\Delta + |\lambda|^{\frac{1}{m}})^{\frac{2m-m_j}{2}} g_j|_{L_1(\mathbb{R}_+^{n+1}; E)} \\ & + \sum_{j=1}^m |(-\Delta + |\lambda|^{\frac{1}{m}})^{\frac{2m-m_j-1}{2}} D_y g_j|_{L_1(\mathbb{R}_+^{n+1}; E)}\Big). \end{aligned}$$

This proves the assertion. □

For a parameter-elliptic operator $A(D)$ of order $2m$ and angle of ellipticity ϕ_A, we define the $L_1(\mathbb{R}^{n+1}_+;E)$-realization A^0_B of the boundary value problem (3.6) with $g=0$ as

$$A^0_B u := \mathcal{A}(D)u \tag{3.9}$$

$$D(A^0_B) := \{u \in W^{2m}_1(\mathbb{R}^{n+1}_+;E); \mathcal{B}_j(D)u = 0 \quad \text{for all} \quad j=1,\dots,m\} \tag{3.10}$$

and set

$$A_B := \overline{A^0_B}.$$

It follows from Theorem 3.1 that $(\lambda + A_B)$ is invertible for all $\lambda \in \Sigma_{\pi-\phi}$ with $\phi > \phi_{\mathcal{A}}$ and that

$$(\lambda + A_B)^{-1} = P(\lambda + A_{\mathbb{R}^{n+1}})^{-1}E_0 + \sum_{j=1}^{m} R^j_\lambda.$$

Theorem 3.1 also implies that $-\sum_{\pi-\phi} \subset \rho(A)$ and that $|\lambda(\lambda + A_B)^{-1}| \leq M$ for $\lambda \in \Sigma_{\pi-\phi}$ with $\phi > \phi_{\mathcal{A}}$. We thus have the following corollary.

Corollary 3.3. *Suppose that $\phi_{\mathcal{A}} < \frac{\pi}{2}$. Then $-A_B$ generates an analytic C_0-semigroup on $L_1(\mathbb{R}^{n+1}_+;E)$.*

References

[ADN59] S. Agmon, A. Douglis, L. Nirenberg, *Estimates near the boundary for solutions of elliptic partial differential equations satisfying general boundary conditions, I.* Comm. Pure Appl. Math. **22** (1959), 623–727. II. Comm. Pure Appl. Math. **17** (1964), 35–92.

[Ama83] H. Amann, *Dual semigroups and second order linear elliptic boundary value problems.* Israel J. Math. **45** (1983), 225–254.

[Ama01] H. Amann, *Elliptic operators with infinite-dimensional state spaces.* J. Evol. Equ. **1** (2001), 143–188.

[AE96] H. Amann, J. Escher, *Strongly continuous dual semigroups.* Ann. Mat. Pura Appl. **CLXXI** (1996), 41–62.

[DHP01] R. Denk, M. Hieber, J. Prüss, *$\mathcal{R}$-boundedness, Fourier multipliers and problems of elliptic and parabolic type.* Memoirs Amer. Math. Soc., to appear.

[DiB91] G. Di Blasio, *Analytic semigroups generated by elliptic operators in L^1 and parabolic equations.* Osaka J. Math. **28** (1991), 367–384.

[Gui93] D. Guidetti, *On elliptic systems in L^1.* Osaka J. Math. **30** (1993), 397–429.

[Tan97] H. Tanabe, *Functional Analytic Methods for Partial Differential Equations.* Marcel Dekker, New York, 1997.

Robert Denk
Universität Regensburg
Naturwissenschaftliche Fakultät I – Mathematik
D-93040 Regensburg, Germany
E-mail address: robert.denk@mathematik.uni-regensburg.de

Matthias Hieber
Technische Universität Darmstadt
Fachbereich Mathematik
Schlossgartenstr. 7
D-64289 Darmstadt, Germany
E-mail address: hieber@mathematik.tu-darmstadt.de

Jan Prüss
Martin-Luther-Universität Halle-Wittenberg
Fachbereich Mathematik und Informatik
Institut für Analysis
Theodor-Lieser-Straße 5
D-06120 Halle, Germany
E-mail address: anokd@volterra.mathematik.uni-halle.de

Progress in Nonlinear Differential Equations
and Their Applications, Vol. 55, 149–156

On Moving Boundaries in Elliptic–Parabolic Systems

Joachim Escher

1. Introduction

Of concern is a simple model describing in vivo cancer growth for a single tumor. The model comprises a reaction–diffusion equation describing the evolution of the nutrient concentration, denoted by u and an elliptic equation for the internal pressure, denoted by p, in the tissue. The cell proliferation rate is denoted by $f(u)$, where $f : \mathbb{R} \to \mathbb{R}$ is assumed to be smooth. At time t the tumor occupies the domain $\Omega(t)$ with the moving boundary $\Gamma(t)$. In dimensionless form p and u satisfy the equations

$$-\Delta p = f(u) \quad \text{in} \quad \Omega(t), \tag{1.1}$$

$$\partial_t u - \Delta u + u = 0 \quad \text{in} \quad \Omega(t). \tag{1.2}$$

To formulate the boundary conditions to be satisfied by p and u, let V denote the normal velocity of the boundary, i.e., the component of the derivative of $t \mapsto \Gamma(t)$ in direction of the outward normal on $\Gamma(t)$. Moreover, we write $H(t)$ for the mean curvature of $\Gamma(t)$ and assume that the nutrient concentration outside the tumor is at a constant value ψ. Then we have the following boundary conditions

$$V = -\partial_\nu p \quad \text{on} \quad \Gamma(t) \tag{1.3}$$

$$p = cH \quad \text{on} \quad \Gamma(t) \tag{1.4}$$

$$u = \psi \quad \text{on} \quad \Gamma(t). \tag{1.5}$$

Here c is a positive constant, called surface tension. Further we use the sign convention that V is positive if the tumor grows and that H is positive if $\Gamma(t)$ is convex with respect to $\Omega(t)$. Finally, we close the system by the initial conditions

$$\Gamma(0) = \Gamma_0 \quad \text{and} \quad u(0,\cdot) = u_0 \ \text{in} \ \Omega_0 \tag{1.6}$$

where Ω_0 denotes the domain initially occupied by the tumor, so that $\Gamma_0 := \partial\Omega_0$ is the initial shape of the tumor. The initial nutrient concentration is denoted by u_0.

In order to give a precise statement of our results, we denote by B^s_{pp}, where $p \in [1,\infty]$ and $s \geq 0$, the usual Besov spaces, cf. [12].

Theorem 1. *Let $n \geq 2$ and $p > n$. Let further Ω_0 be a bounded domain in $\mathbb{R}^n$ of class $B^{4-1/p}_{pp}$ and assume that $u_0 - \psi \in W^2_{p,0}(\Omega_0)$. Then system (1.1)–(1.6) possess* *a unique classical solution* (p,u,Γ) *of class* $W_p^2 \times W_p^2 \times B_{pp}^{4-1/p}$ *on some time interval* $[0,T)$ *with* $T>0$. *This solution is smooth on the time–space manifold*

$$\bigcup_{t\in(0,T)} (\{t\}\times\overline{\Omega}(t)).$$

Let Γ_0 be a compact hypersurface of $\mathbb{R}^n$ which is of class C^4. Then it follows from Proposition 2.3.2 in [12] that $C^4(\Gamma_0)\subset B_{pp}^{4-1/p}(\Gamma_0)$. Thus any bounded C^4–domain in $\mathbb{R}^n$ is within the scope of Theorem 1.

We briefly exhibit the method used to establish the above result. In the first step we introduce suitable diffemorphisms to transform the original domains $\Omega(t)$ and original moving boundary $\Gamma(t)$ onto a fixed domain Ω with a fixed boundary Σ, respectively. Next we perform a reduction of the transformed system by eliminating the pressure. We are then left with a nonlinear system of the form

$$\partial_t Z = F(Z), \quad Z(0) = Z_0, \tag{1.7}$$

where Z and Z_0 denote the transformed pairs (u,Γ) and (u_0,Γ_0), respectively. Besides the transformed differential operators from (1.2) and (1.3) the operator F contains the solution operator for the (transformed) pressure p as well. It turns out, that F is a quasilinear and nonlocal operator. However, it can be shown in that the evolution equation is of parabolic type, in the sense that F contains a leading linear term which generates an analytic semigroup on a suitable product of Sobolev spaces over Ω and over Σ. This makes it possible to apply the well established theory of quasilinear parabolic evolution equations which is based on analytic semigroups, see [1, 11]. System (1.1)–(1.6) was introduced by [9] and [10], see also [3]. The radially symmetric situation was recently investigated in a series of papers by Friedman and Reitich [6, 7, 8]. Existence and uniqueness of nonsymmetric solutions have been shown even more recently in [2] in the physically relevant case of space dimension $n=3$. However, existence of solutions is guaranteed in [2] only under high regularity assumptions on the initial data and under a serious loss of regularity for the solution. More precisely, it is shown in the main result of [2] that, given Ω_0 of class $C^{17+\alpha}$ and u_0 of class $C^{13+\alpha}$, the solution (u,Γ) belongs to $C^{7+\alpha}\times C^{11+\alpha}$. We point out that the regularity properties of the solution constructed by Theorem 1 above are optimal with respect to that aspect.

2. The transformed system

Let Ω_0 be a bounded domain in $\mathbb{R}^n$ such that its boundary Γ_0 is a compact orientable embedded hypersurface of class $C^{2+\alpha}$. we choose a smooth hypersurface Σ in a sufficiently small tubular neighbourhood of Γ_0 and write μ for the outward unit normal field on Σ. Then for $a>0$, but small enough, the mapping

$$X:\Sigma\times(-a,a)\to\mathbb{R}^n, \quad (p,r)\mapsto p+r\mu(p) \tag{2.1}$$

is a smooth diffeomorphism onto its range $\mathcal{R} := \operatorname{im}(X)$, i.e., $X\in \operatorname{Diff}^\infty(\Sigma\times(-a,a),\mathcal{R})$. It is clear that we can choose Σ in such a way that $\Gamma_0\subset\mathcal{R}$. Later on it will be convenient to decompose the inverse of X into $X^{-1} = (P, \Lambda)$ such that

$$P \in C^\infty(\mathcal{R}, \Sigma) \quad \text{and} \quad \Lambda \in C^\infty(\mathcal{R}, (-a, a)).$$

Here P is the metric projection of $\mathcal{R}$ onto Σ, i.e., $P(x)$ is the nearest point on Σ to x, and $\Lambda(x)$ is the signed distance from x to Σ. Clearly, $\mathcal{R}$ is the set of those points with distance less than a to Σ. We call

$$\mathrm{Ad} := \left\{ \rho \in C^2(\Sigma)\,;\; \|\rho\|_{C^1(\Sigma)} < b \right\} \tag{2.2}$$

with $b \in (0, a)$, the set of admissible functions. Given $\rho \in \mathrm{Ad}$, let

$$\theta_\rho : \Sigma \to \mathbb{R}^n, \quad p \mapsto p + \rho(p)\mu(p)$$

and $\Gamma_\rho := \mathrm{im}(\theta_\rho)$. Then Γ_ρ is a C^2–hypersurface which is diffeomorphic to Σ, i. e. $\theta_\rho \in \mathrm{Diff}^2(\Sigma, \Gamma_\rho)$. We write Ω_ρ for the domain enclosed by Γ_ρ. We also need a suitable extension $\Theta_\rho : \mathbb{R}^n \to \mathbb{R}^n$ of θ_ρ to $\mathbb{R}^n$. To this end we introduce a special case of the so–called Hanzawa transformation. Let $b \in (0, a/4)$ be given, and pick $\varphi \in C^\infty(\mathbb{R}, [0, 1])$ such that

$$\varphi(r) = \begin{cases} 1 & \text{if} \quad |r| \le b, \\ 0 & \text{if} \quad |r| > 3b \end{cases} \qquad \text{and} \qquad \sup|\phi'(r)| < 1/b. \tag{2.3}$$

Then, given $\rho \in \mathrm{Ad}$, define

$$\Theta_\rho(x) := \begin{cases} X(P(x), \Lambda(x) + \varphi(\Lambda(x))\rho(P(x)) & \text{if } X \in \mathcal{R}, \\ x & \text{if } X \notin \mathcal{R}. \end{cases}$$

Observe that $|\varphi'(r)\rho(p)| < 1$, for $\rho \in \mathrm{Ad}$ and $r \in \mathbb{R}, p \in \Sigma$. Hence the function $[r \mapsto r + \varphi(r)\rho(p)]$ is strictly increasing for any $\rho \in \mathrm{Ad}$. From this one easily deduces that $\Theta_\rho \in \mathrm{Diff}^2(\mathbb{R}^n, \mathbb{R}^n)$ such that

$$\Theta_\rho \in \mathrm{Diff}^2(\Omega, \Omega_\rho) \quad \text{and} \quad \Theta_\rho|\Sigma = \theta_\rho. \tag{2.4}$$

We consider now time–dependent functions $\rho : J \to \mathrm{Ad}$, where $J := [0, T]$ for some $T > 0$. More precisely, given

$$\rho \in C^1(J, C(\Sigma)) \cap C(J, \mathrm{Ad})$$

we obtain a family

$$\Gamma := \left\{ \Gamma(t) = \Gamma_{\rho(t)}\,;\; t \in J \right\}$$

of hypersurfaces in $\mathbb{R}^n$. The domains enclosed by those hypersurfaces are denoted by $\Omega(t) = \Omega_{\rho(t)}$, $st \in J$. It will be convenient to describe $\Gamma(t)$ as the zero–level set of a suitable time–dependent function. Letting

$$\phi_\rho : J \times \mathbb{R} \to \mathbb{R}, \quad (t, x) \mapsto \Lambda(x) - \rho(t, P(x)),$$

we obviously have that $\Gamma(t) = \phi_\rho(t, \cdot)^{-1}(0)$. Hence the outward unit normal field $\nu(\cdot, t)$ on $\Gamma(t)$ at the point $x = X(s, \rho(t, s))$ can be represented as

$$\nu(t, s) = \frac{\nabla\phi_\rho(t, x)}{|\nabla\phi_\rho(t, x)|_{x = X(s, \rho(t, s))}}, \quad (t, s) \in J \times \Sigma. \tag{2.5}$$

In order to calculate the normal velocity of $[t \mapsto \Gamma(t)]$ at the time $t \in [0,T)$ and at the point $x = X(s,\rho(t,s))$, let us introduce the function ψ by setting

$$\psi(\tau,r) := \Lambda(x + r\nu(t,s)) - \rho(P(x + r\nu(t,s)), t+\tau)$$

for $(\tau,r) \in [0,T-t) \times [0,a-b)$. Clearly $\psi(0,0) = 0$ and

$$\partial_2\psi(0,0) = \nabla\phi_\rho(t,x)\cdot\nu(t,s) = |\nabla\phi_\rho(t,x)| \neq 0.$$

Hence the implicit function theorem implies that there is an $\varepsilon > 0$ and a function $f \in C^1([0,\varepsilon),\mathbb{R})$ such that $\psi(\tau,f(\tau)) = 0$ for $\tau \in [0,\varepsilon)$. Further,

$$f'(0) = \frac{-\partial_1\psi(0,0)}{\partial_2\psi(0,0)} = \frac{\partial_t\rho(t,s)}{|\nabla\phi_\rho(t,x)|}. \tag{2.6}$$

Observe that $x + f(\tau)\nu(t,s)$ belongs to $\Gamma(t+\tau)$. Thus $f(\tau)$ is the increase at $x = X(s,\rho(t,s)) \in \Gamma(t)$ in direction of the outer normal $\nu(t,s)$ so that the normal velocity $V(t,s)$ of $[t \mapsto \Gamma(t)]$ at t and $x = X(s,\rho(t,s))$ is given by $f'(0)$. Therefore (2.6) yields

$$V(t,s) = \frac{\partial_t\rho(t,s)}{|\nabla\phi_\rho(t,x)|_{x=X(s,\rho(t,s))}}, \qquad (t,s) \in J\times\Sigma. \tag{2.7}$$

We can now rewrite the one–phase Stefan condition (1.3) in the form

$$\partial_t\rho(t,s) = -(\nabla p(t,x)|\nabla\phi_\rho(t,x))|_{x=X(s,\rho(t,s))}, \qquad (t,s) \in J\times\Sigma, \tag{2.8}$$

where $(\cdot|\cdot)$ stands for the Euclidean inner product in $\mathbb{R}^n$. We need some further notation. Let

$$\Omega_{\rho,J} := \bigcup_{t\in J}(\{t\}\times\Omega_{\rho(t)}) \quad \text{for} \quad \rho \in C^1(J,C(\Sigma)) \cap C(J,\mathrm{Ad})$$

and define

$$\Phi_\rho : J\times\Omega \to \Omega_{\rho,J}, \quad (t,y) \mapsto (t,\Theta_{\rho(t)}(y)).$$

Then $\Phi_\rho \in \mathrm{Diff}^0(J\times\Omega,\Omega_{\rho,J})$ and hence the pull–back operator Φ_ρ^* defined by

$$\Phi_\rho^* u := u\cdot\Phi_\rho \quad \text{for} \quad u \in BUC(\Omega_{\rho,J}), \tag{2.9}$$

as well as the push–forward operator Φ_*^ρ given by

$$\Phi_*^\rho v := v\cdot\Phi_\rho^{-1} \quad \text{for} \quad v \in BUC(J\times\Omega) \tag{2.10}$$

are well defined. Using this notation we introduce the following transformed differential operators

$$\begin{aligned} A(\rho)v &:= -\Phi_\rho^*\Delta(\Phi_*^\rho v), \\ B(\rho)v &:= \Phi_\rho^*(\nabla(\Phi_*^\rho v)|\nabla\phi_\rho) \end{aligned} \tag{2.11}$$

for $v \in C(J,W_p^2(\Omega))$. Of course we need also the transformed versions of the nonlinear operator f and the mean curvature, respectively:

$$g(\rho,v) := -\Phi_\rho^* f(\Phi_*^\rho v) \quad \text{and} \quad H(\rho) := \Phi_\rho^* H_{\Gamma_\rho}, \tag{2.12}$$

where $H_{\Gamma_{\rho(t)}}$ denotes the mean curvature of $\Gamma_{\rho(t)}$ for $t \in J$. Furthermore, given

$$v \in C(J,BUC^1(\Omega)), \quad q \in C(J,BUC^1(\Omega)), \quad \rho \in C(J,\mathrm{Ad}),$$

we introduce a mapping $R(v,q,\rho)$ on Ω by

$$R(v,q,\rho)(y) := \begin{cases} [\varphi \circ \Lambda] \cdot [(B(\rho)q) \circ P] \cdot [B_\mu(\rho)v](y), & y \in \mathcal{R} \\ 0, & y \notin \mathcal{R} \end{cases} \tag{2.13}$$

where $B_\mu(\rho)v := \Phi^*_\rho(\nabla(\Phi^\rho_* v)|\mu \circ P)$. The transformed initial data is of course:

$$v_0 := u_0 \circ \Theta_{\rho_0}. \tag{2.14}$$

Then it follows from (2.11)–(2.14) that the original system (1.1)–(1.6) can be transformed in the following system for $q := \Phi^*_\rho p$ and $v := \Phi^*_\rho u$:

$$\begin{cases} A(\rho)q &= g(\rho,v) & \text{in} & J \times \Omega \\ \partial_t v + A(\rho)v &= R(v,q,\rho) - v & \text{in} & J \times \Omega \\ \partial_t \rho + B(\rho)q &= 0 & \text{in} & J \times \Omega \\ q &= H(\rho) & \text{on} & J \times \Sigma \\ v &= \psi & \text{on} & J \times \Sigma \\ v(0,\cdot) &= v_0 & \text{in} & \Omega \\ \rho(0,\cdot) &= \rho_0 & \text{on} & \Sigma. \end{cases} \tag{2.15}$$

Let now $\rho_0 \in B^{4-1/p}_{pp}(\Sigma) \cap \mathrm{Ad}$ and $w_0 \in W^2_p(\Omega)$ be given, where $p > n$. We call (v,q,ρ) a classical solution to (2.15) with initial data (v_0,ρ_0) if the following conditions hold:

(i) $\rho \in C(J, B^{4-1/p}_{pp}(\Sigma) \cap \mathrm{Ad}) \cap C^1(J, B^{2-1/p}_{pp}(\Sigma)) \cap C^\infty(\dot{J}, BUC^\infty(\Sigma))$;

(ii) $v \in C^\infty(\dot{J} \times \Omega, \mathbb{R})$;

(iii) $(v,q) \in C(J, W^2_p(\Omega) \times W^2_p(\Omega))$;

(iv) (v,q,ρ) satisfies (2.15) pointwise on $J \times \bar{\Omega}$.

In order to give a precise notion of a classical solution to (1.1)–(1.6), let Γ_0 satisfy the assumption stated at the beginning of this section and let $u_0 \in W^2_p(\Omega_0)$. Given $\rho \in C^1(J, C(\Sigma)) \cap C(J, \mathrm{Ad})$, define

$$\Omega_{\rho,\dot{J}} := \bigcup_{t \in (0,T)} (\{t\} \times \Omega_{\rho(t)}),$$

and observe that $\Omega_{\rho,\dot{J}}$ is an open subset of $\mathbb{R}^{n+1}$. We call (u,p,Γ) a classical solution to (1.1)–(1.6) with initial data (u_0,Γ_0) if there exists a $\rho \in C^1(J, C(\Sigma)) \cap C(J, \mathrm{Ad})$ such that the following conditions hold:

(i) $\Gamma(t) = \Gamma_{\rho(t)}$ for $t \in J$;

(ii) $u \in C^\infty(\Omega_{\rho,\dot{J}}, \mathbb{R})$;

(iii) $(u(t,\cdot), p(t,\cdot)) \in W^2_p(\Omega_{\rho(t)}) \times W^2_p(\Omega_{\rho(t)})$ for $t \in J$;

(iv) (u,p,Γ) satisfies (1.1)–(1.6) pointwise on $\bigcup_{t \in J}(\{t\} \times \bar{\Omega}_{\rho(t)})$.

Then the following result holds true, cf. [5]:

Lemma 1. *Suppose that $\rho_0 \in B_{pp}^{4-1/p}(\Sigma) \cap Ad$ such that $\Gamma_0 = \Gamma_{\rho_0}$ and put $v_0 := \Theta_{\rho_0}^* u_0$. Moreover, assume that (w, q, ρ) is a classical solution to (2.15) with initial data (v_0, ρ_0). Then, letting*

$$u := \Phi_*^\rho v, \quad p := \Phi_*^\rho q, \quad \Gamma(t) := \Gamma_{\rho(t)}, \quad t \in J,$$

we have that (u, p, Γ) is a classical solution to (1.1)–(1.6).

3. The reduced system

In this section we reduce system (2.15) by eliminating the transformed pressure q. In order to do this, we need a suitable representation of the solution operator to the boundary value problem

$$A(\rho)q = f \ \text{ in } \Omega, \qquad \gamma_\Sigma q = g \ \text{ on } \Sigma, \tag{3.1}$$

where γ_Σ stands for the trace operator with respect to Σ. In the following we use the notation $U := B_{pp}^{4-1/p}(\Sigma) \cap \text{Ad}$. Using Theorem 4.3.3 in [12] it can be shown that here exists

$$(S, T) \in C^\infty(U, \mathcal{L}(L_p(\Omega) \times B_{pp}^{2-1/p}(\Sigma), W_p^2(\Omega)))$$

such that, given $\rho \in U$ and $(f, g) \in L_p(\Omega) \times B_{pp}^{2-1/p}(\Sigma)$, the unique solution $q \in W_p^2(\Omega)$ to (3.1) is given by $q = S(\rho)f + T(\rho)g$.

Then it is easily verified that the pair (v, ρ) satisfies the following reduced system

$$\left\{\begin{array}{rcll} \partial_t v + A(\rho)v &=& F_1(v,\rho) & \text{in} \quad J \times \Omega \\ \partial_t \rho + B(\rho)T(\rho)H(\rho) &=& S(\rho)f(v) & \text{in} \quad J \times \Sigma \\ v &=& \psi & \text{on} \quad J \times \Sigma \\ v(0,\cdot) &=& v_0 & \text{in} \quad \Omega \\ \rho(0,\cdot) &=& \rho_0 & \text{on} \quad \Sigma, \end{array}\right. \tag{3.2}$$

where we set for simplicity

$$F_1(v, \rho) := R(v, S(\rho)f(v) + T(\rho)H(\rho), \rho) - v. \tag{3.3}$$

It is well–known that the mean curvature operator carries a quasilinear structure in the following sense:

$$P \in C^\infty(U, \mathcal{L}(B_{pp}^{4-1/p}(\Sigma), B_{pp}^{2-1/p}(\Sigma)) \text{ and } K \in C^\infty(U, B_{pp}^{3-1/p}(\Sigma))$$

such that

$$H(\rho) = P(\rho)\rho + K(\rho) \quad \text{for} \quad \rho \in U.$$

see Lemma 3.1 in [4]. Hence the second equation in (3.2) becomes

$$\partial_t \rho + B(\rho)T(\rho)P(\rho)\rho = B(\rho)[S(\rho)f(v) - T(\rho)K(\rho)] \quad \text{in} \quad J \times \Sigma.$$

Let us now introduce the following nonlinear operator $F := (F_1, F_2)$ for the pair $Z := (v, \rho)$, where F_1 is given in (3.3) and

$$F_2(Z) := B(\rho)[S(\rho)f(v) - T(\rho)K(\rho)].$$

We also introduce a quasilinear operator

$$\mathbb{A}(Z)Z := (A(\rho)v, B(\rho)T(\rho)P(\rho)\rho). \tag{3.4}$$

Observe that for Z_1 fixed, the mapping $[Z \to \mathbb{A}(Z_1)Z]$ is linear. Now the system (3.2) can formally be re-expressed in the form

$$Z' + \mathbb{A}(Z)Z = F(Z), \qquad Z(0) = Z_0, \tag{3.5}$$

where, of course, $Z_0 := (v_0, \rho_0)$.

It can be shown that for fixed Z_1 the linear operator $-\mathbb{A}(Z_1)$ generates a strongly continuous analytic semigroup in the product space $L_p(\Omega) \times B_{pp}^{1-1/p}(\Sigma)$ with domain $W^2_{p,0}(\Omega) \times B_{pp}^{4-1/p}(\Sigma)$, cf. [5]. Using this generation result and a carful analysis of the nonlinear mappings $[Z \mapsto \mathbb{A}(Z)]$ and $[Z \mapsto F(Z)]$, one can show the abstract quasilinear evolution equation (3.5) is uniquely solvable by the theory developed by H. Amann [1]. The full detail of this procedure, including a bootstapping argument to get the smoothness of the solution, will be given in the forthcoming paper [5].

References

[1] H. Amann, *Nonhomogeneous linear and quasilinear elliptic and parabolic boundary value problems.* In: H. J. Schmeisser, H. Triebel, eds., Function Spaces, Differential Operators and Nonlinear Analysis, Teubner, Stuttgart, Leipzig, 1993, 9–126.

[2] B. V. Bazaliy, A. Friedman, *A free boundary problem for an elliptic-parabolic system: application to a model of tumor growth.* Preprint.

[3] H. M. Byrne, M. A. J. Chaplain, *Growth of necrotic tumors in the presence and absence of inhibitors.* Math. Biosciences **135** (1996), 187–216.

[4] J. Escher, G. Simonett, *Classical solutions for Hele-Shaw models with surface tension.* Adv. Diff. Equs. **2** (1997), 619–642.

[5] J. Escher, *Classical solutions to moving boundary problems for elliptic–parabolic systems.* Preprint.

[6] A. Friedman, F. Reitich, *Analysis of a mathematical model for growth of tumor.* J. Math. Biology **38** (1999), 262–284.

[7] A. Friedman, F. Reitich, *Symmetry-breaking bifurcation of analysic solutiond to free boundary problems: an application to a model of tumor growth.* Trans. Amer. Math. Soc., to appear.

[8] A. Friedman, F. Reitich, *Nonlinear stability of quasi-static Stefan problems with surface tension: a continuation approach.* Ann. Scu. Norm. Super. Pisa, to appear.

[9] H. P. Greenspan, *Models for the growth of a solid tumor by diffusion.* Stud. Appl. Math. **LI** (1972), 317–340.

[10] H. P. Greenspan, *On the growth and stability of cell cultures and solid tumors.* J. Theor. Biology **56** (1976), 229–242.

[11] A. Lunardi, *Analytic Semigroups and Optimal Regularity in Parabolic Problem,* Birkhäuser, Basel, 1995.

[12] H. Triebel, *Theory of Function Spaces*, Birkhäuser, Basel, 1983.

Joachim Escher
Institute for Applied Mathematics
University of Hannover
Postfach 6009
D–30060 Hannover, Germany
E-mail address: escher@ifam.uni-hannover.de

Progress in Nonlinear Differential Equations
and Their Applications, Vol. 55, 157–168

Spectral Mapping Theorems and Spectral Space-Independence

Eva Fašanga

1. Introduction

Motivated by the interest in spectral mapping theorems for $\mathcal{C}_0$-semigroups of bounded linear operators on a Banach space (see, e.g., [8]), we present here a unified approach for spectral mapping theorems for various kinds of functional calculi. As a byproduct of our investigation, and motivated by [7], we obtain that the spectrum of an operator, when defined on two different spaces, is in some circumstances the same.

Many results presented here are known, or even classical, but we collect them for didactical purpuses to show that ideas can be identical in different situations.

The paper is essentially divided into two main parts: Chapter 3 desribing the functional calculi and formulating the spectral mapping theorems, and Chapter 4 being devoted to prove spectral space-independence. The method of proving the spectral mapping theorems is the use of an abstract result from [16] for Banach algebra homomorphisms, where we compute the so-called Arveson spectrum of an appropriate homomorphism. Therefore in Chapter 2 we collect some basic notions from the theory of Banach algebras and cite this abstract theorem.

In Section 3.1 we formulate the (weak) spectral mapping theorems for functional calculus associated to generators of bounded $\mathcal{C}_0$-groups, and more generally, to groups which have non-quasianalytic growth bound. It is called the Phillips functional calculus. The weak spectral mapping theorem is known for a longer time for the Dirac measures (see, e.g., [15] for a proof and references), and even for the sum of an L^1-function and a discrete measure (see [5], [12]). The method we use allows for a weak spectral mapping theorem for discrete measures or L^1-functions (but not sums). Our approach is very close to that of [14].

In Section 3.2 we investigate the discrete analogue of the preceeding situation (replace $(\mathbb{R}, +)$ by $(\mathbb{Z}, +)$), starting from a doubly power bounded operator, and more generally, from an operator with non-quasianalytically growing powers. The associated functional calculus is again well-known. Both the discrete and continuous situation is studied in [12] in the context of non-quasianalytic representations of locally compact abelian groups.

The last section of Chapter 3 is devoted to the Davies functional calculus. Almost all ingredients and results are taken from [6], except of the weak spectral mapping theorem, as formulated in Theorem 3.12 (this is proved in [3]). Note nevertheless that a result very similar to Lemma 3.11 may be found in [7], where it is used to prove the independence of the spectrum on the underlying space. We have been also motivated by the paper [11], where the Davies calculus and the calculus of Galé and Pytlik are described.

Chapter 4 is written after inspiration by the paper [7] where the spectral space-independence is proved for the Davies functional calculus (and is applied to a variety of operators, including Schrödinger operators). We recollect in a little bit rearranged manner the ideas from [7] in such a way, that they can be repeated for the functional calculi for $\mathcal{C}_0$-groups and the discrete analogues. In this way, we obtain (to our best knowledge) new results on the independence of the spectrum on the underlying Banach space for all operators under our investigations.

In our opinion, the main disadvantage of these results is that only operators whose spectrum is equal to the boundary of the spectrum (and hence to the approximate spectrum) are treated. Our approach cannot be verbatim used for, e.g., generators of analytic semigroups, although there is a well-developped functional calculus for them. The reason is, in the context of the Phillips functional calculus, that $(L^1(\mathbb{R},+),\star)$, being a Banach algebra, is not regular.

In [2] the spectral mapping theorem is approached through the theory of spectral distributions. We will not analyze in detail the relation between such distributions and our homomorphisms here. In the direction of spectral space-dependence and independence see also the work [1], and in the context of semigroups and their growth bound the work [4, Section 2].

The author would like to thank J. Milota for many stimulating discussions and to W. Desch for the motivating example, where the spectrum is space-dependent.

2. Preliminaries: Spectrum and Gelfand transform

A Banach algebra $\mathcal{A}$ is a Banach space over $\mathbb{C}$ with multiplication "$\cdot$" satisfying $\|a \cdot b\| \leq \|a\| \, \|b\|$. Its Gelfand space is $\Delta\mathcal{A} := \{\varphi : \mathcal{A} \to \mathbb{C}$; nonzero, continuous, linear, multiplicative $\}$ equipped with the w*-topology. The Gelfand transform of $a \in \mathcal{A}$ is defined by $\widehat{a} : \Delta\mathcal{A} \to \mathbb{C}$, $\widehat{a}(\varphi) = \varphi(a)$, hence we get a mapping from $\mathcal{A}$ into the set of (continuous) functions defined on $\Delta\mathcal{A}$. If $e \in \mathcal{A}$ is the unit and $a \in \mathcal{A}$, then we set $\sigma(a) := \mathbb{C} \setminus \{\lambda \in \mathbb{C}, (\lambda e - a)^{-1} \in \mathcal{A}\}$, the spectrum of a.

A commutative Banach algebra $\mathcal{A}$ is called semisimple if for any $a, b \in \mathcal{A}$, $a \neq b$, there exists a $\varphi \in \Delta\mathcal{A}$ such that $\varphi(a) \neq \varphi(b)$ (i.e., the Gelfand map is injective). It is called regular if for any $F \subset \Delta\mathcal{A}$ closed and any $\varphi \in \Delta\mathcal{A} \setminus F$ there exists an $a \in \mathcal{A}$ such that $\widehat{a} = 0$ on F and $\widehat{a}(\varphi) \neq 0$.

The following result is a kind of Gelfand theorem (that the spectrum of a may be computed as the range of $\widehat{a}$, in the commutative case), but the situation here is more complicated by having a homomorphism and a noncommutative Banach algebra.

Theorem 2.1. *Let $\mathcal{A}$ be a commutative, semisimple, regular Banach algebra, $\mathcal{B}$ be a Banach algebra with unit, $\Pi : \mathcal{A} \to \mathcal{B}$ be a homomorphism (i.e., continuous, linear, multiplicative). Then for any $a \in \mathcal{A}$ we have*

$$\sigma(\Pi(a)) = \overline{\widehat{a}(\operatorname{Sp}\Pi)},$$

where we will call

$$\operatorname{Sp}\Pi := \bigcap_{b\in\ker\Pi} \ker\widehat{b}$$

the Arveson spectrum of Π.

This abstract result is proved in [16] when Π is injective, and a hint for the noninjective case is given there, too. See also [10], [3] for the proof.

3. Spectral mapping theorems

In this chapter X will be a Banach space and A a linear operator on X. We will prove that $\sigma(g(A)) = \overline{g(\sigma(A))}$, where $g(A)$ is a function of A. Our approach is based on the abstract Theorem 2.1 and on the close relation between the Arveson spectrum and $\sigma(A)$ (this will be formulated always as a lemma preceeding the main theorem).

3.1. The Phillips functional calculus for strongly continuous groups

3.1.1. L^1-FUNCTIONS Let A be a linear densely defined closed operator on X. Assume that A generates a bounded C_0-group $\{e^{At}\}_{t\in\mathbb{R}}$ (observe that this implies that $\sigma(A) \subset i\mathbb{R}$, and so $\sigma(A)$ equals to its boundary and hence to the approximate spectrum $\sigma_{\mathrm{ap}}(A)$). We associate to A a mapping Π defined in the following diagram:

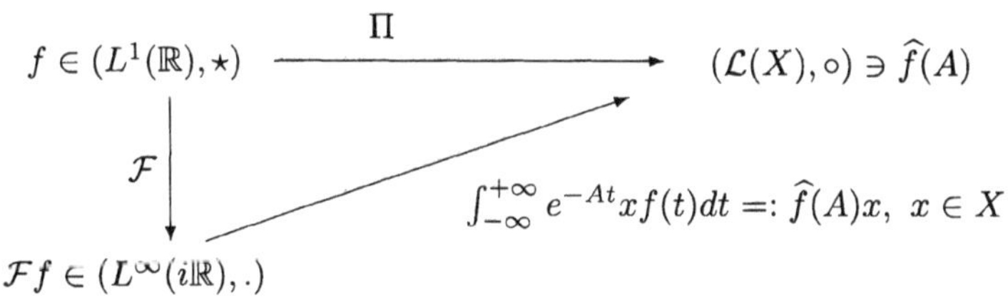

Here $\mathcal{F}$ is the Fourier transform, which for our purposes is defined by $(\mathcal{F}f)(i\varrho) := \int_{-\infty}^{\infty} e^{-i\varrho t} f(t)dt$, so that $\mathcal{F}f$ is a function defined on $i\mathbb{R} \supset \sigma(A)$. The mapping $\Pi : f \mapsto \widehat{f}(A)$ is the so-called Phillips functional calculus.

The space $L^1(\mathbb{R})$ with convolution forms a Banach algebra, which is commutative, regular and semisimple, its Gelfand space is isomorphic to $i\mathbb{R}$ (we identify them) and the Gelfand transform is $\widehat{f}(i\varrho) = (i\varrho)(f) = \mathcal{F}f(i\varrho)$. The space $L^\infty(i\mathbb{R})$ with pointwise multiplication is again a commutative Banach algebra, and $\mathcal{L}(X)$ with composition is a (noncommutative) Banach algebra with unit. The mappings

Π and $\mathcal{F}$ are algebra homomorphisms, and $\Pi \circ \mathcal{F}^{-1}$ on its domain is a functional calculus in the sense that the function $e^{\cdot\lambda_0}\chi_{(-\infty,0)}(.)$ is mapped to $(\lambda_0 - .)^{-1}$ by $\mathcal{F}$, and to $(\lambda_0 - A)^{-1}$ by Π, for any $\Re\lambda_0 > 0$.

The Arveson spectrum of Π is $\operatorname{Sp}\Pi = \cap\{\ker(\mathcal{F}f);\ f \in L^1,\ \widehat{f}(A) = 0\}$ (we use the notation $\ker g := \{x;\ g(x) = 0\} = g_{-1}(\{0\})$ for a mapping g). We collect the following results for purposes of comparison. The following lemma was discovered in [9] (see also [8, Proposition IV.3.18] or [15, Theorem 5.4.1]).

Lemma 3.1. $\operatorname{Sp}\Pi = \sigma(A)$.

Theorem 3.2. $\sigma(\widehat{f}(A)) = \overline{\widehat{f}(\sigma(A))}$.

Proof. By Lemma 3.1 and Theorem 2.1 we have:

$$\sigma(\widehat{f}(A)) = \sigma(\Pi(f)) = \overline{\widehat{f}(\operatorname{Sp}\Pi)} = \overline{\widehat{f}(\sigma(A))}.$$

□

3.1.2. Discrete measures Let A and X be as in the previous paragraph. We replace only $L^1(\mathbb{R})$ by the space of discrete measures:

$$M_d(\mathbb{R}) := \{\sum_{-\infty}^{\infty} c_n\delta_{t_n};\quad c_n \in \mathbb{C},\ t_n \in \mathbb{R},\ \sum |c_n| < \infty\},$$

i.e., consider

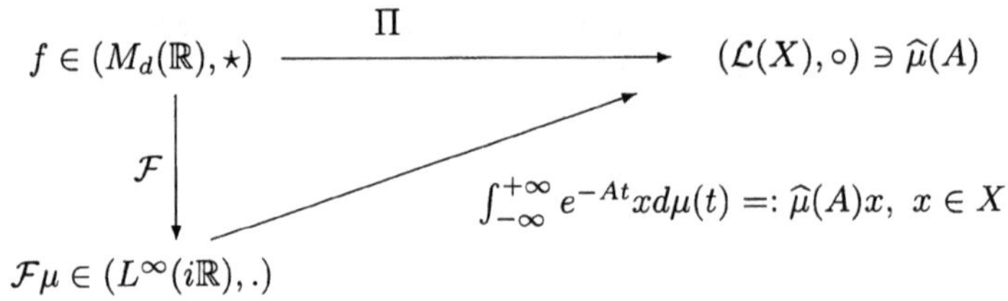

Here the Fourier transform of a measure is defined analogously by $(\mathcal{F}\mu)(i\varrho) := \int_{-\infty}^{\infty} e^{-i\varrho t}d\mu(t)$. Here again, $M_d(\mathbb{R})$ with convolution is a commutative regular semisimple Banach algebra, its Gelfand space is isomorphic to $\widehat{\mathbb{R}}^B$, the Bohr compactification of $\mathbb{R}$, and hence also to $i\widehat{\mathbb{R}}^B$ and after their identification the Gelfand transform is the extension of the Fourier transform (more precisely, the unique continuous extension of an almost periodic function to the Bohr compactification, factored by i):

$$\widehat{\mu}(iz) = \left(\sum c_n e^{-it_n\cdot}\right)(z), \quad z \in \widehat{\mathbb{R}}^B, \quad \mu \in M_d.$$

(One can look at $(M_d(\mathbb{R}), \star)$ as at the group algebra associated to the locally compact abelian group $(\mathbb{R}, +)$ with the discrete topology.) We again have homomorphisms of algebras and the name functional calculus is justified by the following example: the measure δ_{-t_0} is mapped to $e^{\cdot t_0}$ by $\mathcal{F}$ and to e^{At_0} by Π, for any $t_0 \in \mathbb{R}$. The Arveson spectrum is $\operatorname{Sp}\Pi = \cap\{\ker\widehat{\mu};\ \mu \in M_d,\ \widehat{\mu}(A) = 0\}$.

We shall prove the following lemma, from which the main result of this paragraph follows easily with the help of Theorem 2.1 (see also [12, Corollary 1.3.16]). For more detailed proofs one can consult [10].

Lemma 3.3. $\operatorname{Sp}\Pi = \overline{\sigma(A)}^{i\widehat{\mathbb{R}}^B}$

Proof.

$\supset$: Here we use the following result: "If $\mu \in M_d$ is such that $\widehat{\mu}(A) = 0$ then $\sigma_{\text{ap}}(A) \subset \ker\mathcal{F}\mu$", which can be proved similarly as in [8, proof of Proposition IV.3.18]. We have then

$$\sigma(A) = \sigma_{\text{ap}}(A) \subset \bigcap_{\widehat{\mu}(A)=0} \ker\mathcal{F}\mu \subset \operatorname{Sp}\Pi,$$

and take the closure in $i\widehat{\mathbb{R}}^B$ (observe that the last set is closed).

$\subset$: Here the regularity of the Banach algebra M_d and two approximation properties (namely that L^1 contains an approximate unit and also one with elements whose Fourier transforms have compact supports) are essentially used.

We shall, moreover, use the property that if $f \in L^1$ is such that $\mathcal{F}f = 0$ on a (euclidean) neighbourhood of $\sigma(A)$ then $\widehat{f}(A) = 0$. This is proved, e.g., in [8, Lemma IV.3.17] with the additional assumption that the support of $\mathcal{F}f$ is compact, but this assumption can be omitted by choosing a sequence $(\varrho_n) \subset L^1$ such that $\operatorname{supp}\mathcal{F}\varrho_n$ is compact and $\varrho_n \star f \to f$ for all $f \in L^1$.

Now we turn to the proof of the second inclusion by contradiction. Let $z \in \operatorname{Sp}\Pi \setminus \overline{\sigma(A)}^{i\widehat{\mathbb{R}}^B}$. We may choose a compact neigbourhood (in the Bohr topology) U of $\overline{\sigma(A)}^{i\widehat{\mathbb{R}}^B}$ such that $z \notin U$. The regularity assures the existence of a $\mu \in M_d$ such that $\widehat{\mu} = 0$ on U, $\widehat{\mu}(z) \neq 0$. Take $\varrho_n(.) = 2n\chi_{[-\frac{1}{n},\frac{1}{n}]}(.)$, an approximate unit. Then $\varrho_n \star \mu \in L^1$, $\mathcal{F}(\varrho_n \star \mu) = \mathcal{F}\varrho_n\mathcal{F}\mu = 0$ on $U \cap i\mathbb{R}$, hence $\mathcal{F}(\varrho_n \star \mu) = 0$ on a neighbourhood of $\sigma(A)$ (euclidean topology). From the above-cited result we get that $\widehat{\varrho_n \star \mu}(A) = 0$. To prove that then $\widehat{\mu}(A) = 0$, it is enough to check the following computation for any $x \in X$:

$$\begin{aligned}
\|\widehat{\varrho_n \star \mu}(A)x - \widehat{\mu}(A)x\| &= \|\int_{\mathbb{R}} e^{-At}x \int_{t-\frac{1}{n}}^{t+\frac{1}{n}} 2n\, d\mu(s)\, dt - \int_{\mathbb{R}} e^{-At}x\, d\mu(t)\| = \\
&= \|\int_{\mathbb{R}} 2n \int_{s-\frac{1}{n}}^{s+\frac{1}{n}} e^{-At}x\, dt\, d\mu(s) - \int_{\mathbb{R}} e^{-As}x\, d\mu(s)\| = \\
&= \|\int_{\mathbb{R}} 2n \int_{-\frac{1}{n}}^{\frac{1}{n}} \left(e^{-A(s+\tau)}x - e^{-As}x\right) d\tau\, d\mu(s)\| \leq \\
&\leq \sup_{s\in\mathbb{R}} \|e^{-As}\|\, \|\mu\| \sup_{z\in[-\frac{1}{n},\frac{1}{n}]} \|e^{-Az}x - x\| \to 0,
\end{aligned}$$

Therefore the choice of z implies that $\widehat{\mu}(z) = 0$, which is a contradiction. □

Theorem 3.4. $\sigma(\widehat{\mu}(A)) = \overline{\widehat{\mu}(\sigma(A))}$.

Proof. By Theorem 2.1 and Lemma 3.3, using the continuity of Gelfand transform, we get:

$$\sigma(\widehat{\mu}(A)) = \sigma(\Pi(\mu)) = \overline{\widehat{\mu}(\operatorname{Sp}\Pi)} = \overline{\widehat{\mu}(\overline{\sigma(A)}^{i\widehat{\mathbb{R}}^B})} = \overline{\widehat{\mu}(\sigma(A))}.$$

□

Corollary 3.5. $\sigma(e^{t_0 A}) = \overline{e^{t_0 \sigma(A)}}$, $t_0 \in \mathbb{R}$.

Proof. Take $\mu = \delta_{-t_0}$. □

3.1.3. Non-quasianalytic bounds The situation is the same as in the previuos paragraphs but the group generated by A is not bounded. Let $w : \mathbb{R} \to [1, \infty)$ be measurable and such that $w(t+s) \leq w(t)w(s)$. We shall suppose that $\|e^{At}\| \leq w(t)$ for $t \in \mathbb{R}$. In our diagram we have to consider a weighted space (a subspace of $L^1(\mathbb{R})$):

$$\begin{array}{ccc} f \in (L^1(\mathbb{R}, w), \star) & \xrightarrow{\quad\Pi\quad} & (\mathcal{L}(X), \circ) \ni \widehat{f}(A) \\ \Big\downarrow \mathcal{F} & \nearrow & \\ & & \int_{-\infty}^{+\infty} e^{-At} x f(t) dt =: \widehat{f}(A)x, \; x \in X \\ \mathcal{F}f \in (L^\infty(i\mathbb{R}), .) & & \end{array}$$

The spaces are Banach algebras and the mappings homomorphisms. It is known that $(L^1(\mathbb{R}, w), \star)$ is regular if and only if w is non-quasianalytic, i.e.,

$$\int_{-\infty}^{\infty} \frac{\ln w(t)}{1+t^2} dt < \infty,$$

and then the Gelfand space is isomorphic to $i\mathbb{R}$ (cf., e.g., [13, Section 5.1]). All the necessary ingredients to prove corresponding results are true in the non-quasianalytic situation (see [10] for the references), hence we could repeat the arguments. We can also replace L^1 by discrete measures (again weighted). We summarize the results.

Theorem 3.6. *Let A be the generator of a C_0-group with non-quasianalytic growth w and $f \in L^1(\mathbb{R}, w)$ or $\mu \in M_d(\mathbb{R}, w)$. Then*

$$\sigma(\widehat{f}(A)) = \overline{\widehat{f}(\sigma(A))}, \qquad \sigma(\widehat{\mu}(A)) = \overline{\widehat{\mu}(\sigma(A))}.$$

In particular, for $\mu = \delta_{-t_0}$ we get $\sigma(e^{t_0 A}) = \overline{e^{t_0 \sigma(A)}}$. We emphasize that this particular case has been proved by many authors and our approach was inspired by that of [14] and by the paper [16].

One could consider $\Pi : M(\mathbb{R}) \to \mathcal{L}(X)$ defined for all (bounded resp. weighted) Borel measures. For μ in $L^1(\mathbb{R}) + M_d(\mathbb{R})$ (or even in a greatest regular closed subalgebra of $(M(\mathbb{R}), \star)$) it has been proved in [12, Theorem 1.4.2] (see also [5]) that $\sigma(\widehat{\mu}(A)) = \overline{\widehat{\mu}(\sigma(A))}$. This statement contains our Theorems 3.2, 3.4, 3.6 as particular cases. However, to repeat verbatim our arguments in this more general context one should compute the Gelfand space of the algebra which may not be possible.

3.2. Calculus for discrete groups of operators

3.2.1. DOUBLY POWER BOUNDED OPERATORS Let A be a bounded linear operator on X. Assume that A is doubly power bounded, i.e., $\|A^n\| \leq M$ for all $n \in \mathbb{Z}$. Observe that this implies that $\sigma(A) \subset \partial B(0,1)$, the unit circle. With A we associate the homomorphism Π by the following diagram:

$$\begin{array}{ccc}
(a_n) \in (l^1(\mathbb{Z}), \star) & \xrightarrow{\quad\Pi\quad} & (\mathcal{L}(X), \circ) \ni f(A) \\
\Big\downarrow \mathcal{F} & \nearrow & \\
& & \sum_{n=-\infty}^{+\infty} a_n A^n x =: f(A)x, \ x \in X \\
f(z) := \sum_{n=-\infty}^{\infty} a_n z^n \in (L^\infty(\partial B(0,1)), .) & &
\end{array}$$

Observe that for $|\lambda_0| \neq 1$ the function $\frac{1}{\lambda_0 - .}$ lies in the range of $\mathcal{F}$ and is mapped to $(\lambda_0 - A)^{-1}$ by Π (since $\frac{1}{\lambda_0 - z} = \sum_{n=0}^{\infty} \frac{1}{\lambda_0^{n+1}} z^n$ for $|\lambda_0| > 1$). The space $l^1(\mathbb{Z})$ with convolution is a (commutative, semisimple) regular Banach algebra, its Gelfand space may be identified with $\partial B(0,1)$ and the Gelfand transform is $\widehat{(a_n)}(z) = z((a_n)) = \sum a_n z^n$, $z \in \partial B(0,1)$. The Arveson spectrum is $\operatorname{Sp}\Pi = \cap\{\ker f;\ f(A) = \sum_{-\infty}^{\infty} a_n A^n = 0\}$. The following lemma is proved in [12] (see also [8, Proposition IV.3.20]) and implies a spectral mapping theorem.

Lemma 3.7. $\operatorname{Sp}\Pi = \sigma(A)$

Theorem 3.8. *$\sigma(f(A)) = f(\sigma(A))$ for $f(z) = \sum_{n=-\infty}^{\infty} a_n z^n$ with $\sum |a_n| < \infty$.*

Proof. Apply Theorem 2.1 and the above lemma with the observation that the compact set $\sigma(A)$ is mapped by the continuous function f onto a closed set:

$$\sigma(f(A)) = \sigma(\Pi((a_n)) = \overline{\widehat{(a_n)}(\operatorname{Sp}\Pi)} = \overline{\widehat{(a_n)}(\sigma(A))} = \overline{f(\sigma(A))} = f(\sigma(A)).$$

□

3.2.2. THE WEIGHTED CASE Let A be as in the previuos paragraph, but we allow a growth $\|A^n\| \leq w(n)$, $n \in \mathbb{Z}$, where $w : \mathbb{Z} \to [1, \infty)$ is such that $w(n+m) \leq w(n)w(m)$, $n, m \in \mathbb{Z}$. We may introduce a homomorphism Π associated to A on the weighted Banach algebra $(l^1(\mathbb{Z},w),\star)$ as in the continuous case. The algebra $l^1(\mathbb{Z},w)$ is regular if and only if w is non-quasianalytic (see [13]), i.e.,

$$\sum_{n=-\infty}^{\infty} \frac{\ln w(n)}{1+n^2} < \infty,$$

and this fact yields a spectral mapping theorem in this case. Compare the proof of the following lemma to those of Lemmas 3.3, 3.1 resp. 3.7.

Lemma 3.9. $\operatorname{Sp}\Pi = \sigma(A)$

Proof.

$\supset$: Let $t \in \sigma(A)$. Since $\sigma(A) = \partial\sigma(A) = \sigma_{\mathrm{ap}}(A)$, we can find approximate eigenvectors (x_k), $\|x_k\| = 1$, $Ax_k - tx_k \to 0$. Now if $f(A) = \sum a_n A^n = 0$ then

$$|f(t)| = \|f(t)x_k\| = \|\sum(a_n t^n x_k - f(A)x_k)\| \le \sum |a_n|\ \|t^n x_k - A^n x_k\| \to 0$$

where the convergence is justified by the Lebesgue theorem. So $t \in \ker f$.

$\subset$: Here we need the auxiliary statement: if $f = 0$ on a neighbourhood of $\sigma(A)$ (euclidean topology restricted to $\partial B(0,1)$) then $f(A) = 0$. To prove this, we repeat the computation from [12]:

$$\begin{aligned} f(A)x &= \sum a_n A^n x = \lim_{r\to 1-} A^{-1} \sum a_n r^{|n+1|} A^{n+1} x = \\ &= \lim_{r\to 1-} A^{-1} \frac{1}{2\pi i} \int_{|z|=1} \sum_{m\in\mathbb{Z}} a_m z^m \sum_{n\in\mathbb{Z}} z^{-n} r^{|n|} A^n x \, dz = \\ &= \lim_{r\to 1-} A^{-1} \frac{1}{2\pi i} \int_{|z|=1} f(z) \frac{z}{r} \left((zr^{-1} - A)^{-1}x - r^2(zr - A)^{-1}x\right) dz = 0, \end{aligned}$$

for $x \in X$, where in the last limit we have used the Lebesgue theorem.

We turn now to the proof of the inclusion by contradiction. Let $t \in \operatorname{Sp}\Pi \setminus \sigma(A)$. Then we can find $(a_n) \in l^1(\mathbb{Z},w)$ such that $f(z) := \sum a_n z^n = 0$ on a neighbourhood of $\sigma(A)$ and $f(t) \neq 0$ (here we use the regularity of $l^1(\mathbb{Z},w)$). The auxiliary statement gives us $f(A) = 0$, which leads to the contradiction $f(t) = 0$ □

Theorem 3.10. *Let A be a bounded linear operator with non-quasianalytic power bound w. Let $f(z) = \sum_{n=-\infty}^{\infty} a_n z^n$ with $\sum |a_n| w(n) < \infty$. Then $\sigma(f(A)) = f(\sigma(A))$.*

3.3. Davies functional calculus

In this section A will be a closed densely defined linear operator on X. We assume that $\sigma(A) \subset \mathbb{R}$ and that for some $\alpha \geq 0$ the following resolvent-estimate holds:

$$\sup_{\lambda\in\mathbb{C}\setminus\mathbb{R}} \frac{\|(\lambda - A)^{-1}\|}{(\sqrt{1+|\lambda|^2})^\alpha} |\Im\lambda|^{\alpha+1} < \infty \tag{1}$$

Define $A_{n+1}(\mathbb{R})$ to be the completion of $\mathcal{C}_c^\infty(\mathbb{R})$ with respect to the norm

$$\|f\|_{n+1} := \sum_{k=0}^{n+1} \int_{\mathbb{R}} |f^{(k)}(x)| \left(\sqrt{1+|x|^2}\right)^{k-1} dx.$$

With A we associate a functional calculus for functions from A_{n+1}. It is proved in [6] that the following mapping (based on the so-called Helffer-Sjöstrand formula) defined on $\mathcal{C}_c^\infty$ extends to an algebra homomorphism for $n > \alpha$:

$$f \in (A_{n+1}(\mathbb{R}), .) \xrightarrow{\quad \Pi \quad} (\mathcal{L}(X), \circ) \ni -\tfrac{1}{\pi} \int_{\mathbb{C}} \tfrac{\partial \widetilde{f}}{\partial \bar{z}}(z)(z-A)^{-1} dz =: f(A)$$

Here $\widetilde{f}$ is an almost analytic extension of f to $\mathbb{C}$, see [6] for the details. The space $A_{n+1}(\mathbb{R})$ is a Banach algebra with pointwise multiplication, it is semisimple and regular, its Gelfand space is isomorphic to $\mathbb{R}$ (see [3] for example). Since the Gelfand map is the identity: $\widehat{f}(t) = t(f) = f(t)$, $t \in \mathbb{R}$ and the function $\frac{1}{\lambda_0 - .}$ is mapped to $(\lambda_0 - A)^{-1}$ for $\lambda \in \mathbb{C}\backslash\mathbb{R}$, we need not introduce another homomorphism to get a functional calculus in the usual sense. The Arveson spectrum is $\operatorname{Sp}\Pi = \cap\{\ker f;\ f \in \mathcal{A}_{n+1},\ f(A) = 0\}$ and we have (see [3]) the following results.

Lemma 3.11. $\operatorname{Sp}\Pi = \sigma(A)$

Theorem 3.12. $\sigma(f(A)) = \overline{f(\sigma(A))}$, $f \in \mathcal{A}_{n+1}$.

We note that in the context of the calculus of Galé and Pytlik, Lemma 3.11 is analogous to [11, Proposition 2.5].

4. Spectral space-independence

In this section we consider one operator which is defined on two different Banach spaces, examples can be differential operators acting on $L^p(\Omega)$ with appropriate domains, for different values of p. We prove that in a number of situations the spectrum is independent on the space provided some natural assumptions are satisfied. These results, combined with the spectral mapping theorem, yield that the spectrum of a function of these operators is independent on the underlying Banach space, too.

In the whole chapter X_1, X_2 are Banach spaces such that $X_1 \cap X_2$ is dense in both X_1 and X_2, A_1, A_2 are linear operators on X_1, X_2, respectively, such that $A_1 x = A_2 x$ for $x \in X_1 \cap X_2$.

Remark 4.1. The spectrum does, in general, depend on the space. Consider, for an $\alpha > 0$ fixed, the Banach space $X_\alpha := \{x := (x_n)_{n=0}^\infty;\ x_n \in \mathbb{C},\ \|x\|_\alpha := \sum |x_n| \alpha^n < \infty\}$ and the shift operator $A_\alpha : X_\alpha \to X_\alpha$, $(x_0, x_1, \dots) \mapsto (x_1, x_2, \dots)$. It is easy to check that $\|A_\alpha\| \le \frac{1}{\alpha}$, and that $\lambda \in \sigma_{\mathrm{ap}}(A_\alpha)$ for any $|\lambda| < \frac{1}{\alpha}$ (in fact, for $x^k := (1, \lambda, \lambda^2, \dots, \lambda^k, 0, \dots)$ we have $\|Ax^k - \lambda x^k\|_\alpha \to 0$, $\|x^k\|_\alpha \to \frac{1}{1-\alpha|\lambda|} \neq 0$, as $k \to \infty$), so $\sigma(A_\alpha) = B(0, \frac{1}{\alpha})$, the ball of radius $\frac{1}{\alpha}$.

4.1. Phillips functional calculus for continuous groups

We assume that A_i is the generator of a bounded C_0-group on X_i (more generally, we may require that $\|e^{A_i t}\| \leq w(t)$ with $w : \mathbb{R} \to [1, \infty)$ measurable, submultiplicative and non-quasianalytic), i=1,2, and finally we assume that $e^{A_1 t}x = e^{A_2 t}x$ for $x \in X_1 \cap X_2$ and $t \in \mathbb{R}$. Consider the situations described in Section 3.1 with the notation $\Pi_i(\mu) = \widehat{\mu}(A_i)$. Then the following results hold both for L^1-functions and discrete measures, we formulate them for measures.

Lemma 4.2. *$\widehat{\mu}(A_1) = 0$ if and only if $\widehat{\mu}(A_2) = 0$.*

Proof. For $x \in X_1 \cap X_2$, a dense set, we may write:

$$0 = \widehat{\mu}(A_1)x = \int_{-\infty}^{\infty} e^{-A_1 t}x \, d\mu(t) = \int_{-\infty}^{\infty} e^{-A_2 t}x \, d\mu(t) = \widehat{\mu}(A_2)x.$$

□

Theorem 4.3. $\sigma(A_1) = \sigma(A_2)$

Proof.

$$\sigma(A_1) = \operatorname{Sp}\Pi_1 = \bigcap_{\widehat{\mu}(A_1)=0} \ker \widehat{\mu} = \bigcap_{\widehat{\mu}(A_2)=0} \ker \widehat{\mu} = \operatorname{Sp}\Pi_2 = \sigma(A_2)$$

□

Corollary 4.4. $\sigma(\widehat{\mu}(A_1)) = \sigma(\widehat{\mu}(A_2))$

4.2. Discrete group of operators

Here let A_1, A_2 be bounded linear operators. Note that in this situation the assumptions mean that A_1 and A_2 are essentially the same, i.e., A_1 can be extended to X_2 in a unique way and will be equal to A_2. We will suppose that A_i is doubly power bounded on X_i (or more generally, the powers may have a common non-quasianalytic growth bound w), i=1,2, and finally we assume that $A_1^n x = A_2^n x$ for $x \in X_1 \cap X_2$ and $n \in \mathbb{Z}$ (for n positive, this follows from the first assumptions automatically). For the situations described in Section 3.2 the following results are valid.

Lemma 4.5. *Assume that $(a_n) \in l^1(\mathbb{Z}, w)$. Then $\sum a_n A_1^n = 0$ if and only if $\sum a_n A_2^n = 0$.*

Theorem 4.6. $\sigma(A_1) = \sigma(A_2)$

Proof.

$$\begin{aligned} \sigma(A_1) &= \operatorname{Sp}\Pi_1 = \bigcap_{\sum a_n A_1^n = 0} \ker(\sum a_n z^n) = \bigcap_{\sum a_n A_2^n = 0} \ker(\sum a_n z^n) \\ &= \operatorname{Sp}\Pi_2 = \sigma(A_2) \end{aligned}$$

□

Corollary 4.7. *$\sigma(f(A_1)) = \sigma(f(A_2))$ for $f(z) = \sum a_n z^n$ with $\sum |a_n| w(n) < \infty$.*

4.3. Davies functional calculus

Here we present the results from [7] for purposes of comparison. We assume that A_1, A_2 satisfy the resolvent estimate (1) with a common $\alpha \geq 0$, $i = 1, 2$, and that $(\lambda - A_1)^{-1}x = (\lambda - A_2)^{-1}x$ for $x \in X_1 \cap X_2$ and $\lambda \in \mathbb{R} \setminus \mathbb{C}$. Consider the situation described in Section 3.3 with $n > \alpha$ and the notation $\Pi_i(f) = f(A_i)$.

Lemma 4.8. *$f(A_1) = 0$ if and only if $f(A_2) = 0$, whenever $f \in C_c^\infty(\mathbb{R})$.*

Proof. For $x \in X_1 \cap X_2$, a dense set, we may write:

$$\begin{aligned} 0 = f(A_1)x &= \frac{-1}{\pi} \int_{\mathbb{C}} \frac{\partial \widetilde{f}}{\partial \overline{z}}(z)(z - A_1)^{-1}x \, dz \\ &= \frac{-1}{\pi} \int_{\mathbb{C}} \frac{\partial \widetilde{f}}{\partial \overline{z}}(z)(z - A_2)^{-1}x \, dz = f(A_2)x. \end{aligned}$$

□

Theorem 4.9. $\sigma(A_1) = \sigma(A_2)$

Proof.

$$\begin{aligned} \sigma(A_1) &= \operatorname{Sp} \Pi_1 = \bigcap_{f(A_1)=0,\ f \in \mathcal{A}_{n+1}} \ker f \subset \\ &\subset \bigcap_{f(A_1)=0,\ f \in C_c^\infty} \ker f = \bigcap_{f(A_2)=0,\ f \in C_c^\infty} \ker f \subset \sigma(A_2), \end{aligned}$$

here it remains to prove the last inclusion. Suppose by contradiction that $t \in \cap\{\ker f;\ f \in C_c^\infty,\ f(A_2) = 0\} \setminus \sigma(A_2)$. Then we can find a function $f \in C_c^\infty$ such that $f(t) = 1$ and $f = 0$ on a neighbourhood of $\sigma(A_2)$. The properties of the calculus (namely: "if $f = 0$ on a neighbourhood of $\sigma(A)$ then $f(A) = 0$", see [6]) imply that $f(A_2) = 0$, hence $f(t) = 0$, a contradiction. □

Corollary 4.10. *$\sigma(f(A_1)) = \sigma(f(A_2))$ for $f \in \mathcal{A}_{n+1}$, $n > \alpha$.*

Acknowledgment

Partially supported by the grants No. MSM 113200007, GAUK No. 166/99, and GAČR No. GP201/01/D094.

References

[1] W. Arendt, *Gaussian estimates and interpolation of the spectrum in L^p*. Diff. and Int. Equs. **7** (1994), 1153–1168.

[2] M. Balabane, H. Emamirad, M. Jazar, *Spectral Distributions and generalization of Stone's theorem.* Acta Appl. Math. **31** (1993), 275–295.

[3] A. Bátkai, E.Fašanga, *The spectral mapping theorem for Davies' functional calculus.* Dept. Math. Anal. Preprint Series No. MATH-KMA-2001/57, Charles University, Prague, 2001.

[4] R. F. Curtain, H. Logemann, S. Townley, H. Zwart, *Well-posedness, stabilizability, and admissibility for Pritchard-Salamon systems.* J. of Mathematical Systems, Estimation, and Control **7** (1997), 439–476.

[5] C. D'Antoni, R. Longo, L. Zsidó, *A spectral mapping theorem for locally compact groups of operators.* Pacific J. Math. **103** (1981), 17–24.

[6] E.-B. Davies, *The functional calculus.* J. London Math. Soc. **52** (1995), 166–176.

[7] E.-B. Davies, *L^p spectral independence and L^1 analyticity.* J. London Math. Soc. **52** (1995), 177–184.

[8] K. J. Engel, R. Nagel, *One-Parameter Semigroups for Linear Evolution Equations.* Springer, 2000.

[9] D. E. Evans, *On the spectrum of a one-parameter strongly continuous group representation.* Math. Scand. **39** (1976), 80–82.

[10] E. Fašangová, *A Banach algebra approach to the weak spectral mapping theorem for C_0-groups.* Ulmer Seminare, 2000, 174–181.

[11] J. E. Galé, P.J.Miana, T.Pytlik, *Spectral properties and norm estimates associated to the $C_c^{(k)}$-functional calculus.* Publicaciones del seminario Matemático Preprint No. 4, Universidad de Zaragoza, Spain, 2000.

[12] S. Z. Huang, *Non-quasianalytic representations of locally compact abelian groups.* Ph.D. Thesis, Universität Tübingen, 1996.

[13] Y. I. Lyubich, *Introduction to the Theory of Banach Representations of Groups.* Birkhäuser Verlag, 1988 (original Russian edition 1985).

[14] R. Nagel, S. Z. Huang, *Spectral mapping theorems for C_0-groups satisfying non-quasianalytic growth conditions.* Math. Nachr. **169** (1994), 207–218.

[15] J. M. A. M. van Neerven, *The asymptotic behaviour of semigroups of linear operators.* Birkhäuser, 1996.

[16] H. Seferoğlu, *Spectral mapping theorem for representations of measure algebras.* Proc. Edinburgh Math. Soc. **40** (1997), 261–266.

Eva Fašanga
Department of Mathematical Analysis
Charles University
Sokolovská 83
186 75 Praha 8, Czech Republic
E-mail address: `fasanga@karlin.mff.cuni.cz`

Progress in Nonlinear Differential Equations
and Their Applications, Vol. 55, 169–180

The Laplacian with Generalized Wentzell Boundary Conditions

Angelo Favini, Gisèle Ruiz Goldstein, Jerome A. Goldstein,
Enrico Obrecht, and Silvia Romanelli

1. Introduction

The regularity of the solutions of the heat equation

$$\frac{\partial u}{\partial t} = \Delta u$$

with suitable boundary conditions in different types of function spaces is an important issue in many applications to problems coming from Physics and Engineering.

The best type of regularity is achieved when the Laplacian Δ generates a (C_o) semigroup of operators which is analytic on the considered space (for the concept of analytical semigroup and the related properties we refer to [3], [10], [14]).

Here we focus on the realization of the Laplacian in the Sobolev space $H^1(\Omega)$, with generalized Wentzell boundary condition given by

$$\Delta u + \beta \frac{\partial u}{\partial n} + \gamma u = 0 \quad on \quad \partial\Omega, \tag{1.1}$$

where Ω is a bounded subset of $\mathbf{R}^N$ with sufficiently regular boundary $\partial\Omega$.

The functions β and γ are taken nonnegative in $C^1(\partial\Omega)$, with β strictly positive, while $n(x)$ denotes the unit outer normal at x.

In a similar order of ideas as in [8], where operators in the divergence form with generalized Wentzell boundary condition were considered on suitable L^p spaces, the tools used here are based essentially on the energy methods and on the $a - priori$ estimates method due to Agmon, Douglis and Nirenberg (see, e.g., Lunardi [11, Chapter 3], Pazy [14, Chapter 7], Tanabe [16, Chapter 3]).

Let us notice that results concerning analyticity of the (C_o) semigroup generated by realizations of the Laplacian with Wentzell boundary condition (i.e., $\Delta u = 0$ on $\partial\Omega$) can be obtained following the methods used in [2], [4]–[7] and [9], while for the Laplacian with Robin boundary condition we refer, e.g., to [11] and to [15].

Different approaches involving the Laplacian with inhomogeneous boundary conditions in $C(\overline{\Omega})$ were recently studied by Arendt in [1].

2. The Main Result

Let A represent the Laplacian Δ acting on functions defined on $\overline{\Omega}$, where Ω is a bounded subset of $\mathbf{R}^N$ with a smooth boundary, so that the divergence theorem holds on Ω. We consider the elliptic equation

$$\lambda u - Au = h \quad \text{in} \quad \Omega \tag{2.1}$$

with $Re\,\lambda > 0$ and $h \in H^1(\Omega)$ given, with the generalized Wentzell boundary condition

$$Au + \beta \frac{\partial u}{\partial n} + \gamma u = 0 \quad \text{on} \quad \partial\Omega. \tag{2.2}$$

Here β, γ are sufficiently smooth functions on $\partial\Omega$, with γ nonnegative and $\beta \geq \epsilon_o > 0$, and in the expression $\frac{\partial u}{\partial n}$, the vector $n = n(x)$ is the unit outer normal to $\partial\Omega$ at x.

From (2.1) we have $Au = \lambda u - h$, so that plugging this into (2.2) we obtain

$$\beta \frac{\partial u}{\partial n} + (\gamma + \lambda) u = h \quad \text{on} \quad \partial\Omega. \tag{2.3}$$

Of course, to be more precise, we should replace h by $h|_{\partial\Omega}$ in (2.3) etc., but the meaning of (2.3) should be clear. To solve the elliptic boundary value problem (2.1), (2.3) we shall use a weak formulation and the Riesz representation theorem. Elliptic boundary value problem theory gives us a unique solution $u \in C^{2+\epsilon}(\overline{\Omega})$ if h, β, γ and $\partial\Omega$ are smooth enough (see, e.g., Lunardi [11]). But we wish to find explicit bounds describing how the solution depends on λ and these bounds come naturally from energy methods.

Multiply (2.1) by $\overline{v} \in H^1(\Omega)$ and integrate over Ω. Then the result is

$$\lambda \int_\Omega u\overline{v}\,dx - \int_\Omega (\Delta u)\overline{v}\,dx = \int_\Omega h\overline{v}\,dx, \tag{2.4}$$

which by the divergence theorem gives

$$\lambda \int_\Omega u\overline{v}\,dx + \int_\Omega \nabla u \cdot \nabla\overline{v}\,dx - \int_{\partial\Omega} \frac{\partial u}{\partial n}\overline{v}\,d\sigma = \int_\Omega h\overline{v}\,dx.$$

Using $\frac{\partial u}{\partial n} = \frac{h-(\gamma+\lambda)u}{\beta}$ from (2.3) and $\beta \geq \epsilon_0 > 0$, we deduce

$$\lambda \int_\Omega u\overline{v}\,dx + \int_\Omega \nabla u \cdot \nabla\overline{v}\,dx + \int_{\partial\Omega} (\lambda+\gamma) u\overline{v}\frac{d\sigma}{\beta} = \int_\Omega h\overline{v}\,dx + \int_{\partial\Omega} h\overline{v}\frac{d\sigma}{\beta}. \tag{2.5}$$

Let $L(u,v)$ be the left hand side of (2.5) and let $F(v)$ be the corresponding right hand side. Notice that h is fixed. Let $\mathcal{H}$ be the closure of $C^1(\overline{\Omega})$ in the norm

$$\|u\|_{\mathcal{H}} := \left(\int_\Omega |u|^2\,dx + \int_\Omega |\nabla u|^2\,dx + \int_{\partial\Omega} |u|^2 \frac{d\sigma}{\beta}\right)^{\frac{1}{2}}.$$

Observe that in [12], [13] Maz'ja stated that

$$\|u\|^2_{L^{\frac{N}{N-1}}(\Omega)} \leq C(N, Vol(\Omega))(\|\nabla u\|^2_{L^2(\Omega)} + \|u|_{\partial\Omega}\|^2_{L^2(\partial\Omega)}) \tag{2.6}$$

and hence the norm $\|\cdot\|_{\mathcal{H}}$ is equivalent to the $H^1(\Omega)$-norm.

Since

$$
\begin{aligned}
|L(u,v)| &\le \max\{|\lambda|,1\}\|u\|_{H^1(\Omega)}\|v\|_{H^1(\Omega)} + (|\lambda| + \|\gamma\|_\infty)\|u\|_{L^2(\partial\Omega,\frac{d\sigma}{\beta})}\|v\|_{L^2(\partial\Omega,\frac{d\sigma}{\beta})} \\
&\le c_1(\lambda)\|u\|_{\mathcal{H}}\|v\|_{\mathcal{H}}, \\
|F(v)| &\le \|h\|_{\mathcal{H}}\|v\|_{\mathcal{H}},
\end{aligned}
$$

provided that $h \in H^1(\Omega)$, as we assumed, we have that L is a bounded sesquilinear form on $\mathcal{H}$ and F is a bounded conjugate linear functional on $\mathcal{H}$. Also

$$
\begin{aligned}
Re\, L(u,u) &\ge Re\,\lambda\|u\|^2_{L^2(\Omega)} + \|\nabla u\|^2_{L^2(\Omega)} + Re\,\lambda\|u\|^2_{L^2(\partial\Omega,\frac{d\sigma}{\beta})} \\
&\ge c_2(\lambda)\|u\|^2_{\mathcal{H}}.
\end{aligned}
$$

By the Riesz representation theorem (or the Lax-Milgram lemma), for all $h \in \mathcal{H}$ there is a unique $u \in \mathcal{H}$ such that $L(u,v) = F(v)$ holds for all $v \in \mathcal{H}$, that is (2.5) holds and this u is our weak solution of (2.1)–(2.2). Moreover, as before mentioned, $u \in C^{2+\epsilon}(\Omega)$, whenever $\beta, \gamma, h, \partial\Omega$ are smooth enough. We shall show that, in fact, the following result holds.

Theorem 2.1. *Assume that Ω is a bounded subset of $\mathbf{R}^N$ with boundary $\partial\Omega$ of class C^2, β,γ are nonnegative functions in $C^1(\partial\Omega)$, with $\beta > 0$. Then the operator $\mathcal{A}$ defined by*

$$
\begin{aligned}
D(\mathcal{A}): &= \{u \in H^1(\Omega) : Au \in H^1(\Omega),\ \tfrac{\partial u}{\partial n} \text{ exists in } L^2(\partial\Omega), \\
&\qquad Au + \beta\tfrac{\partial u}{\partial n} + \gamma u = 0 \text{ on} \partial\Omega = 0\} \\
\mathcal{A}u: &= Au, \quad u \in D(\mathcal{A}),
\end{aligned}
$$

generates an analytic semigroup on $H^1(\Omega)$.

Notice that the operator $\mathcal{A}$ is obviously different from the closure of the realization of A in the suitable space $L^2(\overline{\Omega}, d\mu)$ studied in [8] in a more general situation.

To accomplish this, it seems convenient to discuss in advance in detail the one-dimensional case, and then to parallel the proof for any dimension by the divergence theorem.

3. The One-Dimensional Case

Let us consider $\Omega = (0,1)$ and define

$$\beta_j := (-1)^{j+1}\beta(j), \qquad \gamma_j := \gamma(j) \qquad j = 0,1.$$

Taking into account that

$$\frac{\partial u}{\partial n}(j) = (-1)^{j+1}\frac{du}{dx}(j), \qquad j = 0,1$$

here the generalized Wentzell boundary condition reads as

$$Au(j) + \beta_j u'(j) + \gamma_j u(j) = 0, \qquad j = 0,1$$

where β_j, $\gamma_j \in \mathbf{R}$, $j = 0, 1$ with

$$\beta_o < 0,\ \beta_1 > 0,\ \gamma_o, \gamma_1 \geq 0. \tag{3.1}$$

The first step consists into finding the *a-priori* bounds for $(\lambda - \mathcal{A})u$.

Let $Re\,\lambda > 0$ and consider

$$\lambda u - \mathcal{A}u = h \in H^1(0,1) \tag{3.2}$$

with the generalized Wentzell boundary conditions

$$Au(j) + \beta_j u'(j) + \gamma_j u(j) = 0, \quad j = 0, 1. \tag{3.3}$$

Let us multiply (3.2) by $\overline{u}$ and integrate over $(0,1)$. The result is

$$\lambda \int_{(0,1)} |u|^2\, dx + \int_{(0,1)} |u'|^2\, dx - [u'\overline{u}]_0^1 = \int_{(0,1)} h\overline{u}\, dx. \tag{3.4}$$

Since, in view of (3.2)–(3.3), we have

$$u'(1)\overline{u}(1) = -\frac{\gamma_1 + \lambda}{\beta_1}|u(1)|^2 + \frac{1}{\beta_1}h(1)\overline{u}(1), \tag{3.5}$$

$$u'(0)\overline{u}(0) = -\frac{\gamma_o + \lambda}{\beta_o}|u(0)|^2 + \frac{1}{\beta_o}h(0)\overline{u}(0), \tag{3.6}$$

we obtain that (3.4) reads equivalently

$$\begin{aligned}&\lambda\|u\|^2_{L^2(0,1)} + \|u'\|^2_{L^2(0,1)} + \tfrac{\gamma_1+\lambda}{\beta_1}|u(1)|^2 - \tfrac{1}{\beta_1}h(1)\overline{u}(1) - \tfrac{\gamma_o+\lambda}{\beta_o}|u(0)|^2 \\ &+\tfrac{1}{\beta_o}h(0)\overline{u}(0) = \textstyle\int_{(0,1)} h\overline{u}\, dx.\end{aligned} \tag{3.7}$$

Taking the real parts in both members of (2.7), we obtain

$$\begin{aligned}&Re\,\lambda\|u\|^2_{L^2(0,1)} + \|u'\|^2_{L^2(0,1)} + \tfrac{\gamma_1+Re\,\lambda}{\beta_1}|u(1)|^2 - \tfrac{\gamma_o+Re\,\lambda}{\beta_o}|u(0)|^2 \\ &= Re\left[\textstyle\int_{(0,1)} h\overline{u}\, dx + \tfrac{1}{\beta_1}h(1)\overline{u}(1) - \tfrac{1}{\beta_o}h(0)\overline{u}(0)\right]\end{aligned} \tag{3.8}$$

and therefore

$$\begin{aligned}&Re\,\lambda\left(\|u\|^2_{L^2(0,1)} + \tfrac{|u(1)|^2}{\beta_1} - \tfrac{|u(0)|^2}{\beta_0}\right) \leq \|h\|_{L^2(0,1)}\|u\|_{L^2(0,1)} \\ &+ \tfrac{|h(1)|}{\sqrt{\beta_1}}\tfrac{|u(1)|}{\sqrt{\beta_1}} + \tfrac{|h(0)|}{\sqrt{-\beta_0}}\tfrac{|u(0)|}{\sqrt{-\beta_0}} \\ &\leq \left(\|h\|^2_{L^2(0,1)} + \frac{|h(1)|^2}{\beta_1} - \frac{|h(0)|^2}{\beta_0}\right)^{\frac{1}{2}} \cdot \left(\|u\|^2_{L^2(0,1)} + \frac{|u(1)|^2}{\beta_1} - \frac{|u(0)|^2}{\beta_0}\right)^{\frac{1}{2}} \\ &\leq \left(\|h\|^2_{L^2(0,1)} + \|h'\|^2_{L^2(0,1)} + \tfrac{|h(1)|^2}{\beta_1} - \tfrac{|h(0)|^2}{\beta_0}\right)^{\frac{1}{2}} \\ &\cdot\left(\|u\|^2_{L^2(0,1)} + \|u'\|^2_{L^2(0,1)} + \tfrac{|u(1)|^2}{\beta_1} - \tfrac{|u(0)|^2}{\beta_0}\right)^{\frac{1}{2}}.\end{aligned} \tag{3.9}$$

Now, taking imaginary parts in (3.7), we have

$$Im\,\lambda\left(\|u\|^2_{L^2(0,1)}+\tfrac{|u(1)|^2}{\beta_1}-\tfrac{|u(0)|^2}{\beta_0}\right)$$
$$=Im\left[\int_{(0,1)}h\overline{u}\,dx+\tfrac{1}{\beta_1}h(1)\overline{u}(1)-\tfrac{1}{\beta_0}h(0)\overline{u}(0)\right],$$

whence

$$|Im\,\lambda|\quad\left(\|u\|^2_{L^2(0,1)}+\tfrac{|u(1)|^2}{\beta_1}-\tfrac{|u(0)|^2}{\beta_0}\right)$$
$$\le\left(\|h\|^2_{L^2(0,1)}+\|h'\|^2_{L^2(0,1)}+\tfrac{|h(1)|^2}{\beta_1}-\tfrac{|h(0)|^2}{\beta_0}\right)^{\frac12}\tag{3.10}$$
$$\cdot\left(\|u\|^2_{L^2(0,1)}+\|u'\|^2_{L^2(0,1)}+\tfrac{|u(1)|^2}{\beta_1}-\tfrac{|u(0)|^2}{\beta_0}\right)^{\frac12}.$$

We also need the corresponding estimates for $\|u'\|_{L^2(0,1)}$. Since $h\in H^1(0,1)$, we have

$$\lambda u'-u'''=h'\quad\text{in}\quad(0,1);\tag{3.11}$$

multiplying (3.11) by $\overline{u}'$ and integrating on $(0,1)$, we get

$$\lambda\|u'\|^2_{L^2(0,1)}-[u''\overline{u}']_0^1+\|u''\|^2_{L^2(0,1)}=\int_{(0,1)}h'\overline{u}'\,dx.\tag{3.12}$$

On the other hand, taking also into account (3.5) and (3.6), we compute

$$\begin{aligned}[u''\overline{u}']_0^1 &=(-\beta_1u'(1)-\gamma_1u(1))\overline{u}'(1)+(\beta_0u'(0)+\gamma_0u(0))\overline{u}'(0)\\ &=-\beta_1|u'(1)|^2+\beta_0|u'(0)|^2-\gamma_1u(1)\overline{u}'(1)+\gamma_0u(0)\overline{u}'(0)\\ &=-\beta_1|u'(1)|^2+\beta_0|u'(0)|^2-\gamma_1\left(\tfrac{1}{\beta_1}\overline{h}(1)u(1)-\tfrac{\gamma_1+\overline{\lambda}}{\beta_1}|u(1)|^2\right)+\\ &\quad+\gamma_0\left(\tfrac{1}{\beta_0}\overline{h}(0)u(0)-\tfrac{\gamma_0+\overline{\lambda}}{\beta_0}|u(0)|^2\right).\end{aligned}$$

Hence

$$\begin{aligned}Re\,\lambda\,\|u'\|^2_{L^2(0,1)}&+\|u''\|^2_{L^2(0,1)}+\beta_1|u'(1)|^2-\beta_0|u'(0)|^2=\tfrac{\gamma_1}{\beta_1}(\gamma_1+Re\,\lambda)|u(1)|^2+\\ &-\tfrac{\gamma_o(\gamma_o+Re\,\lambda)}{\beta_o}|u(0)|^2+Re\left[\int_{(0,1)}h'\overline{u}'\,dx-\tfrac{\gamma_1}{\beta_1}\overline{h}(1)u(1)\right.\\ &\left.+\tfrac{\gamma_0}{\beta_0}\overline{h}(0)u(0)\right].\end{aligned}$$

In view of (3.8),

$$\frac{\gamma_1}{\beta_1}(\gamma_1+Re\,\lambda)\,|u(1)|^2-\frac{\gamma_0}{\beta_0}(\gamma_0+Re\,\lambda)\,|u(0)|^2$$
$$\le(\gamma_0+\gamma_1)\,Re\left(\int_{(0,1)}h\overline{u}\,dx+\frac{1}{\beta_1}h(1)\overline{u}(1)-\frac{1}{\beta_0}h(0)\overline{u}(0)\right).$$

It follows that

$$\begin{aligned}Re\,\lambda\,\|u'\|^2_{L^2(0,1)} &\leq Re\left(\int_{(0,1)} h'\overline{u}'\,dx + (\gamma_0+\gamma_1)\int_{(0,1)} h\overline{u}\,dx\right) + \\ &\quad + Re\left(\tfrac{\gamma_0}{\beta_1}h(1)\overline{u}(1) - \tfrac{\gamma_1}{\beta_0}h(0)\overline{u}(0)\right) \\ &\leq C Re\left(\int_{(0,1)} h'\overline{u}'\,dx + \int_{(0,1)} h\overline{u}\,dx\right) + \\ &\quad + C Re\left(\tfrac{1}{\beta_1}h(1)\overline{u}(1) - \tfrac{1}{\beta_0}h(0)\overline{u}(0)\right) \\ &\leq C\,(\|h\|_{L^2(0,1)}\|u\|_{L^2(0,1)} + \|h'\|_{L^2(0,1)}\|u'\|_{L^2(0,1)} \\ &\quad + \tfrac{|h(1)|}{\sqrt{\beta_1}}\tfrac{|u(1)|}{\sqrt{\beta_1}} + \tfrac{|h(0)|}{\sqrt{-\beta_0}}\tfrac{|u(0)|}{\sqrt{-\beta_0}}\Big) \\ &\leq C\left(\|h\|^2_{L^2(0,1)} + \|h'\|^2_{L^2(0,1)} + \tfrac{|h(1)|^2}{\beta_1} - \tfrac{|h(0)|^2}{\beta_0}\right)^{\frac{1}{2}} \\ &\quad \cdot\left(\|u\|^2_{L^2(0,1)} + \|u'\|^2_{L^2(0,1)} + \tfrac{|u(1)|^2}{\beta_1} - \tfrac{|u(0)|^2}{\beta_0}\right)^{\frac{1}{2}},\end{aligned} \tag{3.13}$$

where C is a suitable positive constant.

Then, from (3.9) and (3.13), it follows that

$$\begin{aligned}&Re\,\lambda\left(\|u\|^2_{L^2(0,1)} + \|u'\|^2_{L^2(0,1)} + \tfrac{|u(1)|^2}{\beta_1} - \tfrac{|u(0)|^2}{\beta_0}\right) \\ &\leq (C+1)\cdot\left(\|h\|^2_{L^2(0,1)} + \|h'\|^2_{L^2(0,1)} + \tfrac{|h(1)|^2}{\beta_1} - \tfrac{|h(0)|^2}{\beta_0}\right)^{\frac{1}{2}} \\ &\quad\cdot\left(\|u\|^2_{L^2(0,1)} + \|u'\|^2_{L^2(0,1)} + \tfrac{|u(1)|^2}{\beta_1} - \tfrac{|u(0)|^2}{\beta_0}\right)^{\frac{1}{2}},\end{aligned}$$

so that

$$\begin{aligned}Re\,\lambda\ &\left(\|u\|^2_{L^2(0,1)} + \|u'\|^2_{L^2(0,1)} + \tfrac{|u(1)|^2}{\beta_1} - \tfrac{|u(0)|^2}{\beta_0}\right)^{\frac{1}{2}} \\ &\leq (C+1)\left(\|h\|^2_{L^2(0,1)} + \|h'\|^2_{L^2(0,1)} + \tfrac{|h(1)|^2}{\beta_1} - \tfrac{|h(0)|^2}{\beta_0}\right)^{\frac{1}{2}}.\end{aligned} \tag{3.14}$$

From (3.12) it also follows that

$$\begin{aligned}&Im\,\lambda\left(\|u'\|^2_{L^2(0,1)} + \tfrac{\gamma_1}{\beta_1}|u(1)|^2 - \tfrac{\gamma_0}{\beta_0}|u(0)|^2\right) \\ &= Im\left[\int_{(0,1)} h'\overline{u}'\,dx - \tfrac{\gamma_1}{\beta_1}\overline{h}(1)u(1) + \tfrac{\gamma_0}{\beta_0}\overline{h}(0)u(0)\right]\end{aligned}$$

and thus

$$\begin{aligned}&|Im\,\lambda|\left(\|u'\|^2_{L^2(0,1)} + \frac{\gamma_1}{\beta_1}|u(1)|^2 - \frac{\gamma_0}{\beta_0}|u(0)|^2\right) \\ &\leq \|h'\|_{L^2(0,1)}\|u'\|_{L^2(0,1)} + \frac{\gamma_1}{\beta_1}|h(1)|\,|u(1)| - \frac{\gamma_0}{\beta_0}|h(0)|\,|u(0)|\end{aligned}$$

$$\leq\left(\|h'\|^2_{L^2(0,1)}+\frac{\gamma_1}{\beta_1}|h(1)|^2-\frac{\gamma_0}{\beta_0}|h(0)|^2\right)^{\frac{1}{2}}\cdot\left(\|u'\|^2_{L^2(0,1)}+\frac{\gamma_1}{\beta_1}|u(1)|^2-\frac{\gamma_0}{\beta_0}|u(0)|^2\right)^{\frac{1}{2}}$$

$$\begin{aligned}&\leq\left(\|h\|^2_{L^2(0,1)}+\|h'\|^2_{L^2(0,1)}+\tfrac{\gamma_1}{\beta_1}|h(1)|^2-\tfrac{\gamma_0}{\beta_0}|h(0)|^2\right)^{\frac{1}{2}}\cdot\\&\left(\|u\|^2_{L^2(0,1)}+\|u'\|^2_{L^2(0,1)}+\tfrac{\gamma_1}{\beta_1}|u(1)|^2-\tfrac{\gamma_0}{\beta_0}|u(0)|^2\right)^{\frac{1}{2}}.\end{aligned}\tag{3.15}$$

So

$$\begin{aligned}|Im\ \lambda|\,\|u'\|^2_{L^2(0,1)}&\leq C_1\left(\|h\|^2_{L^2(0,1)}+\|h'\|^2_{L^2(0,1)}+\tfrac{|h(1)|^2}{\beta_1}-\tfrac{|h(0)|^2}{\beta_0}\right)^{\frac{1}{2}}\cdot\\&\cdot\left(\|u\|^2_{L^2(0,1)}+\|u'\|^2_{L^2(0,1)}+\tfrac{|u(1)|^2}{\beta_1}-\tfrac{|u(0)|^2}{\beta_0}\right)^{\frac{1}{2}}.\end{aligned}\tag{3.16}$$

By (3.10) and (3.16), we get

$$\begin{aligned}&|Im\ \lambda|\left(\|u\|^2_{L^2(0,1)}+\|u'\|^2_{L^2(0,1)}+\frac{|u(1)|^2}{\beta_1}-\frac{|u(0)|^2}{\beta_0}\right)^{\frac{1}{2}}\leq\\&(C_1+1)\left(\|h\|^2_{L^2(0,1)}+\|h'\|^2_{L^2(0,1)}+\frac{|h(1)|^2}{\beta_1}-\frac{|h(0)|^2}{\beta_0}\right)^{\frac{1}{2}}.\end{aligned}\tag{3.17}$$

Combining (3.14) and (3.17), we get the desired estimate

$$\begin{aligned}&|\lambda|\left(\|u\|^2_{L^2(0,1)}+\|u'\|^2_{L^2(0,1)}+\frac{|u(1)|^2}{\beta_1}-\frac{|u(0)|^2}{\beta_0}\right)^{\frac{1}{2}}\leq\\&C_2\left(\|h\|^2_{L^2(0,1)}+\|h'\|^2_{L^2(0,1)}+\frac{|h(1)|^2}{\beta_1}-\frac{|h(0)|^2}{\beta_0}\right)^{\frac{1}{2}}.\end{aligned}$$

The solvability of (3.2) is easily obtained by a direct computation. Hence, we have proved Theorem 2.1 in the one-dimensional case.

4. The General Case

Here we assume that $\beta,\gamma\in C^1(\partial\Omega)$, $\beta\geq\epsilon_o>0$ and $\gamma\geq 0$ on $\partial\Omega$. Let us consider the operator $\mathcal{A}u=\Delta u$ with domain

$$\begin{aligned}D(\mathcal{A}):=\{u\in H^1(\Omega)\quad&:\ \Delta u\in H^1(\Omega),\ \tfrac{\partial u}{\partial n}\text{ exists in }L^2(\partial\Omega),\\&\Delta u+\beta\tfrac{\partial u}{\partial n}+\gamma u=0\quad\text{on}\quad\partial\Omega\}.\end{aligned}\tag{4.1}$$

If $(\lambda I-\mathcal{A})u=h\in H^1(\Omega)$, the divergence theorem gives immediately

$$\lambda\|u\|^2_{L^2(\Omega)}+\int_\Omega|\nabla u|^2\,dx-\int_{\partial\Omega}\frac{\partial u}{\partial n}\,\overline{u}\,d\sigma=\int_\Omega h\overline{u}\,dx$$

so that, since

$$\frac{\partial u}{\partial n}=-\frac{\gamma u+\Delta u}{\beta}=-\frac{\gamma}{\beta}u+\frac{h-\lambda u}{\beta}=-\frac{\lambda+\gamma}{\beta}u+\frac{h}{\beta}\tag{4.2}$$

we obtain

$$\lambda\|u\|^2_{L^2(\Omega)} + \int_\Omega |\nabla u|^2\,dx + \int_{\partial\Omega} \frac{\lambda+\gamma}{\beta}|u|^2\,d\sigma - \int_{\partial\Omega} \frac{h}{\beta}\overline{u}\,d\sigma = \int_\Omega h\overline{u}\,dx. \tag{4.3}$$

Taking real and imaginary parts in both sides of (4.3), we get

$$Re\,\lambda\left(\|u\|^2_{L^2(\Omega)} + \int_{\partial\Omega}\frac{|u|^2}{\beta}\,d\sigma\right) + \int_\Omega |\nabla u|^2\,dx + \int_{\partial\Omega}\frac{\gamma}{\beta}|u|^2\,d\sigma$$

$$= Re\left(\int_\Omega h\overline{u}\,dx + \int_{\partial\Omega}\frac{h\overline{u}}{\beta}\,d\sigma\right) \tag{4.4}$$

and

$$Im\,\lambda\left(\|u\|^2_{L^2(\Omega)} + \int_{\partial\Omega}\frac{|u|^2}{\beta}\,d\sigma\right) = Im\left(\int_\Omega h\overline{u}\,dx + \int_{\partial\Omega}\frac{h\overline{u}}{\beta}\,d\sigma\right), \tag{4.5}$$

so that for some suitable positive constant k one has

$$|\lambda|\left(\|u\|^2_{L^2(\Omega)} + \int_{\partial\Omega}\frac{|u|^2}{\beta}\,d\sigma\right)^{\frac{1}{2}} \le k\left(\|h\|^2_{L^2(\Omega)} + \int_{\partial\Omega}\frac{|h|^2}{\beta}\,d\sigma\right)^{\frac{1}{2}} \tag{4.6}$$

where $h \in H^1(\Omega)$. Since

$$\lambda\frac{\partial u}{\partial x_j} - \frac{\partial}{\partial x_j}(\Delta u) = \frac{\partial h}{\partial x_j}, \quad j = 1,..,N, \tag{4.7}$$

multiplying (4.7) by $\frac{\partial \overline{u}}{\partial x_j}$ and then integrating over Ω the obtained equality, we have by formal integration by parts

$$\lambda\int_\Omega |\frac{\partial u}{\partial x_j}|^2\,dx - \int_{\partial\Omega}\Delta u\frac{\partial\overline{u}}{\partial x_j}n_j\,d\sigma + \int_\Omega \Delta u\frac{\partial^2\overline{u}}{\partial x_j^2}\,dx = \int_\Omega \frac{\partial h}{\partial x_j}\cdot\frac{\partial\overline{u}}{\partial x_j}\,dx, \tag{4.8}$$

i.e.,

$$\lambda\|\nabla u\|^2_{L^2(\Omega)} - \int_{\partial\Omega}\Delta u\frac{\partial\overline{u}}{\partial n}\,d\sigma + \|\Delta u\|^2_{L^2(\Omega)} = \sum_{j=1}^N\int_\Omega\frac{\partial h}{\partial x_j}\frac{\partial\overline{u}}{\partial x_j}\,dx. \tag{4.9}$$

From the boundary condition

$$\Delta u = -\beta\frac{\partial u}{\partial n} - \gamma u \quad \text{on} \quad \partial\Omega, \tag{4.10}$$

we have

$$\begin{aligned} -\textstyle\int_{\partial\Omega}\Delta u\frac{\partial\overline{u}}{\partial n}\,d\sigma = & \textstyle\int_{\partial\Omega}(\beta\frac{\partial u}{\partial n} + \gamma u)\frac{\partial\overline{u}}{\partial n}\,d\sigma \\ = & \textstyle\int_{\partial\Omega}\beta|\frac{\partial u}{\partial n}|^2\,d\sigma + \int_{\partial\Omega}\gamma u\frac{\partial\overline{u}}{\partial n}\,d\sigma \\ & \text{(by (4.10))} \\ = & \textstyle\int_{\partial\Omega}\beta|\frac{\partial u}{\partial n}|^2\,d\sigma - \int_{\partial\Omega}\gamma u\left(\frac{\gamma\overline{u}+\Delta\overline{u}}{\beta}\right)\,d\sigma \\ = & \textstyle\int_{\partial\Omega}\beta|\frac{\partial u}{\partial n}|^2\,d\sigma - \int_{\partial\Omega}\gamma u\left(\frac{\gamma\overline{u}+\overline{\lambda}\overline{u}-\overline{h}}{\beta}\right)\,d\sigma \\ = & \textstyle\int_{\partial\Omega}\beta|\frac{\partial u}{\partial n}|^2\,d\sigma - \int_{\partial\Omega}\gamma\frac{\gamma+\overline{\lambda}}{\beta}|u|^2\,d\sigma + \int_{\partial\Omega}\frac{\gamma}{\beta}\overline{h}u\,d\sigma. \end{aligned} \tag{4.11}$$

If we plug (4.11) in (4.9), we get

$$\lambda\|\nabla u\|^2_{L^2(\Omega)} + \int_{\partial\Omega}\beta|\tfrac{\partial u}{\partial n}|^2\,d\sigma + \quad \|\Delta u\|^2_{L^2(\Omega)} - \int_{\partial\Omega}\gamma\left(\tfrac{\gamma+\bar\lambda}{\beta}\right)|u|^2\,d\sigma + \\ + \int_{\partial\Omega}\tfrac{\gamma}{\beta}\,\overline{h}u\,d\sigma = \sum_{j=1}^n\int_\Omega \tfrac{\partial h}{\partial x_j}\tfrac{\partial \overline{u}}{\partial x_j}\,dx. \tag{4.12}$$

Notice that from (4.3),

$$\int_{\partial\Omega}\frac{\gamma+Re\,\lambda}{\beta}|u|^2\,d\sigma \le Re\left(\int_\Omega h\overline{u}\,dx + \int_{\partial\Omega}\frac{h}{\beta}\,\overline{u}\,d\sigma\right). \tag{4.13}$$

Therefore, from (4.12)

$$\begin{aligned}
Re\,\lambda \quad & \|\nabla u\|^2_{L^2(\Omega)} + \int_{\partial\Omega}\beta|\tfrac{\partial u}{\partial n}|^2\,d\sigma + \|\Delta u\|^2_{L^2(\Omega)} = \\
& = \int_{\partial\Omega}\gamma\,\tfrac{\gamma+Re\,\lambda}{\beta}|u|^2\,d\sigma - Re\int_{\partial\Omega}\tfrac{\gamma}{\beta}\,\overline{h}u\,d\sigma + Re\sum_{j=1}^n\int_\Omega\tfrac{\partial h}{\partial x_j}\tfrac{\partial\overline{u}}{\partial x_j}\,dx \\
& \le k_1\ Re\left(\int_\Omega h\overline{u}\,dx + \int_{\partial\Omega}\tfrac{h}{\beta}\,\overline{u}\,d\sigma\right) + k_2\left(\int_{\partial\Omega}\tfrac{|h|^2}{\beta}\,d\sigma\right)^{\frac12}\left(\int_{\partial\Omega}\tfrac{|u|^2}{\beta}\,d\sigma\right)^{\frac12} \\
& \quad +\|\nabla h\|_{L^2(\Omega)}\,\|\nabla u\|_{L^2(\Omega)} \\
& \le k_3\left[\|h\|_{L^2(\Omega)}\|u\|_{L^2(\Omega)} + \left(\int_{\partial\Omega}\tfrac{|h|^2}{\beta}\,d\sigma\right)^{\frac12}\left(\int_{\partial\Omega}\tfrac{|u|^2}{\beta}\,d\sigma\right)^{\frac12} + \right. \\
& \left. +\|\nabla h\|_{L^2(\Omega)}\,\|\nabla u\|_{L^2(\Omega)}\right].
\end{aligned}$$

Since

$$Re\,\lambda\left(\|u\|^2_{L^2(\Omega)} + \int_{\partial\Omega}\frac{|u|^2}{\beta}\,d\sigma\right) \le Re\left(\int_\Omega h\overline{u}\,dx + \int_{\partial\Omega}\frac{h\overline{u}}{\beta}\,d\sigma\right)$$

as a consequence of (4.4) we get,

$$\begin{aligned}
& Re\,\lambda\left(\|u\|^2_{L^2(\Omega)} + \int_{\partial\Omega}\tfrac{|u|^2}{\beta}\,d\sigma + \|\nabla u\|^2_{L^2(\Omega)}\right) \\
& \le \|h\|_{L^2(\Omega)}\,\|u\|_{L^2(\Omega)} + \left(\int_{\partial\Omega}\tfrac{|u|^2}{\beta}\,d\sigma\right)^{\frac12}\left(\int_{\partial\Omega}\tfrac{|h|^2}{\beta}\,d\sigma\right)^{\frac12} + \\
& +k_3\left[\|h\|_{L^2(\Omega)}\,\|u\|_{L^2(\Omega)} + \left(\int_{\partial\Omega}\tfrac{|u|^2}{\beta}\,d\sigma\right)^{\frac12}\left(\int_{\partial\Omega}\tfrac{|h|^2}{\beta}\,d\sigma\right)^{\frac12} + \right. \\
& \left. +\|\nabla h\|_{L^2(\Omega)}\,\|\nabla u\|_{L^2(\Omega)}\right] \\
& \le k_4\left(\|u\|^2_{L^2(\Omega)} + \int_{\partial\Omega}\tfrac{|u|^2}{\beta}\,d\sigma + \|\nabla u\|^2_{L^2(\Omega)}\right)^{\frac12}\left(\|h\|^2_{L^2(\Omega)} + \int_{\partial\Omega}\tfrac{|h|^2}{\beta}\,d\sigma + \right. \\
& \left. +\|\nabla h\|^2_{L^2(\Omega)}\right)^{\frac12},
\end{aligned}$$

i.e.,

$$Re\,\lambda\left(\|u\|^2_{L^2(\Omega)}+\int_{\partial\Omega}\tfrac{|u|^2}{\beta}\,d\sigma+\|\nabla u\|^2_{L^2(\Omega)}\right)^{\frac{1}{2}}$$
$$\le k_4\left(\|h\|^2_{L^2(\Omega)}+\|\nabla h\|^2_{L^2(\Omega)}+\int_{\partial\Omega}\tfrac{|h|^2}{\beta}\,d\sigma\right)^{\frac{1}{2}}. \tag{4.14}$$

Taking the imaginary parts in both sides of (4.12), we get

$$Im\,\lambda\left(\|\nabla u\|^2_{L^2(\Omega)}+\int_{\partial\Omega}\frac{\gamma}{\beta}|u|^2\,d\sigma\right)=Im\left(\sum_{j=1}^{n}\int_\Omega\frac{\partial h}{\partial x_j}\frac{\partial\overline{u}}{\partial x_j}\,dx-\int_{\partial\Omega}\frac{\gamma}{\beta}\overline{h}u\,d\sigma\right) \tag{4.15}$$

and thus

$$|Im\,\lambda|\left(\|\nabla u\|^2_{L^2(\Omega)}+\int_{\partial\Omega}\tfrac{\gamma}{\beta}|u|^2\,d\sigma\right)\le\|\nabla h\|_{L^2(\Omega)}\|\nabla u\|_{L^2(\Omega)}+$$
$$+\left(\int_{\partial\Omega}\tfrac{\gamma}{\beta}|h|^2\,d\sigma\right)^{\frac{1}{2}}\left(\int_{\partial\Omega}\tfrac{\gamma}{\beta}|u|^2\,d\sigma\right)^{\frac{1}{2}}.$$

It follows that

$$|Im\,\lambda|\left(\|\nabla u\|^2_{L^2(\Omega)}+\int_{\partial\Omega}\frac{\gamma}{\beta}|u|^2\,d\sigma\right)^{\frac{1}{2}}\le\left(\|\nabla h\|^2_{L^2(\Omega)}+\int_{\partial\Omega}\frac{\gamma}{\beta}|h|^2\,d\sigma\right)^{\frac{1}{2}}. \tag{4.16}$$

Hence

$$|Im\,\lambda|\,\|\nabla u\|_{L^2(\Omega)}\le k_5\left(\|\nabla h\|^2_{L^2(\Omega)}+\int_{\partial\Omega}\frac{|h|^2}{\beta}\,d\sigma\right)^{\frac{1}{2}}. \tag{4.17}$$

In view of (4.14), (4.17) and (4.6) we conclude that

$$|\lambda|\left(\|u\|^2_{L^2(\Omega)}+\int_{\partial\Omega}\tfrac{|u|^2}{\beta}\,d\sigma+\|\nabla u\|^2_{L^2(\Omega)}\right)^{\frac{1}{2}}$$
$$\le K'\left(\|h\|^2_{L^2(\Omega)}+\|\nabla h\|^2_{L^2(\Omega)}+\int_{\partial\Omega}\tfrac{|h|^2}{\beta}\,d\sigma\right)^{\frac{1}{2}},$$

as desired. By the discussion preceding Theorem 2.1, for any $h\in H^1(\Omega)$, there exists a unique weak solution $u\in\mathcal{H}$. Applying the usual Agmon-Douglis-Nirenberg estimates (see, for instance, Tanabe [16, p.77]) to the problem

$$(\lambda I-A)u=h\quad in\quad\Omega,$$

$$\beta\frac{\partial u}{\partial n}+(\gamma+\lambda)u=h\quad in\quad\partial\Omega,$$

since $h_{|\partial\Omega}\in H^{\frac{1}{2}}(\partial\Omega)$ we get that $u\in H^2(\Omega)$ and hence $\frac{\partial u}{\partial n}$ exists on the boundary. Furthermore,

$$\Delta u=\lambda u-h\in H^1(\Omega).$$

Therefore Theorem 2.1 is proved.

References

[1] W. Arendt, *Resolvent positive operators and inhomogeneous boundary conditions.* Annali Scuola Norm. Sup. Pisa Cl. Sci. Serie IV **XXIX (3)** (2000), 639–670.

[2] V. Barbu and A. Favini, *Analytic semigroups on $H^1(\Omega)$ generated by degenerate elliptic operators.* Semigroup Forum **62** (2001), 317–328.

[3] K.J. Engel and R. Nagel, *One-parameter Semigroups for Linear Evolution Equations.* Graduate Texts in Mathematics **194**, Springer, Berlin, Heidelberg, New York, 1999.

[4] A. Favini, G.R. Goldstein, J.A. Goldstein and S. Romanelli, *C_o-semigroups generated by second order differential operators with general Wentzell boundary conditions.* Proc. Am. Math. Soc. **128** (2000), 1981–1989.

[5] A. Favini, G.R. Goldstein, J.A. Goldstein and S. Romanelli, *An analytic semigroup associated to a degenerate evolution equation.* Stochastic Processes and Functional Analysis (J.A. Goldstein, N.A. Gretsky and J. Uhl, eds.) M. Dekker, New York, 1997, 85–100.

[6] A. Favini, G.R. Goldstein, J.A. Goldstein and S. Romanelli, *Generalized Wentzell boundary conditions and analytic semigroups in $C[0,1]$.* In: Semigroups of Operators: Theory and Applications, (A.V. Balakrishnan ed.), PLNDE 42, Birkhäuser, Basel, 2000, 125–130.

[7] A. Favini, G.R. Goldstein, J.A. Goldstein and S. Romanelli, *Analytic semigroups on $L^p_w(0,1)$ and on $L^p(0,1)$ generated by some classes of second order differential operators.* Math. Nach. **238** (2002), 78–102.

[8] A. Favini, G.R. Goldstein, J.A. Goldstein and S. Romanelli, *The heat equation with generalized Wentzell boundary condition.* J. Evol. Eqns. **2** (2002), 1–19.

[9] A. Favini and A. Yagi, *Degenerate Differential Equations in Banach Spaces.* M. Dekker, New York, 1999.

[10] J.A. Goldstein, *Semigroups of Linear Operators and Applications.* Oxford University Press, Oxford, New York, 1985.

[11] A. Lunardi, *Analytic Semigroups and Optimal Regularity in Parabolic Problems.* Birkhäuser Verlag, Basel, 1995.

[12] V.G. Maz'ja, *Classes of regions and embedding theorems for function spaces.* Soviet Math. Dokl. **1** 1960, 882–885.

[13] V.G. Maz'ja, *Sobolev Spaces.* Springer Verlag, Berlin, Heidelberg, New York, 1985.

[14] A. Pazy, *Semigroups of Linear Operators and Applications to Partial Differential Equations.* Appl. Math. Sci., vol. 44, Springer Verlag, Berlin, Heidelberg, New York, 1983.

[15] K. Taira, *Analytic Feller semigroups.* Conferenze del Seminario di Matematica, Università di Bari **267** (1997).

[16] H. Tanabe, *Equations of Evolution.* Pitman, London, San Francisco, Melbourne, 1979.

Angelo Favini
Dipartimento di Matematica
Università di Bologna
Piazza di Porta S. Donato 5
I-40126 Bologna, Italy
E-mail address: favini@dm.unibo.it

Gisèle Ruiz Goldstein
CERI
University of Memphis
Memphis, Tennessee 38152, USA
E-mail address: ggoldste@memphis.edu

Jerome A. Goldstein
Department of Mathematical Sciences
University of Memphis
Memphis, Tennessee 38152, USA
E-mail address: jgoldste@memphis.edu

Enrico Obrecht
Dipartimento di Matematica
Università di Bologna
Piazza di Porta S. Donato 5
I-40126 Bologna, Italy
E-mail address: obrecht@dm.unibo.it

Silvia Romanelli
Dipartimento Interuniversitario di Matematica
Università di Bari
via E. Orabona 4
I-70125 Bari, Italy
E-mail address: romans@dm.uniba.it

Progress in Nonlinear Differential Equations
and Their Applications, Vol. 55, 181–193

Consistency and Stabilization of Rational Approximation Schemes for C_0-semigroups

Simone Flory, Frank Neubrander, and Lutz Weis

1. Consistency, Stability and Approximation Order

A rational function $r(\cdot)$ with numerator $p(\cdot)$ of degree p and denominator $q(\cdot)$ of degree $q \geq p$ is a *rational approximation of the exponential of approximation order* m if

$$|r(z) - e^z| \leq C|z|^{m+1} \text{ as } z \to 0.$$

It is called a *rational Padé approximation of the exponential* if $m = p + q$; i.e., if

$$|r(z) - e^z| \leq C|z|^{p+q+1} \text{ as } z \to 0.$$

We note that given p and q the Padé approximations are those rational approximations which have maximal approximation order m. Also, for given p, q there exists exactly one rational Padé approximant of the exponential (if one normalizes $q(\cdot)$ such that $q(0) = 1$; see [HW], p. 50). Furthermore

$$|r(\infty)| = \begin{cases} 1 \text{ if } q = p \\ 0 \text{ if } q > p. \end{cases}$$

A rational approximation is called *A-stable* if $|r(z)| \leq 1$ for $Re z \leq 0$. It is well known that a rational Padé approximation $r(\cdot)$ is A-stable if and only if $q - 2 \leq p \leq q$ (see [HW], p. 60, Theorem 4.12).

Example 1.1. The following are examples for A-stable, rational Padé approximations with $q - 2 \leq p \leq q$. The function $r(z) = \frac{1}{1-z}$ has approximation order $m = 1$, the functions $r(z) = \frac{1+\frac{z}{2}}{1-\frac{z}{2}}$ and $r(z) = \frac{2}{(z-1-i)(z-1+i)}$ have approximation order $m = 2$, the function $r(z) = \frac{6+2z}{(z-2-i\sqrt{2})(z-2+i\sqrt{2})}$ has approximation order $m = 3$, and $r(z) = \frac{(z+3-i\sqrt{3})(z+3+i\sqrt{3})}{(z-3-i\sqrt{3})(z-3+i\sqrt{3})}$ has approximation order $m = 4$.

The Padé-approximation $r(z) = \frac{6}{6-6z+3z^2-z^3}$ has approximation order $m = 3$. Note that it is not A-stable since $p = q-3$. In fact, $|r(\frac{i}{2})| = |\frac{6}{3-5i}| = |\frac{9+15i}{17}| > 1$.

An example for a rational approximation scheme that is not a Padé approximation is the so-called Calahan scheme of approximation order $m = 3$ with $p = q = 2$ given by $r(z) = \frac{6-2\sqrt{3}z-(1+\sqrt{3})z^2}{(2+\sqrt{3})(z-\frac{3+\sqrt{3}}{2+\sqrt{3}})^2}$.

Definition 1.2. *Let $r(\cdot)$ be a rational (Padé) approximation of the exponential and let A be the generator of a C_0-semigroup $T(\cdot)$ on some Banach space X. Then $V(t) := r(tA)$, $t \geq 0$, is called a rational (Padé) approximation scheme for $T(\cdot)$.*

Well-known examples of rational Padé approximation schemes of semigroups are the Backwards Euler scheme $V_{be}(t) := r(tA) = (I - tA)^{-1}$, and the Crank-Nicolson scheme $V_{cn}(t) := r(tA) = (I + \frac{t}{2}A)(I - \frac{t}{2}A)^{-1}$. Notice that rational approximation schemes consist of bounded linear operators commuting with the semigroup $T(t)$ and the resolvent $R(\lambda, A)$. The following generalization of the semigroup formula

$$T(t)x - x = A\int_0^t T(s)x\,ds = tA\int_0^1 T(t(1-s))x\,ds, x \in X$$

can be used to discuss consistency properties of rational approximation schemes. We denote by $\sigma(A)$ the spectrum of the operator A and by $s(A) := \sup\{Re\lambda : \lambda \in \sigma(A)\}$ its spectral bound.

Theorem 1.3 (Padé Representation). *Let A be the generator of a strongly continuous semigroup $T(\cdot)$ on a Banach space X and let $r(\cdot) = \frac{p(\cdot)}{q(\cdot)}$ be a rational approximation of the exponential of order m, where $p(\cdot)$ is of degree $p \leq m$ and $q(\cdot)$ is of degree $q \leq m$ with $q(0) = 1$. Then there is a polynomial $F(\cdot)$ with $F^{(m)}(\cdot) \equiv 1$ such that*

$$q(tA)T(t)x - p(tA)x = (tA)^{p+q+1}\int_0^1 F(s)T(t(1-s))x\,ds \qquad (1.1)$$

for all $x \in D(A^{p+q})$. Moreover, if $r(\cdot)$ has no poles in the left halfplane (this is the case if $q - 4 \leq p \leq q$, see [AS]) and $s(A) \leq 0$, then

$$T(t)x = r(tA)x + (tA)^{p+q+1}(q(tA))^{-1}\int_0^1 F(s)T(t(1-s))x\,ds \qquad (1.2)$$

for all $x \in D(A^p)$.

Proof. Let $F(\cdot)$ be a polynomial of degree m with $F^{(m)} \equiv 1$. Then, for $x \in D(A^m)$,

$$q(A)T(1)x - p(A)x = A^{m+1}\int_0^1 F(s)T(1-s)x\,ds,$$

where

$$p(A) = F(1)A^m + F'(1)A^{m-1} + \cdots + F^{(m)}(1),$$
$$q(A) = F(0)A^m + F'(0)A^{m-1} + \cdots + F^{(m)}(0).$$

This can be seen by induction. In order for $p(\cdot)$ to be of order p and $q(\cdot)$ to be of order q, we choose a polynomial $F(\cdot)$ such that

$$F(1) = F'(1) = \cdots = F^{m-p-1}(1) = 0,$$
$$F(0) = F'(0) = \cdots = F^{m-q-1}(0) = 0.$$

In case $r(\cdot)$ is a Padé approximation; that is, $m = p + q$, a polynomial $F(\cdot)$ that satisfies these conditions is $F(s) = s^p(s-1)^q$. Otherwise, if $q(z) = \sum_{i=0}^{q} b_i z^i$ with

$q(0) = b_0 = 1$, then $F(s) = \sum_{i=0}^{m} b_{m-i}\frac{s^i}{i!}$ (see also [HW], Chap. 4, Thm 3.10). Thus, for $x \in D(A^{p+q})$,

$$q(A)T(1)x - p(A)x = A^{p+q+1}\int_0^1 F(s)T(1-s)x\,ds. \tag{$*$}$$

Now let $X = \mathbb{C}$, $x = 1$ and $A = z$. Then, $T(1)x = e^z$ and

$$e^z - r(z) = \frac{1}{q(z)}z^{p+q+1}\int_0^1 s^p(s-1)^q e^{z(1-s)}\,ds$$

(for a Padé approximation, this is Padé's classical error estimate; see [Pa], Section 66). Replacing A by tA in $(*)$, we obtain (1.1). To see (1.2), we note that all zeros of $q(\cdot)$ are in the right halfplane if $r(\cdot)$ is A-stable. In fact, $r(\cdot)$ has no poles in the left halfplane since $|r(z)| \le 1$ for $Re z \le 0$. Thus, $q(A)$ is invertible and $q(A)^{-1} := \prod_{i=1}^{q} R(\lambda_i, A)$ for some λ_i with $Re\lambda_i > 0$. It follows from (1.1) that

$$T(t)x = r(tA)x + (tA)^{p+q+1}q(tA)^{-1}\int_0^1 F(s)\,T(t(1-s))x\,ds$$

for all $x \in D(A^{p+q})$. Since $A^q\prod_{i=1}^{q} R(\lambda_i, A)$ is a bounded operator on X, the identity holds on $D(A^p)$. □

An approximation scheme $\{V(t), t \in [0,\delta]\}$ of linear operators defined on a dense subset $D \subset D(A)$ is called *consistent* if, for all $x \in D$, $V(0)x = x$ and

$$\frac{V(t)x - x}{t} \to Ax$$

as $t \to 0$. It follows from $\frac{1}{t}[V(t)x - T(t)x] = [\frac{V(t)x-x}{t} - Ax] - [\frac{T(t)x-x}{t} - Ax]$ that an approximation scheme $V(t)$ is consistent if and only if

$$\|V(t)x - T(t)x\| = o(t) \text{ as } t \to 0 \text{ for all } x \in D. \tag{1.3}$$

An approximation scheme $\{V(t), t \in [0,\delta]\}$ is called *consistent of approximation order m* if $V(0) = I$ on D and, for all $x \in D$, there exists a constant $M(x)$ such that

$$\|\frac{V(t)x - T(t)x}{t}\| \le M(x)t^m \text{ for all } t \in [0,\delta]. \tag{1.4}$$

An approximation scheme $\{V(t), t \in [0,\delta]\}$ is called *stable* if there exist δ-dependent positive constants M and ω such that for all $t \in [0,\delta]$ and $n \in \mathbb{N}$,

$$\|V(t)^n\| \le Me^{\omega nt}, \tag{1.5}$$

or, equivalently, there exists an increasing positive function $M(\cdot)$ such that for all $T > 0$, $n \in \mathbb{N}$, $t \in [0,\delta]$ with $nt \in [0,T]$,

$$\|V(t)^n\| \le M(T). \tag{1.6}$$

For a proof see, e.g., [FNZ], Theorem 2.1, p. 7. An approximation scheme V is called *convergent* if for all $T > 0$

$$\sup_{t\in[0,T]} \|V(\frac{t}{n})^n x - T(t)x\| \to 0$$

as $n \to \infty$ for all $x \in X$. The following theorem is usually referred to as the *Lax Equivalence Theorem*. It was proved by P. Lax and R. Richtmyer under a stronger consistency requirement as the one given above. With consistency defined as in this paper, the main implication $(i) \to (ii)$ is due to P. Chernoff. The proof of the implication $(ii) \to (i)$ can be found in [FNZ].

Theorem 1.4. (Lax-Richtmyer 56, Chernoff 74) *Let A be the generator of a C_0-semigroup on a Banach space X and let $\mathcal{V} := \{V(t), t \in [0,\delta]\}$ be a consistent approximation scheme with $V(0) = I$. Then the following are equivalent.*

(i) *$\mathcal{V}$ is stable.*
(ii) *$\mathcal{V}$ is convergent.*

If $\mathcal{V}$ is a rational approximation of a semigroup $T(\cdot)$, then $\mathcal{V}$ and $T(\cdot)$ commute and

$$V(\frac{t}{n})^n x - T(t)x = \frac{t}{n}\sum_{j=0}^{n-1} V(\frac{t}{n})^{n-1-j} T(\frac{t}{n})^j (\frac{V(\frac{t}{n})x - T(\frac{t}{n})x}{\frac{t}{n}}). \tag{1.7}$$

Thus, if $\mathcal{V}$ is stable and consistent of approximation order m on D (see 1.4), then the approximation order m determines the speed of convergence on D; i.e., if for all $x \in D$, there exists a constant $M(x)$ such that $\|\frac{V(t)x-T(t)x}{t}\| \le M(x)t^m$ for all $t \in [0,\delta]$, then for all $T > 0, x \in D$ there exists a constant $M(T,x)$ such that

$$\|V(\frac{t}{n})^n x - T(t)x\| \le M(T,x)\frac{1}{n^m}.$$

Moreover, since D is assumed to be dense in X, the stability of the approximation scheme implies convergence for all $x \in X$. For example, if A generates a strongly continuous semigroup $T(\cdot)$ with $\|T(t)\| \le M$ for all $t \ge 0$ and some $M \ge 1$, then it follows from the Hille-Yosida theorem (see, e.g., [EN], Chap. 2, Thm 3.8.) and $(I - tA)^{-n} = \lambda^n R(\lambda, A)^n$ for $\lambda = \frac{1}{t}$ that $\|V_{be}(t)^n\| = \|(I-tA)^{-n}\| \le M$ for $n \in \mathbb{N}$ and $t \in [0,\infty)$. Moreover, by Theorem 1.3, there exists a constant $C > 0$ such that $\|\frac{T(t)x - V_{be}(t)x}{t}\| \le Ct\|A^2x\|$ for all $x \in D(A^2)$ and $t > 0$. Thus, by (1.7), we obtain that

$$\|V_{be}(\frac{t}{n})^n x - T(t)x\| \le t^2 CM^2 \frac{1}{n}\|A^2x\|$$

for all $t > 0$ and $x \in D(A^2)$. In particular, since the operators $V_{be}(\frac{t}{n})^n$ are uniformly bounded, $V_{be}(\frac{t}{n})^n x \to T(t)x$ as $n \to \infty$ for all $x \in X$. However, for initial data $x \notin D(A^2)$, the convergence can be arbitrarily slow (easy examples show that for $x \notin D(A)$ the convergence can be like $\frac{1}{\ln(\ln(\cdots(\ln(n))\cdots))}$) and the convergence rate $\frac{1}{\sqrt{n}}$ is best possible for $x \in D(A)$.

The following are optimal convergence results for smooth initial data due to P. Brenner and V. Thomée [BT] based on earlier work of R. Hersh and T. Kato [HK]. Instead of employing the Dunford-Schwartz calculus in their proofs, Brenner and Thomée as well as Hersh and Kato use the Hille-Phillips calculus (see [HP], Chapter XV). Note that the following results do not assume the stability of the approximation scheme.

Theorem 1.5 (Brenner-Thomée). *Let A be the generator of a strongly continuous semigroup $T(\cdot)$ with $\|T(t)\| \leq Me^{\omega t}$ for all $t \geq 0$. If $r(\cdot)$ is an A-stable rational approximation of the exponential of order m and $V(t) := r(tA)$, then there exist constants $C, c > 0$ such that*

$$\|V(\frac{t}{n})^n x - T(t)x\| \leq CMte^{c\omega t}(\frac{t}{n})^m \|A^{m+1}x\|$$

for $t \geq 0, n \in \mathbb{N}, x \in D(A^{m+1})$. In particular, for all $T > 0$ there exists $C'_T > 0$ such that

$$\|V(\frac{t}{n})^n x - T(t)x\| \leq C'_T \frac{1}{n^m} \|A^{m+1}x\|$$

for $n \in \mathbb{N}, x \in D(A^{m+1})$ and $t \in [0, T]$.

Theorem 1.6 (Brenner-Thomée). *Let A be the generator of a strongly continuous semigroup $T(\cdot)$ with $\|T(t)\| \leq Me^{\omega t}$ for all $t \geq 0$. If $r(\cdot)$ is an A-stable rational approximation of the exponential of order m and $V(t) := r(tA)$, then, for $s = 0, 1, 2 \cdots, m+1$ with $s \neq \frac{m+1}{2}$,*

$$\|V(\frac{t}{n})^n x - T(t)x\| \leq CMe^{c\omega t} t^{s-\beta(s)} (\frac{t}{n})^{\beta(s)} \|A^s x\|$$

for $t \geq 0, n \in \mathbb{N}, x \in D(A^s)$, where $\beta(s) = s\frac{m}{m+1} + \min(0, \frac{s}{m+1} - \frac{1}{2})$. In particular, for all $T > 0$ there exists $C'_T > 0$ such that

$$\|V(\frac{t}{n})^n x - T(t)x\| \leq C'_T \frac{1}{n^{\beta(s)}} \|A^s x\|$$

for all $t \in [0, T]$, $n \in \mathbb{N}$, and $x \in D(A^s)$.

Theorem 1.6 does not apply for $s = 1$ and $m = 1$. For $p = 0, q = 1$ (Backward-Euler), the following theorem fills the gap.

Theorem 1.7. *Let A be the generator of a strongly continuous semigroup $T(\cdot)$ on a Banach space X with $\|T(t)\| \leq M$ for $t \geq 0$. Then, for all $x \in D(A)$,*

$$\|V_{BE}(\frac{t}{n})^n x - T(t)x\| \leq M(M+2)\frac{t}{\sqrt{n}} \|Ax\|.$$

Proof. Let $Lip_0(\mathbb{R}_+, X)$ denote the Banach space of all $F : [0, \infty) \to X$ with $F(0) = 0$ and $\|F\|_{Lip} := \sup_{t,s} \|\frac{F(t)-F(s)}{t-s}\| < \infty$. By the Riesz-Stieltjes Representation theorem (see [ABHN], Theorem 2.1.1), there exists an isometric isomorphism $Lip_0(\mathbb{R}_+, X) \to \mathcal{L}(L^1(\mathbb{R}_+), X)$ which assigns to each $F \in Lip_0(\mathbb{R}_+, X)$ an operator $T_F \in \mathcal{L}(L^1(\mathbb{R}_+), X)$ such that $T_F(g) = \int_0^\infty g(s)\, dF(s)$ for all continuous

$g \in L^1(\mathbb{R}_+)$ and $T_F\chi_{[0,t]} = F(t)$ for $t \in \mathbb{R}_+$. We will use this to show that for $n \in \mathbb{N}$ and $t > 0$,

$$\|F(t) - (-1)^n \frac{1}{n!}(\frac{n}{t})^{n+1} \frac{d^n}{d\lambda^n}(\frac{r(\lambda)}{\lambda})|_{\lambda=\frac{n}{t}}\| \leq 2\|F\|_{Lip} \frac{t}{\sqrt{n}}, \tag{1.8}$$

where $F \in Lip_0(\mathbb{R}_+, X)$ and $r(\lambda) := \int_0^\infty e^{-\lambda s}\, dF(s) = \lambda \int_0^\infty e^{-\lambda s} F(s)\, ds$. By the Leibniz formula and $r^{(j)}(\lambda) = \int_0^\infty e^{-\lambda s}(-s)^j\, dF(s) = T_F(e^{-\lambda\cdot}(-\cdot)^j)$,

$$\begin{aligned} F(t) &- (-1)^n \tfrac{1}{n!}(\tfrac{n}{t})^{n+1} \tfrac{d^n}{d\lambda^n}(\tfrac{r(\lambda)}{\lambda})|_{\lambda=\frac{n}{t}} \\ &= F(t) - (-1)^n \tfrac{1}{n!}(\tfrac{n}{t})^{n+1} \textstyle\sum_{j=0}^n \binom{n}{j}(-1)^{n-j}(n-j)!(\tfrac{t}{n})^{n-j+1} r^{(j)}(\tfrac{n}{t}) \\ &= F(t) - \textstyle\sum_{j=0}^n (-1)^j \tfrac{1}{j!}(\tfrac{n}{t})^j r^{(j)}(\tfrac{n}{t}) = T_F[\chi_{[0,t]} - \textstyle\sum_{j=0}^n \tfrac{1}{j!}(\tfrac{n}{t})^j (\cdot)^j e^{-\frac{n}{t}\cdot}], \end{aligned}$$

and thus

$$\begin{aligned} \|F(t) &- (-1)^n \tfrac{1}{n!}(\tfrac{n}{t})^{n+1} \tfrac{d^n}{d\lambda^n}(\tfrac{r(\lambda)}{\lambda})|_{\lambda=\frac{n}{t}}\| \\ &\leq \|T_F\| \|\chi_{[0,t]} - \textstyle\sum_{j=0}^n \tfrac{1}{j!}(\tfrac{n}{t})^j(\cdot)^j e^{-\frac{n}{t}\cdot}\|_1 \\ &= \|F\|_{Lip} \|\chi_{[0,t]} - \textstyle\sum_{j=0}^n \tfrac{1}{j!}(\tfrac{n}{t})^j(\cdot)^j e^{-\frac{n}{t}\cdot}\|_1. \end{aligned}$$

Moreover,

$$\begin{aligned} \|\chi_{[0,t]} &- \textstyle\sum_{j=0}^n \tfrac{1}{j!}(\tfrac{n}{t})^j(\cdot)^j e^{-\frac{n}{t}\cdot}\|_1 \\ &= \textstyle\int_0^t |1 - \sum_{j=0}^n \tfrac{1}{j!}(\tfrac{n}{t}s)^j e^{-\frac{n}{t}s}|\, ds + \int_t^\infty |\sum_{j=0}^n \tfrac{1}{j!}(\tfrac{n}{t}s)^j e^{-\frac{n}{t}s}|\, ds \\ &= \textstyle\int_0^t (1 - \sum_{j=0}^n \tfrac{1}{j!}(\tfrac{n}{t}s)^j e^{-\frac{n}{t}s})\, ds + \int_t^\infty \sum_{j=0}^n \tfrac{1}{j!}(\tfrac{n}{t}s)^j e^{-\frac{n}{t}s}\, ds \\ &= \tfrac{t}{n}\textstyle\int_0^n (1 - \sum_{j=0}^n \tfrac{1}{j!}u^j e^{-u})\, du + \tfrac{t}{n}\int_n^\infty \sum_{j=0}^n \tfrac{1}{j!}u^j e^{-u}\, du \\ &= t - \tfrac{t}{n}\textstyle\sum_{j=0}^n \int_0^n \tfrac{1}{j!}u^j e^{-u}\, du + \tfrac{t}{n}\sum_{j=0}^n \int_n^\infty \tfrac{1}{j!}u^j e^{-u}\, du \\ &= t + \tfrac{t}{n}\textstyle\sum_{j=0}^n [\int_n^\infty \tfrac{1}{j!}u^j e^{-u}\, du - \int_0^n \tfrac{1}{j!}u^j e^{-u}\, du], \end{aligned}$$

and $\int \frac{1}{j!}u^j e^{-u}\, du = -e^{-u}\sum_{m=0}^j \frac{1}{m!}u^m$ which can be seen by induction. Therefore,

$$\begin{aligned} \|\chi_{[0,t]} - \textstyle\sum_{j=0}^n \tfrac{1}{j!}(\tfrac{n}{t})^j(\cdot)^j e^{-\frac{n}{t}\cdot}\|_1 &= t + \tfrac{t}{n}\textstyle\sum_{j=0}^n [2e^{-n}\sum_{m=0}^j \tfrac{1}{m!}n^m - 1] \\ &= t - \tfrac{t}{n}(n+1) + \tfrac{t}{n}\textstyle\sum_{j=0}^n 2e^{-n}\sum_{m=0}^j \tfrac{1}{m!}n^m \\ &= -\tfrac{t}{n} + \tfrac{t}{n}2e^{-n}\textstyle\sum_{j=0}^n \sum_{m=0}^j \tfrac{1}{m!}n^m = \tfrac{t}{n}[2e^{-n}\sum_{m=0}^n \sum_{j=m}^n \tfrac{1}{m!}n^m - 1] \\ &= \tfrac{t}{n}[2e^{-n}\textstyle\sum_{m=0}^n (n-m+1)\tfrac{1}{m!}n^m - 1] \\ &= \tfrac{t}{n}[2e^{-n}[\textstyle\sum_{m=0}^n \tfrac{1}{m!}n^{m+1} - \sum_{m=1}^n \tfrac{1}{(m-1)!}n^m + \sum_{m=0}^n \tfrac{1}{m!}n^m] - 1] \\ &= \tfrac{t}{n}[2e^{-n}[\textstyle\sum_{m=0}^n \tfrac{1}{m!}n^{m+1} - \sum_{m=0}^{n-1} \tfrac{1}{m!}n^{m+1} + \sum_{m=0}^n \tfrac{1}{m!}n^m] - 1] \\ &= \tfrac{t}{n}[2e^{-n}\tfrac{1}{n!}n^{n+1} + 2e^{-n}\textstyle\sum_{m=0}^n \tfrac{1}{m!}n^m - 1] \\ &\leq \tfrac{t}{n}[2e^{-n}\tfrac{1}{n!}n^{n+1} + 2e^{-n}e^n - 1] = \tfrac{t}{n}[2e^{-n}\tfrac{1}{n!}n^{n+1} + 1] \leq 2\tfrac{t}{\sqrt{n}} \end{aligned}$$

since, by Stirling's formula,

$$2e^{-n}\frac{1}{n!}n^{n+1} = 2e^{-n}n^n \frac{1}{(n-1)!} = \sqrt{\frac{2}{\pi}}\sqrt{n}e^{-\omega(n)} \leq \sqrt{\frac{2}{\pi}}e^{\frac{1}{12n}}\sqrt{n} \leq \sqrt{n}.$$

This yields (1.8). Now let $x \in D(A)$, $F(t) := T(t)x - x = \int_0^t T(s)Ax\, ds$ and $r(\lambda) = \int_0^\infty e^{-\lambda t}\, dF(t)$. Then $\|F\|_{Lip} \leq M\|Ax\|$ and integration by parts yields

$\frac{r(\lambda)}{\lambda} = \int_0^\infty e^{-\lambda s}(T(s)x - x)\, ds = R(\lambda, A)x - \frac{x}{\lambda}$. Hence,

$$\begin{aligned}
F(t) &- (-1)^n \tfrac{1}{n!}(\tfrac{n}{t})^{n+1} \tfrac{d^n}{d\lambda^n}(\tfrac{r(\lambda)}{\lambda})|_{\lambda=\frac{n}{t}} \\
&= T(t)x - x - (-1)^n \tfrac{1}{n!}(\tfrac{n}{t})^{n+1} \tfrac{d^n}{d\lambda^n}[R(\lambda, A)x - \tfrac{x}{\lambda}]_{\lambda=\frac{n}{t}} \\
&= T(t)x - x - (-1)^n \tfrac{1}{n!}(\tfrac{n}{t})^{n+1} \tfrac{d^n}{d\lambda^n} R(\lambda, A)x|_{\lambda=\frac{n}{t}} + (-1)^n \tfrac{1}{n!}(\tfrac{n}{t})^{n+1} \tfrac{d^n}{d\lambda^n} \tfrac{x}{\lambda}|_{\lambda=\frac{n}{t}} \\
&= T(t)x - x - (\tfrac{n}{t})^{n+1} R(\tfrac{n}{t}, A)^{n+1} x + (-1)^n \tfrac{1}{n!}(\tfrac{n}{t})^{n+1}[(-1)^n \tfrac{n!}{(\frac{n}{t})^{n+1}}]x \\
&= T(t)x - x - (\tfrac{n}{t})^{n+1} R(\tfrac{n}{t}, A)^{n+1} x + x \\
&= T(t)x - (\tfrac{n}{t})^{n+1} R(\tfrac{n}{t}, A)^{n+1} x = T(t)x - V_{BE}(\tfrac{t}{n})^{n+1} x.
\end{aligned}$$

Therefore, by (1.8), $\|T(t)x - V_{BE}(\frac{t}{n})^{n+1}x\| \le 2M\frac{t}{\sqrt{n}}\|Ax\|$ for all $x \in D(A)$. Note that $\|V_{BE}(\frac{t}{n})^n\| \le M$ due to the Hille-Yosida theorem. Thus,

$$\begin{aligned}
\|T(t)x - V_{BE}(\tfrac{t}{n})^n x\| &\le \|T(t)x - V_{BE}(\tfrac{t}{n})^{n+1}x\| + \|V_{BE}(\tfrac{t}{n})^{n+1}x - V_{BE}(\tfrac{t}{n})^n x\| \\
&\le 2M\tfrac{t}{\sqrt{n}}\|Ax\| + \|V_{BE}(\tfrac{t}{n})^n\|\|V_{BE}(\tfrac{t}{n})x - x\| \\
&\le 2M\tfrac{t}{\sqrt{n}}\|Ax\| + M\|\tfrac{n}{t}R(\tfrac{n}{t}, A)x - x\| \\
&\le 2M\tfrac{t}{\sqrt{n}}\|Ax\| + M\|R(\tfrac{n}{t}, A)Ax\| \\
&\le 2M\tfrac{t}{\sqrt{n}}\|Ax\| + M^2\tfrac{t}{n}\|Ax\| \le M(M+2)\tfrac{t}{\sqrt{n}}\|Ax\|. \qquad \square
\end{aligned}$$

Unfortunately, except for the Backward-Euler scheme (and for all schemes with $p = q - 1$, see [AS], Theorem II.1.2) the Lax-Chernoff Equivalence Theorem 1.4 does not apply for generators of strongly continuous semigroups in a Banach space X due to the lack of stability. In fact, for $X = L^1(\mathbb{R})$ and $T(t)f(x) := f(x+t)$, $\|V_{CN}(\frac{t}{n})^n\| \ge C\sqrt{n}$ according to T. Kato (the example can be found in [CHMM], p. 224, Example 3.8). The following are known stability and convergence results for analytic semigroups due to [LTW] and [Ha] which we will modify in the next section using $\mathcal{H}^\infty$-techniques.

Theorem 1.8. (Larsson, Thomée, Wahlbin 91) *Let A generate an analytic C_0-semigroup on a Banach space X with $s(A) < 0$ and let $r(\cdot)$ be an A-stable, rational approximation of the exponential of approximation order m. Then*

(a) $\|V(\frac{t}{n})^n x - T(t)x\| \le M_T \frac{1}{n^j}\|A^j x\|$ *for all* $1 \le j \le m, x \in D(A^j), t \in [0, T]$.

(b) *If* $|r(\infty)| < 1$, *then* $V(t) := r(tA)$ *defines a stable, rational approximation scheme and* $\|V(\frac{t}{n})^n x - T(t)x\| \le M_T \frac{1}{n^m}\|x\|$ *for all* $x \in X, t \in [0, T]$.

If A generates an analytic semigroup, then rational approximation schemes that are not stable can be stabilized by employing first a lower order approximation scheme with $|r(\infty)| = 0$. Most importantly, this technique can also be used to obtain optimal convergence estimates that are valid for all $x \in X$ (and not only for $x \in D(A^{m+1})$ as in Theorem 1.5).

Theorem 1.9. (Hansbo 99) *Let A generate an analytic C_0-semigroup $T(\cdot)$ on a Banach space X with $s(A) < 0$ and let*

$$r_n(z) := \begin{cases} r_s(z)^n & n < N \\ r_s(z)^N r_a(z)^{n-N} & n \ge N \end{cases},$$

where $r_a(\cdot)$ *is an A-stable, rational approximation of the exponential of order* $m \geq 2$, $N \geq 0$, *and* $r_s(\cdot)$ *is A-stable and of order* $m-1$ *with* $r_s(\infty) = 0$. *Furthermore, let* $\frac{m-N}{2} \leq j \leq m$. *Then, for* $n \geq \max\{-j, 1\}$,

$$\|r_n(\frac{t}{n}A)x - T(t)x\| \leq C\frac{t^j}{n^m}\|A^j x\| \text{ for all } x \in D(A^j).$$

2. $\mathcal{H}^\infty$-Calculus

Let $\mathcal{H}^\infty(\Sigma_\phi)$ denote the Banach space of all bounded analytic functions $f : \Sigma_\phi \to \mathbb{C}$ where $0 < \phi < \pi/2$ and $\Sigma_\phi := \{\lambda \in \mathbb{C} : |\arg(\lambda)| > \pi - \phi\}$ is an open left sector. Let A be a closed linear operator on a Banach space X with dense domain $D(A)$ in X such that $0 \notin p\sigma(A) \cup p\sigma(A^*)$; that is,

$$A \text{ is injective and range } A \text{ is dense in } X.$$

Then A is called sectorial of type ω for some $0 < \omega < \pi$ if $\sigma(A) \subset \Sigma_\omega$ and for all $\phi > \omega$ there is a C_ϕ such that $\|\lambda R(\lambda, A)\| \leq C_\phi$ for λ in the open right sector $\{\lambda \in \mathbb{C} : |\arg(\lambda)| < \pi - \phi\}$. Let $\omega(A)$ be the infimum of all angles ω for which A is sectorial of type ω. Let

$$\mathcal{H}_0^\infty(\Sigma_\phi) := \{f : \mathcal{H}^\infty(\Sigma_\phi) \text{ and there exist } C_f, \epsilon_f > 0 \text{ such that } |f(\lambda)| \leq C_f|\tfrac{\lambda}{(1-\lambda)^2}|^{\epsilon_f} \text{ for all } \lambda \in \Sigma_\omega.\}.$$

For $f \in \mathcal{H}_0^\infty(\Sigma_\phi)$ and a sectorial operator A of type ω with $\omega < \phi < \pi$

$$f(A) := \int_\Gamma f(\lambda)R(\lambda, A)\, d\lambda$$

can be defined as a Bochner integral, where $\Gamma := \partial\Sigma_\theta$ and $\omega < \theta < \phi$. We need the growth condition for $f \in \mathcal{H}_0^\infty(\Sigma_\phi)$ in order to guarantee that the above integral converges. For $f \in \mathcal{H}^\infty(\Sigma_\phi)$ the integral does not necessarily converge; thus, we first define an auxiliary $\mathcal{H}_0^\infty(\Sigma_\phi)$-calculus which will be extended to $\mathcal{H}^\infty(\Sigma_\phi)$ later on.

Let A be a sectorial operator of type ω and $\omega < \phi < \pi$. For $f \in \mathcal{H}_0^\infty(\Sigma_\phi)$ define $|||f|||_A := \|f\|_{\mathcal{H}^\infty(\Sigma_\phi)} + \|f(A)\|_{\mathcal{B}(X)}$, and consider the algebra

$$\mathcal{H}_A^\infty(\Sigma_\phi) := \{f \in \mathcal{H}^\infty(\Sigma_\phi) : \text{ there is } \{f_n\} \subset \mathcal{H}_0^\infty(\Sigma_\phi) \text{ with } f_n(\lambda) \to f(\lambda) \text{ for all } \lambda \in \Sigma_\phi \text{ and } \sup_n |||f_n|||_A < \infty\}.$$

Then $D(\Phi) := \mathcal{H}_A^\infty(\Sigma_\phi)$ is the domain of a closed operator $\Phi : \mathcal{H}^\infty(\Sigma_\phi) \supset D(\Phi) \to \mathcal{B}(X)$ defined by $\Phi(f) := f(A)$, where $f(A)x := \lim f_n(A)x$ for $f_n \in \mathcal{H}_0^\infty(\Sigma_\phi)$ with $f_n(\lambda) \to f(\lambda)$ for $\lambda \in \Sigma_\phi$ and $|||f_n|||_A \leq C$ (for a proof of the closedness of Φ, see [We]).

The sectorial operator A *has a bounded* $\mathcal{H}^\infty(\Sigma_\phi)$*-calculus* if the operator Φ is bounded, or, equivalently if $\mathcal{H}^\infty(\Sigma_\phi) = \mathcal{H}^\infty_A(\Sigma_\phi)$. Thus, if A has a bounded $\mathcal{H}^\infty(\Sigma_\phi)$-calculus, then there exists a $C > 0$ such that

$$\|f(A)\|_{\mathcal{B}(X)} \leq C\|f\|_{\mathcal{H}^\infty(\Sigma_\phi)},$$

where $f(A)x := \lim \int_{\partial\Sigma_\phi} f_n(\lambda)R(\lambda, A)x\,d\lambda$ and $f_n \in \mathcal{H}^\infty_0(\Sigma_\phi)$ with $f_n(\lambda) \to f(\lambda)$ for all $\lambda \in \Sigma_\phi$ and $\sup_n |||f_n|||_A < \infty$. Moreover, if A has a bounded $\mathcal{H}^\infty(\Sigma_\phi)$-calculus, then $\Phi : f \to f(A)$ is an algebra homomorphism from $\mathcal{H}^\infty(\Sigma_\phi)$ into $\mathcal{B}(X)$.

Example 2.1. Let $X = \mathcal{H}^\infty(\Sigma_{\frac{\pi}{2}})$ and let A be defined by $Ag(z) = zg(z)$ for $g \in D(A)$. Then A is sectorial of type $\frac{\pi}{2}$ as

$$\|\lambda R(\lambda, A)\| = \sup_{\|g\|\leq 1} \sup_{z\in\Sigma_{\frac{\pi}{2}}} |\lambda(\lambda - z)^{-1}g(z)| \leq \sup_{z\in\Sigma_{\frac{\pi}{2}}} |\lambda(\lambda - z)^{-1}| \leq \frac{1}{\cos(\phi - \frac{\pi}{2})}$$

for all $\frac{\pi}{2} < \phi < \pi$ and $|\arg\lambda| < \pi - \phi$. Notice that $\Sigma_{\frac{\pi}{2}} \subset \sigma(A)$, the spectrum of A, and thus, A is not the generator of an analytic semigroup. However, A has a bounded $\mathcal{H}^\infty(\Sigma_\phi)$-calculus on X for all $\phi > \frac{\pi}{2}$ since

$$\|f(A)\| \leq \sup_{\|g\|\leq 1} \sup_{z\in\Sigma_{\frac{\pi}{2}}} |f(z)g(z)| \leq \|f\|_{\mathcal{H}^\infty(\Sigma_{\frac{\pi}{2}})} \leq \|f\|_{\mathcal{H}^\infty(\Sigma_\phi)}.$$

Note that if an operator A has an $\mathcal{H}^\infty(\Sigma_\phi)$-calculus with $0 < \phi < \frac{\pi}{2}$, it generates an analytic semigroup. However, we do not need to require $s(A) < 0$, which is the case in Theorems 1.8 and 1.9. Examples for generators A with an $\mathcal{H}^\infty(\Sigma_\phi)$-calculus with $0 < \phi < \frac{\pi}{2}$ where $s(A) < 0$ does not hold, are the Laplacian on $L^p(\mathbb{R}^n)$, $1 < p < \infty$, or all densely defined, injective normal operators A with dense range on Hilbert spaces whose spectrum is contained in Σ_ϕ for some $\phi < \frac{\pi}{2}$.

Theorem 2.2. *Suppose* A *has a bounded* $\mathcal{H}^\infty(\Sigma_\phi)$*-calculus with* $0 < \phi < \frac{\pi}{2}$, *and* $r(\cdot)$ *is an A-stable rational approximation scheme of order* m. *Then,*

$$\|V(\frac{t}{n})^n f - T(t)f\| \leq C\frac{t^j}{n^j}\|A^j f\| \textit{ for all } 1 \leq j \leq m, f \in D(A^j), t \in [0, T].$$

Proof. Since A has a bounded $\mathcal{H}^\infty(\Sigma_\phi)$-calculus,

$$\|A^{-j}(V(\frac{t}{n})^n - T(t))\| \leq M\|(r(\frac{t}{n}\cdot)^n - e^{t\cdot})\frac{1}{(\cdot)^j}\|_{\mathcal{H}^\infty(\Sigma_\phi)}.$$

For $Re(z) < 0$ and $|\frac{t}{n}z| > R > 0$, there exists a $C > 0$ such that $|r(\frac{t}{n}z)^n - e^{tz}| \leq 2 \leq C|z|^j \frac{t^j}{n^j}$ for all $1 \leq j \leq m$. Let $z \in \Sigma_\phi$ and let $c = \cos(\omega) > 0$. Then $e^{\mathrm{Re} z} \leq e^{-c|z|}$ and $|r(z)| \leq 1$. Moreover, if $|z| < R$ for small R, then there exists $C > 0$ such that $|r(z) - e^z| \leq C|z|^{m+1}$. Thus, for $z \in \Sigma_\phi$ and $|\frac{t}{n}z| < R$, $\left|r(\frac{t}{n}z)^n - e^{tz}\right| \frac{1}{|z|^j} \leq \frac{1}{|z|^j}\left|r(\frac{t}{n}) - e^{\frac{t}{n}z}\right| \sum_{i=0}^{n-1} \left|r(\frac{t}{n}z)\right|^{n-1-i} \left|e^{\frac{t}{n}z}\right|^i \leq \frac{1}{|z|^j}\left|r(\frac{t}{n}z) - e^{\frac{t}{n}z}\right| \sum_{i=0}^{n-1} \left|e^{\frac{t}{n}z}\right|^i \leq \frac{1}{|z|^j}\left|r(\frac{t}{n}z) - e^{\frac{t}{n}z}\right| \sum_{i=0}^{\infty} \left[e^{-c\left|\frac{t}{n}z\right|}\right]^i \leq \frac{1}{|z|^j}\left|r(\frac{t}{n}z) - e^{\frac{t}{n}z}\right| \frac{1}{1-e^{-c\left|\frac{t}{n}z\right|}} \leq$

$C\frac{1}{|z|^j}\left|\frac{t}{n}z\right|^{m+1}\frac{1}{1-e^{-c\left|\frac{t}{n}z\right|}} \leq C\frac{t^j}{n^j}\left|\frac{t}{n}z\right|^{m-j}\frac{|\frac{t}{n}z|}{1-e^{-c\left|\frac{t}{n}z\right|}} \leq C\frac{t^j}{n^j}R^{m-j}\frac{|\frac{t}{n}z|}{1-e^{-c\left|\frac{t}{n}z\right|}} \leq$ $C\frac{t^j}{n^j}R^{m-j}\frac{1}{c}e^{c\left|\frac{t}{n}z\right|} \leq C\frac{t^j}{n^j}R^{m-j}\frac{1}{c}e^{cR}$ since $\frac{1-e^{-cw}}{w} = \int_0^c e^{-ws}\,ds \geq ce^{-wc}$ for $w > 0$. Thus, there exists a constant $M > 0$ such that $\left|r(\frac{t}{n}z)^n - e^{tz}\right|\frac{1}{|z|^j} \leq M\frac{t^j}{n^j}$. □

Theorem 2.3. *If A has a bounded $\mathcal{H}^\infty(\Sigma_\phi)$-calculus for some $0 < \phi \leq \frac{\pi}{2}$, then all A-stable rational approximation schemes $V(t) := r(tA)$ are stable.*

Proof. Since A has a bounded $\mathcal{H}^\infty(\Sigma_\phi)$-calculus, we have $\|V(t)^n\| = \|r(tA)^n\| \leq M\|r(t\cdot)^n\|_{\mathcal{H}^\infty(\Sigma_\phi)} \leq M$ since $|r(z)| \leq 1$ for $Re z \leq 0$ for all t and n. □

For the next results we will need the following statements; for proofs see, for example, [Ha], Lemma 2.1 and 2.2.

Lemma 2.4. *Let $r(\cdot)$ be A-stable and let $r_n(\cdot), r_a(\cdot), r_s(\cdot)$ be as in Theorem 1.9. Then, for all $R > 0$ there are $C, c > 0$ such that*

(1) $|r(z)| \leq e^{cRez}$ *for* $z \in \Sigma_\phi$. *In particular,* $|r(z)| \leq e^{cRez} \leq e^{-c|z|}$ *for* $z \in \Sigma_\phi, |z| \leq R$.
(2) $|r_n(z)| \leq C|z|^{-m}e^{-c\frac{n-m}{|z|}}$ *for* $z \in \Sigma_\phi, |z| \geq R, n \geq m$.
(3) $|r_a(z)^n - e^{nz}| \leq C|z|^{m+1}ne^{-cn|z|}$ *for* $z \in \Sigma_\phi, |z| \leq R$.
(4) $|r_s(z)^m - e^{mz}| \leq C|z|^m$ *for* $z \in \Sigma_\phi$.

Theorem 2.5. *Suppose A has a bounded $\mathcal{H}^\infty(\Sigma_\phi)$-calculus with $0 < \phi < \frac{\pi}{2}$, and $r(\cdot)$ is an A-stable rational approximation scheme of order m. If $|r(\infty)| < 1$, then*

$$\|V(\frac{t}{n})^n f - T(t)f\| \leq M_T\frac{1}{n^m}\|f\| \text{ for all } f \in X, t \in [0,T].$$

Proof. Since A has a bounded $\mathcal{H}^\infty(\Sigma_\phi)$-calculus,

$$\|V(\frac{t}{n})^n - T(t)\| \leq M\|r(\frac{t}{n}\cdot)^n - e^{t\cdot}\|_{\mathcal{H}^\infty(\Sigma_\phi)}.$$

Let $\epsilon > 0$ such that $\gamma := |r(\infty)| + \epsilon < 1$. Let $R > 0$ such that $|r(\frac{t}{n}z)| \leq \gamma < 1$ for all $|\frac{t}{n}z| > R$. Then, $|r(\frac{t}{n}z)^n - e^{tz}| \leq \gamma^n + e^{-c|tz|} \leq \gamma^n + e^{-cn} \leq \frac{C}{n^m}$ for all $m > 0$. On the other hand, for $|\frac{t}{n}z| \leq R$, we obtain by Lemma 2.4 (1), $e^{\text{Re}z} \leq e^{-c|z|}$, and

$$\sup_{x\geq 0}|x^k e^{-x}| = k^k e^{-k}, \tag{2.1}$$

$|r(\frac{t}{n}z)^n - e^{tz}| \leq |r(\frac{t}{n}z) - e^{\frac{t}{n}z}|\sum_{j=0}^{n-1}|r(\frac{t}{n}z)|^{n-1-j}|e^{\frac{t}{n}z}|^j \leq C|\frac{t}{n}z|^{m+1}ne^{-c(n-1)|\frac{t}{n}z|}$ $= \frac{C}{n^m}|tz|^{m+1}e^{-|ctz|}e^{cR} \leq \frac{C_m}{n^m}$. □

Next we will prove a similar result to Theorem 1.9, replacing the condition $s(A) < 0$.

Theorem 2.6. *Let A be the generator of an analytic semigroup $T(\cdot)$ on a Banach space X that has a bounded $\mathcal{H}^\infty(\Sigma_\phi)$-calculus with $0 < \phi < \frac{\pi}{2}$. Let*

$$r_n(z) := \begin{cases} r_s(z)^n & n < m \\ r_a(z)^{n-m}r_s(z)^m & n \geq m \end{cases},$$

where $r_a(z)$ is an A-stable rational approximation of the exponential of order $m \geq 2$ and $r_s(z)$ is A-stable and of order $m-1$ with $r_s(\infty) = 0$. Then, for $n \in \mathbb{N}$,

$$\|r_n(\frac{t}{n}A)f - T(t)f\| \leq C\frac{1}{n^m}\|f\| \text{ for all } f \in X.$$

Proof. Since A has a bounded $\mathcal{H}^\infty(\Sigma_\phi)$-calculus,

$$\|r_n(\frac{t}{n}A) - T(t)\| \leq M\|r_n(\frac{t}{n}\cdot) - e^{t\cdot}\|_{\mathcal{H}^\infty(\Sigma_\phi)}.$$

We rewrite $r_n(\frac{t}{n}z) - e^{tz} = r_s(\frac{t}{n}z)^m r_a(\frac{t}{n}z)^{n-m} - e^{tz} = f_1(\frac{t}{n}z) + f_2(\frac{t}{n}z)$, where $f_1(\frac{t}{n}z) := r_s(\frac{t}{n}z)^m(r_a(\frac{t}{n}z)^{n-m} - e^{(n-m)\frac{t}{n}z})$ and $f_2(\frac{t}{n}z) := e^{(n-m)\frac{t}{n}z}(r_s(\frac{t}{n}z)^m - e^{m\frac{t}{n}z})$. Then, for $|\frac{t}{n}z| \leq R$, by Lemma 2.4 (3) and (2.1) there exists a constant C_m such that $|f_1(\frac{t}{n}z)| \leq C|\frac{t}{n}z|^{m+1}(n-m)e^{-c(n-m)|\frac{t}{n}z|} \leq \frac{C}{n^{m+1}}|tz|^{m+1}ne^{-c|tz|}e^{cmR} \leq \frac{C_m}{n^m}$. On the other hand, by Lemma 2.4 (2) and $\sup_{x>0}\frac{n^m}{x^m}e^{-c\frac{n^2}{x}} = \frac{m^m}{c^m n^m}e^{-m}$ we find for $|\frac{t}{n}z| > R$ a constant C_m such that $|f_1(\frac{t}{n}z)| \leq |r_n(\frac{t}{n}z)| + |e^{(n-m)\frac{t}{n}z}| \leq \frac{C}{|\frac{t}{n}z|^m}e^{-\frac{cn}{|\frac{t}{n}z|}} \leq \frac{C_m}{n^m}$. Therefore, $|f_1(\frac{t}{n}z)| \leq \frac{C}{n^m}$ for all $z \in \Sigma_\phi$. Also, it follows from (2.1) and Lemma 2.4 (4) that there is a constant C_m such that

$$|f_2(\frac{t}{n}z)| = |e^{(n-m)\frac{t}{n}z}(r_s(\frac{t}{n}z)^m - e^{m\frac{t}{n}z})| \leq e^{-c(n-m)|\frac{t}{n}z|}C|\frac{t}{n}z|^m \leq \frac{C_m}{n^m}. \qquad \square$$

One can prove the following variation of Theorem 2.6 which allows to use less smoothing steps if the initial data is sufficiently smooth. In particular, for $f \in D(A^j)$, j less smoothing steps are needed. For example, Theorem 2.7 will yield

$$\|V_{BE}(\frac{t}{n})^1 V_{CN}(\frac{t}{n})^{n-1}f - T(t)f\| \leq C\frac{t}{n^2}\|Af\|.$$

Theorem 2.7. *Let A be the generator of an analytic semigroup $T(\cdot)$ on a Banach space X that has a bounded $\mathcal{H}^\infty(\Sigma_\phi)$-calculus with $0 < \phi < \frac{\pi}{2}$. For $0 \leq j \leq m$, let*

$$r_n(z) := \begin{cases} r_s(z)^n & n < m-j \\ r_a(z)^{n-m+j}r_s(z)^{m-j} & n \geq m-j \end{cases},$$

where $r_a(z)$ is an A-stable rational approximation of the exponential of order $m \geq 2$ and $r_s(z)$ is A-stable and of order $m-1$ with $r_s(\infty) = 0$. Then, for $f \in D(A^j)$,

$$\|r_n(\frac{t}{n}A)f - T(t)f\| \leq C\frac{t^j}{n^m}\|A^j f\|.$$

Proof. Since A has a bounded $\mathcal{H}^\infty(\Sigma_\phi)$-calculus,

$$\begin{aligned}&\|A^{-j}(V_s(\tfrac{t}{n})^{m-j}V_a(\tfrac{t}{n})^{n-m+j} - T(t))\| \\ &\qquad\leq M\|(r_s(\tfrac{t}{n}\cdot)^{m-j}r_a(\tfrac{t}{n}\cdot)^{n-m+j} - e^{t\cdot})\tfrac{1}{(\cdot)^j}\|_{\mathcal{H}^\infty(\Sigma_\phi)}.\end{aligned}$$

We rewrite $(r_s(\frac{t}{n}z)^{m-j}r_a(\frac{t}{n}z)^{n-m+j} - e^{tz})\frac{1}{z^j} = f_1(\frac{t}{n}z) + f_2(\frac{t}{n}z)$, where $f_1(\frac{t}{n}z) := r_s(\frac{t}{n}z)^{m-j}(r_a(\frac{t}{n}z)^{n-m+j} - e^{(n-m+j)\frac{t}{n}z})\frac{1}{z^j}$ and $f_2(\frac{t}{n}z) := e^{(n-m+j)\frac{t}{n}z}(r_s(\frac{t}{n}z)^{m-j} - e^{(m-j)\frac{t}{n}z})\frac{1}{z^j}$. Then, by estimates similar to the ones in the previous proof, there exists a constant C_m such that $|f_i(\frac{t}{n}z)| \leq C_m\frac{t^j}{n^m}$. $\square$

References

[ABHN] W. Arendt, C. Batty, M. Hieber and F. Neubrander, *Vector-Valued Laplace Transforms and Cauchy Problems.* Monographs in Mathematics, Birkhäuser Verlag, 2001.

[AS] A. Ashyralyev and P. Sobolevskii, *Well-posedness of Parabolic Difference Equations.* Operator Theory. Advances and Applications, Birkhäuser Verlag, 1994.

[BT] P. Brenner and V. Thomée, *On rational approximations of semigroups.* SIAM J. Numer. Anal. **16** (1979), no. 4, 683–694.

[Ch] P. R. Chernoff, *Product Formulas, Nonlinear Semigroups, and Addition of Unbounded Operators.* Memoirs of the American Math. Soc. 140, American Mathematical Society, Providence, Rhode Island, 1974.

[CHMM] A. Chorin, T. Hughes, M. McCracken and J. Marsden, *Product formula and numerical algorithms.* Comm. Pure Appl. Math. **31** (1978), 605–656.

[EN] K.-J. Engel and R. Nagel, *One-Parameter Semigroups for Linear Evolution Equations.* Graduate Texts in Mathematics 194, Springer, 1999.

[FNZ] S. Flory, F. Neubrander and Y. Zhuang, *On the regularization and stabilization of approximation schemes for C_0-semigroups.* Partial Differential Equations and Spectral Theory, edited by M. Demuth and B.-W. Schulze, OT 126, Birkhaeuser Verlag, 2001.

[HW] E. Hairer and G. Wanner, *Solving Ordinary Differential Equations II – Stiff and Differential-Algebraic Problems.* Springer-Verlag, 1991.

[Ha] A. Hansbo, *Nonsmooth data error estimates for damped single step methods for parabolic equations in Banach spaces.* Calcolo **36** (1999), 75–101.

[HK] R. Hersh and T. Kato, *High-accuracy stable difference schemes for well-posed initial value problems.* SIAM J. Numer. Anal. **16** (1979), no. 4, 670–682.

[HP] E. Hille and S. Phillips, *Functional Analysis and Semi-Groups.* American Mathematical Society, Providence, 1957.

[LR] P. Lax and R. Richtmyer, *Survey of stability of linear finite difference equations.* Comm. Pure Appl. Math. **9** (1956), 267–293.

[LTW] S. Larsson, V. Thomée and L. B. Wahlbin, *Finite-element methods for a strongly damped wave equation.* IMA J. Numer. Anal. **II** (1991), 115–142.

[LuR] M. Luskin and R. Rannacher, *On the smoothing property of the Crank-Nicolson scheme.* Applicable Analysis **14** (1982), 117–135.

[Pa] M. H. Padé, *Sur répresentation approchée d'une fonction par des fractionelles.* Ann. de l'Ecole Normale Superieure **9** (1892).

[Th] V. Thomée, *Galerkin Finite Element Methods for Parabolic Problems.* Springer-Verlag, 1997.

[We] L. Weis, *Functional Calculus and Differential Operators.* Internet Seminar, Lecture Notes, University of Karlsruhe, 2002.

Simone Flory
Department of Mathematics
Louisiana State University
Baton Rouge, LA 70808, USA
E-mail address: simone@simoneflory.com

Frank Neubrander
Department of Mathematics
Louisiana State University
Baton Rouge, LA 70808, USA
E-mail address: neubrand@math.lsu.edu

Lutz Weis
Mathematisches Institut I
Universität Karlsruhe
Englerstr. 2
D-76131 Karlsruhe, Germany
E-mail address: lutz.weis@mathematik.uni-karlsruhe.de

Progress in Nonlinear Differential Equations
and Their Applications, Vol. 55, 195–213

Poincaré's Inequality on Stratified Sets and Applications

Alexey A. Gavrilov, Serge Nicaise, and Oleg M. Penkin

1. Introduction

In our days, partial differential equations on multistructures have a width range of applications in continuum mechanics, aerodynamics, neurobiology, electronics and so on (cf. [5]). The most important questions are existence and regularity results for stationary or nonstationary (linear or nonlinear) equations, spectral theory, control theory and numerical analysis, all reveal the specific character of the multistructure in question.

This topic starts with the study of equations on one-dimensional stratified sets (also called topological networks or geometrical graphs) at the end of the 70th's with the works of G. Lumer [16, 17] and Y.V. Pokornyi [32, 25]. Diffusion equations on topological networks were first considered in [16, 17, 18] (see also [4, 7, 9, 14, 19]). Spectral theory on such objects (system composed of strings or beams) is also of great interest and reveals specific features of these structures which help to describe their properties [3, 6, 8, 11, 13, 19, 20, 25, 33, 34].

The Laplace-Beltrami operator Δ is used, for example, to describe small movements of the system composed of strings and membranes under the effect of small forces. Weak solvability of the associated Dirichlet problem has been discussed by numerous authors (see, for example, [4, 9, 19, 27]). Some regularity results may be also expected but are somehow difficult in general due to the presence of singular points, in various cases the subject had been considered in [21, 22].

In this paper, we present a general notion of stratified sets which is a large class of multistructures and is an extension of previous notions like 1-d networks, 2-d networks and other ones (see below). We underline a sufficient condition (firm pairs) which guarantees the validity of the so-called Poincaré's inequality. We give some elementary notion of differential geometry on such objects. This allows the introduction of second order elliptic operators in divergence form. Poincaré's inequality then yields existence and uniqueness results of a weak solution to the associated Dirichlet-Neumann problem. Using standard semigroup theory, we further deduce existence results for the associated diffusion equations. At the end, we shortly show that Poincaré's inequality on stratified sets leads to the so-called Opial inequality (see [1] for the standard case).

2. Some preliminary notions

The general notion of stratified sets may be found in the book [31]. We specialize this notion for our convenience. Namely, a connected set Ω in $\mathbf{R}^n$ is said to be stratified if there exists a finite sequence of closed subsets of $\mathbf{R}^n$

$$\Omega^{k_0} \subset \Omega^{k_1} \subset \cdots \subset \Omega^{k_m} = \Omega \text{ when } k_0 < k_1 < \cdots < k_m, \tag{1}$$

with the following properties:

i) $\Omega^{k_i} \setminus \Omega^{k_{i-1}}$ is a smooth submanifold in $\mathbf{R}^n$ of dimension k_i, with the convention $\Omega^{k_{-1}} = \emptyset$. Its connected components will be called k_i-dimensional strata and will be denoted by $\sigma_{k_i j}$. The second index serves for the numeration of the strata. We shall assume that there is a finite number of strata in Ω and that each of them has a compact closure in $\mathbf{R}^n$. It is important to notice that the boundary of the stratum is piecewise smooth, because it consists of strata. However, it could have some singularities like cracks, cuspidal edges and so on. In order to avoid some serious difficulties we then assume that the boundary of the strata is Lipschitz.

ii) The boundary $\partial\sigma_{ki} = \overline{\sigma}_{ki} \setminus \sigma_{ki}$ of each stratum σ_{ki} with $k \geq 1$ is a union of strata σ_{mj} with $m < k$. We write $\sigma_{mj} \prec \sigma_{ki}$ if $\sigma_{mj} \subset \partial\sigma_{ki}$.

iii) If $\sigma_{k-1,j} \prec \sigma_{ki}$ and $y \in \sigma_{ki}$ tends to $x \in \sigma_{k-1,j}$ along some continuous curve, then the tangent space $T_y\sigma_{ki}$ has a limit position $\lim\limits_{y \to x} T_y\sigma_{ki}$ which contains the tangent space $T_x\sigma_{k-1,j}$.

The sequence (1) is called a stratification of Ω. Each set can be stratified in several ways. For example, a sphere $S^2 \subset \mathbf{R}^3$ has a trivial stratification $\Omega^2 = S^2$. The other way is to define Ω^1 as the equator of S^2 and take Ω^2 as above. More exactly a stratified set is a triple (Ω, S, ϕ), where Ω is an initial set and S is a stratification like (1). The mapping ϕ describes how to construct Ω using all the pieces σ_{ki}. Nevertheless we shall refer to Ω itself as a stratified set (with fixed S and ϕ, of course).

Before going on, let us present some examples of stratified sets:

- One-dimensional networks (see [4, 2, 6, 8, 15, 16, 17, 19, 25]), where 0-d strata are the vertices and 1-d strata are the edges.
- Two-dimensional polygonal topological networks in the sense of [21, 22], in that case, 1-d strata are the edges and 2-d strata are the faces of the network.
- Take the unit cube of $\mathbf{R}^3$ with the following stratification: the vertices are the 0-d strata, the edges are the 1-d strata, the faces are the 2-d strata and finally the interior of the cube is the 3-d stratum.
- Take for 1-d strata two concentric circles of the plane and as 2-d stratum the area between them.
- In the plane, take as 0-d strata the points $\sigma_{04} = (0,0)$, $\sigma_{02} = (1,0)$, $\sigma_{01} = (2,0)$ and $\sigma_{03} = (0,1)$, as 1-d strata the intervals $(\sigma_{04}, \sigma_{02})$, $(\sigma_{02}, \sigma_{01})$, $(\sigma_{02}, \sigma_{03})$ and $(\sigma_{04}, \sigma_{03})$ and finally as 2-stratum the triangle of vertices $\sigma_{02}, \sigma_{03}, \sigma_{04}$.

- Take as n-d stratum ($n \geq 1$) a bounded open set O of $\mathbf{R}^\mathbf{n}$ with a smooth boundary and as $(n-1)$-d stratum the boundary of O.

Together with Ω we shall fix an open and connected subset Ω_0 consisting in some strata of Ω and such that $\overline{\Omega}_0 = \Omega$. The topological terms like "closed", "open" and so on refer to the topology on Ω induced by the embedding $\Omega \subset \mathbf{R}^n$. In this topology the difference $\Omega \setminus \Omega_0$ is the boundary $\partial\Omega_0$ of Ω_0. We shall further assume that the boundary of Ω_0 is decomposed as follows

$$\partial\Omega_0 = \Gamma_D \cup \Gamma_N,$$

where Γ_D and Γ_N are also union of strata of Ω (Γ_D or Γ_N may be empty).

Each σ_{ki} as a submanifold of $\mathbf{R}^n$ is obviously a Riemannian manifold. So we can define a measure μ_{ki} on them by means of

$$\mu_{ki}(\omega) = \int_\omega d\mu_{ki} = \int_\omega \sqrt{g}\, dy^1 \dots dy^k,$$

where $\sqrt{g}$ is a determinant of the metric tensor $G = (g_{\alpha\beta})$ with respect to the local coordinates y^α. Using this metric we can define a measure μ entirely on Ω according to the formula

$$\mu(\omega) = \sum_{\sigma_{ki}} \mu_{ki}(\omega \cap \sigma_{ki}). \tag{2}$$

Here we make a sum on all strata which may have different geometrical dimensions. As a consequence a set ω included in strata of dimension $< k_m$ may have a non zero measure. We say that ω is measurable if the above sum is meaningful. The restriction of μ onto σ_{ki} is μ_{ki}, nevertheless it will be denoted by μ as well.

As usually, a function u will be called measurable if $u^{-1}(M)$ is measurable for each Lebesgue measurable set $M \subset \mathbf{R}$. Moreover one can show that if $u : \Omega \to \mathbf{R}$ is integrable, then each restriction u_{ki} of u onto σ_{ki} is also integrable and

$$\int_\Omega u\, d\mu = \sum_{\sigma_{ki}} \int_{\sigma_{ki}} u_{ki}\, d\mu. \tag{3}$$

All integrals in this right-hand side are Lebesgue integrals.

We now introduce some functional spaces on a stratified set that will be useful later on.

- $C_\sigma(\Omega_0)$ is the set of function with continuous restrictions u_{ki} (such restrictions might also be denoted by $u_{|\sigma_{ki}}$ or $u_{|ki}$).
- $C(\Omega_0)$ and $C(\Omega)$ are the sets of continuous functions on Ω_0 and Ω respectively.
- $C^{0,1}_\sigma(\Omega_0)$ is the space of functions u such that each restriction u_{ki} is Lipschitz continuous in $\bar{\sigma}_{ki}$ (as a consequence by Rademacher's theorem u_{ki} has first order partial derivatives in σ_{ki} (a.e.) which are uniformly bounded)

- $C^{0,1}(\Omega_0) = C^{0,1}_\sigma(\Omega_0) \cap C(\Omega_0)$.
- $C^{0,1}_0(\Omega_0)$ is the set of functions in $C^{0,1}(\Omega_0)$ vanishing on Γ_D.
- $L^2_\mu(\Omega_0)$ is the completion of $C(\Omega_0)$ with respect to the norm in $C(\Omega_0)$ generated by the inner product

$$(u,v) = \int\limits_{\Omega_0} uv\, d\mu.$$

- $\overset{\circ}{H}{}^1_\mu(\Omega_0)$ is the completion of $C^{0,1}_0(\Omega_0)$ with respect to the norm $\|\cdot\|_{<>}$ in $C^{0,1}_0(\Omega_0)$ generated by the inner product

$$< u, v >= \int\limits_{\Omega_0} uv\, d\mu + \int\limits_{\Omega_0} \nabla u \cdot \nabla v\, d\mu.$$

The product $\nabla u \cdot \nabla v$ is the usual inner product of covariant vectors, i.e.

$$\nabla u \cdot \nabla v = g^{\alpha\beta} \frac{\partial u}{\partial y^\alpha} \frac{\partial v}{\partial y^\beta}.$$

- $H^1_\mu(\Omega_0)$ is the completion of $C^{0,1}(\Omega_0)$ with respect to the norm $\|\cdot\|_{<>}$ in $C^{0,1}(\Omega_0)$.

3. Poincaré's inequality.

The first question we want to discuss is the validity of the so-called Poincaré inequality on a stratified set, which here takes the form

$$\int\limits_{\Omega_0} u^2 d\mu \leq C \int\limits_{\Omega_0} |\nabla u|^2 d\mu, \forall u \in \overset{\circ}{H}{}^1_\mu(\Omega_0), \tag{4}$$

for some positive constant C.

It turns out that such an estimate is valid for "firm" stratified sets (see below for the exact definition). The proof of the estimate (4) is based on the following local inequalities (similar to (4)) on each stratum. All our proofs are performed in the two-dimensional case, because they are easily illustrated. Note further that the proofs are quite standard but we give them for the sake of completeness.

Lemma 3.1. *Let $u \in H^1_\mu(\Omega_0)$, $\sigma_{ki} \subset \Omega_0$ and $\sigma_{k-1,j} \prec \sigma_{ki}$. Then there exists a constant C (independent of u) such that*

$$\int\limits_{\sigma_{ki}} u^2 d\mu \leq C \left(\int\limits_{\sigma_{k-1,j}} u^2 d\mu + \int\limits_{\sigma_{ki}} |\nabla u|^2 d\mu \right). \tag{5}$$

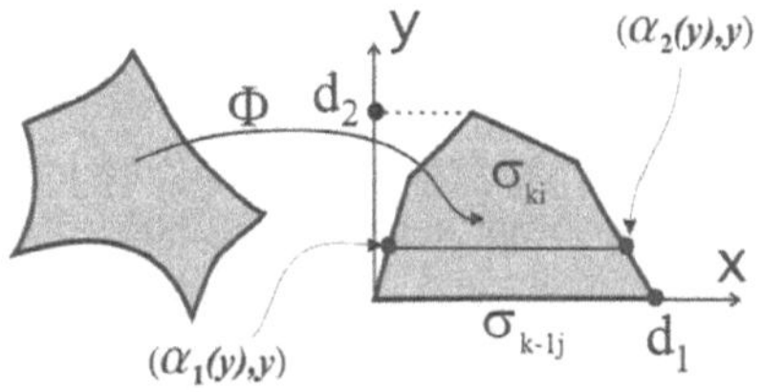

FIGURE 1. The simplification of the initial configuration

Proof. First of all assume that σ_{ki} is a polyhedron. On Fig.1 σ_{ki} is a pentagon.

By density it is sufficient to prove (5) for the functions in $C^{0,1}(\Omega_0)$. In this case we may write

$$u(x,y) = u(x,0) + \int_0^y \frac{\partial u}{\partial y}(x,s)ds.$$

As a consequence, we obtain

$$\begin{aligned} u^2(x,y) &\leq 2u^2(x,0) + 2\left(\int_0^y \frac{\partial u}{\partial s}(x,s)ds\right)^2 \\ &\leq 2u^2(x,0) + 2d_2 \int_0^y \left(\frac{\partial u}{\partial s}\right)^2 (x,s)ds. \end{aligned}$$

Let $\sigma_{ki} \cap \{(x,y) : y = s\} = (\alpha_1(s), \alpha_2(s)) \times \{s\}$. Then integrating the above estimate in x yields

$$\begin{aligned} \int_{\alpha_1(y)}^{\alpha_2(y)} u^2(x,y)dx &\leq 2 \int_{\alpha_1(y)}^{\alpha_2(y)} u^2(x,0)dx + 2d_2 \int_{\alpha_1(y)}^{\alpha_2(y)} \left(\int_0^y \left(\frac{\partial u}{\partial s}\right)^2 (x,s)ds\right) dx \\ &\leq 2 \int_{\sigma_{k-1,j}} u^2 d\mu + 2d_2 \int_{\sigma_{ki}} \left(\frac{\partial u}{\partial y}\right)^2 d\mu. \end{aligned}$$

Integrating on y, we finally get

$$\int_{\sigma_{ki}} u^2 d\mu \leq 2d_2 \int_{\sigma_{k-1,j}} u^2 d\mu + 2d_2^2 \int_{\sigma_{ki}} |\nabla u|^2 d\mu,$$

which implies (5).

If the stratum is not polyhedral, but is diffeomorphic to a polyhedron, i.e., there exists a one-to-one differentiable transformation Φ of this shape into some polyhedron (see the above figure) and $0 < m \leq |\det \Phi'| \leq M$, then (5) is fulfilled with the constant $C \cdot M/m$, where C is the previous constant.

If σ_{ki} is not diffeomorphic to a polyhedron, we will need to divide σ_{ki} into polyhedron-like pieces and apply the next Lemma.

Lemma 3.2. *Under the conditions of the previous Lemma let* $\sigma_{ki} \subset \Omega_0$ *and* $\sigma_{k-1,j} \prec \sigma_{ki}$, $\sigma_{k-1,l} \prec \sigma_{ki}$. *Then there exists a constant* C *such that*

$$\int\limits_{\sigma_{k-1,j}} u^2 d\mu \leq C \left(\int\limits_{\sigma_{k-1,l}} u^2 d\mu + \int\limits_{\sigma_{ki}} |\nabla u|^2 d\mu \right). \tag{6}$$

Proof. We start with the simple case of a straight triangle shown on the next figure.

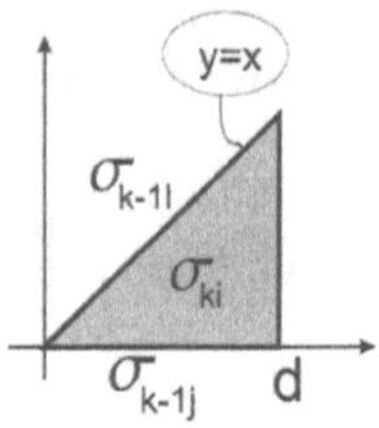

FIGURE 2. A straight triangle

Then we may write

$$u(x,x) = u(x,0) + \int\limits_0^x \frac{\partial u}{\partial y}(x,s)ds.$$

Integrating the square of this inequality on x yields the conclusion.

The proof for a curvilinear triangle is based on preliminary transformation of the initial shape of σ_{ki} into a straight triangle.

For general stratum the classical approach is to use an adequate partition of unity on σ_{ki}. Here we show that the use of a partition of unity is not necessary.

Indeed, let σ_{ki} be an arbitrary stratum. Since the closure of σ_{ki} is compact and its boundary is Lipschitz we can divide it in a finite number of curvilinear triangles (here we still continue to consider two-dimensional case; the differences with general case are insignificant) by means of artificial strata σ_{k-1,j_ν}. Let $\sigma_{k-1,j_1}, \dots, \sigma_{k-1,j_m}$ be a chain with the following properties:

- $\sigma_{k-1,j_{p+1}}$ and σ_{k-1,j_p} are the sides of the triangle σ_{kj_p}.
- $\sigma_{k-1,j_1} \subset \sigma_{k-1,j}$ and $\sigma_{k-1,j_m} \subset \sigma_{k-1,l}$.

We can apply inequality (6) to the triangle σ_{kj_p} and obtain:

$$\int\limits_{\sigma_{k-1,j_p}} u^2 d\mu \leq C_{j_p} \left(\int\limits_{\sigma_{k-1,j_{p+1}}} u^2 d\mu + \int\limits_{\sigma_{kj_p}} |\nabla u|^2 d\mu \right).$$

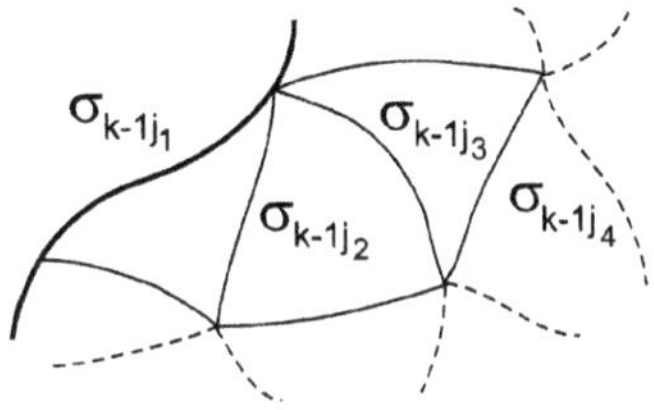

FIGURE 3. σ_{ki} divided into triangular pieces

Consequent combinations of such inequalities for $p = 1, 2, \ldots, m$ lead to

$$\int_{\sigma_{k-1,j_1}} u^2 d\mu \leq C \left(\int_{\sigma_{k-1,j_m}} u^2 d\mu + \sum_{p=1}^{m} \int_{\sigma_{kj_p}} |\nabla u|^2 d\mu \right)$$

with $C = \max_{1 \leq p \leq m} C_{j_1} C_{j_2} \ldots C_{j_p}$.

Estimating the right-hand side we obtain

$$\int_{\sigma_{k-1,j_1}} u^2 d\mu \leq C \left(\int_{\sigma_{k-1,l}} u^2 d\mu + \int_{\sigma_{ki}} |\nabla u|^2 d\mu \right).$$

The artificial stratum σ_{k-1,j_1} is only a part of the stratum $\sigma_{k-1,j}$. But the same inequality is true for other parts of $\sigma_{k-1,j}$. Summing over all the parts we obtain (6). Lemma 3.2 is proved. □

Remark 3.3. *The constant C in (6) is a monotone increasing function of the diameter of σ_{ki}.* □

Let us now come back to the proof of Lemma 3.1 when σ_{ki} is not diffeomorphic to a polyhedron. In this case we should start with an arbitrary artificial stratum σ_{kj_p} of the above-mentioned triangulation. At first we take some chain $\sigma_{kj_p}, \sigma_{kj_{p-1}}, \ldots \sigma_{kj_1}$, where $\sigma_{kj_1} \succ \sigma_{k-1,j_1}$ and apply (5) to curvilinear triangle σ_{kj_p} (above we proved the inequality (5) for curvilinear polyhedrons). As a consequence we have

$$\int_{\sigma_{kj_p}} u^2 d\mu \leq C_{j_p} \left(\int_{\sigma_{k-1,j_p}} u^2 d\mu + \int_{\sigma_{kj_p}} |\nabla u|^2 d\mu \right),$$

where σ_{k-1,j_p} is an artificial stratum lying between σ_{kj_p} and $\sigma_{kj_{p-1}}$. Now we apply (6) to the strata σ_{k-1,j_p} and $\sigma_{k-1,j_{p-1}}$ and so on (all the next steps should be based on the inequality (6)). The final result will be the inequality

$$\int_{\sigma_{kj_p}} u^2 d\mu \le C\left(\int_{\sigma_{k-1,j_1}} u^2 d\mu + \sum_{s=1}^{p} \int_{\sigma_{kj_s}} |\nabla u|^2 d\mu\right),$$

where $C = \max_{1\le p\le m} C_{j_1}C_{j_2}\dots C_{j_p}$.

In the right-hand side we can replace σ_{k-1,j_1} by $\sigma_{k-1,j}$. Summing over all σ_{kj_s} we obtain (5). The proof of Lemma 3.1 is then complete. □

Lemma 3.4. *Under the conditions of Lemma 3.1 there exists a constant C such that*

$$\int_{\sigma_{k-1,j}} u^2 d\mu \le C\left(\int_{\sigma_{ki}} u^2 d\mu + \int_{\sigma_{ki}} |\nabla u|^2 d\mu\right). \tag{7}$$

Proof. As in Lemma 3.2 we start with the case of a straight triangle shown on Fig. 2 and write

$$u(x,0) = u(x,y_1) - \int_0^{y_1} \frac{\partial u}{\partial y}(x,s)\,ds.$$

Therefore we get

$$\int_0^d u^2(x,0)dx \le 2\int_0^d u^2(x,y_1)dx + 2d\int_0^d\int_0^{y_1}\left(\frac{\partial u}{\partial y}\right)^2(x,s)dxds.$$

Integrating this estimate on y_1 yields (7).

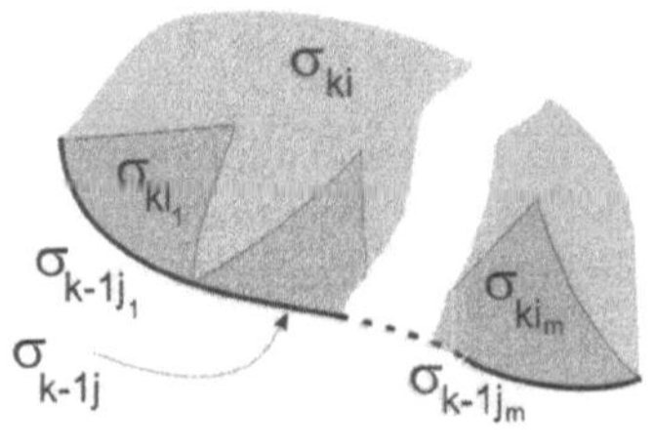

FIGURE 4. To lemma 3.2

As a consequence, this result still holds for curvilinear triangles.

If σ_{ki} is not diffeomorphic to a curvilinear triangle, we use the following procedure: we take small pieces $\sigma_{ki_1}, \dots, \sigma_{ki_m}$ of the stratum σ_{ki} lying along the stratum $\sigma_{k-1,j}$ (see Fig. 4) which are curvilinear triangles. For each piece we apply

(7) and then sum over all such pieces. So we obtain

$$\int_{\sigma_{k-1,j}} u^2 d\mu \le \sum_{p=1}^{m} C_p \left(\int_{\sigma_{ki_p}} u^2 d\mu + \int_{\sigma_{ki_p}} |\nabla u|^2 d\mu \right)$$

$$\le \sum_{p=1}^{m} C_p \left(\int_{\sigma_{ki}} u^2 d\mu + \int_{\sigma_{ki}} |\nabla u|^2 d\mu \right)$$

and Lemma 3.4 is proved. □

Let us now make the following definition:

Definition 3.5. *The pair (Ω_0, Γ_D) is firmly connected (or is a firm pair) if for any stratum σ_{ki} of Ω_0, there exists a stratum σ_{mj} of Γ_D and a firm chain joining σ_{ki} to σ_{mj}, i.e., there exists a connected sequence $\sigma_{k_1 i_1}, \sigma_{k_2 i_2} \dots, \sigma_{k_p i_p}$ with the following properties:*

- *$\sigma_{k_1 i_1} = \sigma_{ki}$ and $\sigma_{k_p i_p} = \sigma_{mj}$;*
- *for each $1 \le q \le p-1$, either $\sigma_{k_q i_q} \prec \sigma_{k_{q+1} i_{q+1}}$ or $\sigma_{k_q i_q} \succ \sigma_{k_{q+1} i_{q+1}}$;*
- *$|k_{q+1} - k_q| = 1$ for each $1 \le q \le p-1$.*

Figure 5 shows examples of firm and not firm pairs. The right example presents a non firm pair (Ω_0, Γ_D), when $\Omega_0 = \Omega \setminus \sigma_{01}$ and $\Gamma_D = \sigma_{01}$, since there exists no firm chain joining σ_{21} to σ_{01}. On the left we can see a firm pair (Ω_0, Γ_D) with Ω_0 and Γ_D as before (the desired chain joining σ_{21} to σ_{01} is here $\sigma_{21}, \sigma_{12}, \sigma_{02}, \sigma_{11}, \sigma_{01}$). If Ω_0 is a 1-d network or a two-dimensional polygonal topological networks Ω_0 with a nonempty boundary $\partial\Omega_0$ (with the stratification described above) and if we choose $\Gamma_D = \partial\Omega_0$, then we easily check that (Ω_0, Γ_D) is a firm pair.

Now we can formulate the main result of this section.

Theorem 3.6. *Let (Ω_0, Γ_D) be a firm pair such that $\Gamma_D \neq \emptyset$. Then there exists a positive constant C such that*

$$\int_{\Omega_0} u^2 \, d\mu \le C \int_{\Omega_0} |\nabla u|^2 d\mu. \tag{8}$$

for any function $u \in \overset{\circ}{H}{}^1_\mu(\Omega_0)$.

Proof. For an arbitrary stratum $\sigma_{ki} \subset \Omega_0$, let us fix a stratum $\sigma_{mj} \subset \Gamma_D$ and a firm chain $\sigma_{ki} = \sigma_{k_1 i_1}, \sigma_{k_2 i_2} \dots, \sigma_{k_p i_p} = \sigma_{mj}$ joining σ_{ki} to σ_{mj}. For $q = 1, \cdots, p-1$, using Lemma 3.1 or Lemma 3.4 according to the case $k_{q+1} - k_q = -1$ or $k_{q+1} - k_q = 1$, we obtain

$$\int_{\sigma_{k_q i_q}} u^2 d\mu \le C_{k_q i_q} \left(\int_{\sigma_{k_{q+1} i_{q+1}}} u^2 d\mu + \int_{\sigma_{k_{q+1} i_{q+1}} \cup \sigma_{k_q i_q}} |\nabla u|^2 d\mu \right).$$

By induction, we get

$$\int_{\sigma_{ki}} u^2 d\mu \le C \left(\int_{\sigma_{mj}} u^2 d\mu + \sum_{q=1}^{p} \int_{\sigma_{k_q i_q}} |\nabla u|^2 d\mu \right).$$

Taking into account that u vanishes on Γ_D, we have

$$\int_{\sigma_{ki}} u^2 d\mu \le C \sum_{q=1}^{p} \int_{\sigma_{k_q i_q}} |\nabla u|^2 d\mu \le C \int_{\Omega_0} |\nabla u|^2 d\mu.$$

Taking the sum on all strata, we obtain (8). □

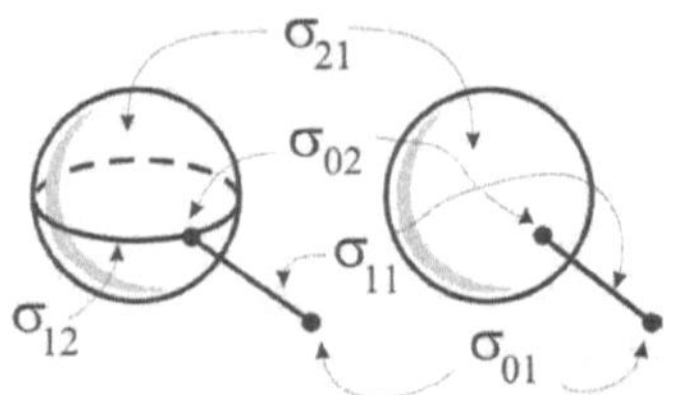

FIGURE 5. Firm and not firm pairs

4. Elliptic operators on stratified sets.

The next two sections are devoted to the study of the elliptic and parabolic operators on stratified sets. Here we start with some elementary differential geometry on stratified sets.

First without loss of generality we may suppose that Ω satisfies the following property: if $\sigma_{k-1,j} \prec \sigma_{ki}$, then for all $x \in \sigma_{k-1,j}$, there exists a neighbourhood $V(x)$ of x such that $V(x) \cap \sigma_{ki}$ is connected in Ω. If this property does not hold for some strata σ_{ki}, we may decompose σ_{ki} into some substrata (as in Figure 6).

A vector field $\vec{F}$ on Ω_0 is said to be tangent to Ω_0 if for each stratum $\sigma_{ki} \subset \Omega_0$ and each $x \in \sigma_{ki}$, the vector $\vec{F}(x)$ belongs to the tangent space $T_x\sigma_{ki}$. We shall say that a tangent vector field $\vec{F}$ is in $PH^1_\mu(\Omega_0)$ (piecewise H^1) if each component of the restriction $\vec{F}_{ki}$ to any stratum σ_{ki} of Ω_0 is in $H^1_\mu(\sigma_{ki})$.

We may now define the divergence operator as follows:

Definition 4.1. *For a tangent vector field $\vec{F}$ in $PH^1_\mu(\Omega_0)$, we define* $\operatorname{div} \vec{F}$ *as the scalar field defined in Ω_0 by*

$$(\operatorname{div} \vec{F})_{|\sigma_{k-1,j}} = \operatorname{div}_{k-1}(\vec{F}_{|\sigma_{k-1,j}}) + \sum_{\sigma_{ki} \succ \sigma_{k-1,j}} \vec{\nu}_{ki} \cdot \vec{F}_{|\overline{\sigma_{ki}}}, \tag{9}$$

for all strata $\sigma_{k-1,j}$ of Ω_0. Here $\operatorname{div}_{k-1}\vec{F}$ is the standard $(k-1)$-dimensional divergence on the stratum $\sigma_{k-1,j}$, the vector $\vec{\nu}_{ki}$ is the unit normal vector along $\sigma_{k-1,j}$ directed inside the stratum σ_{ki} and $\vec{F}_{|\overline{\sigma_{ki}}}$ means the extension of $\vec{F}_{|\sigma_{ki}}$ to $\overline{\sigma_{ki}}$ in the sense of trace.

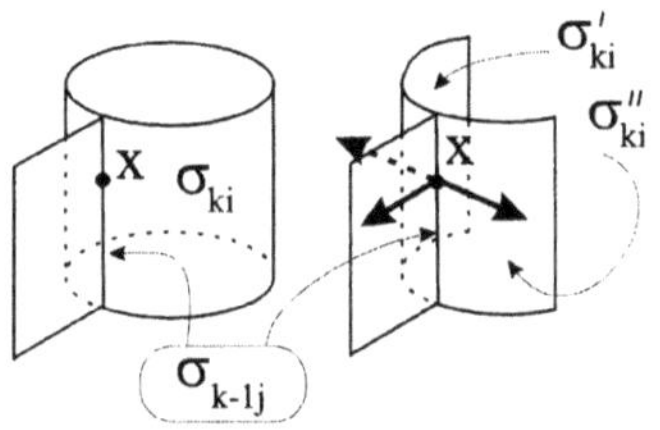

FIGURE 6. Division of the stratum σ_{ki}

Note that the definition (9) of the divergence is justified by the validity of the following divergence theorem (see Theorem 5.1 of [30]):

Theorem 4.2. *For all tangent vector field $\vec{F}$ in $PH^1_\mu(\Omega_0)$, we have*

$$\int_{\Omega_0} \operatorname{div} \vec{F}\, d\mu = -\int_{\partial\Omega_0} (\vec{F})_\nu\, d\mu,$$

where $(\vec{F})_\nu(x)$ for $x \in \sigma_{k-1,j}$ is an abbreviation for

$$\sum_{\substack{\sigma_{ki}\succ\sigma_{k-1,j}\\ \sigma_{ki}\subset\Omega_0}} (\vec{F}\cdot\vec{\nu})_{|\overline{\sigma_{ki}}}(x).$$

Proof. Apply the standard divergence theorem on each stratum σ_{ki} and take the sum. □

Remark further that the operator $-\operatorname{div}$ is the formal adjoint of the gradient operator ∇ (acting on scalar functions) since the above arguments imply that we have

$$\int_{\Omega_0} \nabla u\cdot\vec{F}\, d\mu = -\int_{\Omega_0} u \operatorname{div}\vec{F}\, d\mu - \int_{\partial\Omega_0} u(\vec{F})_\nu\, d\mu, \tag{10}$$

for all $u \in H^1_\mu(\Omega_0)$ and any tangent vector field $\vec{F}$ in $PH^1_\mu(\Omega_0)$.

This suggests the following definition for the Laplace-Beltrami operator: for a function $u \in \overset{\circ}{H}{}^1_\mu(\Omega_0)$ such that $\nabla u \in PH^1_\mu(\Omega_0)$, we take $\Delta u = \operatorname{div}\nabla u$, or equivalently

$$\Delta u(x) = \Delta_{k-1}u(x) + \sum_{\sigma_{ki}\succ\sigma_{k-1,j}} \vec{\nu}_{ki}\cdot(\nabla u)_{|\overline{\sigma_{ki}}}(x), \tag{11}$$

for all $x \in \sigma_{k-1,j}$, when Δ_{k-1} is the classical Laplace-Beltrami operator on $\sigma_{k-1,j}$.

Together with Δ we shall also consider the operator $(\Delta_p)(\cdot) = \operatorname{div}(p\nabla(\cdot))$, for some function $p \in L^2_\mu(\Omega_0)$.

5. Weak solvability of the Dirichlet-Neumann problem.

The next problem is the analogue of the classical Dirichlet-Neumann problem:

$$Lu(x) = \Delta_p u(x) - qu(x) = f(x) \qquad x \in \Omega_0, \tag{12}$$

$$u = 0 \qquad \text{on } \Gamma_D, \tag{13}$$

$$(p\nabla u)_\nu = 0 \qquad \text{on } \Gamma_N. \tag{14}$$

Here we assume that p is a function in $L^2_\mu(\Omega_0)$ which is strictly positive in Ω_0, i.e., there exists $\alpha > 0$ such that

$$p(x) \geq \alpha, \forall \text{ a.e. } x \in \Omega_0.$$

We further assume that $q \in L^2_\mu(\Omega_0)$ satisfies $q + \frac{\alpha}{C}$ is strictly positive in Ω_0, where C is the constant in the Poincaré inequality, this means that there exists $\beta > 0$ such that

$$q(x) + \frac{\alpha}{C} \geq \beta, \forall \text{ a.e. } x \in \Omega_0. \tag{15}$$

Note that the values of p at 0-dimensional strata are not used. This means that we can weaken the condition of the strict positiveness of p but it is of less importance.

We say that u is a weak solution of the boundary value problem (12)–(14) if the following identity holds for any v (in a specified space, see below):

$$\int_{\Omega_0} (p\nabla u \nabla v + quv) d\mu = \int_{\Omega_0} fv \, d\mu. \tag{16}$$

As usual, this definition is motivated by the following analogue of the classical Green formula for stratified sets.

Theorem 5.1. *Let $u \in L^2_\mu(\Omega_0)$ be such that $p\nabla u \in PH^1_\mu(\Omega_0)$ and $v \in H^1_\mu(\Omega_0)$. Then*

$$\int_{\Omega_0} v\Delta_p u \, d\mu = -\int_{\Omega_0} p\nabla v \nabla u \, d\mu - \int_{\partial\Omega_0} v(p\nabla u)_\nu d\mu. \tag{17}$$

Proof. Direct consequence of the identity (10). □

As a consequence we get the second Green formula.

Theorem 5.2. *Let $u, v \in L^2_\mu(\Omega_0)$ be such that $p\nabla u, p\nabla v \in PH^1_\mu(\Omega_0)$ Then*

$$\int_{\Omega_0} (vLu - uLv) d\mu = \int_{\partial\Omega_0} (u(p\nabla v)_\nu - v(p\nabla u)_\nu) d\mu. \tag{18}$$

Similar Green's formula may be found in [28, 29].

Following the standard way let us define a new inner product $[\cdot,\cdot]$ in $C^{0,1}_0(\Omega_0)$ by means of the left-hand side of (16). The completion of $C^{0,1}_0(\Omega_0)$ for the norm

$\|\cdot\|_{[]}$, generated by this inner product is a Hilbert space, denoted by $\overset{\circ}{H}{}^1_\mu(\Omega_0, p, q)$. Note that this argument is meaningful since by Theorem 3.6, we have

$$\int_{\Omega_0} \{p|\nabla v|^2 + q|v|^2\}\, d\mu \geq \int_{\Omega_0} (q + \frac{\alpha}{C})|v|^2\, d\mu \geq \beta \int_{\Omega_0} |v|^2\, d\mu, \forall v \in C_0^{0,1}(\Omega_0). \quad (19)$$

Now we can prove the following existence and uniqueness result.

Theorem 5.3. *If the pair (Ω_0, Γ_D) is firmly connected and Γ_D is not empty. Then for any $f \in L^2_\mu(\Omega_0)$ the problem (12)–(14) has a unique weak solution $u \in \overset{\circ}{H}{}^1_\mu(\Omega_0, p, q)$, i.e., satisfying (16) for all $v \in \overset{\circ}{H}{}^1_\mu(\Omega_0, p, q)$.*

Proof. It suffices to show that the functional F defined by the right-hand side of (16) is bounded on $\overset{\circ}{H}{}^1_\mu(\Omega_0, p, q)$. For that purpose, let us denote

$$C_f = \left(\int_{\Omega_0} f^2 d\mu \right)^{\frac{1}{2}}.$$

For any $v \in C_0^{0,1}(\Omega_0)$, by Cauchy-Schwarz's inequality and the estimate (19), we may write

$$|F(v)|^2 \leq C_f^2 \int_{\Omega_0} v^2 d\mu \leq \frac{C_f^2}{\beta} \|v\|^2_{[]}.$$

This means that F is bounded on $\overset{\circ}{H}{}^1_\mu(\Omega_0, p, q)$ by density.

The conclusion follows from the Riesz representation theorem. □

Remark 5.4. *When p and q are bounded, the space $\overset{\circ}{H}{}^1_\mu(\Omega_0, p, q)$ coincides with $\overset{\circ}{H}{}^1_\mu(\Omega_0)$, because, in this case, the norm $\|\cdot\|_{[]}$ is equivalent to the norm $\|\cdot\|_{<>}$. Therefore in this case the above constructed solution belongs to $\overset{\circ}{H}{}^1_\mu(\Omega_0)$. For general p, we may say that u satisfies (12) in Friedrichs' sense. More exactly, the operator L admits a selfadjoint extension $\widetilde{L}$ (in $L^2_\mu(\Omega_0)$) and $\widetilde{L}u = f$. In the next section we will show that such an extension is always possible. In any case u is a solution of (12) in the sense of distributions on each stratum.* □

Remark 5.5. *For 1-d networks with Γ_D equal to the exterior boundary, Theorem 5.3 recovers standard results of* [3, 4, 9, 19, 22]. *Theorem 5.3 also covers the case of dynamical interface problems studied in* [10]. □

6. Diffusion equations on stratified sets.

In this section we keep the previous conditions on p and q. We start with the density of $C_0^{0,1}(\Omega_0)$ into $L^2_\mu(\Omega_0)$ which is based on the next result.

Lemma 6.1. *Let $u_{ki} \in L^2_\mu(\sigma_{ki})$ with $\sigma_{ki} \subset \Omega_0$ and $\varepsilon > 0$. Then there exists a function $v^{ki} \in C_0^{0,1}(\Omega_0)$ with the following properties:*

$$\int_{\sigma_{ki}} |u_{ki} - v^{ki}_{|\sigma_{ki}}|^2 \, d\mu \leq \varepsilon, \tag{20}$$

$$\int_{\Omega_0 \setminus \sigma_{ki}} |v^{ki}|^2 \, d\mu \leq \varepsilon. \tag{21}$$

Proof. By a standard density result, there exists $\varphi_{ki} \in \mathcal{D}(\sigma_{ki})$ (smooth function with a compact support in σ_{ki}) such that

$$\int_{\sigma_{ki}} |u_{ki} - \varphi_{ki}|^2 \, d\mu \leq \varepsilon. \tag{22}$$

Using a finite number of local charts and a partition of unity $(\psi_j)_{j=1}^J$, the function $\psi_j \varphi_{ki}$ is transformed into $\varphi \in \mathcal{D}(V')$, where V' is bounded domain of $\mathbf{R}^{k-1}$. Take a function $\eta \in \mathcal{D}(]-1,1[^{n-k+1})$ such that $\eta(0) = 1$ and for any $0 < \delta < 1$, set

$$w(y', y) = \eta(\frac{y}{\delta})\varphi(y'), \forall y' \in V', y \in]-1,1[^{n-k+1}.$$

Then clearly $w \in \mathcal{D}(\mathbf{R^n})$ and satisfy

$$w(y', 0) = \phi(y'), \forall y' \in V',$$
$$\text{supp } w \subset \{x \in \mathbf{R^n} : \mathbf{d}(\mathbf{x}, \mathbf{S}) < \delta\},$$

where S is the support of φ. Going back to σ_{ki}, we get $w^{ki} \in \mathcal{D}(\mathbf{R^n})$ such that

$$w^{ki}_{|\sigma_{ki}} = \phi_{ki},$$
$$\text{supp } w^{ki} \subset \{x \in \mathbf{R^n} : \mathbf{d}(\mathbf{x}, \mathbf{S_{ki}}) < \mathbf{C}\delta\},$$

for some positive constant C and S_{ki} is the support of φ_{ki}.

Now for $0 < \delta < 1$ small enough we take

$$v^{ki} = w^{ki}_{|\Omega_0}.$$

Indeed from the construction, v^{ki} clearly belongs to $C_0^{0,1}(\Omega_0)$ and satisfies (for δ small enough)

$$v^{ki}_{|\sigma_{ki}} = \phi_{ki},$$
$$\text{supp } v^{ki} \subset \{x \in \Omega_0 : d(x, S_{ki}) < C\delta\},$$
$$v^{ki}_{|\sigma_{k'i'}} = 0, \text{ for } k' < k \text{ or } (k' = k \text{ and } i' \neq i).$$

The first property and the estimate (22) yield (20). The third property implies that

$$\int_{\Omega_0 \setminus \sigma_{ki}} |v^{ki}|^2 \, d\mu = \sum_{\substack{k' > k \\ \sigma_{k'i'} \subset \Omega_0}} \int_{\sigma_{k',i'}} |v^{ki}|^2 \, d\mu.$$

The second property then yields (21) for δ small enough, since for $k' > k$

$$\mu(\text{ supp } v^{ki} \cap \sigma_{k',i'}) \to 0 \text{ as } \delta \to 0.$$

□

Theorem 6.2. *Let $u \in L^2_\mu(\Omega_0)$. Then for any $\varepsilon > 0$ there exists a function $v \in C_0^{0,1}(\Omega_0)$ such that*

$$\int_{\Omega_0} |u - v|^2 \, d\mu \leq \epsilon.$$

Proof. We may write

$$u = \sum_{\sigma_{ki} \subset \Omega_0} \tilde{u}_{ki},$$

where $\tilde{u}_{ki}$ is the extension of u_{ki} by zero outside σ_{ki}. Moreover we remark that (20) and (21) simply mean that

$$\int_{\Omega_0} |\tilde{u}_{ki} - v^{ki}|^2 \, d\mu \leq 2\varepsilon.$$

Taking

$$v = \sum_{\sigma_{ki} \subset \Omega_0} v^{ki},$$

the conclusion follows from the triangular inequality and the fact that the number of strata is finite. □

Corollary 6.3. *The space $\overset{\circ}{H}{}^1_\mu(\Omega_0, p, q)$ is dense in $L^2_\mu(\Omega_0)$.*

We now consider the triple (a, V, H), with $H = L^2_\mu(\Omega_0)$, $V = \overset{\circ}{H}{}^1_\mu(\Omega_0, p, q)$ and a the the bilinear form defined as the left-hand side of (16). Since V is dense in H and a is continuous, symmetric and strongly coercive in V, its Friedrichs' extension A defined by

$$\begin{aligned} D(A) &= \{u \in H : \exists g \in H : a(u, v) = \int_{\Omega_0} gv \, d\mu, \forall v \in V\}, \\ Au &= g, \end{aligned}$$

is a positive selfadjoint operator on H.

To see the relationship with problem (12)–(14), we define the operator L as follows:

$$\begin{aligned} D(L) &= \{u \in \overset{\circ}{H}{}^1_\mu(\Omega_0, p, q) : p\nabla u \in PH^1_\mu(\Omega_0) \text{ satisfying (14)}\}, \\ Lu &= -\Delta_p u + qu, \forall u \in D(L). \end{aligned}$$

Theorem 6.4. *The operator A is an extension of L, in other words, for all $u \in D(L)$, one has*

$$Au = Lu.$$

Proof. Fix $v \in C_0^{1,0}(\Omega_0)$. Then by Theorem 5.1, we have

$$a(u, v) = \int_{\Omega_0} \{-\Delta_p u \, v + qu \, v\} \, d\mu = \int_{\Omega_0} Lu \, v \, d\mu.$$

By density we conclude that $Au = Lu$. □

Corollary 6.5. *The operator L admits a positive selfadjoint Friedrichs' extension $\tilde{L} = A$.*

Remark 6.6. *If $p \in PH^1_\mu(\Omega_0)$, then Lemma 6.1 and Theorem 6.2 show that the domain $D(L)$ is dense in $L^2_\mu(\Omega_0)$ (since the builded functions v_{ki} actually belong to $D(L)$ since they are smooth on each stratum) and therefore, in that case, L is a positive symmetric operator.*

Theorem 6.7. *The operator $-A$ generates an analytic semigroup on $L^2_\mu(\Omega_0)$.*

Proof. This result is a consequence of Theorem IV.6.B of [35] (see also [36, 24]) and of the above properties of the triple (a, V, H). □

This theorem allows to obtain in natural way a weak solvability for the following diffusion equation (see, for example, [24]): for $\phi \in H$ and $f \in C^{0,\alpha}([0,T]; H)$, with $0 < \alpha \leq 1$, there exists $u \in C([0,T], H) \cap C^1(]0,T]; H)$ solution of

$$\frac{\partial u}{\partial t} - \tilde{L}u = f, \forall t \in]0,T],$$
$$u(0,x) = \phi(x), x \in \Omega_0.$$

By Corollary 6.5, this system is formally equivalent to

$$\frac{\partial u_{k-1,j}}{\partial t} - \operatorname{div}_{k-1}(p_{k-1,j}\nabla u_{k-1,j})$$
$$+ \sum_{\sigma_{ki} \succ \sigma_{k-1,j}} \vec{\nu}_{ki} \cdot (p\nabla u)_{|\sigma_{ki}} = f \text{ in }]0,T] \times \sigma_{k-1,j},$$
$$u = 0 \text{ on }]0,T] \times \Gamma_D,$$
$$(p\nabla u)_\nu = 0 \text{ on }]0,T] \times \Gamma_N,$$
$$u(0,x) = \phi(x), x \in \Omega_0.$$

7. Opial's inequality.

Let us mention one more application of Poincaré's inequality, namely Opial's inequality. A lot of works are devoted to the study of Opial's inequality (for example, the significant one – [1]). The first version of the inequality proved by Opial himself is as follows [23]:

$$\int_0^h u|u'|dt \leq \frac{h}{4} \int_0^h |u'|^2 dt,$$

where u is a positive function on $[0,h]$ such that $u(0) = u(h) = 0$. This inequality finds its application in the proofs of some nonlinear boundary-value problems. In [1] the authors generalized this inequality to the case of vector-valued functions, functions of many variables and so on. As far as we know the case of functions on stratified sets has not yet been considered. Here is the analogue:

Theorem 7.1. *Let (Ω_0, Γ_D) be a firm pair. Then there exists a positive constant C such that*

$$\int_{\Omega_0} |u||\nabla u| d\mu \leq C \int_{\Omega_0} |\nabla u|^2 d\mu, \forall u \in \overset{\circ}{H}{}^1_\mu(\Omega_0).$$

Proof. Direct consequence of Cauchy-Schwarz's and Poincaré's inequalities:

$$\int_{\Omega_0} |u||\nabla u| d\mu \leq \left(\int_{\Omega_0} |u|^2 d\mu \right)^{\frac{1}{2}} \left(\int_{\Omega_0} |\nabla u|^2 d\mu \right)^{\frac{1}{2}} \leq C \int_{\Omega_0} |\nabla u|^2 d\mu.$$

□

References

[1] R.P. Agarwal and P.Y.H. Pang, *Opial Inequalities with Applications in Differential and Difference Equations.* Kluwer Academic Publishers, 1995.

[2] F. Ali Mehmeti, *A characterisation of generalized C^∞ notion on nets.* Integral Eq. and Operator Theory **9** (1986), 753–766.

[3] F. Ali Mehmeti, *Regular solutions of transmission and interaction problems for wave equations.* Math. Meth. Appl. Sc. **11** (1989), 665–685.

[4] F. Ali Mehmeti, *Nonlinear Waves in Networks.* Akademie-Verlag, 1994.

[5] F. Ali Mehmeti, J. von Below, and S. Nicaise, *Partial differential equations on multi-structures.* Proceedings of the conference held at the CIRM in Luminy, France, 1999, Marcel Dekker, (2001).

[6] J. von Below, *A characteristic equation associated to an eigenvalue problem on c^2-networks.* Linear Algebra and appl. **71** (1985), 309–325.

[7] J. von Below, *Classical solvability of linear parabolic equations on networks.* J. Differential Equations **72** 1988, 316–337.

[8] J. von Below, *Sturm-Liouville eigenvalue problems on networks.* Math. Meth. Appl. Sc. **10** (1988), 383–395.

[9] J. von Below, *Parabolic Network Equations.* Habilitation Thesis, Eberhard-Karls-Universität Tübingen, 1993.

[10] J. von Below, and S. Nicaise, *Dynamical interface transition with diffusion in ramified media.* Communications in Partial Differential Equations **21** (1996), 255–279.

[11] A. Borovskikh, R. Mustafokulov, K. Lazarev, and Yu. Pokornyi, *A class of fourth-order differential equations on a spatial net.* Doklady Math. **52** (1995), 433–435.

[12] B. Dekoninck and S. Nicaise, *Control of networks of Euler-Bernoulli beams.* ESSAIM-COCV **4** (1999), 57–82.

[13] B. Dekoninck and S. Nicaise, *The eigenvalue problem for networks of beams.* Linear Algebra and its Applications **314** (2000), 165–189.

[14] M.I. Kamensky, O.M. Penkin and Yu.V. Pokornyi, *On the semigroup of the diffusion process in the space network.* Doklady Academii Nauk **368** (1999), 157–159.

[15] J.E. Lagnese, G. Leugering, and E.J.P.G. Schmidt, *Modeling, analysis and control of dynamic elastic multi-link structures.* Birkhäuser, Boston, 1994.

[16] G. Lumer, *Espaces ramifiés et diffusions sur les réseaux topologiques.* C. R. Acad. Sc. Paris, Série A **291** (1980), 627–630.

[17] G. Lumer, *Connecting of local operators and evolution equation on network.* L. N. in Math. **787**, Springer Verlag, (1980), 219–234.

[18] G. Lumer, *Equations de diffusion générales sur des réseaux infinis.* L. N. in Math. **1061**, Springer Verlag, (1984), 230–243.

[19] S. Nicaise, *Diffusion sur les espaces ramifiés.* Thesis, Université de Mons, 1986.

[20] S. Nicaise, *Spectre des réseaux topologiques finis.* Bull. Sc. Math., 2ème série **111** (1987), 401–413.

[21] S. Nicaise, *Le laplacien sur les réseaux deux-dimensionnels polygonaux topologiques.* J. Math. pures et appl. **67** (1988), 93–113.

[22] S. Nicaise, *Polygonal interface problems.* Methoden und Verfahren Math. Physik **39**, Peter Lang Verlag, (1993).

[23] Z. Opial, *Sur une inégalité.* Ann. Polon. Math. **8** (1960), 61–63.

[24] A. Pazy, *Semigroups of Linear Operators and Applications to Partial Differential Equations.* Springer-Verlag, 1983.

[25] O.M. Penkin, Yu.V. Pokornyi, and E.N. Provotorova, *On a vectorial boundary value problem.* Boundary value problems, Interuniv. Collect. sci. Works, Perm', 1983, 64–70.

[26] O.M. Penkin, *Some qualitative properties of the boundary values problems on graphs.* Thesis, Voronezh, 1988.

[27] O.M. Penkin, E.M. Bogatov, and Ju.M. Kashkarov, *Solvability of elliptic boundary value problems on stratified sets.* Differential equations **34** (1998), 1291–1293.

[28] O.M. Penkin and Yu.V. Pokornyi, *On incompatible inequalities for elliptic equations on stratified sets.* Differential equations **34** (1998), 1111–1117.

[29] O.M. Penkin and Yu.V. Pokornyi, *On differential inequalities on manifolds of hybrid type.* Russian Mathematical Surveys **53** (1998), 788–789.

[30] O.M. Penkin, *About a geometrical approach to multistructures and some qualitative properties of solutions.* In F. Ali Mehmeti, J. von Below and S. Nicaise eds., Partial differential equations on multistructures, Marcel Dekker, 2001.

[31] F. Pham, *Introduction à l'étude topologique des singularités de Landau.* Gauthier-Villars, Paris, 1967.

[32] Yu.V. Pokornyi, *Oscillatory properties of the spectrum of one class of vectorial boundary value problems.* VIII-th school in the operator theory in function spaces, Riga, 1983, 46–48.

[33] Yu.V. Pokornyi and O. Penkin, *Sturm theorems for equations on graphs.* Soviet Math. Dokl. **40** (1990), 640–642.

[34] J.P. Roth, *Le spectre du laplacien sur un graphe.* L. N. in Math. **1096**, Springer-Verlag, 1984, 521–539.

[35] R.E. Showalter, *Hilbert space methods for partial differential equations.* Monographs and Studies in Math. **1**, Pitman, 1977.

[36] K. Yosida, *Functional Analysis.* Springer-Verlag, 1964.

Alexey A. Gavrilov
Voronezh State University
Universitetskaja pl., 1
394000 Voronezh, Russia

Serge Nicaise
Université de Valenciennes et du Hainaut Cambrésis
MACS
Institut des Sciences et Techniques de Valenciennes
F-59313 Valenciennes Cedex 9, France

Oleg M. Penkin
Voronezh State University
Universitetskaja pl., 1
394000 Voronezh, Russia

Progress in Nonlinear Differential Equations
and Their Applications, Vol. 55, 215–225

Variational Principles for Granular Matter

Karl P. Hadeler and Christina Kuttler

1. Introduction

In the last few years research on so-called soft materials, as opposed to classical states like solids, fluids, and gases, has attracted much attention. Among various types of foams, polymers, etc., granular materials appear as most interesting in view of a variety of surprising experiments, novel connections to mathematical fields like geometry, partial differential equations, particle systems, and also industrial applications [9] [16] [15]. It is surprising that granular matter became an attractive research field only recently since basic experiments on equilibrium configurations have been performed a century ago [1]. In some situations granular matter can be seen as a granular gas (Boltzmann approach for discrete particles with free path length [20]) or as a granular fluid (Savage-Hutter model for height and speed of avalanches [22] [23], an adaptation of the shallow water or St. Venant equations). If large amounts of sand are slowly accumulated then continuum models are appropriate which use the "geometry of sand" and the distinction of a standing layer and a rolling layer [8] [5] [6] [10] [11] [12] [13]. These models are based on the angle of repose of the material considered and two functional laws which describe how the speed of rolling grains is related to the slope of the standing material and how grains are deposited or start moving on a slope. Also in stationary situations with a non-vanishing rolling layer these two parameter functions play a role [12]. Models with instantaneous deposition have been studied in [2] [3] [19]. Another approach to stationary problems has been designed in [21].

Granular matter is distinguished from fluids and gases by an additional scale parameter, the grain size. Grain size must be seen in comparison to the total amount of matter or number of grains. A granular material is a collection of grains where typically every grain is in contact with several other grains. Grains interact by impact and by rolling and frictional contact. There are some standardized grains for experiments like spherical glass beads, but the grains in most granular materials like sand, coal, wheat are highly irregular or have a complicated shape. If these grains are piled up then the individual shapes of grains lead to static frictional forces which are averaged and produce the angle of repose, as has been first formulated by Coulomb 1773. Thus, the angle of repose is a macroscopic parameter. In describing static configurations of granular matter no further material parameter is needed. Usually the angle of repose is given by its tangent which is called $\alpha > 0$. For most materials the angle of repose ranges between 20° and 45°. One can normalize $\alpha = 1$ but we prefer to keep the parameter α in all formulae.

We study equilibrium configurations. As a first model situation we consider the deposition of dry sand on a horizontal table, i.e., a flat surface $\mathbb{R}^d$. In general we will be interested in the two-dimensional situation $d = 2$. Physicists call the problem of forming a sand heap on a two-dimensional table a three-dimensional problem, with some justification, as we shall see in the case of a general obstacle. For the moment we prefer to speak of two dimensions because of close analogies with the two-dimensional Dirichlet problem of the Laplacean. We also consider the case $d = 1$ of a one-dimensional table modelled by a straight line $\mathbb{R}$. This case can be experimentally realized by pouring dry sand between parallel vertical glass plates. According to our model, a sand heap is a bounded non-negative function $u : \mathbb{R}^d \to \mathbb{R}$ which is Lipschitz continuous with Lipschitz constant α. The most obvious special case is the circular cone with slope α which describes the heap generated by a point source above the table. Of course the tip of the cone is right under the source.

If the function u is differentiable in some open domain then it satisfies the inequality $|\operatorname{grad} u| \leq \alpha$. In view of the eikonal equation $|\operatorname{grad} u| = \alpha$ we call these functions subeikonal functions (or "solutions").

If instead we assume that the table is some bounded domain $\Omega \subset \mathbb{R}^d$ (or interval, respectively) then boundary conditions have to be imposed. Simple experiments tell how to formulate boundary conditions which are physically meaningful. Think of a flat table with a possibly rather irregular rim (a tray) onto which dry sand is deposited. Sand is leaving the table where the rim is low, and sand is held back where the rim is high. Hence we assume a boundary function $\phi : \partial\Omega \to [0, \infty]$ describing the height of the rim. We assume that the function ϕ is non-negative and that there is $\xi \in \partial\Omega$ with $\phi(\xi)$ finite. The function ϕ could describe a rim with narrow vertical slits which at some places is infinitely high. Again motivated by physical intuition, we pose a formal boundary value problem for the eikonal equation

$$|\operatorname{grad} u| \leq \alpha \tag{1}$$

$$u(x) \leq \phi(x) \quad \text{for} \quad x \in \partial\Omega. \tag{2}$$

Now we must give some meaning to the inequality (1). As before, we try to interpret the inequality (1) as a Lipschitz condition $|u(x) - u(y)| \leq \alpha|x - y|$ for all $x, y \in \Omega$. For convex domains Ω this interpretation is meaningful. However, for non-convex domains this inequality is too restrictive. What we really want, are functions which have a local Lipschitz constant α but not necessarily a global Lipschitz constant α with respect to the Euclidean metric of $\mathbb{R}^d$ into which Ω is embedded.

A way out is to define an intrinsic metric of the domain which is closely related to the rays along which granular matter is transported. This approach needs some smoothness properties of the boundary. In particular, we need that each boundary point can be reached from the interior by a straight line segment. Then one can define the path metric as follows. For $x, y \in \bar{\Omega}$ let $d(x, y)$ be the infimum of the lengths of all paths connecting x and y. Here a path is a continuous, piecewise linear path in Ω with finitely many segments. With this definition, one can interpret equation (1) as

$$|u(x) - u(y)| \leq \alpha\, d(x, y) \quad \text{for} \quad x, y \in \bar{\Omega}. \tag{3}$$

We consider non-negative functions satisfying (3) and (2) as models for sand heaps on the tray.

One can ask for the maximal volume of any sand heap on the tray. We shall show that there is a unique solution $\bar{u}$ to this problem. In the classical case $\phi \equiv 0$ we do not need any condition on the boundary $\partial\Omega$, because if $C \subset \mathbb{R}^d$ is any compact set, then $\delta : \mathbb{R}^d \to \mathbb{R}$, $\delta(x) = \operatorname{dist}(x, C)$ is Lipschitz continuous with Lipschitz condition 1 with respect to the Euclidean distance in $\mathbb{R}^d$. Hence in the case $\phi = 0$ the solution of the maximum volume problem is given by $\bar{u}(x) = \alpha \operatorname{dist}(x, \partial\Omega)$ for any type of boundary. The classical case is somewhat misleading: The distance function is Lipschitz near the boundary with respect to the Euclidean metric (even for points which are geometrically close but far from each other along the boundary) because it vanishes anyway on $\partial\Omega$.

Next we study a table with an uneven surface which we may call a distributed obstacle against the flow of granular matter. Again there is a bounded domain $\bar{\Omega} \subset \mathbb{R}^d$ and a non-negative function $\psi : \bar{\Omega} \to \mathbb{R}$ (the obstacle). We assume that ψ is Lipschitz continuous with some Lipschitz constant K. In general K will be larger than α, i.e., the slope of ψ may exceed the angle of repose. In a physical experiment sand will be deposited at locations where ψ is rather flat, and steep slopes will remain bare of sand. The table has a rim described by a bounded function $\phi : \partial\Omega \to [0, \infty]$ as before such that $\phi \geq \psi$ on $\partial\Omega$. Then the model sand heaps are functions $u : \bar{\Omega} \to \mathbb{R}$ with the property that u has local Lipschitz constant α wherever $u > \psi$ and $u \leq \phi$ on $\partial\Omega$. Also for this problem we can ask for the heap with maximal volume. The special case of an uneven table without a rim, $\phi = \psi$ on $\partial\Omega$, is of particular interest.

Then we study the case of a general obstacle which is a compact set $\Psi \subset \mathbb{R}^{d+1}$. Also in this case we try to characterize the maximal amount of sand that can be deposited on the obstacle.

We shall study these problems, we present and explain the necessary mathematical framework and, in particular, we discuss a rather strong analogy with the standard Dirichlet and obstacle problems for the Laplace equation. Finally we shall discuss some findings for stationary solutions with prescribed sources. For proofs, in particular for the equivalence of the variational and the transport path concepts, we refer to [14].

2. Comparison with the Laplace Equation

Let $\Omega \subset \mathbb{R}^d$ be a bounded domain and let a boundary function $\phi : \partial\Omega \to \mathbb{R}$ be given. The Dirichlet problem (e.g., [17]) asks for a continuous function $u : \bar{\Omega} :\to \mathbb{R}$ which satisfies the Laplace equation $\Delta u = 0$ in the interior Ω and which assumes the boundary data, $u(x) = \phi(x)$ for $x \in \partial\Omega$. In general ($d \geq 2$) the problem has no solution. From the maximum principle it follows that the solution is unique (provided it exists). The problem has a solution if the boundary $\partial\Omega$ and the boundary data ϕ are sufficiently smooth. Examples show clearly the separate effects of $\partial\Omega$ and of ϕ. If a point on the boundary cannot be reached from the interior by sufficiently "many" sequences or paths then the solution can fail to assume the datum at this point. If the boundary itself is smooth but the function ϕ is not continuous at some boundary point then trivially the solution cannot be continuous at this point.

There are at least four different approaches to the Dirichlet problem. i) maximizing the Dirichlet integral $\int_\Omega |\operatorname{grad} u|^2 dx$ over all sufficiently smooth functions satisfying the boundary data [17]. This approach asks for solutions in a strict sense and it defers the problem of smoothness of the boundary and the boundary data to the choice of appropriate function (e.g., Sobolev) spaces. ii) Perron's method of subharmonic functions or the method of balayage [17]. Here the value of the solution at $x \in \Omega$ is presented as the supremum over all subharmonic functions v satisfying $v \leq \phi$ on the boundary. This method is based on a weaker concept of solution. It has the advantage that, in principle, it yields a unique candidate for the solution and one can check for individual boundary points whether the solution assumes the boundary values. In [7] a general approach to Perron methods for viscosity solutions has been designed. iii) In the Brownian motion approach the value at the point $x \in \Omega$ is obtained by sampling over Brownian paths running from x to the boundary [4]. In a discretized version this approach leads to a Monte Carlo procedure. iv) Solving analogous problems on square grids and letting the mesh width go to zero. This idea is the basis for many numerical algorithms.

In the standard obstacle problem for the Laplace or Poisson equation there is an obstacle $\psi : \bar{\Omega} \to \mathbb{R}$ (with $\max \psi(x) \geq 0$ and $\psi(x) < 0$ for $x \in \partial\Omega$), a function $f : \bar{\Omega} \to \mathbb{R}$, and a boundary condition $u = 0$. The goal is to find a function $u : \bar{\Omega} \to \mathbb{R}$ with $u \geq \psi$ which satisfies the boundary condition and which satisfies the Poisson equation $-\Delta u = f$ equation where $u > \psi$. This problem can also be formulated as a variational inequality [18]. Let $K = \{v \in H_0^1(\Omega) : v \geq \psi\}$. Find $u \in K$ with $\int \operatorname{grad} u \operatorname{grad}(v-u)dx \geq \int f(v-u)dx$ for all $v \in K$.

Although the Laplace equation is linear and the eikonal equation is highly non-linear, there is a strong analogy between these two equations with respect to the formulation of the Dirichlet problem and the four approaches to its solution. While discussing these analogies we must keep in mind the hyperbolic character of the eikonal equation which does not admit sufficiently many smooth solutions. If one relaxes regularity (e.g., to piecewise smooth solutions) then there are too many solutions. Hence one needs, in comparison with the Laplace equation, an additional constraint. We choose this constraint to be maximal volume.

There is one further important difference. Since the Laplacean is linear, the superposition principle holds: the sum of two solutions is again a solution. Also the eikonal equation has a nonlinear superposition principle (with respect to the distinguished solutions we are looking for). However, this superposition principle uses the maximum or minimum of functions rather than the sum or the integral [10]. This phenomenon can be explained in physical terms: For a (linear) loaded membrane the sum of two loads produces the sum of deflections, in granular matter the sum of two sources produces (in equilibrium!) the maximum of the two heaps [10]. For each of the geometric settings related to the eikonal equation we try to formulate a variational problem, a Perron approach, and an approach based on paths.

3. Boundary Properties and Seminorms

For granular matter problems a uniformly Lipschitz boundary $\partial\Omega$ is appropriate, see [14] for a detailed definition. A C^2 boundary is sufficient in any case. Then the following is true. For any two points $x, y \in \bar{\Omega}$ there is a piece-wise linear continuous path in Ω connecting x and y. The function $d : (x, y) \mapsto d(x, y)$ is well-defined and it defines the path metric on $\bar{\Omega}$. There is a constant $\kappa < \infty$ such that for all $x, y \in \bar{\Omega}$

$$|x - y| \leq d(x, y) \leq \kappa |x - y|.$$

Hence the metric d and the Euclidean metric of $\mathbb{R}^d$ are equivalent. In particular $(\bar{\Omega}, d)$ is a complete and also compact metric space. For functions $u : \bar{\Omega} \to \mathbb{R}$ we can define two seminorms as

$$L(u) = \sup_{\substack{x,y\in\bar{\Omega}\\ x\neq y}} \frac{|u(x) - u(y)|}{|x - y|}, \quad \operatorname{Lip} u = \sup_{\substack{x,y\in\bar{\Omega}\\ x\neq y}} \frac{|u(x) - u(y)|}{d(x, y)}.$$

Then

$$\operatorname{Lip} u \leq L(u) \leq \kappa \operatorname{Lip} u,$$

the two seminorms are equivalent. Of course $\operatorname{Lip} u$ is the seminorm adapted to the granular matter problems.

Let $C^d(\bar{\Omega})$ be the space of functions u for which $\operatorname{Lip} u$ is finite, with norm $\|u\|_\infty + \operatorname{Lip} u$. This space is topologically equivalent to the standard space $C^{0,1}(\bar{\Omega})$.

Finally we consider local Lipschitz constants. Let $u \in C(\bar{\Omega})$ and $x \in \bar{\Omega}$. We say that L is a local Lipschitz constant at x for u if there is a neighborhood U of x in $\bar{\Omega}$ (i.e., there is a neighborhood V of x in $\mathbb{R}^d$ and $U = V \cap \bar{\Omega}$) such that $|u(y) - u(z)| \leq L\, d(y, z)$ for $y, z \in U$. The infimum of these constants L we call $\operatorname{Lip}_x u$.

4. The Maximum Volume Characterization

Here we characterize a solution to the Dirichlet problem (flat tray problem) by maximizing the volume of the sand heap. Let M be the set of functions $u \in C^d(\bar{\Omega})$ which satisfy $u \geq 0$ in $\bar{\Omega}$, $u \leq \phi$ on $\partial\Omega$, and $\operatorname{Lip} u \leq \alpha$. The functions in M are considered as models for sand heaps. The set M is compact in $C(\bar{\Omega})$. The volume of the heap u is

$$V(u) = \int_{\Omega} u dx. \tag{4}$$

The functional $V : C(\bar{\Omega}) \to \mathbb{R}$ is continuous. Hence it assumes its maximum at some function $\bar{u} \in M$. The function $\bar{u}$ is unique. Suppose there is another function $v \in M$ with $V(v) = V(\bar{u})$. Then form $w = \max(v, \bar{u}) \in M$. Then $V(w) \leq V(\bar{u})$ since $w \in M$. On the other hand $w \geq \bar{u}$ in $\bar{\Omega}$ and $w > \bar{u}$ on some open subset of $\bar{\Omega}$ and hence $V(w) > V(\bar{u})$. Hence we have a contradiction.

5. The Perron Type Characterization

Here we proceed as in the classical Dirichlet problem: Fix a point and take the supremum over a suitable set of functions. Choose $x \in \bar{\Omega}$. Define

$$\tilde{u}(x) = \max\{v(x) : v \in M\}. \tag{5}$$

Hence the subeikonal functions in M play the role that subharmonic functions play for the Dirichlet problem of the Laplacean. With the required properties of the boundary one can show that the solution so obtained is the same as before, $\tilde{u} = \bar{u}$. We can rephrase the statement as the local maximum property

$$u \in M,\ x \in \bar{\Omega} \quad \Rightarrow \quad u(x) \leq \bar{u}(x). \tag{6}$$

One can relax the conditions on the boundary and the function ϕ and still get a solution in the "Perron sense", see [14].

6. Characterization by Transport Paths

Here we construct a solution via transport paths. The underlying idea is the following. Assume there is a heap of sand on the table and its height at a point $x \in \Omega$ is z. Then draw a smooth or at least rectifiable path χ from x to any boundary point ξ. Let $\ell(\chi)$ be the length of the path and let $\phi(\xi)$ be the height of the rim at ξ. If the difference $z - \phi(\xi)$ is greater than $\alpha\, \ell(\chi)$ then a grain of sand at (x, z) can roll down along the path and hence $z > u(x)$ where $u(x)$ is the maximal height. Hence z cannot be greater than the infimum of all $\phi(\xi) + \alpha\ell(\chi)$. If Ω is convex then it is sufficient to consider straight connections from x to the boundary. In the case of non-convex domains straight transport rays are not sufficient.

At the boundary there is another problem. If x is a boundary point and $\phi(x)$ is large in comparison to $\phi(\xi)$ at boundary points ξ close by, then there will be a path χ connecting x to ξ such that $\phi(x) - \phi(\xi)$ exceeds $\alpha\, \ell(\chi)$. Then clearly the solution cannot assume the boundary datum at x.

Consider paths $\chi : [0, 1] \to \mathbb{R}^d$ which are continuous and piecewise linear. For any $x \in \bar{\Omega}$ let $\mathcal{R}(x)$ be the set of such paths with $\chi(0) = x$, $\chi(1) = \xi \in \partial\Omega$, and $\chi(t) \in \Omega$ for $0 < t < 1$. For $\chi \in \mathcal{R}(x)$ let $\ell(\chi)$ be the length of the path. Then define a function

$$\hat{u}(x) = \inf\{\phi(\chi(1)) + \alpha\,\ell(\chi) : \ \chi \in \mathcal{R}(x)\}. \tag{7}$$

The infimum exists. Using the path metric, we can write the function $\bar{u}$ in the form

$$\hat{u}(x) = \inf\{\phi(\xi) + \alpha\, d(x,\xi) : \ \xi \in \partial\Omega\} \tag{8}$$

With the above properties of the boundary, one can show that this function is the one constructed before, $\hat{u} = \bar{u}$.

7. Assumption of the Boundary Values

We say that the boundary datum is assumed at a boundary point $x \in \partial\Omega$ if $\bar{u}(x) = \phi(x)$. The preceding paragraphs yield criteria for assumption of the boundary datum. The first uses the idea of a barrier function [17]. The proof follows from (6).

Barrier criterion: Suppose there is a function $v \in M$ such that $v(x) = \phi(x)$. Then $\bar{u}(x) = \phi(x)$.

The second criterion uses transport paths.

Path criterion: Suppose $\xi \in \partial\Omega$ with $d(\xi, x) \leq \phi(x)/\alpha$ implies $\phi(\xi) \geq \phi(x) - \alpha\, d(\xi, x)$. Then $\bar{u}(x) = \phi(x)$.

The proof follows from (8): If ξ is far from x then $\alpha\, d(x,\xi) \geq \phi(x)$. If ξ is close to x then use the hypothesis. This criterion is useful, e.g., on straight parts of the boundary. Then $|\phi(x) - \phi(\xi)| \leq \alpha\,|x - \xi|$ ensures that the boundary value is assumed.

We conclude this section with a comment on the "meaning" of the question whether the boundary values are assumed or not. In contrast to the Dirichlet problem of the Laplace equation, for boundary value problems for granular matter it is quite natural that boundary values are not assumed. If, for example, the rim of the tray has a narrow indentation or gap, then the upper part of the rim will stay free of sand on both sides of the gap.

8. The Uneven Table: Maximal Volume Approach

The function $\psi : \bar{\Omega} \to \mathbb{R}$ (the distributed obstacle) is non-negative and Lipschitz continuous with some constant $\operatorname{Lip}\psi = K$, in general $K > \alpha$. The boundary function $\phi : \partial\Omega \to \mathbb{R}$ is non-negative, $\phi(x) \geq \psi(x)$ for $x \in \partial\Omega$. Finally we assume that there is $\xi \in \partial\Omega$ with $\phi(\xi)$ finite.

We define M as the set of all functions with the following properties: $u \in C(\bar{\Omega})$, $u \geq \psi$, $u(x) \leq \phi(x)$ for $x \in \partial\Omega$, finally: If $x \in \Omega$ and $u(x) > \psi(x)$ then $\operatorname{Lip}_x u \leq \alpha$. With some effort one can show that a function u with these properties is Lipschitz continuous with Lipschitz constant $\tilde{\alpha} = \max(\alpha, K)$, see [14]. Having this property, it is easy to see that the set M is compact in $C(\bar{\Omega})$. Hence the volume functional $V(u) = \int_\Omega (u - \psi)dx$ (which is essentially (4)) assumes its maximum on M at some function $\bar{u}$. As in the case of the flat table it follows that $\bar{u}$ is uniquely determined.

9. Characterization by a Perron Principle

As in the case of a flat table the values of the solution can be explicitly expressed as the maximum over a certain set of functions, the set M in the present case. The formulation is exactly the same as in (5) and (6). Again $\tilde{u} = \bar{u}$.

10. Characterization by Transport Paths

Think of the paths χ as "support paths" for some level of matter at position $x \in \bar{\Omega}$ rather than transport paths. Let $x \in \bar{\Omega}$. Choose $\chi \in \mathcal{R}(x)$. Then $\chi(0) = x$, $\chi(1) = \xi \in \partial\Omega$, and $\chi(t) \in \Omega$ for $0 < t < 1$. Choose any $t \in (0,1)$. Then $\psi(\chi(t))$ is the height of the obstacle at $\chi(t)$. If a grain starting at x would roll down the path and arrive at $(\chi(t), \psi(\chi(t)))$ then it would come to rest. Taking the argument in reverse, we get that the height of the obstacle at $(\chi(t), \psi(\chi(t)))$ supports a height $\psi(\chi(t)) + \alpha \int_0^t |\dot{\chi}(s)|ds$ at position x. But then there is also the boundary datum. It supports grains at a height $\phi(\chi(1)) + \alpha\,\ell(\chi)$. Hence the height that is supported at x along the path χ is

$$h_\chi(x) = \max\{\max_{0 \le t \le 1} [\psi(\chi(t)) + \alpha \int_0^t |\dot{\chi}(s)|ds],\, \phi(\chi(1)) + \alpha\,\ell(\chi)\}. \tag{9}$$

Grains at position x will roll down any transport path if they can. Hence we define

$$\hat{u}(x) = \inf\{h_\chi(x) : \chi \in \mathcal{R}(x)\}. \tag{10}$$

In the special case of the flat table we have $\psi \equiv 0$ and the formula simplifies to (5). For Lipschitz boundaries one can show that $\hat{u} = \bar{u}$, see [14].

11. General Obstacle

Now we consider obstacles which are not necessarily graphs. In that case the desired solutions will not be graphs either but can be characterized as sets. We have a d-dimensional coordinate x which we assume being horizontal and a one-dimensional coordinate z which is assumed to point upwards. We assume that gravity is pointing downward. The obstacle is a compact set $\Psi \subset \mathbb{R}^{d+1}$. We can assume that Ψ is contained in the half-space $z > 0$. Now there are several interpretations of the problem which are not equivalent. We can imagine that dry sand is poured from high above down onto Ψ and forms a heap (or perhaps many heaps, but we shall always speak of "a heap"), superfluous sand will leave the obstacle and fall in the direction of the negative z axis. Then some parts of the obstacle on which sand could be placed will not be covered by sand. On the other hand, we could pose the problem to deposit a maximum of sand on the obstacle, in whatever manner. An example would be a hollow sphere. Can sand get into the interior or not? In the present case it can.

The figure tells that already in the one-dimensional case the solution can look rather unexpected. Even with the adopted interpretation, there may be domains

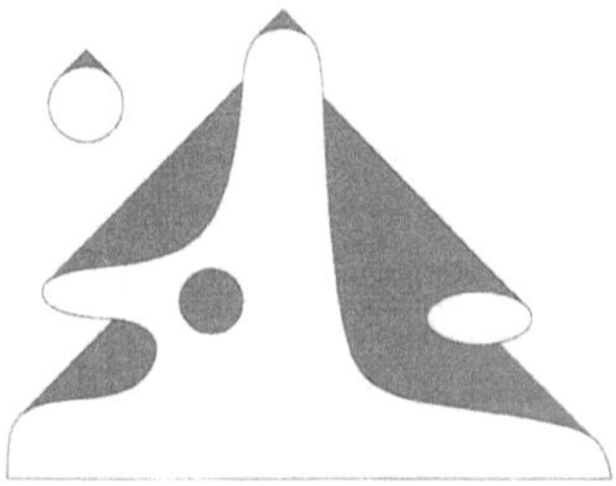

free of sand below components of Ψ which are covered with sand. One way to define the problem properly (and solve it at the same time, by construction) is the following. Suppose $(x, z) \notin \Psi$ with $z > 0$. Consider a continuous, piecewise linear path $\sigma : [0, 1] \to \mathbb{R}^{d+1}$, $\sigma(t) = (\chi(t), \vartheta(t))$ with the following properties

i) $\sigma(0) = (x, z)$,

ii) $\sigma(1) = (\xi, 0)$ with some $\xi \in \mathbb{R}^d$,

iii) $-\dot{\vartheta}(t) \geq \alpha |\dot{\chi}(t)|$ for $0 < t < 1$ with finitely many exceptions, where the derivative is not defined,

iv) $\sigma(t) \notin \Psi$ for $0 < t < 1$.

Hence σ is a steep path which starts at (x, z), passes Ψ and ends below Ψ. This path we consider an admissible transport path for granular matter. If such a path exists, then we call the point (x, z) unsupported (by Ψ). The set of unsupported points is called U_0. It is easy to see that U_0 is open. Then form $U = U_0 \cup \{(x, z) : z \leq 0\}$. This set is also open, hence $W = \mathbb{R}^{d+1} \setminus U$ is closed. It is easy to see that all points in $z > 0$ for which z or $|x|$ is large belong to U. Hence the set W is compact and contains Ψ. Next define the set H as the closure of $W \setminus \Psi$. The compact set H describes the heap of grains supported by the obstacle. If the obstacle Ψ has the form of a table with rim as before then H is the set $H = \{(x, z) : x \in \bar{\Omega},\ \psi \leq z \leq \bar{u}\}$. Hence our construction looks useful.

Even without further assumptions on the topology of Ψ and the regularity of its boundary one can say something about the set H. Physical intuition tells that the surface of H should be smooth where sand covers the obstacle. In fact, this view is correct, see [14]: If a point (x, z) is a boundary point of H but not a boundary point of Ψ then there is a neighborhood of this point where the boundary of H is a Lipschitz continuous graph with Lipschitz constant α.

12. Prescribed Sources

In [10] and [13] we have developed some ideas how stationary solutions look like if sand is poured from distributed sources onto a table or obstacle. We have used the observation from experiments that a point source produces a cone which grows until its foot reaches the edge of the obstacle. After that all additional matter runs over the edge. Another observation tells that in this experimental setting the maximal heaps corresponding to several sources do not add up but the resulting stationary solution is obtained by forming the maximum. Using these principles, we have designed the concept of a Green's function for the case of a flat table without a rim, here the special case $\psi \equiv 0$, $\phi \equiv 0$. Define first $\delta(y) = \text{dist}(y, \partial\Omega)$. Then define a function $\Gamma(x, y)$ of two variables, where y is the position of the point source and x is the variable point at which the height of the heap is observed,

$$\Gamma(x,y) = \begin{cases} \alpha(\delta(y) - |x-y|) & , \ |x-y| < \delta(y) \\ 0 & , \ \text{otherwise.} \end{cases} \tag{11}$$

If the function $f : \bar{\Omega} : \mathbb{R}_+$ describes the sources then

$$u(x) = \max_{y \in \text{supp}(f)} \Gamma(x,y) \tag{12}$$

where $\text{supp}(f)$ is the support of f.

These ideas can be easily carried over to the case of an obstacle ψ provided $\text{Lip}\,\psi < \alpha$. There are principal difficulties in the case $d \geq 2$, $\text{Lip}\,\psi > \alpha$, see [13].

References

[1] F. Auerbach, *Die Gleichgewichtsfiguren pulverförmiger Massen.* Annalen der Physik **5** (1901), 170–219.

[2] G. Aronsson, *A mathematical model in sand mechanics: presentation and analysis.* SIAM J. Math. **22** (1972), 437–458.

[3] G. Aronsson, L.C. Evans, Y. Wu, *Fast/slow diffusion and growing sandpiles.* J. Diff. Equ. **131** (1996), 304–335.

[4] H. Bauer, *Probability Theory.* Walter de Gruyter, Berlin, New York, 1996.

[5] J.-P. Bouchaud, M.E. Cates, J.R. Prakash, S.F. Edwards, *A model for the dynamics of sandpile surfaces.* J. Phys. I France **4** (1994), 1383–1410.

[6] J.-P. Bouchaud, M.E. Cates, J.R. Prakash, S.F. Edwards, *Hysteresis and metastability in a continuum sandpile model.* Phys. Rev. Lett. **74** Nr. 11 (1995), 1982–1985.

[7] M.G. Crandall, H. Ishii, P.-L. Lions, *User's guide to viscosity solutions of second order partial differential equations.* Bull. Amer. Math. Soc. **27** (1992), 1–67.

[8] P.G. de Gennes, *Dynamique superficielle d'un matériau granulaire.* C. R. Acad. Sci. **321** [IIb] (1995), 501–506.

[9] P.G. de Gennes, *Granular matter.* Lecture Notes, Varenna Summer School on Complex Systems, 1996, Società Italiana de Fisica.

[10] K.P. Hadeler, C. Kuttler, *Dynamical models for granular matter.* Granular Matter **2** (1999), 9–18.

[11] K.P. Hadeler, C. Kuttler, *Granular matter in a silo.* Granular Matter 3 (2001) 193–197.

[12] K.P. Hadeler, C. Kuttler, *A model for slowly moving granular matter.* In: D. Helbing et al. (eds), Traffic and Granular Flow '99, Springer Verlag 2000, 511–516.

[13] K.P. Hadeler, *Sand, Kies, Dünen - Mathematische Modelle granularer Medien.* Mitt. Math. Ges. Hamburg **20** (2001), 1–14.

[14] K.P. Hadeler, C. Kuttler, I. Gergert, *Dirichlet and obstacle problems for granular matter.* Preprint University of Tübingen, SFB 382 (2002).

[15] D. Helbing, et al. (eds), *Traffic and Granular Flow '99*, Springer Verlag, 2000.

[16] H.J. Herrmann, J.-P. Hovi, S. Luding, (eds), *Physics of dry granular media.* NATO ASI Series E350, Kluwer Acad. Publ., Dordrecht, 1998.

[17] F. John, *Partial differential equations.* 3rd ed. Springer Verlag, New York Heidelberg Berlin, 1978.

[18] D. Kinderlehrer, G. Stampacchia, *An Introduction to Variational Inequalities and Their Applications.* Academic Press, Boston, 1980.

[19] C. Kuttler, *On the competitive growth of two sand heaps.* Submitted.

[20] S. Luding, O. Strauß, *The equation of state for almost elastic, smooth, polydisperse granular gases for arbitrary density.* In: T. Pöschel, S. Luding (eds), Granular gases. Lect. Notes in Physics, Springer Verlag, 1999–2000.

[21] L. Prigozhin, *Variational model of sandpile growth.* Euro. J. of Applied Math. **7** (1996), 225–235.

[22] S.B. Savage, K. Hutter, *The motion of a finite mass of granular material down a rough incline.* J. Fluid Mech. **199** (1989), 177–215.

[23] S.B. Savage, K. Hutter, *The dynamics of avalanches of granular materials from initiation to runout, I: Analysis.* Acta Mech. **86** (1991), 201–223.

Karl P. Hadeler
Biomathematics
University of Tübingen
Auf der Morgenstelle 10
D-72076 Tübingen, Germany
E-mail address: k.p.hadeler@uni-tuebingen.de

Christina Kuttler
Biomathematics
University of Tübingen
Auf der Morgenstelle 10
D-72076 Tübingen, Germany
E-mail address: christina.kuttler@uni-tuebingen.de

Progress in Nonlinear Differential Equations
and Their Applications, Vol. 55, 227–251

Homogeneous Dynamical Systems and the Age-Structured SIR Model with Proportionate Mixing Incidence

Mimmo Iannelli and Maia Martcheva

1. Introduction

The simplest model of a population is Malthus model. It states that the rate of change of the number of individuals in a population is proportional to the number of individuals so that populations modeled by this law grow or decay exponentially. Malthus model describes an unstructured population, that is a population in which individuals are not distinguishable. Models for structured populations may also display an overall exponential behavior. This is the case with the Lotka-McKendrick model for age-structured populations, where the so called persistent solutions occur, namely solutions that can be expressed as a product of a function of age (the stable distribution) and an exponentially growing function of time. The Lotka-McKendrick model has played an important role in demography, due to its simplicity and relatively good fit to data on a short time scale. Consequently, it remains the preferred theoretical paradigm in the study of human demography.

The Malthus and the Lotka-McKendrick models are, however, linear. Usually non-linear population models are designed to avoid unlimited growth. In fact, through non-linear rates, it is possible to introduce those mechanisms that, in nature, produce containment of the population size.

Nevertheless, in some situations, non-linear models still present exponential growth and persistent solutions. This is, for example, the case when we model diseases both on epidemiological and immunological level, or simply incorporate the sexual structure of the population. In these cases, and others, the model necessarily involves interaction terms reflecting the mutual and reciprocal influence of one subclass over another. These terms as a rule are nonlinear. Although models with mixing subpopulations are unavoidably nonlinear, they still can have exponentially increasing or decreasing solutions and behave somewhat like linear models. Such nonlinear models have homogeneous of degree one right-hand side and are referred to as homogeneous models.

Homogeneous models can be viewed as an analogues of linear models among models with interacting subclasses and have received some interest by demographers because of the possibility to support exponential solutions. In fact, this way the concept of stable distribution has an analogue in the non-linear setting.

One of the main applications of homogeneous models is in the construction of "two-sex" models, that incorporate gender and nonlinear interaction between the sexes that results from pairing. A "two-sex" population model, which is of theoretical interest in mathematical demography, is given by a homogeneous (of degree 1) system of ODEs. It has been completely analyzed by Hadeler and coworkers [8, 9, 10].

Obviously age structure is critical to the process of pair formation in humans. Hence an age-structured two-sex population model has been developed [7, 14] in the early 1970's. The mathematical complexity of that model has been hindering its analysis and only recently results on wellposedness of the model [17, 20] and existence of exponential solutions [25, 19] have appeared.

Homogeneous systems are also gradually gaining recognition as an important modeling tool in mathematical epidemiology. They are more often used to investigate mathematically problems modeled with ODEs. In the structured case difficulties related to the analysis of these models have kept researchers away from them. For models with predominantly linear terms two mechanisms mainly are being used to avoid homogeneity – constant influx of individuals in the population or nonconstant number of contacts per individual in the interaction term of proportionate mixing type. In the ODE models one can assume that the influx is constant and argue favorably since the influx is a general recruitment rate rather than a birth rate. This is not the case in the migration-free age-structured epidemic models where the influx term consists of the total birth rate only. In the linear case, that is when the age-specific per capita birth rate does not depend on the population density, the total birth rate of the population is a linear (and therefore homogeneous) operator. The general form of the mixing term in epidemic modeling has been thoroughly discussed in [4]. In the same article special attention is paid to the most commonly used mixing terms in age-structured epidemiological modeling – proportionate mixing and uniform selectivity mixing. The discussion in [4] demonstrates that assuming all parameters independent of the population densities the resulting force of infection term is homogeneous. Consequently, it is imperative for the majority of the age-structured epidemiological models with age-only dependent coefficients to be homogeneous. If the resulting exponential growth or decay represents a problem from a modeling point of view, nonlinear effects should be incorporated through letting some of the coefficients depend on the population size. Consistency requires that the vital rates (birth and/or death rate) be assumed dependent on the population size. A nonlinear epidemiological SIS model of this type is considered in [16] but their thorough understanding is a matter of future research.

Recognizing the role that homogeneous models play in population dynamics and epidemiology, we consider the problem of analyzing structured homogeneous models. More precisely, we extend the results on homogeneous ODE systems obtained in [8, 9, 10] to structured homogeneous PDE systems. To make the nature of the advances in understanding homogeneous systems clearer we elaborate further. In particular, homogeneous models are usually investigated by projecting on a manifold. The projected system is nonlinear and not homogeneous and can be treated with standard methods. The trade-off is that the projected system is mathematically more complicated. The contribution of the proposed approach consists in the fact that instead of analyzing the eigenvalues of the linearized operator of the projected system, one can analyze the linearized operator of the original system and draw conclusions about stability and instability of both. This approach is particularly suitable for homogeneous epidemiological models since, ideally, we would like to know the behavior of both the proportions of the infected people in the total population as well as their total numbers.

This paper is structured as follows. In the next section we discuss infinite-dimensional homogeneous systems. We consider both the homogeneous system and its projection on a hyperplane and we show that the solution of each can be reconstructed from the solution of the other. Next we demonstrate that the spectrum of the linear operator obtained from the linearization of the projected system is obtained from the spectrum of the linear operator (without the growth rate) obtained from the linearization of the homogeneous system shifted with the growth rate of the population. We illustrate the method on a homogeneous SIR model with proportionate mixing incidence in Section 3. For the SIR model we show that there is always a disease-free equilibrium which is locally stable under the usual threshold condition. If this condition is violated, there is also a unique endemic equilibrium which is stable under an additional condition (see [18]). In Section 4 we summarize our results.

2. Homogeneous dynamical systems

A homogeneous dynamical system is a dynamical system in which the nonlinear term is a homogeneous operator. More precisely, let X be a Banach space, endowed with the norm $\|\cdot\|$, and $K \subset X$ be a closed convex cone. We consider the abstract dynamical system on K, defined by the Cauchy problem:

$$\begin{aligned} u'(t) &= Au + \mathcal{F}(u) \\ u(0) &= u_0 \end{aligned} \tag{2.1}$$

where $u_0 \in K$, $A : \mathcal{D}(A) \subset X \to X$ is a linear operator which is a generator of a C_0-semigroup on X and $\mathcal{F} : K \subset X \to X$ is a (nonlinear) continuous operator, *homogeneous* of degree one (or simply homogeneous), i.e.,

$$\mathcal{F}(\alpha z) = \alpha \mathcal{F}(z) \qquad \forall z \in K, \alpha > 0. \tag{2.2}$$

We suppose that A and $\mathcal{F}$ satisfy some basic assumptions in order that, for every $u_0 \in K$, the system (2.1) to have a unique *mild* solution $u(t) \in C(0,T;X)$, such that $u(t) \in K$, that is, u is the unique solution of the integral equation:

$$u(t) = e^{At}u_0 + \int_0^t e^{A(t-s)}\mathcal{F}(u(s))\,ds. \tag{2.3}$$

Namely, we assume that the semigroup generated by A leaves K invariant:

$$e^{At}(K) \subset K \tag{2.4}$$

and, concerning $\mathcal{F} : K \subset X \to X$, we suppose that it is Lipschitz continuous and Fréchet differentiable on K, that is, we assume that $\forall z \in K, z \neq 0$ there exists a bounded linear operator

$$\mathcal{F}'(z) : X \to X$$

such that

$$\begin{aligned} &\frac{\|\mathcal{F}(z+v) - F(z) - \mathcal{F}'(z)v\|}{\|v\|} \leq o(\|v\|) \quad \forall v \in K \\ &||\mathcal{F}'(z)|| \leq L \quad \forall z \in K, z \neq 0 \end{aligned} \tag{2.5}$$

where $L > 0$.
Moreover, we assume that for all $R > 0$ there exists $\alpha_R > 0$ such that

$$z + \alpha_R \mathcal{F}(z) \in K \quad \forall z \in K, ||z|| \leq R. \tag{2.6}$$

The previous conditions are sufficient to treat the problems we have in mind, but could also be extended to more general cases. In particular these conditions imply that the solution $u(t)$ to problem (2.3) can be approximated in $C(0,T;X)$ by the sequence $u_n(t) \in C^1(0,T;X)$ that satisfies the problems

$$\begin{aligned} &u_n'(t) = A_n u_n + \mathcal{F}(u_n) \\ &u(0) = u_0 \end{aligned} \tag{2.7}$$

where $A_n = n^2(n-A)^{-1} - n$ are the Yosida approximants of A.

We call the system (2.1) *homogeneous* because the operator $\mathcal{F}$ is *homogeneous*.

Homogeneous systems are in some sense half-linear systems since homogeneity is one of the two properties satisfied by a linear operator. The dynamical behavior of homogeneous systems is closer to the dynamical behavior of linear systems, than to nonlinear systems. In particular, homogeneous dynamical systems (like linear systems) in general do not have time-independent solutions. Instead, they have solutions which grow or decay exponentially in time (persistent solutions), and it can be expected that other solutions approach in time a persistent solution.

2.1. Reformulation of the homogeneous problem

Let φ be a bounded linear real functional on X, that is

$$\varphi : X \to \Re.$$

We will use the notation $\langle \varphi, z \rangle = \varphi(z)$. We note that for our abstract discussion we do not need φ to be linear but only homogeneous; however, since in most applications in population dynamics φ is linear we will assume that here. In addition we assume that

$$\begin{aligned} &\text{i)} \ \ \langle \varphi, Az \rangle \ \ \text{is a bounded operator on } X, \quad \text{i.e.,} \quad \varphi \in \mathcal{D}(A^*), \\ &\qquad \text{where } A^* \text{ is the adjoint of } A, \\ &\text{ii)} \ \ \forall u_0 \in K - \{0\} \quad \text{the solution } u(t) \text{ of (2.1) is such} \\ &\qquad \text{that } \langle \varphi, u(t) \rangle > 0. \end{aligned} \tag{2.8}$$

The first of these assumptions is technical and is necessary to justify the linearization procedure that we will perform later for the purpose of investigating stability, the second is important in order to consider the significant quantity whose behavior we want to investigate. In fact, if $u(t)$ is the solution of problem (2.1), we set

$$w(t) = \frac{u(t)}{\langle \varphi, u(t) \rangle}. \tag{2.9}$$

This new variable belongs to K and to the hyperplane

$$\mathcal{P} = \{z \in X; \langle \varphi, z \rangle = 1\}. \tag{2.10}$$

It formally satisfies the dynamical system:

$$\begin{aligned} w'(t) &= Aw + \mathcal{F}(w) - \langle \varphi, Aw(t) + F(w) \rangle w(t) \\ w(0) &= w_0 \in Y \end{aligned} \tag{2.11}$$

where $Y = K \cap \mathcal{P}$.

Actually (2.11) is a semilinear problem (notice that the last term on the right is continuous because of assumption (2.8)) and can be put in the following mild form:

$$w(t) = e^{At} w_0 + \int_0^t e^{A(t-s)} \left[\mathcal{F}(w(s)) - \langle \varphi, Aw(s) + F(w(s)) \rangle w(s)\right] ds. \tag{2.12}$$

On the other hand, $w(t)$ defined in (2.9) satisfies (2.12). In fact we consider the approximating sequence $u_n(t) \in C^1(0,T;X)$ defined in (2.7) and we set (note that $\langle \varphi, u_n(t) \rangle \neq 0$ for $t \in [0,T]$ and n sufficiently large)

$$w_n(t) = \frac{u_n(t)}{\langle \varphi, u_n(t) \rangle}. \tag{2.13}$$

Then we have

$$\begin{aligned} w_n'(t) &= A_n w_n + \mathcal{F}(w_n) - \langle \varphi, A_n w_n(t) + F(w_n) \rangle w_n(t) \\ w_n(0) &= w_0 \in Y \end{aligned} \tag{2.14}$$

so that, going to the limit in the integrated mild form of (2.14), we have that $w(t)$ defined in (2.9) is a solution of (2.12) and $w(t) \in Y$.

Moreover, we note that if we define a continuous function

$$\lambda(t) = \langle \varphi, Aw(t) + F(w(t)) \rangle \tag{2.15}$$

then (2.12) is equivalent to the equation (this can be proved by the same approximating procedure):

$$w(t) = e^{-\int_0^t \lambda(\tau) d\tau} e^{At} w_0 + \int_0^t e^{-\int_s^t \lambda(\tau) d\tau} e^{A(t-s)} \mathcal{F}(w(s)) ds. \tag{2.16}$$

From this, compared with (2.3) we also see that

$$u(t) = e^{\int_0^t \lambda(s) ds} w(t). \tag{2.17}$$

Concerning possible steady states $w^* \in Y$ of (2.11) we note that they must satisfy

$$\begin{aligned} Aw^* + \mathcal{F}(w^*) &= \lambda^* w^* \\ \langle \varphi, w^* \rangle &= 1 \end{aligned} \tag{2.18}$$

where $\lambda^* = \langle \varphi, Aw^* + F(w^*) \rangle$. Observe that λ^* is real since φ is a real-valued functional. Therefore, λ^* is a real eigenvalue corresponding to the eigenvector $w^* \in Y$ of the nonlinear homogeneous operator $A + \mathcal{F}$. We note that by (2.17) the steady state w^* corresponds to an exponential solution with growth rate λ^* of (2.1) with the form

$$u(t) = \alpha_0 e^{\lambda^* t} w^*. \tag{2.19}$$

In fact, it can be seen that every separable solution of (2.1) has the form above. These solutions are referred to in the literature as *persistent solutions.*

2.2. The Linearized problem

We are now concerned with the stability of the exponential solution of (2.1) and the stability of the time-independent solution of (2.11). Following Hadeler [8, 9, 10] we say that the exponential solution of (2.1) is stable if the corresponding time-independent solution of the projected system (2.11) is stable in classical sense.

The analysis of the stability of the solution to problem (2.11) will be performed by linearization, so we must determine the differentiated problem of interest. To this purpose we note that, since $\mathcal{F}(z)$ is homogeneous of degree one with respect to positive constants, we have that it satisfies the following properties:

$$\begin{aligned} \mathcal{F}'(z)z &= \mathcal{F}(z) \\ \mathcal{F}'(\alpha z) &= \mathcal{F}'(z) \end{aligned} \tag{2.20}$$

for all $z \in X$ and $\alpha > 0$. The first of the two equalities above and (2.18) imply

$$\begin{aligned} Aw^* + \mathcal{F}'(w^*)w^* &= \lambda^* w^* \\ \langle \varphi, w^* \rangle &= 1 \end{aligned} \tag{2.21}$$

Thus, λ^* is also an eigenvalue of the linear operator $A + \mathcal{F}'(w^*)$. Actually, we use $B : \mathcal{D}(B) \subset X \to X$ to denote that operator:

$$\begin{aligned} B &= A + \mathcal{F}'(w^*) \\ \mathcal{D}(B) &= \mathcal{D}(A) \end{aligned} \tag{2.22}$$

which is a bounded perturbation of A.

Let $w(t)$ be any solution of (2.12). Define the perturbation of this solution from w^*:

$$\chi(t) = w(t) - w^* \in \ker \varphi$$

where $\ker \varphi$ is the kernel of φ. We observe that, since w^* satisfies

$$w^* = e^{At} w^* + \int_0^t e^{A(t-s)} \left[\mathcal{F}(w^*) - \langle \varphi, Aw^* + F(w^*) \rangle w^* \right] ds \tag{2.23}$$

then $\chi(t)$ satisfies

$$\chi(t) = e^{At}\chi_0 + \int_0^t e^{A(t-s)}\left[\mathcal{F}'(w^*)\chi(s) - \lambda^*\chi(s) - \langle\varphi, B\chi(s)\rangle w^* + \mathcal{G}\left(\chi(s)\right)\right] ds$$
$$\langle\varphi, \chi(t)\rangle = 0 \tag{2.24}$$

where $\mathcal{G}(\cdot)$ is given by

$$\begin{aligned} \mathcal{G}(z) = & - \quad \mathcal{F}'(w^*)z + \mathcal{F}(w^* + z) - \mathcal{F}(w^*) \\ & - \quad \langle\varphi, -\mathcal{F}'(w^*)z + \mathcal{F}(w^* + z) - \mathcal{F}(w^*)\rangle w^* \\ & - \quad \langle\varphi, -\mathcal{F}'(w^*)z + \mathcal{F}(w^* + z) - \mathcal{F}(w^*)\rangle z - \langle\varphi, Az + \mathcal{F}'(w^*)z\rangle z \end{aligned} \tag{2.25}$$

where $z \in \ker\varphi$. We note that $\mathcal{G}$ is a second order term and includes the quadratic term given by

$$\langle\varphi, B\chi(t)\rangle\chi(t)$$

because of assumption (2.8).

Now we define the following linear operator on the space X:

$$\begin{aligned} Sz &= Bz - \langle\varphi, Bz\rangle w^* \\ \mathcal{D}(S) &= \mathcal{D}(B). \end{aligned} \tag{2.26}$$

We note that since the term $\langle\varphi, Bz\rangle w^*$ is a bounded perturbation of B (see 2.8), then S generates a C_0-semigroup on X and (2.24) is equivalent to

$$\chi(t) = e^{-\lambda^* t}e^{St}\chi_0 + \int_0^t e^{-\lambda^*(t-s)}e^{S(t-s)}\mathcal{G}(\chi(s))ds \tag{2.27}$$

In fact, $Sz = Az + \mathcal{F}'(w^*)z - \langle\varphi, Bz\rangle w^*$ and the equivalence of (2.24) and (2.27) can be stated approximating A with the Yosida approximants and proceeding as in Section 2.1.

Equations (2.24) and (2.27) are equations for the perturbation in mild form. We need to show that the perturbation $\chi(t)$ approaches zero as t approaches infinity. That will establish stability for the problem (2.11). In order to analyze the behavior of the solution of (2.24) (or (2.27) respectively), we need to investigate the spectral properties of S.

2.3. Analysis of the spectrum of S

For any linear operator T we denote by $\rho(T)$ the resolvent set of T. In addition, we denote by $\sigma_P(T)$ the point spectrum of T, by $\sigma_R(T)$ the residual spectrum of T, and by $\sigma_C(T)$ the continuous spectrum of T ([28], Section 5.3, p. 264). Then, we have

Theorem 2.1. *Assume λ^* is an eigenvalue of B with algebraic multiplicity one. Then,*

$$(\sigma_P(B) - \lambda^*) \cup 0 = \sigma_P(S). \tag{2.28}$$

In particular we have that:

1. *Zero is an eigenvalue of S which corresponds to eigenvector w^*. Moreover, if $\lambda^* \neq 0$ then it is not an eigenvalue of S.*

2. *If $\alpha \neq \lambda^*$ is an eigenvalue of B corresponding to eigenvector $\mathbf{v} \in \mathcal{D}(B)$, then α is an eigenvalue of S corresponding to eigenvector $\mathbf{v} - \langle \varphi, \mathbf{v} \rangle w^*$.*
3. *If $\alpha \neq 0$ is an eigenvalue of S corresponding to eigenvector $\mathbf{v} \in \mathcal{D}(S)$, then α is an eigenvalue of B corresponding to eigenvector*
$$\mathbf{v} + \frac{\langle \varphi, B\mathbf{v} \rangle}{(\alpha - \lambda^*)} w^*. \tag{2.29}$$

Proof.

1. Direct checking shows that 0 is an eigenvalue of S corresponding to eigenvector w^*. To see that $\lambda^* \neq 0$ is not an eigenvalue of S we assume there exists a vector z such that
$$Bz - \langle \varphi, Bz \rangle w^* = \lambda^* z. \tag{2.30}$$
Applying φ to both sides of the above equation leads to the conclusion that $z \in \ker \varphi$. Now, we consider the projection defined as follows ([28], Section 5.8, p. 305):
$$Qz = \tfrac{1}{2\pi i} \oint (\lambda - B)^{-1} z \, d\lambda \tag{2.31}$$
where the integral is along a closed simple curve around λ^*. We first show that the projection Q is a projection on the space generated by w^*, that is, for $z \in X$ there exists a constant c such that
$$Qz = cw^*.$$
To see that, consider
$$(\lambda^* - B)Qz = \frac{1}{2\pi i} \oint (\lambda^* - B)(\lambda - B)^{-1} z \, d\lambda \tag{2.32}$$
$$= \frac{1}{2\pi i} \oint (\lambda^* - \lambda)(\lambda - B)^{-1} z \, d\lambda + \frac{1}{2\pi i} \oint z \, d\lambda = 0 \tag{2.33}$$
Each of the integrals on the last line above is equal to zero since the integrand is an analytic function. Hence, $Qz \in \ker(\lambda^* - B)$ and therefore Qz must be proportional to the unique eigenvector w^*. To see that Q leaves the elements of the space generated by w^* unchanged, it is enough to see that $Qw^* = w^*$. This last equality can be checked directly from the definition of Q (2.31) observing that
$$(\lambda - B)^{-1} w^* = \frac{1}{\lambda - \lambda^*} w^*. \tag{2.34}$$
Equality (2.34) can be obtained from the identity:
$$\lambda(\lambda - B)^{-1} w^* = w^* + \lambda^*(\lambda - B)^{-1} w^*.$$
Note also that
$$QBz = BQz = \lambda^* Qz \tag{2.35}$$
and applying Q to equality (2.30) we obtain
$$QBz - \langle \varphi, Bz \rangle w^* = \lambda^* Qz. \tag{2.36}$$

Consequently, we have $\langle \varphi, Bz \rangle = 0$ and

$$Bz - \lambda^* z = 0.$$

From $z \neq 0$ it follows that we must have $z = c \cdot w^*$ which is impossible since $z \in \ker \varphi$.

2. Let $\alpha \neq \lambda^*$ be an eigenvalue of B corresponding to eigenvector $\mathbf{v}$, that is $B\mathbf{v} = \alpha\mathbf{v}$. We recall that $Sw^* = 0$. Then

$$\begin{aligned} S(\mathbf{v} - \langle \varphi, \mathbf{v} \rangle w^*) &= B\mathbf{v} - \langle \varphi, B\mathbf{v} \rangle w^* \\ &= B\mathbf{v} - \alpha \langle \varphi, \mathbf{v} \rangle w^* \\ &= \alpha(\mathbf{v} - \langle \varphi, \mathbf{v} \rangle w^*). \end{aligned} \tag{2.37}$$

3. If $\alpha \neq 0$ is an eigenvalue of S corresponding to an eigenvector $\mathbf{v}$, that is $S\mathbf{v} = \alpha\mathbf{v}$, then applying B to the vector (2.29) we obtain

$$\begin{aligned} B\left(\mathbf{v} + \frac{\langle \varphi, B\mathbf{v} \rangle}{\alpha - \lambda^*} w^*\right) &= B\mathbf{v} - < \varphi, B\mathbf{v} > w^* + \left(1 + \frac{\lambda^*}{\alpha - \lambda^*}\right) \langle \varphi, B\mathbf{v} \rangle w^* \\ &= S\mathbf{v} + \frac{\alpha}{\alpha - \lambda^*} \langle \varphi, B\mathbf{v} \rangle w^* \\ &= \alpha\left(\mathbf{v} + \frac{\langle \varphi, B\mathbf{v} \rangle}{\alpha - \lambda^*} w^*\right). \end{aligned} \tag{2.38}$$

□

In other words, Theorem 2.1 states that $\lambda^*, \lambda_1, \lambda_2, \ldots, \lambda_n, \ldots$ are the eigenvalues of B if and only if $0, \lambda_1, \lambda_2, \ldots, \lambda_n, \ldots$ are the eigenvalues of S. Consequently, it extends the results in [8, 9, 10] to the point spectrum of the infinite-dimensional operators obtained from the linearization. Concerning the residual and the continuous spectrum of B and S the following theorem shows that they are actually the same.

Theorem 2.2. *Assume λ^* is an eigenvalue of B with algebraic multiplicity one. Then,*

$$\sigma_R(B) - \{0\} = \sigma_R(S) - \{\lambda^*\} \quad \textit{and} \quad \sigma_C(B) - \{0\} = \sigma_C(S) - \{\lambda^*\}. \tag{2.39}$$

Proof. To see (2.39) we first notice that, since $\lambda w^* - Bw^* = (\lambda - \lambda^*)w^*$, $\lambda \neq \lambda^*$ implies $w^* \in Range\,(\lambda - B)$ and, similarly, since $\lambda w^* - Sw^* = \lambda w^*$, then $\lambda \neq 0$ implies $w^* \in Range\,(\lambda - S)$. Consequently, by the equality

$$\lambda z - Sz = (\lambda z - Bz) + \langle \varphi, Bz \rangle w^*$$

we see that if $\lambda \neq 0$ and $\lambda \neq \lambda^*$ we have

$$Range\,(\lambda - S) = Range\,(\lambda - B). \tag{2.40}$$

This equality is enough to prove the first of (2.39), by the definition itself of residual spectrum. To complete the proof we remark that, standing (2.40) we need only to show that $(\lambda - S)^{-1}$ is continuous if and only if $(\lambda - B)^{-1}$ is. To see this, suppose first that $(\lambda - S)^{-1}$ is continuous. Then the equation

$$\lambda z - Bz = y$$

has one and only one solution for any $y \in Range\,(\lambda - B)$. In fact this solution is necessarily given by

$$z = \frac{1}{\lambda}\langle\varphi, Bz\rangle w^* + (\lambda - S)^{-1} y$$

where the scalar $\langle\varphi, Bz\rangle$ can be explicitly computed from

$$\langle\varphi, Bz\rangle = \frac{\lambda^*}{\lambda}\langle\varphi, Bz\rangle + \langle\varphi, B(\lambda - S)^{-1} y\rangle$$

Thus for all $y \in Range(\lambda - B)$ we have

$$(\lambda - B)^{-1} y = \frac{1}{\lambda - \lambda^*}\langle\varphi, B(\lambda - S)^{-1} y\rangle w^* + (\lambda - S)^{-1} y$$

from which it follows that $(\lambda - B)^{-1}$ is continuous. Similarly one can prove that if $(\lambda - B)^{-1}$ so is $(\lambda - S)^{-1}$ and the proof is complete. □

2.4. Consequences on stability

Now we consider equation (2.27) and we note that the semigroup e^{St} and the nonlinear term $\mathcal{G}(\cdot)$ leave the space

$$X_\varphi = \ker\varphi$$

invariant. Actually, it is easy to check by direct computation that

$$\langle\varphi, \mathcal{G}(z)\rangle = 0 \qquad \forall z \in \ker\varphi.$$

Moreover, we can check that for $\lambda \in \rho(S)$ we have

$$(\lambda - S)^{-1} X_\varphi \subset X_\varphi.$$

In fact, from the equation

$$\lambda z - Bz + \langle\varphi, Bz\rangle w^* = f$$

and $\langle\varphi, f\rangle = 0$ we get

$$\lambda\langle\varphi, z\rangle = 0.$$

As a consequence of these remarks and the fact that $\chi_0 \in X_\varphi$, the equation (2.27) can be embedded in X_φ. Let S_φ be the part of S in X_φ. We note that, since w^* does not belong to X_φ, $\lambda = 0$ is not an eigenvalue for S_φ. Instead, all the other eigenvalues of S are eigenvalue of S_φ; in fact if $\bar\lambda \neq 0$ is an eigenvalue of S with eigenvector $\bar u$, then

$$\bar\lambda\bar u - S\bar u = 0$$

and, since $<\varphi, S\bar u>= 0$, then $\bar\lambda <\varphi, \bar u>= 0$ so that $\bar u \in X_\varphi$.

Namely, we have

$$\sigma_P(S_\varphi) = \sigma_P(S) - \{0\}, \quad \sigma_R(S_\varphi) \subset \sigma_R(S), \quad \sigma_C(S_\varphi) \subset \sigma_C(S)$$

Finally

Theorem 2.3. *Let*

$$\|e^{S_\varphi t}\| \leq M e^{\omega t} \qquad \textit{with} \qquad \omega < \lambda^* \tag{2.41}$$

then

$$\lim_{t\to\infty} \chi(t) = 0. \tag{2.42}$$

Proof. Let

$$\epsilon < \frac{\lambda^* - \omega}{M}.$$

Since $\mathcal{G}$ is a higher order perturbation, then for some R we have

$$\|\mathcal{G}(z)\| \leq \epsilon \|z\| \qquad \forall z, \|z\| \leq R.$$

For any $\|\chi_0\| < R/M$ let

$$\tau = \sup\{t : \|\chi(t)\| \leq R\}.$$

Then we have,

$$\|\chi(t)\| \leq M e^{(\omega-\lambda^*)t}\|\chi_0\| + M\epsilon \int_0^t e^{(\omega-\lambda^*)(t-s)}\|\chi(s)\|\, ds$$

and therefore,

$$\|\chi(t)\| \leq M e^{(\omega-\lambda^*+M\epsilon)t}\|\chi_0\| \leq R \qquad 0 < t < \tau$$

which implies $\tau = \infty$ and (2.42) is valid. □

In the next Corollary we specify conditions under which the assumptions of Theorem 2.3 may hold. Consequently, these conditions lead to local stability of the equilibria of the projected system or, equivalently, to the local stability of the exponential solutions.

Corollary 2.4. *Let S_φ generate an eventually compact semigroup and suppose that*

$$\sup_{\lambda\in\sigma(B),\lambda\neq\lambda^*} \Re\lambda < \lambda^*.$$

Then the conclusion of Theorem 2.3 holds.

Proof. Eventual compactness of the semigroup generated by S_φ implies the so called *spectrum determined growth condition* (see [5], p. 214) that is the bound (2.41) with $\omega < \lambda^*$. □

We note that eventual compactness of the semigroup generated by S_φ is only one of the properties that lead to relation between the long term behavior of the semigroup and the spectrum of its generator. Other properties that can ensure that are quasi-compactness and eventual uniform continuity.

3. A homogeneous SIR model

In this section we use the abstract results of the previous section to investigate a homogeneous SIR model. SIR models are the prototype of the modeling process in mathematical epidemiology. For more than twenty years SIR models have been a field of active research. Age-structured models of SIR type have been investigated [18, 29] in the case when the population is already in steady state, that is, assuming that the growth rate is zero. In this case the demographic process is ruled out and the corresponding normalized system is of comparable complexity with the original one. Models with varying population size are usually treated by normalization. However, the normalized system is more complex than the original one since the normalization adds additional terms in the equation (see 2.11). The theoretical development in the previous section provides a way for the reduction to a great extent of the analysis of the normalized system to the analysis of the original one.

3.1. Setting of the homogeneous SIR model

Let a denote the age and t denote the time. We denote by $a_\dagger$ the maximal age of the population and we assume $a_\dagger < \infty$.

If $S(a,t)$ is the density of susceptible individuals, $I(a,t)$ is the density of infective individuals and $R(a,t)$ is the density of removed (recovered) individuals, then a general form of the age-structured SIR model is given by the following set of equations:

$$\begin{aligned} &S_t(a,t) + S_a(a,t) = -\lambda(a,t)S(a,t) - \mu(a)S(a,t) \\ &I_t(a,t) + I_a(a,t) = \lambda(a,t)S(a,t) - (\mu(a) + \gamma(a))I(a,t) \\ &R_t(a,t) + R_a(a,t) = \gamma(a)I(a,t) - \mu(a)R(a,t) \end{aligned} \tag{3.1}$$

where $\mu(a)$ is the natural age-dependent death rate, $\gamma(a)$ is the recovery rate and $\lambda(a,t)$ is the age-specific transmission rate. The first equation in the model says that a cohort in the susceptible class decreases by the individuals who get infected and the individuals who die. The second equation says that a cohort in the infective class increases by those who become infected, and decreases by those who die and those who recover. The last equation says that a cohort in the recovered class increases by those who recover and decreases by those who die. Therefore, this is a model of a disease which imparts permanent immunity.

The system (3.1) is endowed with boundary conditions. The boundary conditions model the dynamics of the newborns. We assume that there is no vertical transmission. Hence, all individuals are born susceptible and no-one is born infective or immune.

$$\begin{aligned} &S(0,t) = \int_0^{a_\dagger} \beta(a)U(a,t)\,da \\ &I(0,t) = 0, \\ &R(0,t) = 0 \end{aligned} \tag{3.2}$$

where $\beta(a)$ is the age-specific birth rate and

$$U(a,t) = S(a,t) + I(a,t) + R(a,t)$$

is the age-specific density of the population. A key element in model (3.1) is the infection rate $\lambda(a,t)$. Various forms of the infection rate are possible and have been considered so far. We will discuss model (3.1) with the following choice of the infection rate, similar to one in [4]:

$$\lambda(a,t) = \frac{k(a)}{N(t)} \int_0^{a_\dagger} h(\sigma) I(\sigma,t) d\sigma.$$

where

$$N(t) = \int_0^{a_\dagger} U(a,t) da. \tag{3.3}$$

is the total population size. We denote the intracohort infection transmission term by

$$H(t) = \int_0^{a_\dagger} h(a) I(a,t) da. \tag{3.4}$$

Another form of the force of infection which is sometimes used is given by

$$\lambda(a,t) = k(a) \int_0^{a_\dagger} h(\sigma) \frac{I(\sigma,t)}{U(\sigma,t)} d\sigma.$$

We note that the system (3.1) with this choice of the infection rate is not convenient to be treated as discussed in the previous section, that is, it is not convenient to normalize with a functional. Normalizing with $U(a,t)$ one can obtain a model similar to the one discussed in [18, 29] and treat it with similar techniques.

Adding the three equations in (3.1) we obtain the system that describes the dynamics of the cohorts in the total population:

$$\begin{aligned} U_t(a,t) + U_a(a,t) &= -\mu(a) U(a,t) \\ U(0,t) &= \int_0^{a_\dagger} \beta(a) U(a,t)\, da \end{aligned} \tag{3.5}$$

This is the classical Lotka-McKendrick model of the population dynamics. We note that this equation does not depend on S, I and R because in this epidemic model we have not included differential mortality due to the disease; thus the demography of the population is not affected by the occurrence of the epidemics and the total population density is governed by an equation of its own. Equation (3.5) has been vastly investigated [6, 15] (see chapters I, II in [15]) and the behavior of the solutions is well known: there exists a number r called the growth rate of the population and separable solutions of the form

$$U(a,t) = u(a) e^{rt}$$

such that any solution of (3.5) approaches a separable solution. Since the model (3.5) is not only homogeneous but also linear, various treatments are possible and, in fact, more than one are already available.

Before we put the the system (3.1) in the form of an abstract evolution system (2.1) we observe that none of the equations depends on R except the one that describes the dynamics of the recovered class. This suggest that we can consider instead the system

$$\begin{aligned} &S_t(a,t) + S_a(a,t) = -\lambda(a,t)S(a,t) - \mu(a)S(a,t) \\ &I_t(a,t) + I_a(a,t) = \lambda(a,t)S(a,t) - (\mu(a)+\gamma(a))I(a,t) \\ &U_t(a,t) + U_a(a,t) = -\mu(a)U(a,t) \end{aligned} \tag{3.6}$$

with the boundary conditions

$$\begin{aligned} S(0,t) &= \int_0^{a_\dagger} \beta(a)U(a,t)\,da \\ I(0,t) &= 0, \\ U(0,t) &= \int_0^{a_\dagger} \beta(a)U(a,t)\,da \end{aligned} \tag{3.7}$$

We consider the system above with the following conditions on the parameters:

$$\begin{aligned} &\beta(a), \gamma(a), h(a), k(a) \in L^\infty(0,a_\dagger), \\ &\mu(a) \in L^1_{\text{loc}}(0,a_\dagger) \text{ and } \int_0^{a_\dagger} \mu(a)\,da = \infty, \\ &\mu(a)\pi(a) \text{ is bounded a.e. on } (0,a_\dagger) \end{aligned} \tag{3.8}$$

where $\pi(a) = e^{-\int_0^a \mu(\tau)\,d\tau}$ is the probability of surviving to age a. We assume that all parameters are nonnegative and $\beta(a)$ is not zero a.e. on $(0,a_\dagger)$.

We now define the abstract setting from section 2. Let $\mathcal{X} = L^1(0,a_\dagger) \times L^1(0,a_\dagger) \times L^1(0,a_\dagger)$. We introduce the Banach space

$$X = \{f \in \mathcal{X} : \frac{f(\cdot)}{\pi(\cdot)} \in \mathcal{X}\}$$

with the norm

$$\|f\|_X = \sum_{i=1}^{3} \int_0^{a_\dagger} \frac{|f_i(a)|}{\pi(a)}\,da.$$

We note that

$$X \subset \mathcal{X}.$$

Next, we define the cone K:

$$K = \{f \in X : f_i(a) \geq 0 \quad (i=1,2,3), f_1(a) + f_2(a) \leq f_3(a) \text{ a.e. } (0,a_\dagger)\}.$$

Let the operator $A : \mathcal{D}(A) \to \mathcal{X}$ be defined as follows

$$A = \begin{pmatrix} -\frac{\partial}{\partial a} - \mu(a) & 0 & 0 \\ 0 & -\frac{\partial}{\partial a} - \mu(a) - \gamma(a) & 0 \\ 0 & 0 & -\frac{\partial}{\partial a} - \mu(a) \end{pmatrix}$$

with domain

$$\begin{aligned} \mathcal{D}(A) = \{f \in X : \frac{f_i(\cdot)}{\pi(\cdot)} &\text{ is absolutely continuous, } f_1(0) = f_3(0) \\ &= \int_0^{a_\dagger} \beta(a) f_3(a)\,da, f_2(0) = 0\}. \end{aligned}$$

It is not hard to see that the operator A defined above is a generator of a C_0-semigroup on X, such that K is left invariant (see 2.4). Namely, by standard techniques (see [15]) it is possible to represent the solution of the Cauchy problem relative to A through the solution of the linear renewal equation associated to the Lotka-McKendrick equation. By this representation it is then possible to prove the properties of the generated semigroup.

We also notice that the functions defined in (3.4) and (3.3) induce linear functionals on the space X, as follows:

$$\begin{aligned} \mathcal{H}: X \to \Re \qquad & \mathcal{H}(f) := \int_0^{a_\dagger} h(a) f_2(a)\, da, \\ \mathcal{N}: X \to \Re \qquad & \mathcal{N}(f) := \int_0^{a_\dagger} f_3(a)\, da. \end{aligned} \tag{3.9}$$

Consequently, the nonlinear homogeneous operator $\mathcal{F}$ is defined as follows

$$\mathcal{F}(f)(a) = \begin{pmatrix} -k(a) f_1(a) \dfrac{\mathcal{H}(f)}{\mathcal{N}(f)} \\ k(a) f_1(a) \dfrac{\mathcal{H}(f)}{\mathcal{N}(f)} \\ 0 \end{pmatrix} \qquad \forall f \neq 0, f \in K \tag{3.10}$$

and $\mathcal{F}(0) = 0$. Clearly, the operator $\mathcal{F}$ is homogeneous, Lipschitz continuous, Fréchet differentiable (except at zero) and satisfies (2.6).

Next, we define the functional φ as follows:

$$< \varphi, f >:= \mathcal{N}(f) = \int_0^{a_\dagger} f_3(a)\, da. \tag{3.11}$$

For an element $f \in X$, condition (2.8.i) is satisfied since

$$\mu(a)\pi(a) \text{ is bounded a.e. on } (0, a_\dagger).$$

and (2.8.ii) is clearly valid for φ defined as in (3.11) and $f \in K$.

Next, let $f = (S, I, U)^T$, that is f is a vector column. Then the system (3.6) can be written as an abstract evolution equation as in (2.1).

3.2. The nonlinear eigenvalue problem

We search for solutions in the system (2.18). They satisfy a nonlinear eigenvalue problem of the form:

$$\begin{aligned} \lambda S^*(a) + S^*_a(a) &= -k(a) S^*(a) \frac{H^*}{N^*} - \mu(a) S^*(a) \\ \lambda I^*(a) + I^*_a(a) &= k(a) S^*(a) \frac{H^*}{N^*} - (\mu(a) + \gamma(a)) I^*(a) \\ \lambda U^*(a) + U^*_a(a) &= -\mu(a) U^*(a) \end{aligned} \tag{3.12}$$

with

$$\begin{aligned} S^*(0) &= U^*(0) = \int_0^{a_\dagger} \beta(a)U^*(a)\,da \\ I^*(0) &= 0 \\ H^* &= \int_0^{a_\dagger} h(a)I^*(a)\,da \\ N^* &= \int_0^{a_\dagger} U^*(a)\,da \end{aligned} \tag{3.13}$$

Actually, problem (3.12) is equivalent to the problem of existence of persistent solutions of the system (2.1). The last equation of (3.12) can be solved independently of the other two and gives

$$U^*(a) = b_0 e^{-\lambda^* a}\pi(a) \tag{3.14}$$

where λ^* is the unique real solution of the Lotka characteristic equation, that is λ^* satisfies

$$\int_0^{a_\dagger} \beta(a)\pi(a)e^{-\lambda a}\,da = 1 \tag{3.15}$$

The solutions must be in the hyperplane $\mathcal{P}$ (see 2.18) with

$$N^* = \int_0^{a_\dagger} U^*(a)\,da = 1$$

which results the following condition for b_0:

$$b_0 \int_0^{a_\dagger} \pi(a)e^{-\lambda^* a}\,da = 1.$$

The other two equations in the system (3.12) have 2 solutions:

1. The disease-free equilibrium (DFE):

$$\begin{aligned} &S^0(a) = U^0(a) = U^*(a), \\ &I^0(a) = 0. \end{aligned} \tag{3.16}$$

   In this case we have $H^* = 0$.
2. The endemic equilibrium (EE):

$$\begin{aligned} S^*(a) &= b_0 e^{-\lambda^* a}\pi(a)e^{-H^* \int_0^a k(\eta)d\eta} \\ I^*(a) &= b_0 H^* e^{-\lambda^* a}\pi(a)\int_0^a k(\sigma)\Gamma(\sigma,a)e^{-H^* \int_0^\sigma k(\eta)\,d\eta}\,d\sigma \\ U^*(a) &= b_0 e^{-\lambda^* a}\pi(a) \end{aligned} \tag{3.17}$$

   where $\Gamma(\sigma,a) = e^{-\int_\sigma^a \gamma(\eta)\,d\eta}$.

The EE exists if and only if the equation

$$b_0 \int_0^{a_\dagger} h(a)e^{-\lambda^* a}\pi(a)\int_0^a k(\sigma)\Gamma(\sigma,a)e^{-H^* \int_0^\sigma k(\eta)\,d\eta}\,d\sigma da = 1 \tag{3.18}$$

has a positive solution for H^*. Consequently, (3.18) defines a threshold parameter for the existence of an endemic equilibrium of the proportions. We will refer to it as the reproductive number.

$$\mathcal{R}_0 = b_0 \int_0^{a_\dagger} h(a)e^{-\lambda^* a}\pi(a) \int_0^a k(\sigma)\Gamma(\sigma, a)\, d\sigma da. \tag{3.19}$$

In fact, if $\mathcal{R}_0 > 1$ the equation (3.18) has a unique positive solution which we denote by H^* and then use it to determine the EE explicitly from (3.17). If $\mathcal{R}_0 < 1$ the equation (3.18) has no positive solutions and the system only has the DFE. We have the following theorem.

Theorem 3.1. *If $\mathcal{R}_0 < 1$ the nonlinear eigenvalue problem (3.12) has a unique solution (S^0, I^0, U^0) which corresponds to the real eigenvalue λ^*. If $\mathcal{R}_0 > 1$ the nonlinear eigenvalue problem (3.12) has two solutions (S^0, I^0, U^0) and (S^*, I^*, U^*) both corresponding to the unique real eigenvalue λ^*.*

Next, we consider the case of local stability of the solutions of (3.12). Let B_0 be the linear operator corresponding to linearization around the DFE and B be the linear operator corresponding to linearization around the EE, namely B_0 and B are the operators defined in (2.22).

3.3. The spectrum of the operators B_0 and B

First we investigate the spectrum of B_0. To determine the eigenvalues of B_0 we seek nontrivial solutions of the system

$$(B_0 - \lambda)\mathbf{v} = 0 \tag{3.20}$$

where $\mathbf{v} = (s, i, u)^T$. Explicitly, the system (3.20) has the form:

$$\begin{aligned} s' &= -\lambda s - \mu(a)s - \hat{x}k(a)S^0(a) \\ i' &= -\lambda i - \mu(a)i - \gamma(a)i + \hat{x}k(a)S^0(a) \\ u' &= -\lambda u - \mu(a)u \end{aligned} \tag{3.21}$$

with the conditions

$$\begin{aligned} u(0) = s(0) &= \int_0^{a_\dagger} \beta(a)u(a)\, da \\ i(0) &= 0 \\ \hat{x} &= \int_0^{a_\dagger} h(a)i(a)\, da \\ \hat{n} &= \int_0^{a_\dagger} u(a)\, da. \end{aligned} \tag{3.22}$$

Let λ^* be the real solution and α_j, for $j = 1, 2, \dots$ be the complex solutions of the characteristic equation (3.15). We consider two cases:

Case 1: If $\lambda = \lambda^*$ or $\lambda = \alpha_j$ for some j, then the last equation in (3.21) combined with the first boundary condition in (3.22) has a nonzero solution and, therefore, the system (3.21) has at least the nontrivial solution $(u, 0, u)$.

Case 2: Let $\lambda \neq \lambda^*$ and $\lambda \neq \alpha_j$ for $j = 1, 2, \dots$. Then the last equation in (3.21), with the last condition in (3.22), has only the trivial solution. Thus we have $u(0) = 0$, $u(a) = 0$, $\hat{n} = 0$ and, solving the first two equations, we have

$$\begin{aligned} s(a) &= -\hat{x} b_0 \pi(a) \int_0^a e^{-\lambda(a-\sigma)} e^{-\lambda^* \sigma} k(\sigma)\, d\sigma \\ i(a) &= \hat{x} b_0 \pi(a) \int_0^a e^{-\lambda(a-\sigma)} e^{-\lambda^* \sigma} \Gamma(\sigma, a) k(\sigma)\, d\sigma. \end{aligned} \tag{3.23}$$

In order to have a nontrivial solution we need $\hat{x} \neq 0$. From the definition of $\hat{x}$ we obtain the following characteristic equation in λ

$$b_0 \int_0^{a_\dagger} h(a)\pi(a) \int_0^a e^{-\lambda(a-\sigma)} e^{-\lambda^* \sigma} \Gamma(\sigma, a) k(\sigma)\, d\sigma da = 1. \tag{3.24}$$

Any solution of this equation is an eigenvalue of B_0. We note that there is a unique real solution $\tilde{\lambda}$ and a sequence of complex solutions $\tilde{\alpha}_j$ such that $\Re\tilde{\alpha}_j < \tilde{\lambda}$. Moreover, $\tilde{\lambda} > \lambda^*$ (respectively $\tilde{\lambda} < \lambda^*$) if $\mathcal{R}_0 > 1$ (respectively $\mathcal{R}_0 < 1$).

Considering the resolvent of B_0 reveals that B_0 has only eigenvalues in its spectrum. We summarize the above in the following proposition.

Proposition 3.2. *The spectrum of the operator B_0 consists of the eigenvalues*

$$\lambda^*, \alpha_j, \tilde{\lambda}, \tilde{\alpha}_j.$$

Moreover,

1. *If $\mathcal{R}_0 < 1$, then*
$$\Re\alpha_j < \lambda^* \qquad \text{and} \qquad \Re\tilde{\alpha}_j < \tilde{\lambda} < \lambda^*.$$
2. *If $\mathcal{R}_0 > 1$, then*
$$\tilde{\lambda} > \lambda^*.$$

Next, we investigate the spectrum of B. We only have to consider the case $\mathcal{R}_0 > 1$. To determine the eigenvalues of B we seek nontrivial solutions to the system:

$$\begin{aligned} s' &= -\lambda s - \mu(a)s - H^* k(a)s - (\hat{x} - H^*\hat{n})k(a)S^*(a) \\ i' &= -\lambda i - \mu(a)i - \gamma(a)i + k(a)H^* s + (\hat{x} - H^*\hat{n})k(a)S^*(a) \\ u' &= -\lambda u - \mu(a)u \end{aligned} \tag{3.25}$$

with the conditions given in (3.22).

Case 1: If $\lambda = \lambda^*$ or $\lambda = \alpha_j$ for some j, then the last equation in (3.25) combined with the first boundary condition in (3.22) has a nonzero solution and, therefore, the system (3.25) has at least the nontrivial solution $(u, 0, u)$.

Case 2: Let $\lambda \neq \lambda^*$ and $\lambda \neq \alpha_j$ for $j = 1, 2, \ldots$. Then the third equation in (3.25) combined with the first boundary condition in (3.22) has only the trivial solution. Thus we have $u(0) = 0$, $u(a) = 0$, $\hat{n} = 0$. Solving the first two equations, we have

$$s(a) = -\hat{x}b_0\pi(a)e^{-H^*\int_0^a k(\eta)\,d\eta}\int_0^a e^{-\lambda(a-\sigma)}e^{-\lambda^*\sigma}k(\sigma)\,d\sigma$$

$$\begin{aligned} i(a) = \hat{x}b_0\pi(a)&\int_0^a e^{-\lambda(a-\sigma)}e^{-\lambda^*\sigma}\Gamma(\sigma,a)k(\sigma)e^{-H^*\int_0^\sigma k(\eta)\,d\eta}\,d\sigma \\ &- \hat{x}b_0H^*\pi(a)\int_0^a e^{-\lambda(a-\sigma)}\Gamma(\sigma,a)k(\sigma)e^{-H^*\int_0^\sigma k(\eta)\,d\eta} \\ &\times\int_0^\sigma e^{-\lambda(\sigma-\tau)}e^{-\lambda^*\tau}k(\tau)\,d\tau d\sigma. \end{aligned} \tag{3.26}$$

Changing the order of integration in the second integral in i we have

$$\begin{aligned} i(a) = \hat{x}b_0\pi(a)\int_0^a e^{-\lambda(a-\sigma)}e^{-\lambda^*\sigma}k(\sigma)\Big[&\Gamma(\sigma,a)e^{-H^*\int_0^\sigma k(\eta)\,d\eta} \\ &-H^*\int_\sigma^a k(\tau)\Gamma(\tau,a)e^{-H^*\int_0^\tau k(\eta)\,d\eta}\,d\tau\Big]\,d\sigma \end{aligned} \tag{3.27}$$

and, substituting in the definition of $\hat{x}$ we obtain an equation for λ:

$$G(\lambda) = 1 \tag{3.28}$$

where

$$\begin{aligned} G(\lambda) = b_0\int_0^{a_\dagger} h(a)\pi(a)\int_0^a e^{-\lambda(a-\sigma)}e^{-\lambda^*\sigma}k(\sigma)\Big[&\Gamma(\sigma,a)e^{-H^*\int_0^\sigma k(\eta)\,d\eta} \\ &-H^*\int_\sigma^a k(\tau)\Gamma(\tau,a)e^{-H^*\int_0^\tau k(\eta)\,d\eta}\,d\tau\Big]\,d\sigma da. \end{aligned} \tag{3.29}$$

This equation provides the eigenvalues of B but it is not easy to analyze their location because the expression in parentheses is not in general defined in sign. Actually this is a typical occurrence in the age structured epidemic SIR model and is related to the possibility of Hopf type bifurcation. As a matter of fact, if we assume the following condition similar to the one considered in [18]:

$$\frac{I^*(a_\dagger)}{U^*(a_\dagger)} \leq e^{-\int_0^{a_\dagger}\gamma(\eta)\,d\eta} \tag{3.30}$$

we see that in (3.29) the expression in the parenthesis is nonnegative and therefore, $G(\lambda)$ is a decreasing function of λ for real λ. Moreover,

$$G(\lambda^*) < b_0\int_0^{a_\dagger} h(a)\pi(a)e^{-\lambda^* a}\int_0^a k(\sigma)\Gamma(\sigma,a)e^{-H^*\int_0^\sigma k(\eta)\,d\eta}d\sigma da = 1$$

where we have used (3.18). Hence (3.28) has a unique real solution $\hat{\lambda}$ and complex solutions $\hat{\alpha}_j$ for $j = 1, 2, \ldots$ such that

$$\Re\hat{\alpha}_j < \hat{\lambda} < \lambda^*.$$

Considering the resolvent of B shows that B has only eigenvalues in its spectrum and we summarize the above in the following proposition:

Proposition 3.3. *The spectrum of the operator B consists of the eigenvalues*

$$\lambda^*, \alpha_j, \hat{\lambda}, \hat{\alpha}_j.$$

Moreover, if (3.30) is satisfied then

$$\Re\alpha_j < \lambda^* \qquad \text{and} \qquad \Re\hat{\alpha}_j < \hat{\lambda} < \lambda^*. \qquad \square$$

As we have already noticed, this result depends upon the sufficient condition (3.30) that in some sense fixes an upper bound to the endemic state in order to have stability (as we are going to state in the subsequent section). Consequently this condition provides some range for the parameter, outside of which we may have instability and sustained obscillations of the solution.

3.4. Local stability of the disease-free equilibrium and the endemic equilibrium

The results of the previous subsections can be applied to the analysis of the local stability of DFE and EE through Theorem 2.3 and Corollary 2.4. In particular, Propositions 3.2 and 3.3 give the spectrum of the operators B_0 and B and, consequently, of the operators S_0 and S. The final step is to establish that the semigroups generated by the operators $S_{0\varphi}$ and S_φ are eventually compact.

We show that the semigroup generated by S_φ is eventually compact. The argument for the semigroup generated by the operator $S_{0\varphi}$ is a special case.

First we note that from (2.26) we have $S_\varphi = A + \mathcal{F}'(w^*) - \langle\varphi, B\rangle w^*$ where w^* denotes the EE. For the explicit form of $A + \mathcal{F}'(w^*)$ see equations (3.25). We rewrite the operator S_φ in the form $S_\varphi = \mathcal{A} + T$ where $\mathcal{A}$ is the following modification of the operator A:

$$\mathcal{A} = \begin{pmatrix} -\frac{\partial}{\partial a} - \mu(a) - H^*k(a) & 0 & 0 \\ H^*k(a) & -\frac{\partial}{\partial a} - \mu(a) - \gamma(a) & 0 \\ 0 & 0 & -\frac{\partial}{\partial a} - \mu(a) \end{pmatrix}$$

and has the same domain as A. At the DFE when $H^* = 0$ the operator $\mathcal{A}$ coincides with the operator A. To define the operator T let $v = (s, i, u)^T$.

$$(Tv)(a) = \begin{pmatrix} -(\hat{x} - H^*\hat{n})k(a)S^*(a) - \langle\varphi, Bv\rangle S^*(a) \\ (\hat{x} - H^*\hat{n})k(a)S^*(a) - \langle\varphi, Bv\rangle I^*(a) \\ -\langle\varphi, Bv\rangle U^*(a) \end{pmatrix} \tag{3.31}$$

We note that the operator T is compact so that to prove that S_φ is eventually compact it is enough to prove that $\mathcal{A}$ generates an eventually compact semigroup.

To this aim let $w^0 = (f_1, f_2, f_3)^T$. Then for $t > 2a_\dagger$ the semigroup generated by $\mathcal{A}$ is given explicitly by

$$(e^{\mathcal{A}t}w^0)(a) = \begin{pmatrix} \mathcal{B}(t-a)\pi(a)e^{-H^*\int_0^a k(\eta)d\eta} \\ H^*\mathcal{B}(t-a)\pi(a)\displaystyle\int_0^a k(\sigma)\Gamma(\sigma, a)e^{-H^*\int_0^\sigma k(\eta)d\eta}d\sigma \\ \mathcal{B}(t-a)\pi(a) \end{pmatrix} \tag{3.32}$$

where $\mathcal{B}(t)$ is referred to as the *total birth rate* and is the solution of the renewal equation:

$$\mathcal{B}(t) = F(t) + \int_0^t \mathcal{K}(s)\mathcal{B}(t-s)\,ds \quad t \geq 0 \tag{3.33}$$

where

$$F(t) = \int_0^{a_\dagger} \mathcal{K}(s+t)\frac{f_3(s)}{\pi(s)}\,ds.$$

and the kernel $\mathcal{K}(a) = \beta(a)\pi(a)$. We note that $\mathcal{K}(a) = 0$ for $a \geq a_\dagger$. Furthermore, $F(t) = 0$ for $t \geq a_\dagger$ and

$$|F(t)| \leq \bar{\beta}\|f_3\|_{L^1_\pi}$$

where $\bar{\beta} = \|\beta\|_{L^\infty}$.

Consequently, from (3.33), by the Gronwall inequality we see that the total birth rate obeys the following growth bound

$$|\mathcal{B}(t)| \leq \bar{\beta}e^{\bar{\beta}t}\|f_3\|_{L^1_\pi}.$$

Furthermore, for $t \geq 0$ and $h > 0$ small enough we have

$$\begin{aligned}\mathcal{B}(t+h) - \mathcal{B}(t) = F(t+h) - F(t) &+ \textstyle\int_t^{t+h} \mathcal{K}(s)\mathcal{B}(t+h-s)\,ds \\ &+ \textstyle\int_0^t \mathcal{K}(s)[\mathcal{B}(t+h-s) - \mathcal{B}(t-s)]\,ds.\end{aligned}$$

Denote by $\nu(t) = |F(t+h) - F(t)|$. We note that $\nu(t) = 0$ for $t \geq a_\dagger$. With this notation we have

$$|\mathcal{B}(t+h) - \mathcal{B}(t)| \leq \nu(t) + Ch\|f_3\|_{L^1_\pi} + \bar{\beta}\int_0^t |\mathcal{B}(s+h) - \mathcal{B}(s)|\,ds$$

where C is a constant which depends on $\bar{\beta}$ but does not depend on t and h. Applying Gronwall's inequality we have

$$|\mathcal{B}(t+h) - \mathcal{B}(t)| \leq \nu(t) + Che^{\bar{\beta}t}\|f_3\|_{L^1_\pi} + \bar{\beta}\int_0^t e^{\bar{\beta}(t-s)}\nu(s)\,ds.$$

For $t \geq a_\dagger$ we have $\nu(t) = 0$ and the above inequality becomes

$$|\mathcal{B}(t+h) - \mathcal{B}(t)| \leq Che^{\bar{\beta}t}\|f_3\|_{L^1_\pi} + \bar{\beta}\int_0^t e^{\bar{\beta}(t-s)}\nu(s)\,ds. \tag{3.34}$$

On the other hand, the integral on the right hand side of the above inequality can be estimated as follows:

$$\begin{aligned}\int_0^t e^{\bar{\beta}(t-s)}\nu(s)\,ds &\leq e^{\bar{\beta}t}\int_0^{a_\dagger}\int_0^{a_\dagger} |\mathcal{K}(\sigma+s+h) - \mathcal{K}(\sigma+s)|\frac{|f_3(\sigma)|}{\pi(\sigma)}\,d\sigma ds \\ &\leq e^{\bar{\beta}t}\int_0^{a_\dagger}\int_\sigma^{a_\dagger} |\mathcal{K}(s+h) - \mathcal{K}(s)|\frac{|f_3(\sigma)|}{\pi(\sigma)}\,dsd\sigma \\ &\leq \mathcal{O}(h)e^{\bar{\beta}t}\|f_3\|_{L^1_\pi}\end{aligned} \tag{3.35}$$

where

$$\mathcal{O}(h) = \int_0^{a_\dagger} |\mathcal{K}(s+h) - \mathcal{K}(s)|\,ds \to 0 \qquad \text{as } h \to 0$$

independently of f_3. Consequently, by (3.34), for $t > a_\dagger$ we have

$$|\mathcal{B}(t+h) - \mathcal{B}(t)| \leq \mathcal{O}_1(h) e^{\bar{\beta} t} \|f_3\|_{L^1_\pi}.$$

with $\mathcal{O}_1(h) = \bar{\beta}\mathcal{O}(h) + Ch$. Thus, for $t > 2a_\dagger$

$$\int_0^{a_\dagger} |\mathcal{B}(t+h-a) - \mathcal{B}(t-a)|\, da \leq \mathcal{O}_2(h) \|w^0\|_X \tag{3.36}$$

where $\mathcal{O}_2(h) = a_\dagger \mathcal{O}_1(h) e^{\bar{\beta} t}$.

Now we are ready to show that $e^{\mathcal{A}t}$ is compact for $t > 2a_\dagger$. To see this we need to establish that the vector-column of the family of functions obtained for various values of w^0 (see (3.32) is compact in X. Thus, it is sufficient to see that the quotients of all functions with $\pi(a)$ form a compact family in $\mathcal{X}$. We apply the Fréchet-Kolmogorov criterion for compactness in L^1 (see, e.g., [30], p.275). Consider a set of initial values which is bounded in X. For $h > 0$ small enough, using (3.36), the first component satifies (see (3.32):

$$\int_0^{a_\dagger} |\mathcal{B}(t-a) e^{-H^* \int_0^a k(\eta)d\eta} - \mathcal{B}(t-a+h) e^{-H^* \int_0^{a-h} k(\eta)d\eta}|da \tag{3.37}$$

$$\leq \int_0^{a_\dagger} |\mathcal{B}(t-a) - \mathcal{B}(t-a+h)|da + H^* \bar{k} h \int_0^{a_\dagger} \mathcal{B}(t-a)da \leq \mathcal{O}_3(h) \|w^0\|_X$$

where $\mathcal{O}_3(h)$ goes to zero as h goes to zero and does not depend on w^0. Similar estimates hold for the second and third component, so that we can apply the criterion proving that the semigroup generated by $\mathcal{A}$ is compact for $t > 2a_\dagger$.

As a consequence of the previous considerations we have the following theorem.

Theorem 3.4. 1. *Let $\mathcal{R}_0 < 1$. Then the disease-free steady state (3.16) is locally asymptotically stable.*

2. *Let $\mathcal{R}_0 > 1$. Then the disease-free steady state (3.16) is unstable and, if condition (3.30) is satisfied, then the EE is locally asymptotically stable.*

4. Discussion

In this paper we give a rigorous mathematical justification to a simple criterion for local stability of persistent solutions of homogeneous equations which is similar to the one known for steady states of nonlinear equations. In particular, we show that if all eigenvalues of the linearized operator have real part smaller than the growth rate of the population the corresponding exponential solution is locally stable. Thus we extend the result, known to hold for ODEs, to infinite-dimensional systems.

Age-structured epidemiological models are widely discussed in the literature (see for instance [18, 29]). The majority of the age-structured models analyzed so far assume that the population is at steady state, that is the growth rate is zero or, equivalently, the reproductive number of the population is exactly one. This assumption excludes the demographic processes and is typically adequate for modeling diseases with short span like most of the childhood diseases and influenza.

For diseases with lasting presence in the population and relatively long span on individual level, like HIV and Hepatitis C, ignoring the demographic processes in the population is not realistic. The SIR model discussed here considers both demography and epidemics in an age-structured population. Understanding this relatively simple model brings us closer to investigating more realistic models of diseases like HIV and Hepatitis C, involving age and sex structure and interacting demography and epidemics. Age-structured models incorporating demographics have been considered and partially analyzed in [21, 22, 13, 11, 12]. Consequently, these models are homogeneous.

The main purpose of epidemiological models is to extract and provide formal conditions for the eradication of the disease. In populations with constant or asymptotically constant size eradicating the disease means creating conditions for the total number of infecteds to decline to zero or, equivalently, the proportion of infecteds to decline to zero. However, in exponentially growing populations the situation is more complex since the proportion of infecteds may be declining to zero while the number of infecteds is exponentially growing. Hence, in varying populations both the proportions and the densities of the population subclasses, and in particular those responsible for the transmission of the disease, should be monitored.

Infinite-dimensional homogeneous systems have been investigated before in [26]. Webb considers a direct approach, that is, approach which treats the homogeneous system only, without comparing it with the normalized system. This treatment may be more adequate for analyzing systems where normalizing with a functional is not required by the problem. Our framework is more advantageous for the case we are interested in, namely, the local stability of both the original and the projected system. From Theorem 1 in [26], the condition that a rank one projection exists, satisfying certain additional properties, is the hardest one to check. Technically we need a similar condition and we have constructed such a projection when the growth rate can be uniquely determined and there is a single eigenvector corresponding to it. The questions what happens when more than one eigenvectors correspond to a given eigenvalue or when the nonlinear characteristic equation has more than one real solutions remain open. The latter is of interest since this might be the case with the age-structured pair-formation models [19].

We believe that the approach of investigating systems through normalization can be useful in some nonlinear, nonhomogeneous systems. In particular, we have in mind the analysis in [1] where the age profile, that is the proportion of the density of individuals of a given age in the total population, has been used to investigate a version of the Gurtin and MacCamy's nonlinear population model with separable death rate.

Acknowledgment. This work in its larger part was performed while the authors were visiting Purdue University. The authors thank the faculty and staff of the Department of Mathematics for their kind hospitality. The authors also thank Carlos Castillo-Chavez for his comments on the introduction.

References

[1] S. Busenberg and M. Iannelli, *Separable models in age-dependent population dynamics.* J. Math. Biol. **22** (1985), 145–173.

[2] M.E. Gurtin and R.C. MacCamy, *Nonlinear age-dependent population dynamics.* Archive for Rational Mechanics and Analysis **54** (1974), 281–300.

[3] M.E. Gurtin and R.C. MacCamy, *Some simple models for nonlinear age-dependent population dynamics.* Math. Biosci. **43** (1979), 199–211.

[4] S. Busenberg, C. Castillo-Chavez, *A General solution of the problem of mixing of subpopulations and its application to Risk and age-structured epidemic models for the spread of AIDS.* IMA J. Math. Appl. Med. Biol. **8** (1991), 1–29.

[5] Ph. Clément, H. Heijmans, S. Angenent, C. van Duijn, B. de Pagter, *One parameter semigroups.* North-Holland, Amsterdam, New York, Oxford, Tokio, 1987.

[6] W. Feller, *On the integral equation of renewal theory.* Ann. Math. Stat. **12** (1941), 243–267.

[7] A. Fredrickson, *A Mathematical theory of age structure in sexual populations: random mating and monogamous marriage models.* Math. Biosci. **10** (1971), 117–143.

[8] K. Hadeler, R. Waldstätter. A. Wörz-Busekros, *Models for pair formation in bisexual populations.* J. Math. Biol. **26** (1988), 635–649.

[9] K. Hadeler, *Pair formation with maturation period.* J. Math. Biol. **32** (1993), 1–15.

[10] K. Hadeler, *Periodic solutions of homogeneous equations.* J. Diff. Eq. **95** (1992), 183–202.

[11] K. Hadeler, J. Müller, *Vaccination in age structured population I: The reproduction number.* Epidemic models: Their structure and relation to data (V.Isham and G.Medley, eds.), Cambridge University Press, Cambridge, UK, 1996.

[12] K. Hadeler, J. Müller, *Vaccination in age structured population II: Optimal vaccination strategies.* Epidemic models: Their structure and relation to data (V.Isham and G.Medley, eds.), Cambridge University Press, Cambridge, UK, 1996.

[13] H. Hethcote, *The Mathematics of infectious diseases.* SIAM Rev. **42** no. 4 (2000), 599–653.

[14] F. Hoppensteadt, *Mathematical Theories of Populations: Demographics, Genetics and Epidemics.* Society for Industrial and Applied Mathematics, Philadelphia, 1975.

[15] M. Iannelli, *Mathematical Theory of Age-Structured Population Dynamics.* Applied Mathematics Monographs, Comitato Nazionale per le Scienze Matematiche, Consiglio Nazionale delle Ricerche (C.N.R.), Vol. 7, Giardini, Pisa, 1995.

[16] M. Iannelli, M.-Y. Kim, E.-J. Park, *Asymptotic behavior for an S-I-S epidemic model and its approximation.* Nonlin.Anal. **35** (1999), 797–814.

[17] M. Iannelli, M. Martcheva, *A Semigroup approach to the wellposedness of the two-sex population model.* Dynam. Sys. Appl. **5** (1997), 353–370.

[18] H. Inaba, *Threshold and stability results for an age- structured epidemic model.* J. Math. Biol. **28** (1990), 411–434.

[19] M. Martcheva, *Exponential growth in age-structured two-sex populations.* Math. Biosci. **157** (1999), 1–22.

[20] M. Martcheva, F. Milner, *A two-sex age-structured populations model: wellposedness.* Math. Popul. Studies **7(2)** (1999), 111–129.

[21] J. Müller, *Optimal vaccination patterns in age-structured populations.* SIAM J. Appl. Math. **59(1)** (1998), 222–241.

[22] J. Müller, *Optimal vaccination patterns in age-structured populations: endemic case.* Proceedings of the Conference on Dynamical Systems in Biology and Medicine (Veszprém, 1997). Math. Comput. Modelling **31** no. 4-5 (2000), 149–160.

[23] J. Prüss, *Equilibrium solutions of age specific population dynamics of several species.* J. Math. Biol. **11** (1981), 65–84.

[24] J. Prüss, *On the qualitative behaviour of populations with age-specific interactions.* Comput. Math. Appl. **9** (1983), 327–339.

[25] J. Prüss, W. Schappacher, *Persistent age-distributions for a pair-formation model.* J. Math. Biol. **33** (1994), 17–33.

[26] G. Webb, *Asynchronous exponential growth in differential equations with homogeneous nonlinearities.* In: Differential equations models in Banach spaces, Bologna, 1991 (Lecture Notes in Pure and Applied Mathematics, 148), Dekker, New York, 1993.

[27] G. Webb, *Theory of Nonlinear Age-Dependent Population Dynamics.* Marcel Dekker, New York, 1985.

[28] A. Taylor, *Introduction to functional analysis.* John Wiley & Sons, New York, 1958.

[29] H. Thieme, *Stability change of the endemic equilibrium in age-structured models for the spread of $S-I-R$ type infectious diseases.* In: Differential equations models in Biology, Epidemiology and Ecology (Lecture notes in Biomathematics 92), Springer-Verlag, New York, 1991.

[30] K. Yosida, *Functional Analysis.* (second edition) Springer-Verlag, Berlin, Heidelberg, New York, 1968.

Mimmo Iannelli
Dipartimento di Matematica
Università di Trento
I-38050 Povo (Trento), Italy
phone: 39-0461-88-16-57
fax: 39-0461-88-16-24
E-mail address: `iannelli@alpha.science.unitn.it`

Maia Martcheva
Department of Mathematics
Polytechnic University
Brooklyn, NY 11201, USA
phone: 718-260-3294
fax: 718-260-3660
E-mail address: `mayam@magnus.poly.edu`

Progress in Nonlinear Differential Equations
and Their Applications, Vol. 55, 253–273

Asymptotic Behaviour of Solutions of Nonlinear Volterra Equations

Volker G. Jakubowski and Wolfgang M. Ruess

1. Introduction

We study the asymptotic behaviour of generalized solutions u in the sense of Gripenberg [Gri85, CGL96] of the nonlinear Volterra equation

$$\frac{d}{dt}\left(\kappa(u(t)-u_0)+\int_0^t k(t-s)(u(s)-u_0)\,ds\right)+Au(t)\ni f(t), \qquad t\in[0,\infty). \tag{1}$$

Here, $A : X \to 2^X$ is an m-accretive operator in a real Banach space X, $u_0 \in \overline{D(A)}$, and $f \in L^1_{\mathrm{loc}}(\mathbb{R}_+; X)$ with $\mathbb{R}_+ := [0,\infty)$. Moreover, we assume that

$$\begin{aligned}&\kappa \geq 0, \text{ and } k : (0,\infty) \to \mathbb{R} \text{ is nonnegative, nonincreasing such that}\\ &k \in L^1_{\mathrm{loc}}(\mathbb{R}_+).\end{aligned} \tag{2}$$

Note that equations of the form (1) include as special cases both the common nonlinear Cauchy problem (initial-value problem without history): $k \equiv 0$, and the one for fractional derivatives: $\kappa = 0$, $k(t) = \dfrac{t^{-\gamma}}{\Gamma(1-\gamma)}$, with $\gamma \in (0,1)$. In applications, they particularly arise when modeling nonlinear heat flow in materials with memory (see, e.g., [CN81]), as well as in modeling diffusion of fluids in porous media, see Section 5.

Qualitative properties of solutions to (1) have already been considered in [CN79] and [CM88], where the main purpose was to obtain positive solutions. Moreover, for regularity properties of generalized solutions of (1), we refer to [JWb], where existence of strong solutions is shown for m-completely accretive operators A in Banach spaces $X \subset L^1(\Omega)$ without the Radon-Nikodym property.

In this paper, we investigate the inheritance of asymptotic properties by the solution u of (1) from the 'forcing' function f. In this respect, we concentrate on asymptotic almost periodicity and weak almost periodicity in the sense of Eberlein, and give conditions under which these properties go over from f to u.

Our results extend those of [Kre92] and [Kre96] on mild solutions of nonlinear Cauchy problems, as well as those of [CN81] on nonlinear Volterra equations. See also [ALR90, BC81, Hir83] for related questions, and [PR93] for the linear case.

In Section 2, we set the context – with regard to the pair (κ, k) – for our investigation of equation (1), and recall some basic facts about generalized solutions. Most importantly, we derive integral inequalities for these solutions (Propositions 3 and 5) that are the basis for the results on their asymptotic behaviour in the subsequent section.

Section 3 is devoted to the main results of this paper on the asymptotic behaviour of generalized solutions to the abstract Volterra equation (1). We show that asymptotic almost periodicity is inherited by the solution u from the forcing function f (Theorem 8), as well as Eberlein-weak almost periodicity under various circumstances (Theorems 9 and 10). Applications to ergodic theorems for generalized solutions will be included.

In Section 4, we construct a solution of a certain limit equation on the whole real line characterizing the almost periodic part of an asymptotically almost periodic, respectively Eberlein-weak almost periodic solution of (1).

In Section 5, we close the paper by a concrete example for (1) that arises from modeling fluids in porous media.

Notation: As usual, given a Banach space X, and letting $J \in \{\mathbb{R}_+, \mathbb{R}\}$, we shall denote the spaces of continuous, of bounded and continuous, and of bounded and uniformly continuous functions from J into X by $C(J;X)$, $C_b(J;X)$, and $BUC(J;X)$, respectively; the latter two equipped with the sup-norm. Also, $C_0(\mathbb{R}_+;X)$ will denote the subspace of $BUC(\mathbb{R}_+;X)$ of functions vanishing at infinity.

Furthermore, we define the bracket

$$[x,y]_+ = \lim_{\lambda \to 0+} \frac{\|x + \lambda y\| - \|x\|}{\lambda}$$

as the right-hand Gateaux derivative of the norm $\|\cdot\|$ of X at x in direction y, and recall that

$$[x,y]_+ = \sup_{x^* \in \mathbf{J}(x)} \langle y, x^* \rangle,$$

where the duality map $\mathbf{J} : X \to 2^{X^*}$ is defined by

$$\mathbf{J}(x) := \left\{ x^* \in X^* \middle| \, \langle x, x^* \rangle = \|x\| \text{ and } \|x^*\| \le 1 \right\}.$$

Finally, we recall that an operator $A \subset X \times X$ is called *accretive* if $[x-\tilde{x}, y-\tilde{y}]_+ \ge 0$ for all $(x,y), (\tilde{x},\tilde{y}) \in A$, and that an accretive operator A is called m-*accretive* if $R(A + \lambda I) = X$ for all $\lambda > 0$.

2. Integral inequalities for generalized solutions

We start by specifying the conditions on the pair (κ, k) for the results to follow. As both asymptotically almost periodic and Eberlein-weakly almost periodic functions are bounded and uniformly continuous, we shall restrict our study to pairs (κ, k) that guarantee boundedness and continuity of generalized solutions u to (1) for any bounded and continuous forcing function f.

It is shown in [Gri85] that for pairs (κ, k) satisfying (2) and

$$\kappa > 0 \quad \text{or} \quad k(0+) = \infty, \tag{3}$$

and for $u_0 \in \overline{D(A)}$, $f \in C(\mathbb{R}_+; X)$ implies $u \in C(\mathbb{R}_+; X)$. Next, we recall the following well-known facts, see, e.g., [CP90].

Lemma 1. *Let $\kappa \geq 0$, k satisfy* (2). *Then there exists a unique nonnegative Radon measure α on $\mathbb{R}_+$ such that*

$$\kappa\alpha([0,t]) + \int_0^t k(t-s)\alpha([0,s])\,ds = t \qquad \textit{for all} \qquad t > 0. \tag{4}$$

The measure α is called the resolvent of the first kind of the pair (κ, k). Moreover,

$$\alpha(\mathbb{R}_+) < \infty \Leftrightarrow k(\infty) := \lim_{t\to\infty} k(t) > 0. \tag{5}$$

In this case $\alpha(\mathbb{R}_+) = k(\infty)^{-1}$.

Assuming (2),(3), it is well known (c.f. [Clé80]) that for two generalized solutions u_i, $i \in \{1, 2\}$, of (1) for data $u_0^{(i)} \in \overline{D(A)}$ and $f_i \in C(\mathbb{R}_+; X)$ one obtains for all $t \geq 0$

$$\|u_1(t) - u_2(t)\| \leq \|u_0^{(1)} - u_0^{(2)}\| + \int_0^t \|f_1(t-s) - f_2(t-s)\|\,d\alpha(s). \tag{6}$$

As, for any $x \in D(A)$, the function $u(t) \equiv x$ is a generalized solution of (1) for initial value $u_0 = x$, and $f(t) \equiv y \in Ax$, the above inequality implies boundedness of generalized solutions for bounded measures α. Thus, in order to guarantee boundedness of solutions for bounded right hand sides, we will always assume that

$$k(\infty) := \lim_{t\to\infty} k(t) > 0. \tag{7}$$

As an aside, we note that assuming (7) one can rewrite (1) as

$$\frac{d}{dt}\left(\kappa(u(t) - u_0) + \int_0^t \tilde{k}(t-s)(u(s) - u_0)\,ds\right) + (A + \omega I)u(t) \ni f(t) + \omega u_0, \quad t \geq 0. \tag{8}$$

Here, $\omega := k(\infty) > 0$, and $\tilde{k}(t) := k(t) - k(\infty)$. Thus, assumption (7) is equivalent to assuming that $A - \omega I$ is accretive for some $\omega > 0$.

From this point on, we shall always assume that (κ, k) satisfy (2), (3) and (7). Moreover, u will always denote the generalized solution of the Volterra equation (1). We recall that u is the $L^1_{\text{loc}}(\mathbb{R}_+; X)$-limit of a sequence of approximate solutions $\{u_n\}_{n\in\mathbb{N}}$. For an m-accretive operator A, the function u_n is the unique strong solution of

$$\frac{d}{dt}\left(\int_0^t k_n(t-s)(u_n(s) - u_0)\,ds\right) + Au_n(t) \ni f(t), \tag{9}$$

where $\{k_n\}_{n\in\mathbb{N}}$ is an arbitrary sequence of functions satisfying (2) and $k_n(0+) < \infty$ such that

$$\int_0^t k_n(s)\,ds \to \kappa + \int_0^t k(s)\,ds \qquad \text{for all } t > 0 \text{ as } n \to \infty. \tag{10}$$

As the generalized solution does not depend on the choice of the sequence $\{u_n\}$ of approximate solutions with (9) and (10), one might choose a particular sequence $\{u_{\lambda_n}\}_{n\in\mathbb{N}}$ of approximate solutions in order to simplify calculations. Therefore, let α always denote the resolvent of the first kind of the pair (κ, k), given by (4). For $\lambda > 0$, the resolvent ϱ_λ of the measure α is defined as the unique Radon measure on $[0,\infty)$ satisfying

$$\lambda\varrho_\lambda([0,t]) + \int_{[0,t]} \varrho_\lambda([0,t-s])\,d\alpha(s) = \alpha([0,t]) \qquad \text{for all } t \geq 0. \tag{11}$$

Thus, we can define

$$k_\lambda(t) := \frac{1}{\lambda}\left(1 - \varrho_\lambda([0,t])\right) \qquad \text{for all } t > 0. \tag{12}$$

Due to the results of [CN81, Proposition 2.1] and [Gri85, Lemma 3.1], we obtain the following lemma.

Lemma 2. *Let κ, k satisfy (2) and (3). Let α be the resolvent of the first kind of (κ, k). Then, for all $\lambda > 0$, the function k_λ defined by (12) is nonnegative nonincreasing, with $k_\lambda(0+) = \frac{1}{\lambda}$, and*

$$\int_0^t k_\lambda(s)\,ds \to \kappa + \int_0^t k(s)\,ds \qquad \textit{for all } t > 0 \textit{ as } \lambda \to 0+.$$

The resolvent of the first kind α_λ of the function k_λ defined by

$$\int_{[0,t]} k_\lambda(t-s)\,d\alpha_\lambda(s) = 1, \qquad \textit{for all } t \geq 0$$

is given by $\alpha_\lambda = \lambda\delta_0 + \alpha$, where δ_0 is the Dirac measure at the point 0.

Moreover, if k satisfies (7), then

$$k_\lambda(\infty) := \lim_{t\to\infty} k_\lambda(t) = \frac{1}{\lambda + \alpha([0,\infty))}.$$

According to the above lemma, we will now choose sequences $\{u_{\lambda_n}\}_{n\in\mathbb{N}}$ of approximate solutions corresponding to the functions k_{λ_n} for $\lambda_n \to 0+$ as $n \to \infty$. Since our results will not depend on the choice of the particular sequence $\lambda_n \to 0+$, we will in the following statements always speak of the sequence $\lambda \to 0+$, meaning that the particular statement holds for all choices of sequences $\{\lambda_n\}_{n\in\mathbb{N}}$ with $\lambda_n \to 0+$ as $n \to \infty$.

We now state the fundamental integral inequalities for the study of the asymptotic behaviour of generalized solutions.

Proposition 3. *Let κ, k satisfy (2), (3) and (7), and let A be an m-accretive operator in a Banach space X. Let $u_0 \in \overline{D(A)}$, and $f, g \in L^\infty(\mathbb{R}_+; X)$. Moreover, let u be the generalized solution of (1), and let v be the generalized solution of (1) with f replaced by g. Then there exists a constant $C > 0$ such that*

$$\begin{aligned}
&\|u(t+\tau) - v(t+\sigma)\| \\
&\quad \le \int_{[0,t]} [u(t+\tau-s) - v(t+\sigma-s), f(t+\tau-s) - g(t+\sigma-s)]_+ \, d\alpha(s) \\
&\qquad + C \frac{\alpha((t,\infty))}{\alpha([0,\infty))}
\end{aligned}$$

for $t, \tau, \sigma \in \mathbb{R}_+$.

Proof. For all $\lambda > 0$, let $u_\lambda \in L^1_{\text{loc}}(\mathbb{R}_+, X)$ be the unique strong solution of

$$\frac{d}{dt}\left(\int_0^t k_\lambda(t-s)(u_\lambda(s) - u_0)\, ds\right) + Au_\lambda(t) \ni f(t) \qquad \text{for } t \ge 0. \tag{13}$$

Moreover, let $v_\lambda \in L^1_{\text{loc}}(\mathbb{R}_+; X)$ be the unique strong solution of (13) with f replaced by g. Then, by inequality (6), and by the fact that $\alpha_\lambda(\mathbb{R}_+) = \lambda + \alpha(\mathbb{R}_+) < \infty$, there exists a constant $C > 0$ such that for all $0 < \lambda < 1$

$$\|u_\lambda - u_0\|_\infty + \|v_\lambda - u_0\|_\infty \le C. \tag{14}$$

Since, for $\lambda > 0$, u_λ and v_λ are strong solutions, by accretivity of A, we obtain almost everywhere for $t \in \mathbb{R}_+$, and for $0 < \sigma < \tau$

$$\begin{aligned}
0 &\le \big[u_\lambda(t+\tau) - v_\lambda(t+\sigma), f(t+\tau) - g(t+\sigma) - k_\lambda(0+)(u_\lambda(t+\tau) - v_\lambda(t+\sigma)) \\
&\quad - \int_{(0,t+\tau]} (u_\lambda(t+\tau-s) - u_0)\, dk_\lambda(s) + \int_{(0,t+\sigma]} (v_\lambda(t+\sigma-s) - u_0)\, dk_\lambda(s)\big]_+ \\
&\le \big[u_\lambda(t+\tau) - v_\lambda(t+\sigma), f(t+\tau) - g(t+\sigma)\big]_+ \\
&\quad + \big[u_\lambda(t+\tau) - v_\lambda(t+\sigma), -k_\lambda(t+\tau)(u_\lambda(t+\tau) - v_\lambda(t+\sigma))\big]_+ \\
&\quad + \big[u_\lambda(t+\tau) - v_\lambda(t+\sigma), \\
&\quad \int_{(0,t+\sigma]} \{u_\lambda(t+\tau) - v_\lambda(t+\sigma) - (u_\lambda(t+\tau-s) - v_\lambda(t+\sigma-s))\}\, dk_\lambda(s)\big]_+ \\
&\quad + \big[u_\lambda(t+\tau) - v_\lambda(t+\sigma), \\
&\quad \int_{(t+\sigma,t+\tau]} \{u_\lambda(t+\tau) - v_\lambda(t+\sigma) - (u_\lambda(t+\tau-s) - u_0)\}\, dk_\lambda(s)\big]_+ \\
&\le \big[u_\lambda(t+\tau) - v_\lambda(t+\sigma), f(t+\tau) - g(t+\sigma)\big]_+ \\
&\quad - k_\lambda(0+)\|u_\lambda(t+\tau) - v_\lambda(t+\sigma)\| \\
&\quad - \int_{(0,t]} \|u_\lambda(t+\tau-s) - v_\lambda(t+\sigma-s)\|\, dk_\lambda(s) + C(k_\lambda(t) - k_\lambda(\infty)).
\end{aligned}$$

Here, we used the properties of the bracket $[\cdot,\cdot]_+$, in particular the continuity in the second variable, and the fact that $-dk$ is a nonnegative Radon measure on

$(0,\infty)$. Note that, since k_λ is nonincreasing, we estimated $k_\lambda(t+\tau)$ and $k_\lambda(t+\sigma)$ by $k_\lambda(\infty)$. Since the resolvent of the first kind $\alpha_\lambda = \lambda\delta_0 + \alpha$ of k_λ is a nonnegative Radon measure on $[0,\infty)$, the convolution of the above inequality with α_λ yields

$$\begin{aligned} &\|u_\lambda(t+\tau) - v_\lambda(t+\sigma)\| \\ &\quad \le \int_{[0,t]} \big[u_\lambda(t+\tau-s) - v_\lambda(t+\sigma-s), f(t+\tau-s) \\ &\quad -g(t+\sigma-s)\big]_+ \, d\alpha_\lambda(s) + C\frac{\alpha((t,\infty))}{\lambda+\alpha(\mathbb{R}_+)} \end{aligned} \tag{15}$$

almost everywhere for $t \in \mathbb{R}_+$, and for $0 < \sigma < \tau$. Here, we used the fact that, by Lemma 2,

$$\int_{[0,t]} (k_\lambda(t-s) - k_\lambda(\infty))\, d\alpha_\lambda(s) = 1 - \frac{\lambda + \alpha([0,t])}{\lambda + \alpha(\mathbb{R}_+)} = \frac{\alpha((t,\infty))}{\lambda+\alpha(\mathbb{R}_+)} \quad \text{for all } t > 0.$$

By the existence result for generalized solutions (see [CGL96, Theorem 1] or [Gri85, Theorem 1]), we know that the sequence $\{u_\lambda\}_{\lambda>0}$ converges in $L^1_{\mathrm{loc}}(\mathbb{R}_+;X)$ to the generalized solution u of (1) as $\lambda \to 0+$, and $v_\lambda \to v$ in $L^1_{\mathrm{loc}}(\mathbb{R}_+;X)$ as $\lambda \to 0+$. Therefore, we can assume that $u_\lambda(t) \to u(t)$ and $v_\lambda(t) \to v(t)$ almost everywhere for $t \in \mathbb{R}_+$ as $\lambda \to 0+$. Thus, we may take the limit in equation (15) as $\lambda \to 0+$. By continuity of u, v, f and g, and by upper-semicontinuity of the bracket, this gives the assertion, as the roles of τ and σ can be interchanged. □

As a direct consequence of the preceding proposition, we note that uniform continuity is inherited by the solution u from the forcing function f.

Corollary 4. *Let A be an m-accretive operator in a Banach space X, let (2), (3) and (7) be satisfied, and let $u_0 \in \overline{D(A)}$, $f \in BUC(\mathbb{R}_+;X)$. Then the generalized solution u of (1) is uniformly continuous.*

For the proof of Theorem 10 in the subsequent section, it will be convenient to note a version of the integral inequality of Proposition 3 for the right-hand Gateaux derivative of $\frac{1}{2}\|\cdot\|^2$:

$$\langle x, y\rangle_+ := \lim_{\lambda\to 0+} \frac{\|x+\lambda y\|^2 - \|x\|^2}{2\lambda} = \|x\|[x,y]_+ = \sup_{x^*\in \mathbf{F}(x)} \langle y, x^*\rangle$$

for all $x, y \in X$. Here, the corresponding duality map $\mathbf{F}: X \to 2^{X^*}$ is given by

$$\mathbf{F}(x) := \left\{ x^* \in X^* \;\middle|\; \|x\|^2 = \|x^*\|^2 = \langle x, x^*\rangle \right\}.$$

In this context, Proposition 3 takes on the following form.

Proposition 5. *Let κ, k satisfy (2), (3) and (7), and let A be an m-accretive operator in a Banach space X. Let $u_0 \in \overline{D(A)}$ and $f, g \in L^\infty(\mathbb{R}_+;X)$. Moreover,* *let u be the generalized solution of* (1), *and let v be the generalized solution of* (1) *with f replaced by g. Then there exists a constant $C > 0$ such that*

$$\begin{aligned}&\|u(t+\tau) - v(t+\sigma)\|^2 \\ &\quad \le 2\int_{[0,t]} \langle u(t+\tau-s) - v(t+\sigma-s), f(t+\tau-s) - g(t+\sigma-s)\rangle_+ \, d\alpha(s) \\ &\quad + C\frac{\alpha((t,\infty))}{\alpha([0,\infty))}\end{aligned}$$

for $t, \tau, \sigma \in \mathbb{R}_+$.

3. Almost periodicity properties of solutions

In this section it is shown that various almost periodicity properties are inherited by the generalized solution u to the abstract Volterra equation (1) from the forcing function f. We concentrate on asymptotic almost periodicity and on Eberlein-weak almost periodicity.

For $J \in \{\mathbb{R}, \mathbb{R}_+\}$, and $f \in C(J; X)$, we denote by

$$O(f) := \{f_\tau := f(\tau + \cdot)|_J \mid \tau \in J\}$$

the set of all translates of f.

First, recall that a function $f \in C_b(\mathbb{R}; X)$ is *almost periodic* – a.p. for short, if $O(f)$ is relatively compact in $(C_b(\mathbb{R}; X), \|\cdot\|_\infty)$. As usual, we shall denote the space of all almost periodic functions by $AP(\mathbb{R}; X)$.

For the case of the half-line, $f \in C_b(\mathbb{R}_+; X)$ is said to be *asymptotically almost periodic* if $O(f)$ is relatively compact in $(C_b(\mathbb{R}_+; X), \|\cdot\|_\infty)$ (c.f. [AP71], [Fré41], and [RS88]). The space of all such functions on $\mathbb{R}_+$ will be denoted by $AAP(\mathbb{R}_+; X)$.

It is well known that both almost and asymptotically almost periodic functions are uniformly continuous, and that, according to the DeLeeuw-Glicksberg-Decomposition Theorem ([DG61a, DG61b]), any asymptotically almost periodic function uniquely decomposes into the sum of an almost periodic function and a C_0–function:

$$AAP(\mathbb{R}_+; X) = AP(\mathbb{R}; X)|_{\mathbb{R}_+} \oplus C_0(\mathbb{R}_+; X). \tag{16}$$

Of all equivalent descriptions of the above two types of functions (c.f. [AP71], [Fré41], and [RS88]), the following ones will be important for the proofs to follow.

Definition 6. *Let $J \in \{\mathbb{R}, \mathbb{R}_+\}$, then a set $M \subset J$ is called* relatively dense in J, *if there exists a constant $l > 0$ such that $[t, t+l] \cap M \neq \emptyset$ for all $t \in J$.*

Proposition 7. *Let X be a Banach space.*

(i) *A function $f \in C_b(\mathbb{R}; X)$ is almost periodic if and only if for all $\varepsilon > 0$ there exists a relatively dense set $P_\varepsilon \subset \mathbb{R}$ such that*

$$\|f(t+\tau) - f(t)\| \leq \varepsilon \qquad \text{for all } t \in \mathbb{R} \text{ and all } \tau \in P_\varepsilon.$$

The elements τ of P_ε are called ε-almost periods of f.

(ii) *A function $f \in C_b(\mathbb{R}_+; X)$ is asymptotically almost periodic if and only if for all $\varepsilon > 0$ there exists $T_\varepsilon > 0$ and a relatively dense set $P_\varepsilon \subset \mathbb{R}_+$ such that*

$$\|f(t+\tau) - f(t)\| \leq \varepsilon \qquad \text{for all } t \geq T_\varepsilon \text{ and all } \tau \in P_\varepsilon.$$

The elements τ of P_ε are called asymptotic ε-almost periods of f.

We can now prove the inheritance of asymptotic almost periodicity by the solution u to the abstract Volterra equation (1) from the forcing function f.

Theorem 8. *Let A be an m-accretive operator in a Banach space X, let* (2), (3) *and* (7) *be satisfied, and let $u_0 \in \overline{D(A)}$, $f \in AAP(\mathbb{R}_+; X)$. Then the generalized solution u of* (1) *is asymptotically almost periodic.*

Proof. As shown before, the generalized solution u of (1) is an element of $BUC(\mathbb{R}_+; X)$. Given $\varepsilon > 0$, asymptotic almost periodicity of f allows to choose $T_\varepsilon > 0$ and a relatively dense set $P_\varepsilon \subset \mathbb{R}_+$ such that

$$\|f(t+\tau) - f(t)\| \leq \frac{\varepsilon}{2\alpha(\mathbb{R}_+)} \qquad \text{for all } t \geq T_\varepsilon \text{ and all } \tau \in P_\varepsilon.$$

Since, by the continuity of the measure α from above, $\alpha((t,\infty)) \to 0$ as $t \to \infty$, we can choose $T > 0$ large enough such that

$$\alpha((T,\infty)) \leq \frac{\varepsilon}{4\|f\|_\infty + 2C_1}.$$

Here $C_1 := C/\alpha(\mathbb{R}_+)$, where $C > 0$ is the constant given by Proposition 3. Again, by Proposition 3, we conclude for all $t \geq T + T_\varepsilon$, and for all $\tau \in P_\varepsilon$

$$\begin{aligned} \|u(t+\tau) - u(t)\| &\leq \int_{[0,T]} \|f(t+\tau-s) - f(t-s)\| \, d\alpha(s) \\ &\quad + \alpha((T,\infty)) \left(2\|f\|_\infty + \frac{C}{\alpha(\mathbb{R}_+)} \right) \\ &\leq \varepsilon. \end{aligned}$$

Thus, u is asymptotically almost periodic. □

While the preceding result turned out a rather direct application of the integral inequality of Proposition 3, the results on Eberlein-weak almost periodicity will require much more subtle methods of proof.

We start by recalling that a function $f \in C_b(\mathbb{R}_+; X)$ is said to be *Eberlein-weakly almost periodic* – E.-w.a.p. for short – if its set $O(f)$ of translates is weakly relatively compact in $(C_b(\mathbb{R}_+; X), \|\cdot\|_\infty)$, see [Ebe49] and [RS89].

The space of all Eberlein-weakly almost periodic functions on $\mathbb{R}_+$ will be denoted by $W(\mathbb{R}_+;X)$, while its closed linear subspace of all $f \in W(\mathbb{R}_+;X)$ for which the constant 0 function is in the weak$-(C_b(\mathbb{R}_+;X), \|\cdot\|_\infty)$-closure of $O(f)$ is denoted by $W_0(\mathbb{R}_+;X)$.

Every $f \in W(\mathbb{R}_+;X)$ is uniformly continuous ([RS92, Proposition 2.1]), and uniquely decomposes into the sum of an almost periodic function and a $W_0(\mathbb{R}_+;X)-$function ([RS92, Proposition 2.4]):

$$W(\mathbb{R}_+;X) = AP(\mathbb{R};X)|_{\mathbb{R}_+} \oplus W_0(\mathbb{R}_+;X). \tag{17}$$

The space $W(\mathbb{R}_+;X)$ plays an important role in ergodic theory; in fact, every $f \in W(\mathbb{R}_+;X)$ is *uniformly ergodic* ([Ebe49], [RS92, Proposition 2.4]): there exists $z_f \in X$ such that

$$\lim_{T\to\infty} \sup_{h\geq 0} \left\| \frac{1}{T} \int_h^{T+h} f(\tau)\, d\tau - z_f \right\|_X = 0.$$

Returning to the asymptotic behaviour of solutions to (1), even for the special case of the nonlinear Cauchy problem it is unclear whether Eberlein-weak almost periodicity does in general go over from the forcing function f to the solution u (cf. [Kre92]). Positive results, however, can be achieved for those E.-w.a.p. functions that have relatively compact range. The space of those functions will be denoted by $WRC(\mathbb{R}_+;X)$:

$$WRC(\mathbb{R}_+;X) := \{f \in W(\mathbb{R}_+;X) \mid f \text{ has relatively compact range}\}.$$

We can now state the main results on Eberlein-weakly almost periodic solutions to the abstract Volterra equation (1).

Theorem 9. *Let A be an m-accretive operator in a Banach space X, let* (2), (3) *and* (7) *be satisfied, and let $u_0 \in \overline{D(A)}$, $f \in WRC(\mathbb{R}_+;X)$. Then the generalized solution $u \in BUC(\mathbb{R}_+;X)$ of* (1) *is Eberlein-weakly almost periodic on $\mathbb{R}_+$ with relatively compact range.*

For the case of a Banach space X with uniformly convex dual, a corresponding result can be achieved for general $f \in W(\mathbb{R}_+;X)$ (not necessarily having relatively compact range), but with the added assumption that the solution u has relatively compact range.

Theorem 10. *Let A be an m-accretive operator in a Banach space X, with X^* being uniformly convex. Let* (2), (3) *and* (7) *be satisfied, and let $u_0 \in \overline{D(A)}$, $f \in W(\mathbb{R}_+;X)$, and assume that the generalized solution u of* (1) *has relatively compact range. Then u is Eberlein-weakly almost periodic.*

For the special case of the nonlinear Cauchy problem of $k \equiv 0$, the corresponding results have been shown in [Kre92, Kre96], while corresponding results in the general case with A linear are contained in [PR93].

We now turn to the proofs of Theorems 9 and 10. Obviously, the basic ingredient is a criterion for weak relative compactness of the orbit $O(f)$ of a function $f \in C_b(\mathbb{R}_+;X)$. Based on Grothendieck's general double-limits criterion for weak relative compactness in spaces of bounded continuous (scalar-valued) functions on completely regular spaces ([Gro52]), such a criterion has been given in [Mil80, Theorem3] and [RS89, Theorem 2.1]:

Proposition 11. *A function $f \in C_b(\mathbb{R}_+;X)$ is Eberlein-weakly almost periodic if and only if the following double-limits criterion holds: for any sequences $\{t_m, x_m^*\}_{m\in\mathbb{N}} \subset \mathbb{R}_+ \times B_{X^*}$ and $\{\tau_n\}_{n\in\mathbb{N}} \subset \mathbb{R}_+$,*

$$\lim_{n\to\infty}\lim_{m\to\infty}\langle f(t_m+\tau_n), x_m^*\rangle = \lim_{m\to\infty}\lim_{n\to\infty}\langle f(t_m+\tau_n), x_m^*\rangle,$$

whenever both iterated limits exist.

Here, B_{X^*} denotes the closed unit ball of X^*. It is not hard to see that, for $f \in BUC(\mathbb{R}_+;X)$, this double-limits criterion is trivially fulfilled if one of the sequences $\{t_m\}_{m\in\mathbb{N}} \subset \mathbb{R}_+$ or $\{\tau_n\}_{n\in\mathbb{N}} \subset \mathbb{R}_+$ is bounded. Since, in the following, we shall always be dealing with bounded uniformly continuous functions, we can always assume that those sequences tend to infinity in $\mathbb{R}_+$.

For functions with relatively compact range, the following result of [Kre92, Lemma 4.1] (see also [Kre96, Lemma 4.1]) will be needed.

Lemma 12. *Let $\{f_n\}_{n\in\mathbb{N}} \subset C_b(\mathbb{R}_+;X)$ be a sequence such that, for a compact set $K \subset X$,*

$$\{f_n(t) \mid n \in \mathbb{N}, t \in \mathbb{R}_+\} \subset K \quad \text{and} \quad f_n \rightharpoonup 0 \text{ in } C_b(\mathbb{R}_+;X).$$

Then $\|f_n\| \rightharpoonup 0$ in $C_b(\mathbb{R}_+)$.

(Here, the symbol $\rightharpoonup$ denotes weak convergence in the respective Banach space.)

Proof of Theorem 9. In the first step, we show that the generalized solution u of (1) has relatively compact range. Therefore, let $\{t_m\}_{m\in\mathbb{N}}$ be a sequence in $\mathbb{R}_+$. Since f, and thus u, is uniformly continuous (by Corollary 4), we may assume that $\{t_m\}_{m\in\mathbb{N}}$ is strictly increasing. We now define

$$\tau_{m,n} := \begin{cases} t_m - t_n & \text{for } m \ge n, \\ 0 & \text{for } m < n \end{cases} \qquad \text{for all } m,n \in \mathbb{N}.$$

Then, $\tau_{m,n} \to \infty$ as $m \to \infty$ for all $n \in \mathbb{N}$ fixed. Note that, w-cl$(O(f))$ is weakly compact and contained in a closed separable linear subspace of $C_b(\mathbb{R}_+;X)$. Thus, the weak topology on w-cl$(O(f))$ is metrizable. Since $\{f_{\tau_{m,n}} = f(\tau_{m,n}+\cdot) \mid m,n \in \mathbb{N}\} \subset O(f)$, we can apply a diagonalization argument to obtain a common subsequence of $\{\tau_{m,n}\}_{m\in\mathbb{N}}$ for all $n \in \mathbb{N}$, without loss of generality again denoted by $\{\tau_{m,n}\}_{m\in\mathbb{N}}$, such that

$$w\text{-}\lim_{m\to\infty} f_{\tau_{m,n}} = g_n \in C_b(\mathbb{R}_+;X) \qquad \text{for all } n \in \mathbb{N}.$$

Moreover, since f has relatively compact range, we can apply Lemma 12 and obtain

$$\|f_{\tau_{m,n}} - g_n\| \rightharpoonup 0 \quad \text{as } m \to \infty \text{ for all } n \in \mathbb{N}.$$

For $\varepsilon > 0$ given, there exists $N \in \mathbb{N}$ such that for all $n \geq N$

$$C\frac{\alpha((t_n,\infty))}{\alpha(\mathbb{R}_+)} \leq \frac{\varepsilon}{3},$$

where the constant $C > 0$ is given by Proposition 3. For N fixed, the convolution evaluated at t_N is a bounded linear functional on $C_b(\mathbb{R}_+)$. Thus, there exists $M = M(N)$ such that for all $m \geq M$

$$\int_{[0,t_N]} \|f_{\tau_{m,N}}(t_N - s) - g_N(t_N - s)\| \, d\alpha(s) \leq \frac{\varepsilon}{3}.$$

For all $i, j \geq N + M$ one concludes by Proposition 3

$$\begin{aligned}
\|u(t_i) - u(t_j)\| &\leq \|u(t_N + \tau_{i,N}) - u(t_N + \tau_{j,N})\| \\
&\leq \int_{[0,t_N]} \|f_{\tau_{i,N}}(t_N - s) - g_N(t_N - s)\| \, d\alpha(s) \\
&\quad + \int_{[0,t_N]} \|f_{\tau_{j,N}}(t_N - s) - g_N(t_N - s)\| \, d\alpha(s) \\
&\quad + C\frac{\alpha((t_N,\infty))}{\alpha(\mathbb{R}_+)} \\
&\leq \varepsilon.
\end{aligned}$$

This implies that $\{u(t_m)\}_{m\in\mathbb{N}}$ is a Cauchy sequence.

In the second step, we show that u is Eberlein-weakly almost periodic. In order to apply Proposition 11, let $\{t_m\}_{m\in\mathbb{N}} \subset \mathbb{R}_+$, $\{\tau_n\}_{n\in\mathbb{N}} \subset \mathbb{R}_+$, and $\{x_m^*\}_{m\in\mathbb{N}} \subset B_{X^*}$ with $t_m \nearrow \infty$ and $\tau_n \nearrow \infty$ be given such that the iterated limits

$$\lim_{m\to\infty}\lim_{n\to\infty} \langle u(t_m + \tau_n), x_m^*\rangle = \beta \quad \text{and} \quad \lim_{n\to\infty}\lim_{m\to\infty} \langle u(t_m + \tau_n), x_m^*\rangle = \gamma$$

both exist. We have to show that $\beta = \gamma$. Since $f \in WRC(\mathbb{R}_+; X)$, we can assume that $f_{\tau_n} \rightharpoonup g$ for some $g \in C_b(\mathbb{R}_+; X)$ as $n \to \infty$. Moreover, by Lemma 12,

$$\|f_{\tau_n}(\cdot) - g(\cdot)\| \rightharpoonup 0 \qquad \text{in } C_b(\mathbb{R}_+) \text{ as } n \to \infty.$$

By Proposition 3, we have for all $k, m, n \in \mathbb{N}$

$$\begin{aligned}
|\langle u(t_m + \tau_n) - u(t_m + \tau_k), x_m^*\rangle| &\leq \|u(t_m + \tau_n) - u(t_m + \tau_k)\| \\
&\leq \int_{[0,t_m]} \|f_{\tau_n}(t_m - s) - g(t_m - s)\| \, d\alpha(s) \\
&\quad + \int_{[0,t_m]} \|f_{\tau_k}(t_m - s) - g(t_m - s)\| \, d\alpha(s) \\
&\quad + C\,\frac{\alpha((t_m,\infty))}{\alpha(\mathbb{R}_+)}.
\end{aligned}$$

Define the bounded linear functional α_m on $C_b(\mathbb{R}_+)$ for all $m \in \mathbb{N}$ by

$$\alpha_m(h) := \int_{[0,t_m]} h(t_m - s)\, d\alpha(s) \qquad \text{for } h \in C_b(\mathbb{R}_+).$$

Then, since $V := \overline{\operatorname{span}}\{\|f_{\tau_n} - g\| \mid n \in \mathbb{N}\}$ is separable, there exists a subsequence of $\{\alpha_m\}_{m\in\mathbb{N}}$, again denoted by $\{\alpha_m\}_{m\in\mathbb{N}}$, such that $\{\alpha_m|_V\}_{m\in\mathbb{N}}$ is weak* convergent in V^*, i.e., $\alpha_m|_V \rightharpoonup^* \alpha_\infty$ (weak*-convergence) in V^* for some $\alpha_\infty \in V^*$. Thus, we conclude

$$\begin{aligned}
&\lim_{k\to\infty} \lim_{m\to\infty} \lim_{n\to\infty} |\langle u(t_m + \tau_n) - u(t_m + \tau_k), x_m^* \rangle| \\
&\quad \le \lim_{k\to\infty} \lim_{m\to\infty} \lim_{n\to\infty} \langle \alpha_m, \|f_{\tau_n} - g\| \rangle_{V^*,V} + \lim_{k\to\infty} \lim_{m\to\infty} \langle \alpha_m, \|f_{\tau_k} - g\| \rangle_{V^*,V} \\
&\qquad + \lim_{m\to\infty} C \frac{\alpha((t_m, \infty))}{\alpha(\mathbb{R}_+)} \\
&\quad = \lim_{k\to\infty} \langle \alpha_\infty, \|f_{\tau_k} - g\| \rangle_{V^*,V} = 0,
\end{aligned}$$

which implies that the double-limits condition holds. This completes the proof. □

For the proof of Theorem 10, the following result of [Kre92, Lemma 4.6] (see also [Kre96, Lemma 4.6]) will be needed.

Lemma 13. *Let K be a compact set in the dual space X^* of the Banach space X. Moreover, let $\{f_n\}_{n\in\mathbb{N}}$ be a sequence in $C_b(\mathbb{R}_+; X)$ such that $f_n \rightharpoonup 0$ as $n \to \infty$. Then*

$$\sup_{x^* \in K} |\langle f_n(\cdot), x^* \rangle| \rightharpoonup 0 \qquad \text{in } C_b(\mathbb{R}_+) \text{ as } n \to \infty.$$

Proof of Theorem 10. In order to apply Proposition 11, let $\{t_m\}_{m\in\mathbb{N}} \subset \mathbb{R}_+$, $\{\tau_n\}_{n\in\mathbb{N}} \subset \mathbb{R}_+$ and $\{x_m^*\}_{m\in\mathbb{N}} \subset B_{X^*}$ with $t_m \nearrow \infty$ and $\tau_n \nearrow \infty$ be given such that the iterated limits

$$\lim_{m\to\infty} \lim_{n\to\infty} \langle u(t_m + \tau_n), x_m^* \rangle = \beta \quad \text{and} \quad \lim_{n\to\infty} \lim_{m\to\infty} \langle u(t_m + \tau_n), x_m^* \rangle = \gamma$$

both exist. We have to show that $\beta = \gamma$. By assumption, u has relatively compact range. Thus, there exists a compact set $K_1 \subset X$ such that

$$\{u_{\tau_n}(t) - u_{\tau_k}(t) \mid n, k \in \mathbb{N},\ t \in \mathbb{R}_+\} \subset K_1.$$

As X^* is assumed to be uniformly convex, the duality mapping $\mathbf{F}$ is single-valued and norm-continuous, thus $K := \mathbf{F}(K_1) \subset X^*$ is compact. Since $f \in W(\mathbb{R}_+; X)$, we can assume that $f_{\tau_n} \rightharpoonup g$ for some $g \in C_b(\mathbb{R}_+; X)$ as $n \to \infty$. Moreover, by Lemma 13,

$$\sup_{x^* \in K} |\langle f_{\tau_n}(\cdot) - g(\cdot), x^* \rangle| \rightharpoonup 0 \qquad \text{in } C_b(\mathbb{R}_+) \text{ as } n \to \infty.$$

By Proposition 5, we have for all $k, m, n \in \mathbb{N}$

$$
\begin{aligned}
|\langle u(t_m+\tau_n) - u(t_m+\tau_k), x_m^*\rangle|^2 &\leq \|u(t_m+\tau_n) - u(t_m+\tau_k)\|^2 \\
&\leq 2\int_{[0,t_m]} \sup_{x^*\in K} |\langle f_{\tau_n}(t_m-s) \\
&\quad -g(t_m-s), x^*\rangle| \, d\alpha(s) \\
&\quad +2\int_{[0,t_m]} \sup_{x^*\in K} |\langle f_{\tau_k}(t_m-s) \\
&\quad -g(t_m-s), x^*\rangle| \, d\alpha(s) \\
&\quad +C\, \frac{\alpha((t_m,\infty))}{\alpha(\mathbb{R}_+)}.
\end{aligned}
$$

Define the bounded linear functional α_m on $C_b(\mathbb{R}_+)$ for all $m \in \mathbb{N}$ by

$$
\alpha_m(h) := \int_{[0,t_m]} h(t_m-s)\, d\alpha(s) \qquad \text{for } h \in C_b(\mathbb{R}_+).
$$

Then, since $V := \overline{\text{span}}\{\, \sup_{x^*\in K} |\langle f_{\tau_n} - g, x^*\rangle| \;\big|\; n \in \mathbb{N}\}$ is separable, there exists a subsequence of $\{\alpha_m\}_{m\in\mathbb{N}}$, again denoted by $\{\alpha_m\}_{m\in\mathbb{N}}$, such that $\{\alpha_m\big|_V\}_{m\in\mathbb{N}}$ is weak* convergent in V^*, i.e., $\alpha_m\big|_V \rightharpoonup^* \alpha_\infty$ for some $\alpha_\infty \in V^*$. Thus, we conclude

$$
\begin{aligned}
&\lim_{k\to\infty}\lim_{m\to\infty}\lim_{n\to\infty} |\langle u(t_m+\tau_n) - u(t_m+\tau_k), x_m^*\rangle|^2 \\
&\qquad\leq\quad 2\lim_{k\to\infty}\lim_{m\to\infty}\lim_{n\to\infty} \langle \alpha_m, \sup_{x^*\in K} |\langle f_{\tau_n} - g, x^*\rangle|\rangle_{V^*,V} \\
&\qquad\quad + 2\lim_{k\to\infty}\lim_{m\to\infty} \langle \alpha_m, \sup_{x^*\in K} |\langle f_{\tau_k} - g, x^*\rangle|\rangle_{V^*,V} \\
&\qquad\quad + \lim_{m\to\infty} C\frac{\alpha((t_m,\infty))}{\alpha(\mathbb{R}_+)} \\
&\qquad=\quad 2\lim_{k\to\infty} \langle \alpha_\infty, \sup_{x^*\in K} |\langle f_{\tau_k} - g, x^*\rangle|\rangle_{V^*,V} = 0,
\end{aligned}
$$

which implies that the double-limits condition holds. This completes the proof. □

We note, in passing, the following consequence for uniform ergodicity of generalized solutions to (1).

Corollary 14. *Given the assumptions of either Theorem 8, 9 or 10, the generalized solution u to the abstract Volterra equation* (1) *is uniformly ergodic, i.e., there exists $x_u \in X$ such that*

$$
\lim_{T\to\infty}\sup_{h\geq 0} \left\| \frac{1}{T}\int_h^{T+h} u(\tau)\, d\tau - x_u \right\|_X = 0.
$$

4. The limit equation

For the special case of the nonlinear Cauchy problem, and a.a.p.-, respectively, E.-w.a.p.-forcing functions f, and corresponding a.a.p.-, respectively, E.-w.a.p.-solutions u, it is known that the almost periodic part u^∞ of u is the solution to an associated 'limit'-equation on all of $\mathbb{R}$, with the forcing function given by the almost periodic part f^∞ of f ([Kre92, Kre96]).

The purpose of this section is to specify a corresponding 'limit'-Volterra equation on all of $\mathbb{R}$, and to derive the corresponding results on the almost periodic parts in the general context of (1).

We start by specifying such a limit equation, and by defining a corresponding notion of generalized solution.

Given $k \in L^1_{\text{loc}}(\mathbb{R}_+)$, satisfying (2), choose $k_1 \in L^1(\mathbb{R}_+)$ and $k_2 \in BV(\mathbb{R}_+)$ such that

$$k(t) = k_1(t) + k_2(t) \text{ for all } t > 0, \text{ and } k_1, k_2 \text{ satisfy (2).} \tag{18}$$

Then, for an m-accretive operator A in a Banach space X, and $u_0 \in \overline{D(A)}$, and $g \in C_b(\mathbb{R}; X)$, the equation

$$\begin{aligned} &\frac{d}{dt}\Big(\kappa(u(t) - u_0) + \int_{-\infty}^{t} k_1(t-s)(u(s) - u_0)\, ds\Big) \\ &\qquad + k_2(0+)(u(t) - u_0) \qquad\qquad t \in \mathbb{R} \quad (19) \\ &\qquad + \int_{(0,\infty)} (u(t-s) - u_0)\, dk_2(s) + Au(t) \ \ni \ g(t), \end{aligned}$$

is well defined, and will be called the *limit equation* corresponding to the problem (1).

(We note in passing that the limit equation (19) describes a problem on the whole real line $\mathbb{R}$, and is no longer an initial value problem such as (1). The fact that u_0 is still contained in (19) is due to our choice of the right hand side g. Defining $\tilde{g} := k(\infty)u_0 + g$, we would be able to omit the term u_0 in the limit equation (19) by replacing g by $\tilde{g}$.)

As for the notion of a generalized solution to (19), we shall use the following concept, which seems appropriate for the investigation of the asymptotic behaviour of solutions. We start by mentioning that (19) can also be viewed as the limit of initial value problems on intervals $[-T, \infty)$ with T tending to ∞ : consider the approximations

$$\frac{d}{dt}\left(\kappa(u^{(T)}(t) - u_0) + \int_{-T}^{t} k(t-s)(u^{(T)}(s) - u_0)\, ds\right) + Au^{(T)}(t) \ni g(t), \quad t \geq -T \tag{20}$$

of (19) with $T > 0$. These equations are just shifted versions of the initial value problem (1). If $u^{(T)}$ is a strong solution to (20), then $v^{(T)} := u^{(T)}(t-T)$ is a strong solution of (1) with the right hand side f replaced by $g(\cdot - T)$. Therefore, it is almost obvious how to define the notion of a generalized solution of (20).

Definition 15. *A function $u^{(T)} \in L^1_{\mathrm{loc}}([-T,\infty);X)$ is called a* generalized solution *of* (20) *for data $u_0 \in \overline{D(A)}$ and $g \in L^1_{\mathrm{loc}}([-T,\infty);X)$, if the function $v^{(T)} \in L^1_{\mathrm{loc}}(\mathbb{R}_+;X)$ defined by $v^{(T)}(\cdot) := u^{(T)}(\cdot - T)$ is a generalized solution of* (1) *with the right hand side f replaced by $g(\cdot - T)$.*

We now investigate the convergence of the net $\{u^{(T)}\}_{T>0}$ as $T \to \infty$.

Proposition 16. *Let A be an m-accretive operator in a Banach space X, let κ, k satisfy* (2), (3), *and* (7). *Moreover, assume that $u_0 \in \overline{D(A)}$ and $g \in L^\infty(\mathbb{R};X)$. Then the net $\{u^{(T)}\}_{T>0}$ of generalized solutions $u^{(T)}$ of* (20) *converges in $L^\infty_{\mathrm{loc}}(\mathbb{R};X)$ as $T \to \infty$.*

Definition 17. *The limit function of the net $\{u^{(T)}\}_{T>0}$ obtained in Proposition 16 will be called the* generalized solution *of the limit equation (19).*

Proof of Proposition 16. We show that $\{u^{(T)}\}_{T>0}$ is a Cauchy net in $L^\infty_{\mathrm{loc}}(\mathbb{R};X)$. Here, without loss of generality, we set $u^{(T)}(t) := u_0$ for $t < -T$. Defining $v^{(T)}(\cdot) := u^{(T)}(t-T)$ for all $T > 0$, it is clear that $v^{(T)}$ is a generalized solution of (1) with right hand side $g(\cdot - T)\big|_{\mathbb{R}_+}$. Thus, by Proposition 3, there exists a constant $C > 0$ such that for $t \in \mathbb{R}_+$, and $0 < T_1 < T_2$

$$\begin{aligned}\|v^{(T_1)}(t+T_1) - v^{(T_2)}(t+T_2)\| \\ &\leq \int_{[0,t+T_1]} \|g(t+T_1-T_1-s) - g(t+T_2-T_2-s)\|\, d\alpha(s) \\ &\quad + C\frac{\alpha((t+T_1,\infty))}{\alpha(\mathbb{R}_+)}.\end{aligned}$$

For $M > 0$ and $\varepsilon > 0$, choose $T_0 > 0$ such that

$$\alpha((T_0 - M,\infty)) \leq \frac{\varepsilon}{C}\,\alpha(\mathbb{R}_+).$$

Then, by the above inequality, we conclude for $T_0 < T_1 < T_2$, and for $t \in [-M,M]$,

$$\|u^{(T_1)}(t) - u^{(T_2)}(t)\| = \|v^{(T_1)}(t+T_1) - v^{(T_2)}(t+T_2)\| \leq \varepsilon.$$

Since the roles of T_1 and T_2 can be interchanged, we obtain the assertion. $\square$

We first note two of the basic properties of the generalized solution of the limit equation (19).

Proposition 18. *Let A be an m-accretive operator in a Banach space X, and let κ, k satisfy* (2), (3), *and* (7). *Moreover, let $u_0 \in \overline{D(A)}$ and $g \in L^\infty(\mathbb{R};X)$. Then the generalized solution u^∞ of the limit equation* (19) *satisfies*

(i) *$u^\infty \in L^\infty(\mathbb{R};X)$,*
(ii) *if $g \in BUC(\mathbb{R};X)$, then $u^\infty \in BUC(\mathbb{R};X)$.*

Proof. For almost all $t \in \mathbb{R}$, $(x,y) \in A$ and $\varepsilon > 0$, there is $T > |t|$ such that

$$\begin{aligned} \|u^\infty(t) - x\| &\le \|u^\infty(t) - u^{(T)}(t)\| + \|u^{(T)}(t) - x\| \\ &\le \varepsilon + \|u_0 - x\| + \|g - y\|_\infty \alpha(\mathbb{R}_+). \end{aligned}$$

Thus, u^∞ is essentially bounded. If $g \in BUC(\mathbb{R};X)$, then obviously $u^\infty \in C_b(\mathbb{R};X)$. As for uniform continuity, let $\varepsilon > 0$. By the continuity of u at $t = 0$, there exists $h_0 > 0$ such that for all $0 < h < h_0$ and all $T > 0$

$$\|u^{(T)}(h) - u_0\| \le \frac{\varepsilon}{4}.$$

Moreover, by uniform continuity of g, there exists h_0 small enough such that for all $0 < h < h_0$, and for all $t \in \mathbb{R}$,

$$\|g(t+h) - g(t)\| \le \frac{\varepsilon}{4\alpha(\mathbb{R}_+)}.$$

Thus, for any $t \in \mathbb{R}$, choose $T > |t| + h_0$ such that, for all $0 < h < h_0$,

$$\begin{aligned} \|u^\infty(t+h) - u^\infty(t)\| &\le \|u^\infty(t+h) - u^{(T)}(t+h)\| + \|u^\infty(t) - u^{(T)}(t)\| \\ &\quad + \|u^{(T)}(h) - u_0\| + \int_{[0,\infty)} \|g(t+h-s) - g(t-s)\|\, d\alpha(s) \\ &\le \varepsilon. \end{aligned}$$

□

Turning to our initial question of characterizing the almost periodic part of a generalized solution of the Volterra equation (1), we first prove almost periodicity of generalized solutions to the limit equation (19) for almost periodic forcing functions g.

Theorem 19. *Let A be an m-accretive operator in a Banach space X, and let κ, k satisfy* (2), (3), *and* (7). *Moreover, let $u_0 \in \overline{D(A)}$ and $g \in AP(\mathbb{R};X)$. Then the generalized solution u^∞ of the limit equation* (19) *is almost periodic.*

Proof. We already know that u^∞ is bounded and uniformly continuous. Let $\{u^{(T)}\}_{T>0}$ denote the net of generalized solutions $u^{(T)}$ of (20), and let $\varepsilon > 0$. Then there exists a relatively dense set $P_\varepsilon \subset \mathbb{R}$ such that

$$\|g(t+\tau) - g(t)\| \le \frac{\varepsilon}{4\alpha(\mathbb{R}_+)} \qquad \text{for all } t \in \mathbb{R} \text{ and all } \tau \in P_\varepsilon.$$

Thus, for all $t \in \mathbb{R}$, and all $\tau \in P_\varepsilon$, choose $T > |t| + |\tau|$ such that

$$\begin{aligned} &\|u^\infty(t+\tau) - u^\infty(t)\| \\ &\quad \le \|u^\infty(t+\tau) - u^{(T)}(t+\tau)\| + \|u^\infty(t) - u^{(T)}(t)\| + \|u^{(T)}(t+\tau) - u^{(T)}(t)\| \\ &\quad \le \frac{\varepsilon}{2} + \int_{[0,\infty)} \|g(t+\tau-s) - g(t-s)\|\, d\alpha(s) + C\frac{\alpha((t+T,\infty))}{\alpha(\mathbb{R}_+)} \\ &\quad \le \varepsilon, \end{aligned}$$

with a constant $C > 0$ given by Proposition 3. This implies that u^∞ is almost periodic. □

We are now in the position of characterizing the almost periodic part of an a.a.p.-solution to the Volterra equation (1) by the solution to the limit equation (19) with a suitable forcing function.

Theorem 20. *Let A be an m-accretive operator in a Banach space X, and let κ, k satisfy* (2), (3), *and* (7). *Moreover, let $u_0 \in \overline{D(A)}$, $f \in AAP(\mathbb{R}_+; X)$, and let $f^\infty \in AP(\mathbb{R}; X)$ be the almost periodic part of f.*
Then the generalized solution u of the Volterra equation (1) *is asymptotically almost periodic, and the almost periodic part $u^\infty \in AP(\mathbb{R}; X)$ of u is the generalized solution of the limit equation* (19) *with forcing function f^∞.*

Proof. Let u^∞ denote the generalized solution of the limit equation (19) with forcing function f^∞. Then, by Theorem 19 and Theorem 8, we already know that $u^\infty \in AP(\mathbb{R}; X)$ and $u \in AAP(\mathbb{R}_+; X)$. Thus, it remains to show that $\|u^\infty(t) - u(t)\| \to 0$ as $t \to \infty$. Let $u^{(T)}$ be the generalized solution of (20) with right hand side f^∞ for all $T > 0$, and let $\varepsilon > 0$. Then, for all $t \in \mathbb{R}$, choose $T > |t|$ such that

$$\begin{aligned} \|u^\infty(t) - u(t)\| &\leq \|u^\infty(t) - u^{(T)}(t)\| + \|u^{(T)}(t) - u(t)\| \\ &\leq \frac{\varepsilon}{2} + \int_{[0,t]} \|f^\infty(t-s) - f(t-s)\| \, d\alpha(s) + C\frac{\alpha((t,\infty))}{\alpha(\mathbb{R}_+)}, \end{aligned}$$

with a constant $C > 0$ given by Proposition 3. Since $\|f^\infty(t) - f(t)\| \to 0$ as $t \to \infty$, we conclude

$$\lim_{t\to\infty} \|u^\infty(t) - u(t)\| \leq \varepsilon.$$

Thus, u^∞ is the almost periodic part of u. □

We note in passing that, in Theorem 19, if $g \equiv f^\infty \in X$ is constant, then so is u^∞. Combined with the preceding theorem, this gives the result of [CN81, Theorem 3.2]. Thus, Theorem 20 extends [CN81, Theorem 3.2] from functions having a limit at infinity to the class of all asymptotically almost periodic functions.

The analogue of Theorem 20 does also hold for the almost periodic parts of the Eberlein-weak almost periodic solutions u as in Theorems 9 and 10. Once again, the corresponding proofs are subject to the more subtle methods required for those theorems, but also very close to them. We thus omit the proofs, and only state the results.

Theorem 21. *Let A be an m-accretive operator in a Banach space X, and let κ, k satisfy* (2), (3), *and* (7). *Moreover, let $u_0 \in \overline{D(A)}$, $f \in W(\mathbb{R}_+; X)$, and let $f^\infty \in AP(\mathbb{R}; X)$ be the almost periodic part of f, and let u be the generalized solution of the Volterra equation* (1).
Moreover, let either

(i) *f have relatively compact range, or let*
(ii) *X^* be uniformly convex, and u have relatively compact range.*

Then u is Eberlein-weakly almost periodic, and the almost periodic part $u^\infty \in AP(\mathbb{R}; X)$ of u is a generalized solution of the limit equation (19) *with forcing function f^∞.*

5. Example

Equation (1) provides a fairly general model of diffusion of fluids in porous media. Indeed, the law of conservation of mass for a fluid in a porous medium situated in some bounded domain $\Omega \subset \mathbb{R}^3$ is given by

$$m_t + \mathrm{div} q = h. \tag{21}$$

Here, $m = b(p)$ denotes the mass of the fluid depending on the pore pressure p of the fluid, where $b : \mathbb{R} \to \mathbb{R}$ is assumed to be continuous, nondecreasing and normalized by $b(0) = 0$. Moreover, q is the vector field of the flux, and h denotes an external source term. Under a wide range of conditions, Darcy's classical empirical law, stating the dependence of the flux q on the permeability θ and the gradient of the pore pressure, can be applied. Since the permeability may depend on the pore pressure, we obtain the classical porous media equation

$$b(p)_t - \mathrm{div} a(p, Dp) = h. \tag{22}$$

Here, $a : \mathbb{R} \times \mathbb{R}^3 \to \mathbb{R}^3$ is choosen such that the operator $\mathcal{A} : W_0^{1,q}(\Omega) \ni v \mapsto -\mathrm{div} a(v, Dv) \in W^{-1,q/(q-1)}(\Omega)$ is bounded, coercive and pseudo-monotone for some $1 < q < \infty$.

However, in geothermal regions it is often observed that the fluid precipitates minerals in the pores of the medium. Thus the pores are obstructed.

In particular, steam wells used for heat extraction are often self-sealed in relatively short time. If one considers the case when the permeability diminishes with time, the effect of the fluid pressure at the boundary of the flow of the fluid through the medium is delayed and occurs as if the medium had a memory. Thus, in [Cap99] and [Cap00], a modification of Darcy's classical law is suggested using fractional derivatives in time.

The modified Darcy's law now states for some $0 \leq \gamma < 1$

$$q = -\frac{\partial^\gamma}{\partial t^\gamma} a(p, Dp). \tag{23}$$

Here, we use the following definition of a fractional derivative of order $0 < \gamma < 1$ of a function $h : \mathbb{R}_+ \to \mathbb{R}$:

$$\frac{d^\gamma}{dt^\gamma} h(t) := \frac{d}{dt} \left(\int_0^t \frac{(t-s)^{-\gamma}}{\Gamma(1-\gamma)} h(s) \right). \tag{24}$$

Thus, the general porous media equation for $0 \leq \gamma < 1$ is given by

$$\frac{\partial^{1-\gamma}}{\partial t^{1-\gamma}} b(p) - \mathrm{div} a(p, Dp) = \tilde{h}. \tag{25}$$

This leads to an equation of the form (1) for $u = b(p)$ in the space $X = L^1(\Omega)$, if we define the operator A for Dirichlet boundary conditions by

$$(u,w) \in A :\Longleftrightarrow \begin{cases} \text{There exists } v : \Omega \to \mathbb{R} \text{ measurable such that } b(v) = \\ u, w \in L^1(\Omega), \text{ and } T_K(v) \in W_0^{1,q}(\Omega) \text{ for all } K > 0, \\ \text{satisfying} \\ \displaystyle\int_\Omega a(v, Dv) DT_K(u - \phi) \le \int_\Omega w T_K(u - \phi) \\ \text{for all } \phi \in W_0^{1,q}(\Omega) \cap L^\infty(\Omega) \text{ and all } K > 0. \end{cases} \tag{26}$$

Here, the truncation function $T_K : \mathbb{R} \to \mathbb{R}$ used in the above definition is given by $T_K(r) := \max(\min(r, K), -K)$ for $K > 0$.

Note that the above definition of the operator A is motivated by the definition of entropy solutions for elliptic equations (cf. [BBG$^+$95]). As shown in [BW96] the operator A is m-accretive. Moreover, we refer to [JWa] for the problem of existence and uniqueness of entropy solutions of the history dependent elliptic-parabolic problem given by (25).

References

[ALR90] S. Aizicovici, S.O. Londen, and S. Reich, *Asymptotic behavior of solutions to a class of nonlinear Volterra equations.* Diff. Int. Eqns **3** (1990), 813–825.

[AP71] L. Amerio and G. Prouse, *Almost periodic functions and functional equations.* Van Nostrand Reinhold Company, New York, 1971.

[BBG$^+$95] Ph. Bénilan, L. Boccardo, T. Gallouët, R. Gariepy, M. Pierre, and J. L. Vazquez, *An L^1-theory of existence and uniqueness of solutions of nonlinear elliptic equations.* Ann. Sc. Norm. Super. Pisa **22** (2) (1995), 241–273.

[BC81] J.B. Baillon and Ph. Clément, *Ergodic theorems for non-linear Volterra equations in Hilbert space.* Nonlin. Anal.: TMA **5** (1981), 789–801.

[BW96] Ph. Bénilan and P. Wittbold, *On mild and weak solutions of elliptic-parabolic problems.* Adv. Differ. Equ. **1** (6) (1996), 1053–1073.

[Cap99] M. Caputo, *Diffusion of fluids in porous media with memory.* Geothermics **28** (1999), 113–130.

[Cap00] M. Caputo, *Models of flux in porous media with memory.* Water Resources Res. **36**(3) (2000), 693–705.

[CGL96] B. Cockburn, G. Gripenberg, and S.-O. Londen, *On convergence to entropy solutions of a single conservation law.* J. Differ. Equations **128** (1) (1996), 206–251.

[Clé80] Ph. Clément, *On abstract Volterra equations with kernels having a positive resolvent.* Israel J. Math. **36** (3-4)(1980), 193–200.

[CM88] Ph. Clément and E. Mitidieri, *Qualitative properties of solutions of Volterra equations in Banach spaces.* Israel J. Math. **64** (1) (1988), 1–24.

[CN79] Ph. Clément and J. A. Nohel, *Abstract linear and nonlinear Volterra equations preserving positivity.* SIAM J. Math. Anal. **10** (1979), 365–388.

[CN81] Ph. Clément and J. A. Nohel, *Asymptotic behavior of solutions of nonlinear Volterra equations with completely positive kernels.* SIAM J. Math. Anal. **12** (1981), 514–535.

[CP90] Ph. Clément and J. Prüss, *Completely positive measures and Feller semigroups.* Math. Ann. **287** (1990), 73–105.

[DG61a] K. DeLeeuw and I. Glicksberg, *Almost periodic functions on semigroups.* Acta Math. **105** (1961), 99–140.

[DG61b] K. DeLeeuw and I. Glicksberg, *Applications of almost periodic compactifications.* Acta Math. **105** (1961), 63–97.

[Ebe49] W. F. Eberlein, *Abstract Ergodic Theorems and Weak Almost Periodic Functions.* Trans. Am. Math. Soc. *67* (1949), 217–240.

[Fré41] M. Fréchet, *Les fonctions asymptotiquement presque-périodiques.* Rev. Sci. **79** (1941), 341–354.

[Gri85] G. Gripenberg, *Volterra integro-differential equations with accretive nonlinearity.* J. Differ. Equations **60** (1985), 57–79.

[Gro52] A. Grothendieck, *Critères de compactité dans les espaces fonctionelles généraux.* Amer. J. Math. **74** (1952), 168–186.

[Hir83] N. Hirano, *Asymptotic behavior of solutions of nonlinear Volterra equations.* J. Diff. Equations **47** (1983), 163–179.

[JWa] V. G. Jakubowski and P. Wittbold, *On a nonlinear elliptic-parabolic integro-differential equation with L^1-data.* In preparation.

[JWb] V. G. Jakubowski and P. Wittbold, *Regularity of solutions of nonlinear Volterra equations.* In preparation.

[Kre92] J. Kreulich, *Eberlein weak almost periodicity and differential equations in Banach spaces.* Dissertation, Universität Essen, 1992.

[Kre96] J. Kreulich, *Eberlein-weakly almost-periodic solutions of evolution equations in Banach spaces.* Differ. Integral Equ. **9** (5) (1996), 1005–1027.

[Mil80] P. Milnes, *On vector-valued weakly almost periodic functions.* J. Lond. Math. Soc. **22** (2) (1980), 467–472.

[PR93] J. Prüss and W. M. Ruess, *Weak almost periodicity of convolutions.* J. Integral Equations Appl. **5**(4) (1993), 519–530.

[RS88] W. M. Ruess and W. H. Summers, *Compactness in spaces of vector valued continuous functions and asymptotic almost periodicity.* Math. Nachr. **135** (1988), 7–33.

[RS89] W. M. Ruess and W. H. Summers, *Integration of asymptotically almost periodic functions and weak asymptotic almost periodicity.* Diss. Math. **279** (1989), 35 p.

[RS92] W. M. Ruess and W. H. Summers, *Ergodic theorems for semigroups of operators.* Proc. Am. Math. Soc. **114** (2) (1992), 423–432.

Volker G. Jakubowski
Fachbereich Mathematik
Universität Essen
D-45117 Essen, Germany
E-mail address: `volker.jakubowski@uni-essen.de`

Wolfgang M. Ruess
Fachbereich Mathematik
Universität Essen
D-45117 Essen, Germany
E-mail address: `w.ruess@uni-essen.de`

Progress in Nonlinear Differential Equations
and Their Applications, Vol. 55, 275–292

Semigroups for Initial-Boundary Value Problems

Marjeta Kramar, Delio Mugnolo, and Rainer Nagel

1. Semigroups Everywhere

In the foreword of what later became the classic in semigroup theory, E. Hille [Hi48] wrote in 1948:

"The analytical theory of semi-groups is a recent addition to the ever-growing list of mathematical disciplines. It was my good fortune to take an early interest in this discipline and to see it reach maturity. It has been a pleasant association: I hail a semi-group when I see one and

I seem to see them everywhere!"

With this quite general (and provocative) statement he probably expressed his conviction that behind every parabolic or hyperbolic (linear and autonomous) partial differential equation there is a *semigroup* yielding its solutions. In the meantime this has been confirmed in many ways and is presented in excellent form, e.g., in [Go85] or [Pa83]. We briefly recall one of the standard examples.

Example 1.1. Let Ω be a bounded open domain in $\mathbb{R}^n$ with smooth boundary $\partial\Omega$. The heat equation with Dirichlet boundary conditions is

$$\begin{cases} \frac{\partial}{\partial t}u(t,x) &= \sum_{i=1}^n \frac{\partial^2}{\partial x_i^2}u(t,x), & t \geq 0,\ x=(x_1,\dots,x_n)\in\Omega, \\ u(t,z) &= 0, & t\geq 0,\ z\in\partial\Omega, \\ u(0,x) &= f(x), & x\in\Omega. \end{cases} \tag{HE}$$

If we now choose an appropriate function space, e.g., $X := L^2(\Omega)$ and define an operator

$$Au := \Delta u := \sum_{i=1}^n \frac{\partial^2}{\partial x_i^2}u \quad \text{with domain} \quad D(A) := H^2(\Omega)\cap H_0^1(\Omega),$$

the system (HE) can then be re-written as an abstract Cauchy problem

$$\begin{cases} \dot u(t) = Au(t), & t\geq 0, \\ u(0) = f, & x\in\Omega. \end{cases} \tag{ACP}$$

It is well-known (see, e.g., [Pa83, Thm. 7.2.7]) that $(A, D(A))$ generates a strongly continuous (and analytic) semigroup $(T(t))_{t\geq 0}$ on X such that

$$u(t,x) := T(t)f(x), \qquad t \geq 0,\, x \in \Omega,$$

are the (mild) solutions of (HE).

Once a partial differential equation is solved in this way by a semigroup, there exist by now powerful tools to describe the qualitative behavior of the solutions. We only mention the spectral theory for semigroups (see [EN00, Chap. IV]) yielding various Liapunov-type stability theorems or the recent results on maximal regularity of the inhomogeneous version of (ACP) (see, e.g., [DHP01], [We01], or [We01b]).

On the other hand, even when one admits the conceptual lucidity and the generality of the semigroup approach, scepticism may remain, and Hille continues his above quote:

"Friends have observed, however, that there are mathematical objects which are not semi-groups."

Indeed, already for our heat equation (HE) semigroups seem to be inappropriate as soon as we add an inhomogeneous term on the boundary.

Example 1.2. For the heat equation on Ω as in Example 1.1, we consider inhomogeneous boundary values given by $u(t,z) = \phi(t,z)$ for $t \geq 0$, $z \in \partial\Omega$. As we will see below, it is more convenient to write this in the form

$$\begin{cases} \frac{\partial}{\partial t}u(t,x) &= \sum_{i=1}^n \frac{\partial^2}{\partial x_i^2}u(t,x), \qquad t \geq 0,\, x = (x_1,\dots,x_n) \in \Omega, \\ \frac{\partial}{\partial t}u(t,z) &= \psi(t,z), \qquad t \geq 0,\, z \in \partial\Omega, \\ u(0,x) &= f(x), \qquad x \in \Omega, \\ u(0,z) &= g(z), \qquad z \in \partial\Omega, \end{cases} \tag{iHE}$$

where $\psi(t,z) := \frac{\partial}{\partial t}\phi(t,z)$. □

Such a situation typically occurs in all kinds of boundary control problems (see, e.g., [LT00]). The obstacle to use semigroups consists in the fact that the inhomogeneous term

$$t \mapsto \psi(t)$$

does not map into the state space chosen in Example 1.1.

However, as for many other types of equations (e.g., delay equations, integro-differential equations) initially not having the form (ACP), it is possible to extend the state space and then write (iHE) as an inhomogeneous abstract Cauchy problem: take $X := L^2(\Omega)$, $\partial X := L^2(\partial\Omega)$, and $A := \Delta$ the Laplacian with appropriate domain (see § 3 below for details). On the product space

$$\mathcal{X} := X \times \partial X,$$

we define a new operator

$$\mathcal{A} := \begin{pmatrix} A & 0 \\ 0 & 0 \end{pmatrix}, \qquad D(\mathcal{A}) := \left\{ \begin{pmatrix} u \\ v \end{pmatrix} \in D(A) \times \partial X \,:\, u|_{\partial\Omega} = v \right\}. \tag{1.1}$$

Then (iHE) becomes

$$\begin{cases} \dot{\mathcal{U}}(t) = \mathcal{A}\,\mathcal{U}(t) + \binom{0}{\psi(t)}, & t \geq 0, \\ \mathcal{U}(0) = \mathcal{U}_0, \end{cases} \tag{iACP}$$

for a function $\mathbb{R}_+ \ni t \mapsto \mathcal{U}(t) \in \mathcal{X}$ and with the initial data $\mathcal{U}_0 := \binom{f}{g}$. Hence, it becomes an inhomogeneous abstract Cauchy problem and therefore semigroups should be applicable in this situation.

Remark 1.3. The idea to use a product space in order to convert inhomogeneous boundary conditions into the form (iACP) appears and is used systematically by Arendt in [Ar00], see also [ABHN01, Chap. 6].

This important example, and the quotation of Hille's cited at the beginning, suggest the following.

Program 1.4.

(i) Define an abstract setting for the situation described in Example 1.2.

(ii) Show that well-posedness becomes equivalent to the well-posedness of the corresponding (ACP).

(iii) Apply semigroup theory to obtain existence of the solutions.

In the following, we propose an appropriate abstract setting, develop this program, and discuss one typical application.

2. Abstract Initial-Boundary Value Problems

Let X and ∂X be Banach spaces called *state space* and *boundary space*, respectively. We denote by $\mathcal{X} := X \times \partial X$ their product space, and by π_1 and π_2 the projections onto X and ∂X, respectively.

Assumptions 2.1. We consider the following (linear) operators.

$A : D(A) \subseteq X \to X$, called *maximal operator*.

$L : D(A) \to \partial X$, called *boundary operator*.

$B : D(B) \subseteq X \to \partial X$, called *feedback operator*.

For these operators, we assume that $D(A) \subseteq D(B)$ and consider what we call an *abstract initial–boundary value problem on the state space* X *and the boundary space* ∂X:

$$\begin{cases} \frac{d}{dt}u(t) &= Au(t), & t \geq 0, \\ \frac{d}{dt}Lu(t) &= Bu(t), & t \geq 0, \\ u(0) &= f, \\ Lu(0) &= g. \end{cases} \tag{AIBVP}$$

where $f \in X$, $g \in \partial X$. (If we want to emphasize the dependence on the initial data f, g, we write $(\text{AIBVP})_{f,g}$).

We now make precise what we understand by a solution of (ABVP).

Definition 2.2. Let $(f,g) \in D(A) \times \partial X$ be given initial data. A **classical solution** to $(\text{AIBVP})_{f,g}$ is a function u such that

(CS$_1$) $u \in C^1(\mathbb{R}_+, X)$ and $u(t) \in D(A)$ for all $t \geq 0$,
(CS$_2$) $Lu \in C^1(\mathbb{R}_+, \partial X)$,
(CS$_3$) u satisfies $(\text{AIBVP})_{f,g}$.

Moreover, (AIBVP) is called **well-posed** if

(WP$_1$) $\overline{D(A)} = X$,
(WP$_2$) $(\text{AIBVP})_{f,g}$ admits a unique classical solution $u(\cdot, f, g)$ for all $(f,g) \in D(A) \times \partial X$ satisfying $Lf = g$,
(WP$_3$) for all sequences of initial data $(f_n, g_n)_{n\in\mathbb{N}} \subseteq D(A) \times \partial X$ tending to 0 and such that $Lf_n = g_n$, one has $\lim_{n\to\infty} u(t, f_n, g_n) = 0$ and $\lim_{n\to\infty} Lu(t, f_n, g_n) = 0$ uniformly for t on compact intervals.

A **mild solution** to $(\text{AIBVP})_{f,g}$ for arbitrary $(f,g) \in \mathcal{X}$ is a function u such that

(MS$_1$) $u \in C(\mathbb{R}_+, X)$ and $\int_0^t u(s)ds \in D(A)$ for all $t \geq 0$,
(MS$_2$) $B\left(\int_0^{\cdot} u(s)ds\right) \in C(\mathbb{R}_+, \partial X)$,
(MS$_3$) u satisfies the integrated problem

$$\begin{cases} u(t) & = f + A\int_0^t u(s)ds, \qquad t \geq 0, \\ L\int_0^t u(s)ds & = tg + \int_0^t B\left(\int_0^s u(r)\,dr\right)ds, \qquad t \geq 0. \end{cases}$$

In order to treat (AIBVP) using semigroups we consider the operator matrices on $\mathcal{X}$ given by

$$\mathcal{A} := \begin{pmatrix} A & 0 \\ 0 & 0 \end{pmatrix} \quad \text{with domain} \quad D(\mathcal{A}) := \left\{ \begin{pmatrix} u \\ v \end{pmatrix} \in D(A) \times \partial X \ : \ Lu = v \right\}, \tag{2.1}$$

and

$$\tilde{\mathcal{A}} := \begin{pmatrix} A & 0 \\ B & 0 \end{pmatrix} \quad \text{with domain} \quad D(\tilde{\mathcal{A}}) := D(\mathcal{A}). \tag{2.2}$$

We briefly state some of their properties.

Lemma 2.3. *Let B be relatively A-bounded. Then the operator $\tilde{\mathcal{A}}$ is closed if and only if the operator $\binom{A}{L} : D(A) \subseteq X \to X \times \partial X$ is closed.*

Proof. Assume first that $\binom{A}{L}$ is a closed operator. Let

$$\begin{pmatrix} u_n \\ v_n \end{pmatrix}_{n\in\mathbb{N}} \subseteq D(\mathcal{A}), \ \lim_{n\to\infty} \begin{pmatrix} u_n \\ v_n \end{pmatrix} = \begin{pmatrix} u \\ v \end{pmatrix}, \text{ and } \lim_{n\to\infty} \tilde{\mathcal{A}} \begin{pmatrix} u_n \\ v_n \end{pmatrix} = \lim_{n\to\infty} \begin{pmatrix} Au_n \\ Bu_n \end{pmatrix} = \begin{pmatrix} w \\ z \end{pmatrix}$$

for some $u, w \in X$ and $v, z \in \partial X$. Since

$$\lim_{n\to\infty} \begin{pmatrix} A \\ L \end{pmatrix} u_n = \lim_{n\to\infty} \begin{pmatrix} Au_n \\ v_n \end{pmatrix} = \begin{pmatrix} w \\ v \end{pmatrix}$$

and $\binom{A}{L}$ is closed, we obtain $u \in D(A)$ and $Lu = v$, thus showing that $\binom{u}{v} \in D(\tilde{\mathcal{A}})$. Furthermore, $Au = w$. Finally, the continuity of B with respect to the graph norm of A implies that $z = \lim_{n\to\infty} Bu_n = Bu$, and hence $\tilde{\mathcal{A}}\binom{u}{v} = \binom{w}{z}$.

Assume now $\tilde{\mathcal{A}}$ to be closed. Let

$$(u_n)_{n\in\mathbb{N}} \subseteq D(A), \quad \lim_{n\to\infty} u_n = u, \quad \text{and} \quad \lim_{n\to\infty} \begin{pmatrix} A \\ L \end{pmatrix} u_n = \lim_{n\to\infty} \begin{pmatrix} Au_n \\ Lu_n \end{pmatrix} = \begin{pmatrix} w \\ v \end{pmatrix}$$

for some $u, w \in X$, $v \in \partial X$. It follows that $(Bu_n)_{n\in\mathbb{N}}$ converges to some $z \in \partial X$, and that

$$\begin{pmatrix} u_n \\ Lu_n \end{pmatrix}_{n\in\mathbb{N}} \subseteq D(\mathcal{A}), \ \lim_{n\to\infty} \begin{pmatrix} u_n \\ Lu_n \end{pmatrix} = \begin{pmatrix} u \\ v \end{pmatrix}, \ \lim_{n\to\infty} \tilde{\mathcal{A}} \begin{pmatrix} u_n \\ Lu_n \end{pmatrix} = \lim_{n\to\infty} \begin{pmatrix} Au_n \\ Bu_n \end{pmatrix} = \begin{pmatrix} w \\ z \end{pmatrix},$$

and hence, due to closedness of $\tilde{\mathcal{A}}$,

$$\begin{pmatrix} u \\ v \end{pmatrix} \in D(\mathcal{A}) \quad \text{and} \quad \tilde{\mathcal{A}} \begin{pmatrix} u \\ v \end{pmatrix} = \begin{pmatrix} Au \\ Bu \end{pmatrix} = \begin{pmatrix} w \\ z \end{pmatrix},$$

i.e., $u \in D(A)$, $Lu = v$, and $Au = w$. □

Lemma 2.4. *Assume that L is surjective and $\ker(L)$ is dense in X. Then the operator $\tilde{\mathcal{A}}$ is densely defined.*

Proof. Let $u \in X$, $v \in \partial X$, $\varepsilon > 0$. Surjectivity of L ensures that there exists $w \in D(A)$ such that $Lw = v$. Take $\tilde{u}, \tilde{w} \in \ker(L)$ such that $\|u - \tilde{u}\|_X < \varepsilon$ and $\|w - \tilde{w}\|_X < \varepsilon$. Let $z := \tilde{u} + w - \tilde{w} \in D(A)$. Then

$$\left\| \begin{pmatrix} u \\ v \end{pmatrix} - \begin{pmatrix} z \\ v \end{pmatrix} \right\| \leq \left\| \begin{pmatrix} u - \tilde{u} \\ 0 \end{pmatrix} \right\| + \left\| \begin{pmatrix} w - \tilde{w} \\ 0 \end{pmatrix} \right\| \leq 2\varepsilon.$$

Since $L(z) = L(w) = v$, we obtain $\binom{z}{v} \in D(\tilde{\mathcal{A}})$. □

For the operator $\tilde{\mathcal{A}}$, we consider the abstract Cauchy problem on the product space $\mathcal{X}$

$$\begin{cases} \dot{\mathcal{U}}(t) &= \tilde{\mathcal{A}}\,\mathcal{U}(t), \qquad t \geq 0, \\ \mathcal{U}(0) &= \mathbf{x}, \end{cases} \tag{ACP$_\mathbf{x}$}$$

with initial data $\mathbf{x} \in \mathcal{X}$.

As usual (cf. [EN00, Def. II.6.1]), for $\mathbf{x} \in D(\tilde{\mathcal{A}})$ a function $\mathcal{U}$ is called **classical solution** of the abstract Cauchy problem $(\mathrm{ACP})_\mathbf{x}$ if

- $\mathcal{U} \in C^1(\mathbb{R}_+, \mathcal{X})$,
- $\mathcal{U}(t) \in D(\tilde{\mathcal{A}})$ for all $t \geq 0$,
- $\mathcal{U}$ satisfies $(\mathrm{ACP})_\mathbf{x}$.

As in [EN00, Def. II.6.8], the abstract Cauchy problem (ACP) associated to the closed operator $\tilde{\mathcal{A}}$ is called **well-posed** if

- $D(\tilde{\mathcal{A}})$ is dense in $\mathcal{X}$,
- $(\mathrm{ACP})_\mathbf{x}$ admits a unique classical solution $\mathcal{U}(\cdot, \mathbf{x})$ for all $\mathbf{x} \in D(\tilde{\mathcal{A}})$,

- for every sequence of initial data $(\mathbf{x}_n)_{n\in\mathbb{N}} \subseteq D(\tilde{A})$ tending to 0 there holds $\lim_{n\to\infty} \mathcal{U}(t,\mathbf{x}_n) = 0$ uniformly in t on compact intervals.

Further (cf. [EN00, Def. II.6.3]), for arbitrary $\mathbf{x} \in \mathcal{X}$ a function $\mathcal{U}$ is called **mild solution** of $(\mathrm{ACP})_{\mathbf{x}}$ if

- $\mathcal{U} \in C(\mathbb{R}_+, \mathcal{X})$,
- $\int_0^t \mathcal{U}(s)\, ds \in D(\tilde{A})$ for all $t \geq 0$,
- $\mathcal{U}$ satisfies the integrated problem

$$\mathcal{U}(t) = \mathbf{x} + \tilde{A} \int_0^t \mathcal{U}(s)\, ds, \qquad t \geq 0. \tag{IP}$$

We now apply the semigroup characterization of well-posedness of abstract Cauchy problems (see [EN00, Prop. II.6.2, Prop. II.6.4, Cor. II.6.9]).

Proposition 2.5. *Let $(\tilde{A}, D(\tilde{A}))$ be a closed operator on a Banach space $\mathcal{X}$. Then* (ACP) *is well-posed if and only if $\tilde{A}$ generates a strongly continuous semigroup. Moreover, if $(\tilde{A}, D(\tilde{A}))$ generates a strongly continuous semigroup $(\tilde{\mathcal{T}}(t))_{t\geq 0}$, then the function*

$$\mathbb{R}_+ \ni t \mapsto \mathcal{U}(t) := \tilde{\mathcal{T}}(t)\mathbf{x} \in \mathcal{X} \qquad \text{for } \mathbf{x} \in D(\tilde{A}) \quad (\mathbf{x} \in \mathcal{X}, \ \text{respectively})$$

is the unique classical (mild, respectively) solution to $(\mathrm{ACP})_{\mathbf{x}}$.

We now show that well-posedness of (AIBVP) is equivalent to well-posedness of the corresponding (ACP). To that purpose we relate the (classical and mild) solutions of the two problems.

Lemma 2.6.

(i) *If u is a classical solution to* $(\mathrm{AIBVP})_{f,g}$*, then $\mathcal{U} := \binom{u}{Lu}$ is a classical solution to* $(\mathrm{ACP})_{\binom{f}{g}}$.

(ii) *Conversely, if $\mathcal{U}$ is a classical solution to* $(\mathrm{ACP})_{\mathbf{x}}$*, then $u := \pi_1\mathcal{U}$ is a classical solution to* $(\mathrm{AIBVP})_{\pi_1(\mathbf{x}),\pi_2(\mathbf{x})}$.

Proof. (i) Let u be a classical solution to $(\mathrm{AIBVP})_{f,g}$. It follows that $u \in C^1(\mathbb{R}_+, X)$ and $Lu \in C^1(\mathbb{R}_+, \partial X)$, and therefore $\mathcal{U} \in C^1(\mathbb{R}_+, \mathcal{X})$. Moreover, $Lf = g$ holds, hence $\binom{f}{g} \in D(\tilde{A})$, and also $\mathcal{U}(t) \in D(\tilde{A})$ for all $t \geq 0$. Finally, one can see that $(\mathrm{ACP})_{\binom{f}{g}}$ is fulfilled.

(ii) Assume now $\binom{u}{v} := \mathcal{U}$ to be a classical solution to $(\mathrm{ACP})_{\mathbf{x}}$. Then $\mathcal{U}(t) \in D(\tilde{A})$ for all $t \geq 0$ and hence $u(t) \in D(A)$ and $v(t) = Lu(t)$ for all $t \geq 0$. It also follows from $\mathcal{U} \in C^1(\mathbb{R}_+, \mathcal{X})$ that $u \in C^1(\mathbb{R}_+, X)$ and $Lu \in C^1(\mathbb{R}_+, \partial X)$. Finally, $\pi_2(\mathbf{x}) = v(0) = Lu(0) = L(\pi_1(\mathbf{x}))$, and $(\mathrm{AIBVP})_{\pi_1(\mathbf{x}),\pi_2(\mathbf{x})}$ is fulfilled. □

Lemma 2.7.

(i) *If u is a mild solution to* $(\mathrm{AIBVP})_{f,g}$*, then $\mathcal{U} := \binom{u}{v}$ is a mild solution to* $(\mathrm{ACP})_{\binom{f}{g}}$*, where $v(t) := g + B\left(\int_0^t u(s)ds\right)$, $t \geq 0$.*

(ii) *Conversely, assume that $\mathcal{U}$ is a mild solution to* $(\mathrm{ACP})_{\mathbf{x}}$. *Then* $u := \pi_1 \mathcal{U}$ *is a mild solution to* $(\mathrm{AIBVP})_{\pi_1(\mathbf{x}),\pi_2(\mathbf{x})}$.

Proof. (i) Let u be a mild solution to $(\mathrm{AIBVP})_{f,g}$. It follows that $u \in C(\mathbb{R}_+, X)$ and $B\left(\int_0^{\cdot} u(s)\right) ds \in C(\mathbb{R}_+, \partial X)$, and therefore $\mathcal{U} \in C(\mathbb{R}_+, \mathcal{X})$. Moreover, u fulfills the condition (MS_3), and this implies that $L \int_0^t u(s)ds = \int_0^t v(s)ds$, i.e., $\int_0^t \mathcal{U}(s)\, ds \in D(\tilde{\mathcal{A}})$ for all $t \geq 0$. Finally, there holds

$$u(t) = f + A \int_0^t u(s)\, ds, \qquad \text{and} \qquad v(t) = g + B \int_0^t u(s)\, ds$$

(respectively, by (MS_3) and by definition). These equalities express that (IP) is satisfied.

(ii) Assume now $\binom{u}{v} := \mathcal{U}$ to be a mild solution to $(\mathrm{ACP})_{\mathbf{x}}$. It follows that $\mathcal{U} \in C(\mathbb{R}_+, \mathcal{X})$, and hence $u \in C(\mathbb{R}_+, X)$ and $B\left(\int_0^{\cdot} u(s)ds\right) \in C(\mathbb{R}_+, \partial X)$. Moreover, $\int_0^t \mathcal{U}(s)ds \in D(\tilde{\mathcal{A}})$ for all $t \geq 0$, and this yields

$$\int_0^t u(s)\, ds \in D(A) \quad \text{and} \quad L \int_0^t u(s)\, ds = \int_0^t v(s)\, ds = tg + \int_0^t B\left(\int_0^s u(r)\, dr\right) ds,$$

for all $t \geq 0$. Finally, taking the first coordinate of (IP), it follows that $u(t) = \pi_1(\mathbf{x}) + A \int_0^t u(s)ds$. □

Theorem 2.8.

(i) *If* $(\tilde{\mathcal{A}}, D(\tilde{\mathcal{A}}))$ *generates a strongly continuous semigroup* $(\tilde{\mathcal{T}}(t))_{t\geq 0}$ *on* $\mathcal{X}$, *then* (AIBVP) *is well-posed. In this case, the unique classical solution to* $(\mathrm{AIBVP})_{f,g}$ *is given by* $\pi_1 \tilde{\mathcal{T}}(\cdot)\binom{f}{g}$ *for all initial data* $(f,g) \in D(\tilde{\mathcal{A}})$.

(ii) *Conversely, assume* $(\tilde{\mathcal{A}}, D(\tilde{\mathcal{A}}))$ *to be closed and densely defined. If* (AIBVP) *is well-posed, then* $(\tilde{\mathcal{A}}, D(\tilde{\mathcal{A}}))$ *generates a strongly continuous semigroup* $(\tilde{\mathcal{T}}(t))_{t\geq 0}$ *on* $\mathcal{X}$.

Proof. (i) Let $(\tilde{\mathcal{A}}, D(\tilde{\mathcal{A}}))$ generate a strongly continuous semigroup $(\tilde{\mathcal{T}}(t))_{t\geq 0}$ on $\mathcal{X}$. Then $\tilde{\mathcal{A}}$ is densely defined, which implies that A is densely defined, too. Moreover, $\tilde{\mathcal{A}}$ is closed and, by Proposition 2.5, the associated (ACP) is well-posed. By Lemma 2.6(*ii*), $u(\cdot, f, g) = \pi_1 \mathcal{U}(\cdot, \binom{f}{g}) = \pi_1 \tilde{\mathcal{T}}(\cdot)\binom{f}{g}$ yields a classical solution to $(\mathrm{AIBVP})_{f,g}$ for all $(f,g) \in D(A) \times \partial X$ such that $Lf = g$, i.e., for all $(f,g) \in D(\tilde{\mathcal{A}})$. This classical solution is unique by Lemma 2.6(*i*).

Let now $t_0 > 0$ and $(f_n, g_n)_{n\in\mathbb{N}}$ be a sequence of initial data satisfying $Lf_n = g_n$ and tending to 0. Note that $\binom{f_n}{g_n} \in D(\tilde{\mathcal{A}})$ and $\mathcal{U}(t, \binom{f_n}{g_n}) \in D(\tilde{\mathcal{A}})$ for all $t \geq 0$. Hence we have $\lim_{n\to\infty} \mathcal{U}(t, \binom{f_n}{g_n}) = 0$ uniformly in $[0, t_0]$ if and only if $\lim_{n\to\infty} u(t, f_n, g_n) = 0$ and $\lim_{n\to\infty} Lu(t, f_n, g_n) = 0$ (both uniformly in $[0, t_0]$). Since (ACP) is well-posed, the assertion follows.

(ii) Since $\tilde{\mathcal{A}}$ is closed, it suffices by Proposition 2.5 to show that the associated (ACP) is well-posed. Let $\mathbf{x} \in D(\tilde{\mathcal{A}})$. Well-posedness of (AIBVP) yields, by

Lemma 2.6, existence and uniqueness of a classical solution to $(\mathrm{ACP})_{\mathbf{x}}$. By assumption, $\overline{D(\mathcal{A})} = \mathcal{X}$, therefore it only remains to show the continuous dependence on initial data.

Let $t_0 > 0$ and $(\mathbf{x}_n)_{n\in\mathbb{N}} \subseteq D(\mathcal{A})$ be a sequence of initial data tending to 0. Then, $(\pi_1(\mathbf{x}_n), \pi_2(\mathbf{x}_n))_{n\in\mathbb{N}}$ is a sequence of initial data for (AIBVP) tending to 0 and such that $L\pi_1(\mathbf{x}_n) = \pi_2(\mathbf{x}_n)$, and, by assumptions, there holds $\lim_{n\to\infty} u(t, \pi_1\mathbf{x}_n, \pi_2\mathbf{x}_n) = 0$ and $\lim_{n\to\infty} Lu(t, \pi_1\mathbf{x}_n, \pi_2\mathbf{x}_n) = 0$ (both uniformly in $[0, t_0]$). Also, $\mathcal{U}(\cdot, \mathbf{x}_n) = \binom{u}{Li}(\cdot, \binom{\pi_1\mathbf{x}_n}{\pi_2\mathbf{x}_n})$ is the (unique) classical solution to $(\mathrm{ACP})_{\mathbf{x}_n}$ for each $n \in \mathbb{N}$, and we finally obtain $\lim_{n\to\infty} \mathcal{U}(t, \mathbf{x}_n) = \lim_{n\to\infty} \binom{u}{Lu}(t, \binom{\pi_1\mathbf{x}_n}{\pi_2\mathbf{x}_n}) = 0$ uniformly in $[0, t_0]$. □

In the same way we obtain mild solutions to (AIBVP).

Corollary 2.9. *Let $(\tilde{\mathcal{A}}, D(\tilde{\mathcal{A}}))$ generate a strongly continuous semigroup $(\tilde{\mathcal{T}}(t))_{t\geq 0}$ on $\mathcal{X}$. Then $u(t) := \pi_1 \tilde{\mathcal{T}}(\cdot)\binom{f}{g}$ yields the unique mild solution to $(\mathrm{AIBVP})_{f,g}$ for all initial data $(f, g) \in \mathcal{X}$.*

Sufficient conditions on the operators A, B, and L implying $\mathcal{A}$ and $\tilde{\mathcal{A}}$ to be generators, hence implying the well-posedness of (AIBVP), will be given in the next section.

3. Well-Posed Initial-Boundary Value Problems

We first study the situation from Example 1.2, i.e., we consider the case where $B = 0$.

Assumptions 3.1. In order to obtain well-posedness of (AIBVP), we now impose the following.

(G_1) The restriction $A_0 := A|_{\ker(L)} : D(A_0) := \ker(L) \to X$ is closed and densely defined, and it has non-empty resolvent set.

(G_2) The boundary operator $L : D(A) \to \partial X$ is surjective.

(G_3) The operator $\binom{A}{L} : D(A) \subseteq X \to X \times \partial X$ is closed.

The following key lemma is essentially due to Greiner ([Gr87, Lemma 1.2 and Lemma 1.3], see also [CENN01, Lemma 2.3]).

Lemma 3.2. *Let $\lambda \in \rho(A_0)$. Then the restriction $L|_{\ker(\lambda - A)}$ is invertible and its inverse*

$$\mathcal{D}_\lambda := \left(L|_{\ker(\lambda - A)}\right)^{-1} : \partial X \to \ker(\lambda - A),$$

called **Dirichlet operator**, *is bounded. Moreover, for all $\mu \in \rho(A_0)$, $\mathcal{D}_\mu$ is related to $\mathcal{D}_\lambda$ by*

$$\mathcal{D}_\lambda = (I + (\mu - \lambda)R(\lambda, A_0))\mathcal{D}_\mu. \tag{3.1}$$

These Dirichlet operators play an important role in the following.

Lemma 3.3. *Let A_0 generate a strongly continuous semigroup $(T(t))_{t\geq 0}$. For each $\lambda \in \rho(A_0)$ define a family of operators $(Q_\lambda(t))_{t\geq 0}$ from ∂X to X by*

$$Q_\lambda(t)v := (\lambda - A_0)\int_0^t T(s)\mathcal{D}_\lambda v\, ds, \qquad t \geq 0,\ v \in \partial X.$$

Then, for a given $t \geq 0$, all the operators $Q_\lambda(t)$, $\lambda \in \rho(A_0)$, coincide, and will be denoted by $Q(t)$. In particular, if A_0 is invertible, then

$$Q(t) = (I_X - T(t))\,\mathcal{D}_0 \tag{3.2}$$

for all $t \geq 0$.

Proof. Let $\lambda, \mu \in \rho(A_0)$, $v \in \partial X$. Then

$$\begin{aligned} Q_\lambda(t)v &= (\lambda - A_0)\textstyle\int_0^t T(s)\mathcal{D}_\lambda v\, ds \\ &\overset{(3.1)}{=} (\lambda - A_0)\textstyle\int_0^t T(s)\mathcal{D}_\mu v\, ds \\ &\quad +(\mu-\lambda)(\lambda - A_0)\textstyle\int_0^t e^{-\lambda s}T(s)R(\lambda, A_0)e^{\lambda s}\mathcal{D}_\mu v\, ds \\ &= (\lambda - A_0)\textstyle\int_0^t T(s)\mathcal{D}_\mu v\, ds + (\mu-\lambda)\int_0^t T(s)\mathcal{D}_\mu v\, ds = Q_\mu(t)v \end{aligned}$$

by standard properties of strongly continuous semigroups, cf. [EN00, Lemma II.1.3]. Integrating by part we obtain (3.2). □

We now return to the operator matrix $\mathcal{A}$ defined in (2.1).

Lemma 3.4. *Let $\lambda \in \rho(A_0)$. Then*

$$D(\mathcal{A}) = \left\{ \binom{u}{v} \in D(A)\times \partial X \,:\, Lu = v \right\}$$

coincides with

$$\mathcal{H} := \left\{ \binom{u}{v} \in X \times \partial X \,:\, u - \mathcal{D}_\lambda v \in D(A_0) \right\}. \tag{3.3}$$

Moreover, the identity

$$\mathcal{A} - \lambda = \begin{pmatrix} A_0 - \lambda & 0 \\ 0 & -\lambda \end{pmatrix}\begin{pmatrix} I_X & -\mathcal{D}_\lambda \\ 0 & I_{\partial X} \end{pmatrix} =: \mathcal{A}_\lambda \mathcal{R}_\lambda \tag{3.4}$$

holds.

Proof. To show that $D(\mathcal{A})$ equals $\mathcal{H}$ take a vector $\binom{u}{v} \in \mathcal{X}$. Then, $\binom{u}{v} \in \mathcal{H}$ if and only if $u - \mathcal{D}_\lambda v \in \ker(L)$. But $\mathcal{D}_\lambda v \in \ker(\lambda - A) \subseteq D(A)$, and hence $u - \mathcal{D}_\lambda v \in D(A_0) = \ker(L)$ if and only if $u \in D(A)$ and $L(u - \mathcal{D}_\lambda v) = Lu - v = 0$, i.e., $Lu = v$. Thus $D(\mathcal{A}) = \mathcal{H}$.

It is clear that $\mathcal{H} = D(\mathcal{A}_\lambda \mathcal{R}_\lambda)$ since $\binom{u}{v} \in D(\mathcal{A}_\lambda\mathcal{R}_\lambda)$ if and only if

$$\mathcal{R}_\lambda \binom{u}{v} = \binom{u - \mathcal{D}_\lambda v}{v} \in D(\mathcal{A}_\lambda) = D(A_0)\times \partial X.$$

To show (3.4), we take $\binom{u}{v} \in D(\mathcal{A}_\lambda\mathcal{R}_\lambda)$ and obtain

$$\mathcal{A}_\lambda\mathcal{R}_\lambda\binom{u}{v} = \begin{pmatrix} A_0 - \lambda & 0 \\ 0 & -\lambda \end{pmatrix}\binom{u - \mathcal{D}_\lambda v}{v} = \binom{(A-\lambda)u}{-\lambda v} = (\mathcal{A} - \lambda)\binom{u}{v}. \qquad \square$$

Remark 3.5.

1. The factorization (3.4) shows that $\mathcal{A} - \lambda$ is a so-called *one-sided coupled operator matrix*. For more details about this general theory, we refer the reader to [En99], [En01], or [KMN02].
2. Observe that, under the Assumptions 3.1, and by Lemma 2.3 and 2.4, $(\mathcal{A}, D(\mathcal{A}))$ is closed and densely defined. Therefore, Theorem 2.8 implies that (AIBVP) is well-posed if and only if $\mathcal{A}$ generates a strongly continuous semigroup.

Theorem 3.6. *Under the Assumptions 3.1 the following conditions are equivalent.*

a) *The operator $(\mathcal{A}, D(\mathcal{A}))$ generates a strongly continuous semigroup on $\mathcal{X}$.*
b) *The operator $(A_0, D(A_0))$ generates a strongly continuous semigroup on X.*

If a) holds, the semigroup $(\mathcal{T}(t))_{t\geq 0}$ generated by $\mathcal{A}$ is given by

$$\mathcal{T}(t) = \begin{pmatrix} T(t) & Q(t) \\ 0 & I_{\partial X} \end{pmatrix},$$

for all $t \geq 0$, where $(T(t))_{t\geq 0}$ is the semigroup generated by A_0 and $(Q(t))_{t\geq 0}$ is the family of operators introduced in Lemma 3.3. In the particular case of invertible A_0, $(\mathcal{T}(t))_{t\geq 0}$ is given by

$$\mathcal{T}(t) = \begin{pmatrix} T(t) & (I_X - T(t))\,\mathcal{D}_0 \\ 0 & I_{\partial X} \end{pmatrix}. \tag{3.5}$$

Proof. *a)* $\Rightarrow$ *b)*. Assume that $\mathcal{A}$ is a generator. Observe first that $\rho(\mathcal{A}) = \rho(A_0)\setminus\{0\}$. Indeed, it follows from (3.4) that for all $\lambda \in \rho(A_0)\setminus\{0\}$ the resolvent is given by

$$R(\lambda, \mathcal{A}) = \begin{pmatrix} R(\lambda, A_0) & \frac{1}{\lambda}\mathcal{D}_\lambda \\ 0 & \frac{1}{\lambda} \end{pmatrix}. \tag{3.6}$$

Hence, its powers are

$$R(\lambda, \mathcal{A})^n = \begin{pmatrix} R(\lambda, A_0)^n & * \\ 0 & \frac{1}{\lambda^n} \end{pmatrix} \qquad \text{for all } n \in \mathbb{N}.$$

Therefore, $\|R(\lambda, A_0)^n\|$ can be dominated by $\|R(\lambda, \mathcal{A})^n\|$ and satisfies a Hille–Yosida estimate. The closedness of A_0 follows by the closedness of $\mathcal{A}$. Finally, $D(A_0)$ is dense in X by Assumption (G_1), and hence A_0 is a generator.

b) $\Rightarrow$ *a)*. Assume first A_0 to be invertible, and define $\mathcal{T}(t)$ as in (3.5). We show that the family $(\mathcal{T}(t))_{t\geq 0}$ is a strongly continuous semigroup and then verify that its generator is $\mathcal{A}$.

To prove the semigroup property, take $t, s \in \mathbb{R}_+$ and $u \in X$, $v \in \partial X$. We then have

$$\begin{aligned}
\mathcal{T}(t)\mathcal{T}(s)\binom{u}{v} &= \begin{pmatrix} T(t) & Q(t) \\ 0 & I_{\partial X} \end{pmatrix} \begin{pmatrix} T(s) & Q(s) \\ 0 & I_{\partial X} \end{pmatrix} \binom{u}{v} \\
&= \begin{pmatrix} T(t)T(s)u + T(t)\mathcal{D}_0 v - T(t)T(s)\mathcal{D}_0 v + \mathcal{D}_0 v - T(t)\mathcal{D}_0 v \\ v \end{pmatrix} \\
&= \begin{pmatrix} T(t+s)u + (I_X - T(t+s))\mathcal{D}_0 v \\ v \end{pmatrix} \\
&= \mathcal{T}(t+s)\binom{u}{v}.
\end{aligned}$$

Moreover,

$$\mathcal{T}(0) = \begin{pmatrix} T(0) & Q(0) \\ 0 & I_{\partial X} \end{pmatrix} = \begin{pmatrix} I_X & 0 \\ 0 & I_{\partial X} \end{pmatrix} = I_{\mathcal{X}}.$$

To show strong continuity of $(\mathcal{T}(t))_{t\geq 0}$, i.e.,

$$\lim_{t\to 0^+} \mathcal{T}(t)\begin{pmatrix} u \\ v \end{pmatrix} = \begin{pmatrix} \lim_{t\to 0^+} \left[T(t)u + Q(t)v \right] \\ v \end{pmatrix} = \begin{pmatrix} u \\ v \end{pmatrix}$$

for all $\binom{u}{v} \in \mathcal{X}$, it suffices to verify that the operators $Q(t)$ converge to 0 as $t \to 0^+$. This follows indeed from (3.2) and from the boundedness of $\mathcal{D}_0$.

Finally, in order to show that $\mathcal{A}$ is actually the generator of $(\mathcal{T}(t))_{t\geq 0}$, we note that for λ with large real part the Laplace transform of $\mathcal{T}(\cdot)$ coincides with the resolvent of its generator. In particular, and taking into account the explicit formula (3.6) for the resolvent of $\mathcal{A}$, we obtain for the Laplace transform of the upper right entry $\mathcal{T}(\cdot)_{12}$ of $\mathcal{T}(\cdot)$ the equality

$$\mathcal{L}(\mathcal{T}(\cdot)_{12})(\lambda) = R(\lambda, \mathcal{A})_{12} = \frac{1}{\lambda}\mathcal{D}_\lambda.$$

On the other hand, the convolution theorem for the Laplace transform implies that $\mathcal{L}(Q(\cdot)v)(\lambda) = \frac{1}{\lambda}\mathcal{D}_\lambda v$ for all $v \in \partial X$ and λ sufficiently large. This finally proves that $\mathcal{L}(Q(\cdot)v)(\lambda) = \mathcal{L}(\mathcal{T}(\cdot)_{12}v)(\lambda)$ for all $v \in \partial X$ and λ large. The injectivity of the Laplace transformation then implies that $\mathcal{T}(\cdot)_{12} = Q(\cdot)$.

Finally, the case of $0 \notin \rho(A_0)$ can be discussed as above, by rescaling arguments. □

Corollary 3.7. *Under the Assumptions 3.1, A_0 generates a bounded strongly continuous semigroup if and only if $\mathcal{A}$ generates a bounded strongly continuous semigroup.*

Proof. Assume, without loss of generality, A_0 to be invertible. Hence the formula (3.5) yields the semigroup generated by $\mathcal{A}$, and the claim follows. □

Proposition 3.8. *Under the Assumptions 3.1 $\mathcal{A}$ generates a bounded analytic semigroup on $\mathcal{X}$ if and only if A_0 generates a bounded analytic semigroup on X.*

This result may be shown by the same similarity and perturbation techniques of Proposition 4.3 below. However, the following more direct proof can also be used as suggested to us by the referee whom we thank for this remark.

Proof. Assume A_0 to generate a bounded analytic semigroup. By [EN00, Thm. II.4.6], this is equivalent to assume that (i) A_0 generates a bounded strongly continuous semigroup, and (ii) there exists a constant $C > 0$ such that

$$\|sR(r+is, A_0)\| \leq C \tag{3.7}$$

for all $r > 0$ and $0 \neq s \in \mathbb{R}$.

It hence follows by Theorem 3.6 and Corollary 3.7 that also $\mathcal{A}$ generates a bounded strongly continuous semigroup. Moreover, since $\rho(\mathcal{A}) = \rho(A_0) \setminus \{0\}$ as remarked in the proof of Theorem 3.6, the resolvent operator $R(r+is, \mathcal{A})$ is defined for $r > 0$ and $0 \neq s \in \mathbb{R}$. Taking into account (3.1), (3.6), and (3.7), the resolvent of $\mathcal{A}$ satisfies an estimate of the form

$$\begin{aligned} \|sR(r+is, \mathcal{A})\| &\leq \|sR(r+is, A_0)\| + \|\mathcal{D}_\mu\| + \|sR(r+is, A_0)\mathcal{D}_\mu\| + 1 \\ &\leq C := C + \|\mathcal{D}_\mu\| + C\|\mathcal{D}_\mu\| + 1 \end{aligned}$$

for all $r > 0$ and $0 \neq s \in \mathbb{R}$, and arbitrary $\mu \in \rho(A_0)$. It follows that also $\mathcal{A}$ generates a bounded analytic semigroup. The converse implication can be proven likewise. □

An answer to the problem posed in Example 1.2 can now given by simply using the variation of constants formula for semigroups (see [EN00, § VI.7]).

Proposition 3.9. *Assume that the Assumptions 3.1 hold and that A_0 generates a strongly continuous semigroup, and take $f \in X$ and $g \in \partial X$. If $\psi \in L^1(\mathbb{R}_+, \partial X)$, then the unique (mild, in the sense of [EN00, Def. VI.7.2]) solution to the inhomogeneous problem (iACP) is given by*

$$\mathcal{U} : \mathbb{R}_+ \to \mathcal{X}, \qquad \mathcal{U} : t \mapsto \mathcal{U}(t) := \mathcal{T}(t)\begin{pmatrix} f \\ g \end{pmatrix} + \int_0^t \mathcal{T}(t-s)\begin{pmatrix} 0 \\ \psi(s) \end{pmatrix} ds.$$

In the particular case where A_0 is invertible, $\mathcal{U}$ is given by

$$\mathcal{U}(t) = \begin{pmatrix} T(t)f + \mathcal{D}_0 g - T(t)\mathcal{D}_0 g + \int_0^t (\mathcal{D}_0\psi(s) - T(t-s)\mathcal{D}_0\psi(s))\, ds \\ g + \int_0^t \psi(s)\, ds \end{pmatrix}. \tag{3.8}$$

Moreover, if $f \in D(A)$, $f|_{\partial\Omega} = g$, and $\psi \in W^{1,1}(\mathbb{R}_+, \partial X)$, then $\mathcal{U}$ is a classical solution to (iACP).

This result can now be applied to the inhomogeneous problem stated in Example 1.2.

Example 3.10. We have to show that the setting introduced in Example 1.2 fits the hypotheses yielding the results presented in §§ 2-3.

Take as state and boundary space $X := L^2(\Omega)$ and $\partial X := L^2(\partial\Omega)$, respectively. As operator A we take the Laplacian

$$A := \Delta \quad \text{with domain} \quad D(A) := \left\{ u \in H^{\frac{1}{2}}(\Omega) \cap H^2_{\text{loc}}(\Omega) \,:\, \Delta u \in L^2(\Omega) \right\}$$

and as boundary operator L the trace operator

$$Lu := u\big|_{\partial\Omega} \quad \text{with domain} \quad D(L) := D(A)$$

as introduced in [LM72, §§ 2.5-8]. In particular, A is defined on the maximal domain such that the traces of its elements exist as $L^2(\partial\Omega)$ functions. This is why $(A, D(A))$ can be called the *maximal operator*.

With these definitions, (iHE) takes the form (iACP). In order to apply Proposition 3.9 we have to show the well-posedness of (AIBVP). For this, it suffices, by Theorem 2.8, that $\mathcal{A}$ generates a strongly continuous semigroup on the product space $\mathcal{X}$.

We first note that the operator A_0 is in this case nothing but the Dirichlet Laplacian Δ^D, which, by standard results (see, e.g., [Pa83, Thm. 7.2.7]) generates an analytic semigroup on X. Surjectivity of L follows from [LM72, Thm. 2.7.4], see also [CENN01, Lemma 3.1]. Finally, the proof of the closedness of $\binom{A}{L}$ can be found in [CENN01, Lemma 3.2].

The Dirichlet Laplacian on $L^2(\Omega)$ is invertible, and hence an explicit solution to (iHE) is yielded by (3.8), where $(T(t))_{t\geq 0}$ is the heat semigroup and $\mathcal{D}_0$ the operator that maps each $g \in D(A)$ into the unique harmonic function attaining boundary values g.

Summing up, the Assumptions 3.1 are satisfied and by Proposition 3.8 well-posedness of the homogeneous (HE) is ensured. Existence and uniqueness of the solution to problem (iHE) then follows from Proposition 3.9 under suitable assumptions on the inhomogeneous term ψ (see [EN00, § VI.7].

Remark 3.11. Although the Dirichlet Laplacian is dissipative, the operator $\mathcal{A}$ is *not* dissipative and the semigroup $(\mathcal{T}(t))_{t\geq 0}$ generated by $\mathcal{A}$ is not contractive on $\mathcal{X}$ (but bounded indeed, by Remark 3.7). In [FGGR02] it is shown how weighted norms and suitable feedbacks can be used in this context to obtain contractivity.

4. Abstract Boundary Feedback Systems

We now return to the operator $\tilde{\mathcal{A}}$ defined in (2.2). This means that we replace the inhomogeneous term ψ in (iHE) or in (iACP) by a boundary feedback, i.e., we take

$$\psi(t) := Bu(t), \qquad t \geq 0,$$

for some operator B from X into ∂X. With this interpretation, we call (AIBVP) an *abstract boundary feedback system*. We combine the results obtained in the previous section and perturbation theory for semigroups. The simplest case in which B is bounded follows directly from [EN00, Thm. III.1.3].

Proposition 4.1. *Assume that $B : X \to \partial X$ is bounded. Under the Assumptions 3.1 the operator $\tilde{\mathcal{A}}$ generates a strongly continuous (respectively, analytic) semigroup on $\mathcal{X}$ if and only if A_0 generates a strongly continuous (respectively, analytic) semigroup on X.*

The case of unbounded feedback operators defined on $D(A)$ can be treated by using the results of [En99] on one-sided coupled operator matrices. We briefly sketch the main steps. First, define

$$B_\lambda := B\mathcal{D}_\lambda : \partial X \to \partial X$$

for all $\lambda \in \rho(A_0)$ (observe that B_λ is well defined since $R(\mathcal{D}_\lambda) \subseteq D(A) \subseteq D(B)$). Then, the following lemma holds.

Lemma 4.2. *For $\lambda \in \rho(A_0)$ the operator $\tilde{\mathcal{A}} - \lambda$ can be factorized as*

$$\tilde{\mathcal{A}} - \lambda = \begin{pmatrix} A_0 - \lambda & 0 \\ B & B_\lambda - \lambda \end{pmatrix} \begin{pmatrix} I_X & -\mathcal{D}_\lambda \\ 0 & I_{\partial X} \end{pmatrix} \tag{4.1}$$

with $D(\tilde{\mathcal{A}}) = \mathcal{H}$ as defined in (3.3).

Proof. Taking into account the identity (3.4), a matrix computation yields

$$\tilde{\mathcal{A}} - \lambda = \mathcal{A}_\lambda \mathcal{R}_\lambda + \begin{pmatrix} 0 & 0 \\ B & 0 \end{pmatrix} = \left[\mathcal{A}_\lambda + \begin{pmatrix} 0 & 0 \\ B & 0 \end{pmatrix} \mathcal{R}_\lambda^{-1} \right] \mathcal{R}_\lambda$$
$$= \left[\begin{pmatrix} A_0 - \lambda & 0 \\ 0 & -\lambda \end{pmatrix} + \begin{pmatrix} 0 & 0 \\ B & B_\lambda \end{pmatrix} \right] \mathcal{R}_\lambda = \begin{pmatrix} A_0 - \lambda & 0 \\ B & B_\lambda - \lambda \end{pmatrix} \begin{pmatrix} I_X & -\mathcal{D}_\lambda \\ 0 & I_{\partial X} \end{pmatrix}$$

Note that the domain of $\tilde{\mathcal{A}} - \lambda$ coincides with $\mathcal{H}$ from (3.3). □

Proposition 4.3. *Let the Assumptions 3.1 hold, and assume that B is relatively A_0-bounded, and that B_{λ_0} is bounded for some $\lambda_0 \in \rho(A_0)$. Then the operator $\tilde{\mathcal{A}}$ generates a strongly continuous (resp., analytic) semigroup on $\mathcal{X}$ if and only if the operator $A_0 - \mathcal{D}_{\lambda_0} B$, defined on $D(A_0)$, generates a strongly continuous (resp., analytic) semigroup on X.*

Proof. By (4.1), we have

$$\tilde{\mathcal{A}} = \begin{pmatrix} A_0 & 0 \\ B & B_{\lambda_0} \end{pmatrix} \mathcal{R}_{\lambda_0} + \lambda_0 \begin{pmatrix} 0 & \mathcal{D}_{\lambda_0} \\ 0 & 0 \end{pmatrix} =: \tilde{\mathcal{A}}_{\lambda_0} \mathcal{R}_{\lambda_0} + \lambda_0 \mathcal{P}_{\lambda_0}.$$

By Lemma 3.2, $\mathcal{P}_{\lambda_0}$ is a bounded perturbation, so it suffices to show that $\tilde{\mathcal{A}}_{\lambda_0} \mathcal{R}_{\lambda_0}$ generates a strongly continuous (resp., analytic) semigroup. Due to the invertibility of $\mathcal{R}_{\lambda_0}$, $\tilde{\mathcal{A}}_{\lambda_0} \mathcal{R}_{\lambda_0}$ is similar to

$$\mathcal{R}_{\lambda_0} \tilde{\mathcal{A}}_{\lambda_0} = \begin{pmatrix} A_0 - \mathcal{D}_{\lambda_0} B & -\mathcal{D}_{\lambda_0} B_{\lambda_0} \\ B & B_{\lambda_0} \end{pmatrix} = \begin{pmatrix} A_0 - \mathcal{D}_{\lambda_0} B & 0 \\ B & 0 \end{pmatrix} + \begin{pmatrix} 0 & -\mathcal{D}_{\lambda_0} B_{\lambda_0} \\ 0 & B_{\lambda_0} \end{pmatrix}$$
$$=: \mathcal{M} + \mathcal{N}.$$

Since B_{λ_0} is bounded, also $-\mathcal{D}_{\lambda_0} B_{\lambda_0}$ and $\mathcal{N}$ are bounded. Hence, by bounded perturbation and similarity arguments, we can conclude that $\tilde{\mathcal{A}}$ is a generator if and only if the operator matrix $\mathcal{M}$ is a generator.

Observe further that B is also $(A_0 - \mathcal{D}_{\lambda_0} B)$-bounded and while the operator matrix $\tilde{\mathcal{A}}$ does not have diagonal domain, $D(\mathcal{M}) = D(\tilde{\mathcal{A}}_{\lambda_0}) = D(A_0) \times \partial X$: hence, by well-known results on matrices with diagonal domain (see, e.g., [Na89, Cor. 3.2 and Cor. 3.3]) one can see that $\mathcal{M}$ generates a strongly continuous (respectively, analytic) semigroup on $\mathcal{X}$ if and only if $A_0 - \mathcal{D}_{\lambda_0} B$ does the same on X. □

Remark 4.4. Observe that, in particular, if B is relatively $\binom{A}{L}$-bounded, then B is relatively A_0-bounded and B_{λ_0} is bounded for all $\lambda_0 \in \rho(A_0)$. Indeed, $\|u\|_{\binom{A}{L}} = \|u\|_A$ for all $u \in D(A_0)$, and hence B is relatively A_0-bounded.

Further, the closedness of $\binom{A}{L} : D(A) \to \mathcal{X}$ implies the closedness of the operator matrix

$$\mathcal{L} := \begin{pmatrix} A & 0 \\ L & 0 \end{pmatrix}, \qquad D(\mathcal{L}) := D(A) \times \partial X,$$

on $\mathcal{X}$, and hence we can define the Banach space $\mathcal{Y} := (D(\mathcal{L}), \|\cdot\|_{\mathcal{L}})$, with $\mathcal{Y} \hookrightarrow \mathcal{X}$. Now observe that, by virtue of Lemma 3.2, the operator

$$\mathcal{H}_\lambda := \begin{pmatrix} 0 & \mathcal{D}_\lambda \\ 0 & 0 \end{pmatrix} : \mathcal{X} \to \mathcal{X}$$

is bounded from $\mathcal{X}$ to $\mathcal{X}$ and its range is contained in $\mathcal{Y}$ for all $\lambda \in \rho(A_0)$. It then follows by [EN00, Cor. B.7] that $\mathcal{H}_\lambda$ is also bounded from $\mathcal{X}$ to $\mathcal{Y}$, and since under the above assumptions

$$\mathcal{B} := \begin{pmatrix} 0 & 0 \\ B & 0 \end{pmatrix}, \qquad D(\mathcal{B}) := D(A) \times \partial X$$

is bounded from $\mathcal{Y}$ to $\mathcal{X}$,

$$\mathcal{B}\mathcal{H}_\lambda = \begin{pmatrix} 0 & 0 \\ 0 & B_\lambda \end{pmatrix} : \mathcal{X} \to \mathcal{X}$$

is bounded, and the claim follows.

Example 4.5. We consider the diffusion-transport equation with dynamical boundary conditions

$$\begin{cases} \dot{u}(t,x) &= u''(t,x) + ku'(t,x), & t \geq 0,\ x \in (0,1), \\ \dot{u}(t,0) &= u'(t,0) + cu(t,0), & t \geq 0, \\ \dot{u}(t,1) &= -u'(t,1) + du(t,1), & t \geq 0, \\ u(0,x) &= f(x), & x \in (0,1), \\ u(0,0) &= a, \quad u(0,1) = b. \end{cases} \tag{DT}$$

where $a, b, c, d \in \mathbb{C}$, $k \geq 0$. This system becomes a (AIBVP) if we consider the state space $X := L^2(0,1)$, the boundary space $\partial X := \mathbb{C}^2$, the operator

$$Au := u'' + ku', \qquad D(A) := H^2(0,1),$$

the boundary operator

$$Lu := \begin{pmatrix} u(0) \\ u(1) \end{pmatrix}, \qquad D(L) := D(A),$$

and the feedback operator

$$Bu := \begin{pmatrix} u'(0) + cu(0) \\ -u'(1) + du(1) \end{pmatrix}, \qquad D(B) := D(A).$$

We first show that the Assumptions (G_1)–(G_3) hold. Observe that L is surjective onto $\mathbb{C}^2$. The operator A_0, defined as the restriction of A to $D(A_0) := H^2(0,1) \cap H^1_0(0,1)$, generates an analytic semigroup and, in particular, (G_1) is satisfied. Finally, one can show as in [CENN01, Lemma 3.3] that the operator $\mathcal{L} := \binom{\frac{d^2}{dx^2}}{L}$ is closed on $D(A)$. Since $\binom{k\frac{d}{dx}}{0}$ is relatively $\mathcal{L}$-bounded with $\mathcal{L}$-bound 0 for $k \geq 0$, their sum $\binom{A}{L}$ is also closed (see [EN00, Lemma III.2.4]). Hence, the Assumptions 3.1 are satisfied and we are now in the position to apply Proposition 4.3.

Observe that $0 \in \rho(A_0)$, and the Dirichlet operator $\mathcal{D}_0$ maps, by definition, each pair $(\alpha, \beta) \in \mathbb{C}^2$ into the unique solution of the ordinary differential equation

$$\begin{cases} u''(x) + ku'(x) = 0, & x \in (0,1), \\ u(0) = \alpha, \quad u(1) = \beta. \end{cases}$$

In order to show that Proposition 4.3 applies, we now check that B is relatively $\binom{A}{L}$-bounded. This will then yield, by virtue of Remark 4.4, that B is relatively A_0-bounded, and that B_λ is bounded for all $\lambda \in \rho(A_0)$. Recall that the first derivative on $L^2(0,1)$ is relatively bounded by the second derivative, with relative bound 0 (see, e.g., [EN00, § III.2]), and hence that the graph norm of the second derivative and the graph norm of A are equivalent. It then follows from the embedding $H^1(0,1) \hookrightarrow C([0,1])$ that we can find suitable constants ξ, ξ_1, ξ_2 such that

$$\begin{aligned} \|Bu\| &= \left(|u'(0) + cu(0)|^2 + |u'(1) - du(1)|^2 \right)^{\frac{1}{2}} \\ &\leq \zeta \left(\|u'\|_{C([0,1])} + |Lu|_{\mathbb{C}^2} \right) \\ &\leq \xi \left(\|u''\|_{L^2(0,1)} + \|u'\|_{L^2(0,1)} \right) + \zeta\, |Lu|_{\mathbb{C}^2} \\ &\leq \xi_1 \|u''\|_{L^2(0,1)} + \xi_2 \|u\|_{L^2(0,1)} + \zeta\, |Lu|_{\mathbb{C}^2} \leq \eta \|u\|_{\binom{A}{L}}, \end{aligned}$$

for all $u \in D(A)$, where $\zeta = 2\max\{1, |c|, |d|\}$ and $\eta = \max\{\xi_1, \xi_2, \zeta\}$. Hence, B is relatively $\binom{A}{L}$-bounded.

We finally show that $A_0 - \mathcal{D}_0 B$ generates an analytic semigroup. The boundedness of B from $(D(A_0), \|\cdot\|_{A_0})$ to $\mathbb{C}^2$ shows that B, and hence $\mathcal{D}_0 B$ is relatively A_0-compact. It then follows from perturbation theory (see [EN00, Cor. III.2.17]) that $A_0 - \mathcal{D}_0 B$, defined on $D(A_0)$, is the generator of an analytic semigroup.

Therefore, the operator

$$\mathcal{A} := \begin{pmatrix} A & 0 \\ B & 0 \end{pmatrix}, \qquad D(\mathcal{A}) := \left\{ \begin{pmatrix} u \\ Lu \end{pmatrix} : u \in D(A) \right\}$$

on $\mathcal{X} := X \times \partial X$ generates an analytic semigroup by Proposition 4.3, and the diffusion-transport equation (DT) is well-posed.

Remark 4.6. A question that naturally arises is how the stability of such a system is affected by the feedback operator. Under certain assumptions on the parameters c and d the semigroup generated by $\mathcal{A}$ is positive, and it can be shown that *stability of our boundary feedback system is independent of the internal diffusion.* This means that the spectral bound of $\mathcal{A}$ is negative if and only if the spectral bound of B_0 is negative. This allows to apply Liapunov-type stability results (see [EN00, Chap. V]), but we refer the reader to [KMN02, §§ 7–9] for details.

Acknowledgment

We thank K.-J. Engel for many helpful comments.

References

[1] W. Arendt, C.J.K. Batty, M. Hieber, and F. Neubrander, *Vector-valued Laplace Transforms and Cauchy Problems.* Monographs in Mathematics **96**, Birkhäuser, 2001.

[2] W. Arendt, *Resolvent positive operators and inhomogeneous boundary value problems.* Ann. Scuola Norm. Pisa Cl. Sci. **29** (2000), 639–670.

[3] V. Casarino, K.-J. Engel, R. Nagel, and G. Nickel, *A semigroup approach to boundary feedback systems.* Tübinger Berichte zur Funktionalanalysis **10** (2000/01), 30–43.

[4] R. Denk, M. Hieber, and J. Prüß, *R-boundedness, Fourier multipliers and problems of elliptic and parabolic type.* Preprint.

[5] K.-J. Engel, *Spectral theory and generator property for one-sided coupled operator matrices.* Semigroup Forum **58** (1999), 267–295.

[6] K.-J. Engel, *Operator Matrices and Systems of Evolution Equations.* Book preprint. Preliminary version available at `www.fa.uni-tuebingen.de/people/klen.html`.

[7] K.-J. Engel and R. Nagel, *One-Parameter Semigroups for Linear Evolution Equations.* Graduate Texts in Mathematics **194**, Springer-Verlag, 2000.

[8] A. Favini, G. Ruiz Goldstein, J.A. Goldstein, and S. Romanelli, *The heat equation with generalized Wentzell boundary condition.* J. Evolution Equations **2** (2002), 1–19.

[9] J.A. Goldstein, *Semigroups of Linear Operators and Applications.* Oxford Mathematical Monographs, Oxford University Press, 1985.

[10] G. Greiner, *Perturbing the boundary conditions of a generator.* Houston J. Math. **13** (1987), 213–229.

[11] E. Hille, *Functional Analysis and Semi-Groups.* American Mathematical Society Colloquium Publications **31**, American Mathematical Society, 1948.

[12] M. Kramar, D. Mugnolo, and R. Nagel, *Theory and applications of one-sided coupled operator matrices.* Conf. Semin. Mat. Univ. Bari (2002), to appear.

[13] J.L. Lions and E. Magenes, *Non-Homogeneous Boundary Value Problems and Applications.* vol. 1, Grundlehren der mathematischen Wissenschaften **181**, Springer-Verlag, 1972.

[14] I. Lasiecka and R. Triggiani, *Control Theory for Partial Differential Equations: Continuous and Approximation Theories.* vol. 1: Abstract Parabolic Systems, Cambridge University Press, 2000.

[15] R. Nagel, *Towards a "matrix theory" for unbounded operator matrices.* Math. Z. **201** (1989), 57–68.

[16] A. Pazy, *Semigroups of Linear Operators and Applications to Partial Differential Equations.* Applied Mathematical Sciences **44**, Springer-Verlag, 1983.

[17] L. Weis, *Operator-valued Fourier multiplier theorems and maximal regularity.* Math. Ann. **319** (2001), 735–758.

[18] L. Weis, *A new approach to maximal regularity.* In: L. Weis, G. Lumer (eds.), Evolution Equations and their Applications in Physical and Life Sciences, (Proceedings Bad Herrenalb 1999), Marcel Dekker, 2001, 195–214.

Marjeta Kramar
University of Ljubljana
Faculty of Civil and Geodetic Engineering
Department for Mathematics and Physics
Jamova 2
1000 Ljubljana, Slovenia
E-mail address: mkramar@fgg.uni-lj.si

Delio Mugnolo
Arbeitsbereich Funktionalanalysis
Mathematisches Institut
Universität Tübingen
Auf Der Morgenstelle 10
D-72076 Tübingen, Germany
E-mail address: demu@fa.uni-tuebingen.de

Rainer Nagel
Arbeitsbereich Funktionalanalysis
Mathematisches Institut
Universität Tübingen
Auf Der Morgenstelle 10
D-72076 Tübingen, Germany
E-mail address: rana@fa.uni-tuebingen.de

Progress in Nonlinear Differential Equations
and Their Applications, Vol. 55, 293–305

Maximal L_p-Regularity for Second Order Elliptic Operators with Uniformly Continuous Coefficients on Domains

Peer C. Kunstmann

1. Introduction and Main Result

Maximal regularity of type L_p is an important tool when dealing with quasi-linear equations of parabolic type (see, e.g., [1], [3]). If the closed linear operator A is the generator of a bounded analytic C_0-semigroup (T_t) in a Banach space X and $p \in (1, \infty)$ then A is said to have maximal L_p-regularity (which we denote by $A \in MR_p(X)$) if for any $f \in L_p((0, \infty), X)$ the solution $u = T * f$ of the equation $u' = Au + f$, $u(0) = 0$, satisfies $u' \in L_p((0, \infty), X)$ and $Au \in L_p((0, \infty), X)$. By the closed graph theorem this is equivalent to the existence of a constant $C > 0$ such that

$$\|u'\|_{L_p((0,\infty),X)} + \|Au\|_{L_p((0,\infty),X)} \le C\|f\|_{L_p((0,\infty),X)}. \tag{1}$$

If the type $\omega(A)$ of the semigroup is negative, then also the $L_p((0, \infty), X)$-norm of u is controlled by the right hand side, and (1) is equivalent to

$$\|u\|_{W_p^1((0,\infty),X)} + \|u\|_{L_p((0,\infty),[D(A)])} \le C\|f\|_{L_p((0,\infty),X)}$$

where $[D(A)]$ denotes the domain of A equipped with the graph norm. It is known that $A \in MR_p(X)$ for some $p \in (1, \infty)$ implies $A \in MR_p(X)$ for all $p \in (1, \infty)$ (see [6]). In this paper we always have $X = L_q(\Omega)$ for some $q \in (1, \infty)$. Thus we may (and will) assume $p = q$ without loss of generality.

Recently it has been shown by N.J. Kalton and G. Lancien that the property "$A \in MR_p(X)$ for all generators of bounded analytic semigroups" characterizes the Hilbert spaces in large classes of Banach spaces including L_q-spaces for $1 \le q < \infty$.

There are several approaches to showing maximal L_p-regularity for a given operator A. It is, e.g., known that $A \in MR_p(L_q)$ if A has bounded imaginary powers (BIP) which in turn is the case if A has a bounded H_∞-calculus, but the reverse implications are false in general (see [1], [2], [5], [12], [15]).

Recently, L. Weis gave a characterization for the property $A \in MR_p(X)$ for a large class of Banach spaces X including L_q-spaces which unified the different approaches (see [15], for general UMD-spaces see [16]). For $X = L_p(\Omega)$, the characterization is in terms of the space $(L_2)L_p(\Omega)$ of all (equivalence classes of) measurable functions $f : \mathbb{R} \times \Omega \to \mathbb{C}$ such that

$$\|f(t,\cdot)\|_{(L_2)L_p} := \|\omega \mapsto (\int_{-\infty}^{\infty} |f(t,\omega)|^2\, dt)^{1/2}\|_{L_p(\Omega)} < \infty,$$

and it says that, if A is the generator of a bounded analytic semigroup in $L_p(\Omega)$, then $A \in MR_p(L_p(\Omega))$ if and only if $tR(it, A)$ is an $(L_2)L_p$-*multiplier*, i.e., if and only if there is a constant $C > 0$ such that

$$\|t(R(it,A)f(t,\cdot))\|_{(L_2)L_p} \leq C\|f(t,\cdot)\|_{(L_2)L_p} \tag{2}$$

for all $f \in (L_2)L_p(\Omega)$. The smallest constant C in (2) is called the $(L_2)L_p$-*multiplier norm* of $tR(it, A)$.

Here we use this characterization result and a perturbation theorem from [15] (see Theorem 2.1) to obtain maximal L_p-regularity for second order elliptic operators on domains $\Omega \subset \mathbb{R}^n$ subject to Dirichlet boundary conditions under the assumption that the second order coefficients are bounded and uniformly continuous. The case of Hölder continuous coefficients which also satisfy a Hölder condition at infinity was treated in [12] on the half space and on bounded and exterior domains with C^2-boundary. Those operators were shown to have bounded imaginary powers. The case $\Omega = \mathbb{R}^n$ where boundary conditions do not enter was treated in [2] for elliptic systems under an additional assumption on the modulus of continuity of the highest order coefficients. Those operators were shown to have a bounded H_∞-calculus. The assumption on the modulus of continuity was removed in [5].

As for operators with bounded imaginary powers or with a bounded H_∞-calculus there is also an "operator sum method" connected with $(L_2)X$-multipliers in the spirit of the well-known Dore-Venni theorem (see [15]). For those applications it is interesting to have the "optimal angle" of maximal L_p-regularity $\theta_r(A)$ defined below. We use the following notation. For any $\psi \in [0, \pi)$ let γ_ψ be the path given by $\gamma_\psi(t) := |t|e^{i\psi \operatorname{sgn} t}$ for $t \neq 0$. If A is the generator of a bounded analytic semigroup then the *sectoriality angle* $\theta_\sigma(A)$ is defined as

$$\theta_\sigma(A) := \inf\{\theta \in [0, \pi/2) : M_\theta(A) := \sup_{0\leq\psi\leq\pi-\theta} \sup_{t\neq 0} \|tR(\gamma_\psi(t), A)\|_{L_p\to L_p} < \infty\}$$

and, in case A has maximal regularity, the *regularity angle* $\theta_r(A)$ is defined as

$$\theta_r(A) := \inf\{\theta \in [0, \pi/2) : R_\theta(A) := \sup_{0\leq\psi\leq\pi-\theta} \|tR(\gamma_\psi(t), A)\|_{(L_2)L_p\to(L_2)L_p} < \infty\}$$

Note that $M_\theta(A) \leq R_\theta(A)$ always.

We now state our main result. Let $\Omega \subset \mathbb{R}^n$ be a domain whose boundary $\partial\Omega$ is uniformly $C^{1,1}$ by which we mean that there are constants $r_0, c > 0$ such that, for any open cube Q of side length r_0 which intersects $\partial\Omega$ there is a $C^{1,1}$-diffeomorphism Φ from Q onto an open subset of $\mathbb{R}^n$ such that $\Phi(Q \cap \Omega) \subset \mathbb{R}^n_+ :=$

$\{(y', y_n) : y' \in \mathbb{R}^{n-1}, y_n > 0\}$, $\Phi(Q \cap \Omega^c) \subset \mathbb{R}^n_-$, $\Phi(Q \cap \partial\Omega) \subset \mathbb{R}^{n-1} \times \{0\}$, and

$$\begin{gathered} \|D\Phi\|_\infty + \|D^2\Phi\|_\infty + \|D\Phi^{-1}\|_\infty + \|D^2\Phi^{-1}\|_\infty \le c \\ c^{-1} \le |\det D\Phi(x)| \le c, \quad x \in Q. \end{gathered} \tag{3}$$

Here $D^k\Phi$ denotes the matrix of the k-th partial derivatives of Φ. Note that the assumptions (3) imply that

$$c^{-1} \le |\det D\Phi^{-1}(y)| \le c, \; y \in \Phi(Q). \tag{4}$$

It is obvious that the boundary of the halfspace $\mathbb{R}^n_+$ is uniformly $C^{1,1}$. If Ω is bounded and $\partial\Omega \in C^{1,1}$ (cf. [17], Definition 2.7) then $\partial\Omega$ is uniformly $C^{1,1}$ by compactness. Of course, also the boundary of the exterior domain $\mathbb{R}^n \setminus \overline{\Omega}$ is uniformly $C^{1,1}$.

Let $1 < p < \infty$. We treat L_p-realizations A of differential operators $\mathcal{A}$ of the following type:

$$\mathcal{A} := \sum_{j,k=1}^n a_{jk}\partial_j\partial_k + \sum_{j=1}^n b_j\partial_j + d$$

under Dirichlet boundary conditions, i.e., the domain of A is $W^2_{p,D}(\Omega) := \{u \in W^2_p(\Omega) : u = 0 \text{ on } \partial\Omega\} = W^2_p(\Omega) \cap W^1_{p,0}(\Omega)$. Here $a_{jk} = a_{kj}$ without loss of generality and the precise regularity assumptions on the second order coefficients are as follows.

$$\begin{cases} \text{uniform continuity: } (a_{jk})_{j,k} \in BUC(\Omega, \mathbb{R}^{n\times n}) \text{ and } \sup_{j,k} \|a_{jk}\|_\infty \le M; \\ \text{ellipticity: } \sum_{j,k} a_{jk}(x)\xi_j\xi_k \ge \alpha|\xi|^2 \text{ for all } x \in \Omega,\ \xi \in \mathbb{R}^n. \end{cases} \tag{5}$$

Here $M, \alpha > 0$. For the complex valued lower order terms we require as in [2] and [5]

$$\begin{cases} \text{For } j = 1, \dots, n\text{: } b_j \in L_{q_1,\mathrm{unif}}(\Omega), \text{ and } d \in L_{q_0,\mathrm{unif}}(\Omega) \\ \text{where } q_1 > \max(p, n) \text{ and } q_0 > \max(p, n/2). \end{cases} \tag{6}$$

Here, for $q \in [1, \infty]$, $L_{q,\mathrm{unif}}(\Omega)$ is defined as

$$L_{q,\mathrm{unif}}(\Omega, \mathbb{C}) := \{f \in L_q(\Omega) : \|f\|_{q,\mathrm{unif}} := \sup_{j\in\mathbb{Z}^n} \|f\|_{L_q(\Omega\cap j+(-1,1)^n)} < \infty\}.$$

Observe $L_{\infty,\mathrm{unif}}(\Omega) = L_\infty(\Omega)$ and $L_s(\Omega) \hookrightarrow L_{s,\mathrm{unif}}(\Omega) \hookrightarrow L_{r,\mathrm{unif}}(\Omega) \hookrightarrow L_{r,\mathrm{loc}}(\Omega)$ for $1 \le r \le s \le \infty$. The assumptions imply that the operator $Au(x) := \mathcal{A}(x)u(x)$ with one of the above domains is a closed linear operator in $L_p(\Omega)$. The following is our main result.

Theorem 1.1. *The operator A is the generator of an analytic semigroup in $L_p(\Omega)$. If $\omega(A) < 0$ then $A \in MR_p(L_p(\Omega))$. If in addition $\pi/2 \ge \theta > \theta_\sigma(A)$, then $R_\theta(A) < \infty$, in particular $\theta_r(A) = \theta_\sigma(A)$.*

Remark 1.2. (i) The first statement in Theorem 1.1 is of course known, but the proof is usually based on Agmon's a-priori estimates for elliptic boundary value problems (see [14], [10], [9]). We give a direct proof here by localization and perturbation techniques which has the advantage that maximal L_p-regularity may be obtained at the same time.

(ii) Notice that the second statement in Theorem 1.1 extends [12, Theorem A] where the highest order coefficients were assumed to be Hölder continuous, and that – for the case of second order elliptic equations (not systems) – it extends the results in [2] and [5] which treated the case $\Omega = \mathbb{R}^n$. We want to mention that the same method of proof may be used to reprove maximal regularity on $\mathbb{R}^n$ for the elliptic systems studied in [2], [5], see Section 4.

(iii) Note that by our assumption on the zero order coefficient d the assumption $\omega(A) < 0$ in Theorem 1.1 may be made without loss of generality.

(iv) The last statement in Theorem 1.1 shows that A has maximal L_p-regularity with an optimal angle.

The paper is organized as follows. In Section 2 we prove Theorem 1.1. In Section 3 we combine Theorem 1.1 with results due to J. Prüss and R. Schnaubelt ([11]) to obtain maximal regularity of type L_s-L_p for non-autonomous parabolic equations. In Section 4 we comment on other types of boundary conditions and higher order operators.

2. Proof of Theorem 1.1

The proof will be done in four steps. We first consider constant coefficients on $\mathbb{R}^n$, and then on $\mathbb{R}^n_+$. In the third step we apply a perturbation result from [15] to treat small deviations from constant coefficient operators, and in the fourth we treat the general case by localization.

2.1. Step I: Constant Coefficients on $\mathbb{R}^n$

We assume that $b_j = d = 0$, the a_{jk} are constants, and $\Omega = \mathbb{R}^n$. This case is well known and reduces to an application of, e.g., the Mikhlin multiplier theorem. Let $\pi/2 \geq \theta > 0$ and $0 \leq \psi \leq \pi - \theta$. Note that applying the Mikhlin theorem in $L_p(\mathbb{R} \times \mathbb{R}^n)$ to $(t, \xi) \mapsto t(\gamma_\psi(t) + a(\xi))^{-1}$ where $a(\xi) := \sum_{j,k} a_{jk}\xi_j\xi_k$ is the symbol of the operator A gives a bound on the $(L_2)L_p$-multiplier norm of $tR(\gamma_\psi(t), A)$ which depends only on the constants α and M from (5) and the angle θ (see [15]). Hence we have $\theta_r(A) = 0$ and $R_\theta(A) \leq R_{M,\alpha,\theta}$ for $\theta > 0$. Note that the semigroup generated by A is bounded and $0 \in \sigma(A)$. Moreover A is injective and has dense range.

2.2. Step II: Constant Coefficients on $\mathbb{R}^n_+$

We continue to assume that $b_j = d = 0$ and that the a_{jk} are constants, but we now let $\Omega = \mathbb{R}^n_+$. Just as in [12], we apply the reflection principle. If $a_{jn} = 0$ for $j = 1, \ldots, n-1$ (recall $a_{nj} = a_{jn}$) and $k_\lambda(x - y)$ denotes the kernel belonging to the symbol $(\lambda - a(\xi))^{-1}$ on $\mathbb{R}^n$ then the resolvent $R(\lambda, A)$ has the kernel $k_\lambda(x - y) - k_\lambda(x - y^*)$. This comes from

$$R(\lambda, A)f = \rho_+ \mathcal{F}^{-1}(\lambda - a(\xi))^{-1}\mathcal{F}E_- f$$

where $\mathcal{F}$ denotes the fourier transform in $\mathbb{R}^n$, ρ_+ denotes the restriction from $\mathbb{R}^n$ to $\mathbb{R}^n_+$ and $E_- f$ denotes the odd extension of f to $\mathbb{R}^n$. Thus we see from step I that

A generates a bounded analytic semigroup in $L_p(\mathbb{R}^n_+)$ and that for $\pi/2 \geq \theta > 0$ and $0 \leq \psi \leq \pi - \theta$ the $(L_2)L_p$-multiplier norm of $tR(\gamma_\psi(t), A)$ is bounded by a constant only depending on M, α, and θ.

If $a'_{\bullet n} := (a_{1n}, \dots, a_{n-1n}) \neq 0$ then we define $T(x', x_n) := (x' - a'_{\bullet n} a_{nn}^{-1} x_n, x_n)$. Note that $a_{nn} \neq 0$ by ellipticity. The linear map $T : \mathbb{R}^n_+ \to \mathbb{R}^n_+$ is bijective with $\det T = 1$, and hence $Jf(x) := f(Tx)$ defines an isometry in $L_p(\mathbb{R}^n_+)$ which leaves invariant $W^2_{p,0}(\mathbb{R}^n_+)$. If we define the operator $\tilde{A} := J^{-1}AJ$ then the relations $\nabla(f \circ T) = T^t(\nabla f) \circ T$ and $\operatorname{div}(g \circ S) = \operatorname{div}(Sg) \circ S$ show that $\tilde{A}$ is a differential operator with constant coefficients $(\tilde{a}_{jk}) = T^t(a_{jk})T$, hence $\tilde{a}_{jn} = 0$ for $j = 1, \dots, n$ by the choice of T. Now $tR(\gamma_\psi(t), A) = tJR(\gamma_\psi(t), \tilde{A})J^{-1}$ and since J is just an isometric change of variables in $L_p(\mathbb{R}^n_+)$ we see that the $(L_2)L_p$-multiplier norms of $tR(\gamma_\psi(t), A)$ and $tR(\gamma_\psi(t), \tilde{A})$ coincide (of course, A generates a bounded analytic semigroup as $\tilde{A}$ does).

Hence $\theta_r(A) = 0$ and $R_\theta(A) \leq R_{M,\alpha,\theta}$ for $\pi/2 \geq \theta > 0$. Note again that A is injective and has dense range, the semigroup generated by A is bounded and $0 \in \sigma(A)$.

2.3. Step III: Small Deviations

We will use the following perturbation result from [15]. Here K is a fixed constant relating $(L_2)L_p$-multiplier norms to R-bounds (see [15], [16]).

Theorem 2.1. *Let A be the generator of a bounded analytic semigroup which has maximal L_p-regularity and let $\pi/2 \geq \theta > \theta_r(A)$. Let B be a linear operator satisfying $D(B) \supset D(A)$ and*

$$\|Bx\| \leq a\|Ax\| \quad (x \in D(A)). \tag{7}$$

If $a^{-1} > K(R_\theta(A)+1)$ then $A+B$ is the generator of a bounded analytic semigroup which has maximal L_p-regularity, $\theta_r(A+B) \leq \theta$, and

$$R_\theta(A+B) \leq \frac{R_\theta(A)}{1 - aK(R_\theta(A)+1)}.$$

If A is injective with dense range then also $A+B$ is injective with dense range.

Now we can treat small deviations from constant coefficient operators.

Theorem 2.2. *Given M, α and $\pi/2 \geq \theta > 0$, there are constants $\eta, C > 0$ such that for any constant coefficient operator $A_0 := \sum_{j,k} a^0_{jk}\partial_j\partial_k$ whose coefficients (a^0_{jk}) satisfy the assumption (5) and any operator $A := \sum_{j,k} a_{jk}\partial_j\partial_k$ with measurable coefficients (a_{jk}) satisfying*

$$\|a_{jk} - a^0_{jk}\|_\infty \leq \eta$$

we have that $A \in MR_p(L_p)$, $\theta_r(A) \leq \theta$, and $R_\theta(A) \leq C$.

Here we consider A_0 and A either on $\mathbb{R}^n$ with domain $W^2_p(\mathbb{R}^n)$ or on $\mathbb{R}^n_+$ with domain $W^2_{p,D}(\mathbb{R}^n_+)$.

Proof. From steps I and II we know that $A_0 \in MR_p(L_p)$, $\theta_r(A_0) = 0$, and $R_\theta(A_0) \le R_{M,\alpha,\theta}$ for any $\pi/2 \ge \theta > 0$. If we define $B := A - A_0$ then $D(B) \supset D(A)$ and for any $f \in D(A)$:

$$\|Bf\|_p = \|(A-A_0)f\|_p \le \eta\|D^2 f\|_p \le \eta C_{M,\alpha}\|A_0 f\|_p.$$

Hence we can apply Theorem 2.1 if $\eta < (C_{M,\alpha}K(R_{M,\alpha,\theta}+1))^{-1}$. □

2.4. Step IV: The General Case

We now localize, apply the results of steps I - III, and patch together again. We give full details here, since dealing with $(L_2)L_p$-norms is a rather recent topic in this context and the not necessarily compact boundary $\partial\Omega$ means that finite partitions of unity might not be sufficient.

We want to show first that A generates an analytic semigroup. This will be done by adapting the method developped in [2], Section 3. We obtain maximal regularity at the same time. We then use a different resolvent representation, inspired by [12], to show optimality of the regularity angle.

First we choose η according to step III for $\theta = \pi/2$, but for the constants Mc^2 and $c^{-2}\alpha$ instead of M and α (observe that $c \ge 1$ always). By the assumption (5) we find a $\delta > 0$ such that $|a_{jk}(x) - a_{jk}(y)| \le c^{-2}\eta/2$ if $|x-y|_\infty \le c\delta$. We fix $r \le \min(r_0, \delta)$ and a C^∞-function φ with $0 \le \varphi \le 1$ and $\operatorname{supp}\varphi \subset Q := (-r,r)^n$ such that

$$\sum_{l \in r\mathbb{Z}^n} \varphi_l^2(x) = 1, \quad x \in \mathbb{R}^n, \tag{8}$$

where $\varphi_l(x) := \varphi(x-l)$. We let $Q_l := l + Q$, and denote the set of all $l \in r\mathbb{Z}^n$ such that Q_l intersects Ω by Γ. Then Γ splits in the set Γ_1 of l for which Q_l intersects the boundary $\partial\Omega$ of Ω and the set Γ_0 of l for which $Q_l \subset \Omega$. We also fix a C^∞-function ψ with $\operatorname{supp}\psi \subset Q$, $0 \le \psi \le 1$, and $\psi = 1$ on $\operatorname{supp}\varphi$ and define $\psi_l(x) := \psi(x-l)$ for $l \in \Gamma$. For any $l \in \Gamma_0$ define coefficients a^l_{jk} by

$$a^l_{jk}(x) := \begin{cases} a_{jk}(x) & , x \in Q_l \\ a_{jk}(l) & \text{otherwise.} \end{cases}$$

We define operators A_l with domain $W^2_p(\mathbb{R}^n)$ by $A_l u(x) := \sum_{j,k} a^l_{jk}(x)(\partial_j\partial_k u)(x)$ and from the steps I and III we know that the $(L_2)L_p$-multiplier norms of $tR(it, A_l)$ are bounded by a constant only depending on M and α.

For $l \in \Gamma_1$ we choose a $C^{1,1}$-diffeomorphism Φ_l according to our assumption, i.e., satisfying (3) and flattening the part of the boundary in Q_l. We denote the induced map $f \mapsto f \circ \Phi_l$ by J_l, thus $J_l^{-1} : g \mapsto g \circ \Phi_l^{-1}$ is its inverse. The operator $A'_l := J_l^{-1} A J_l$ is given by the expression

$$\sum_{j,k} \tilde{a}^l_{jk}\partial_j\partial_k + \sum_j \tilde{b}^l_j \partial_j$$

where

$$(\tilde{a}^l_{jk}) := \tilde{a}^l := (D\Phi_l \cdot a \cdot D\Phi_l{}^t) \circ \Phi_l^{-1} \quad \text{and} \quad \tilde{b}^l_j := (\sum_{k,m} a_{km}\partial_k\partial_m(\Phi_l)_j) \circ \Phi_l^{-1}.$$

We choose $x_l \in \Phi_l(Q_l \cap \Omega)$ and define the operator $\tilde{A}_l$ in $L_p(\mathbb{R}^n_+)$ by $\tilde{A}_l v(x) := \sum_{j,k} \tilde{a}^l_{jk}(x)v(x)$, where $\tilde{a}^l_{jk}(x) := \tilde{a}^l_{jk}(x_l)$ for $x \notin \Phi_l(Q_l \cap \Omega)$, with domain $W^2_{p,D}(\mathbb{R}^n_+)$. We also let $\tilde{b}^l_j(x) := 0$ for $x \notin \Phi_l(Q_l \cap \Omega)$.

Observe that the coefficients of $\tilde{A}_l$ satisfy again the assumptions in step III, the constants M and α in (5) being replaced by Mc^2, $c^{-2}\alpha$, respectively: For $|x-y| \le \delta$ we have $|\Phi_l^{-1}(x) - \Phi_l^{-1}(y)| \le c\delta$, hence

$$\begin{aligned}
&\|\tilde{a}^l(x) - \tilde{a}^l(y)\| \\
&\le cMc|\Phi_l^{-1}(x) - \Phi_l^{-1}(y)| + c\|a(\Phi_l^{-1}(x)) - a(\Phi_l^{-1}(y))\|c + c|\Phi_l^{-1}(x) - \Phi_l^{-1}(y)|Mc \\
&\le 2Mc^3\delta + c^2\eta/(2c^2) \le \eta
\end{aligned}$$

by the choice of δ. Thus we know from steps II and III that the $(L_2)L_p$-multiplier norms of $tR(it, \tilde{A}_l)$ are bounded by a constant only depending on M, α and c.

For notational reasons we introduce corresponding counterparts also for $l \in \Gamma_0$, letting $\Phi_l := I$, $J_l := I$, $\tilde{A}_l := A_l$ with coefficients $\tilde{a}^l := a^l$ and $\tilde{b}^l_j := 0$.

We now derive a representation for the resolvents of A by localization. Let $f \in L_p(\Omega)$, $\lambda \in \mathbb{C}$ and $u \in D(A)$ satisfy $(\lambda - A)u = f$. Multiplying with φ_l and letting $u_l := \varphi_l u$, $f_l := \varphi_l f$ we get $(\lambda - A_H)u_l = f_l - B_l u$, $l \in \Gamma$, where

$$B_l u := (\sum_{j,k} a_{jk}\partial_j\partial_k\varphi_l - d\varphi_l)u + (2\nabla\varphi_l^t a - \varphi_l b^t)\nabla u, \tag{9}$$

and A_H denotes the principle part of A. Before we continue we observe that the assumption (6) and Hölder's inequality yields

$$\begin{aligned}
\|\varphi_l b^t \nabla u\|_p &\le C_\varphi \|b^t\nabla u\|_{L_p(Q_l)} \le C_{\varphi,r}\|b\|_{q_1,\text{unif}}\|\nabla u\|_{L_{r_1}(Q_l)} \\
\|\varphi_l d u\|_p &\le C_\varphi \|du\|_{L_p(Q_l)} \le C_{\varphi,r}\|d\|_{q_0,\text{unif}}\|u\|_{L_{r_0}(Q_l)}
\end{aligned}$$

where $p^{-1} = q_1^{-1} + r_1^{-1} = q_0^{-1} + r_0^{-1}$. By the Sobolev embedding theorem and the boundedness of a and all derivatives of φ_l we hence find that B_l is bounded $W^s_p \to L_p$ for some $1 \le s < 2$ (cf. [12], p. 165, and [2], p. 645). The norm only depends on φ, r, and on the norms $\|b\|_{q_1,\text{unif}}$ and $\|d\|_{q_0,\text{unif}}$.

We apply J_l^{-1} to (9) from the left, let $\tilde{u}_l := J_l^{-1}u_l$, $\tilde{f}_l := J_l^{-1}f_l$, and arrive at

$$(\lambda - \tilde{A}_l)\tilde{u}_l = \tilde{f}_l - J_l^{-1}B_l u - (\tilde{b}^l)^t\nabla\tilde{u}_l, \quad l \in \Gamma. \tag{10}$$

Observe that by the assumptions (3) and (4) the map J_l is an isomorphism $L_p(\Phi_l(\Omega \cap Q_l)) \to L_p(Q_l)$ which respects the boundary conditions and induces an isomorphism of the domains of the respective differential operators. The norms of these isomorphism and of their inverses are bounded uniformly with respect to l.

Writing $m \sim l$ if $Q_l \cap Q_m \ne \emptyset$, and recalling (8), we have

$$J_l^{-1}B_l u = \sum_{m\sim l}(J_l^{-1}B_l\psi_l\varphi_m)u_m = \sum_{m\sim l}(J_l^{-1}B_l J_m\tilde{\chi}_{l,m})\tilde{u}_m,$$

where $\tilde{\chi}_{l,m} := J_m^{-1}(\psi_l \phi_m)$. Letting $\tilde{B}_{l,m} := J_l^{-1} B_l J_m \tilde{\chi}_{l,m}$ for $m \neq l$ in Γ and $\tilde{B}_{l,l} := J_l^{-1} B_l J_l \tilde{\chi}_{l,l} + (\tilde{b}^l)^t \nabla$, we hence have

$$(\lambda - \tilde{A}_l)\tilde{u}_l = f_l - \sum_{m \sim l} \tilde{B}_{l,m} \tilde{u}_m, \; l \in \Gamma.$$

For a sequence $(X_n, |\cdot|_n)$ of Banach spaces we denote by $X_1 \oplus_p X_2$ the product space $X_1 \times X_2$ equipped with the norm $\|(x_1, x_2)\|_p := (|x_1|_1^p + |x_2|_2^p)^{1/p}$, and by $l_p - \bigoplus_n X_n$ the space of all sequences (x_n) such that $x_n \in X_n$, $n \in \mathbb{N}$, and $\|(x_n)\|_p := (\sum_n |x_n|_n^p)^{1/p} < \infty$, which is a Banach space for $\|\cdot\|_p$. We let

$$\bar{X} := (l_p - \bigoplus_{l \in \Gamma_0} L_p(\mathbb{R}^n)) \oplus_p (l_p - \bigoplus_{l \in \Gamma_1} L_p(\mathbb{R}^n_+)),$$

and denoting the elements of $\bar{X}$ by $(v_l)_{l \in \Gamma}$ we define the operator $\bar{A}(v_l) := (\tilde{A}_l v_l)$ with domain

$$D(\bar{A}) := (l_p - \bigoplus_{l \in \Gamma_0} W_p^2(\mathbb{R}^n)) \oplus_p (l_p - \bigoplus_{l \in \Gamma_1} W_{p,D}^2(\mathbb{R}^n_+)),$$

Note that by steps I - III all $\tilde{A}_l$ generate analytic semigroups and we have resolvent estimates uniformly in l. That means that $\bar{A}$ generates an analytic semigroup in $\bar{X}$. The operator $\bar{B}(v_l) := (\sum_{m \sim l} \tilde{B}_{l,m} v_m)$ is bounded

$$(l_p - \bigoplus_{l \in \Gamma_0} W_p^s(\mathbb{R}^n)) \oplus_p (l_p - \bigoplus_{l \in \Gamma_1} W_p^s(\mathbb{R}^n_+)) \to \bar{X}$$

since all operators $\tilde{B}_{l,m}$ are bounded $W_p^s(R) \to L_p(R)$ where $R = \mathbb{R}^n$ for $l \in \Gamma_0$ and $R = \mathbb{R}^n_+$ for $l \in \Gamma_0$, their norms are uniformly bounded in (m, l), and each row and each column of the operator $\bar{B}$ contains at most L entries $\neq 0$ where

$$L := \sup_{l \in \Gamma} \#\{m \in \Gamma : m \sim l\} < \infty \tag{11}$$

by construction. Hence also $\bar{A} - \bar{B}$ generates an analytic semigroup. This means that for $\lambda = it$ and $|t|$ sufficiently large we have

$$(\tilde{u}_l) = (\lambda - (\bar{A} - \bar{B}))^{-1}(\tilde{f}_l)$$

and from

$$u = \sum_l \varphi_l J_l \tilde{u}_l = \sum_l J_l \tilde{\varphi}_l \tilde{u}_l =: \bar{C}(\tilde{u}_l),$$

where $\tilde{\varphi}_l := J_l^{-1} \varphi_l$, we get

$$R(\lambda, A)f = \bar{C} R(\lambda, \bar{A} - \bar{B}) \bar{D} f, \tag{12}$$

where $Df := (J_l^{-1} \varphi_l f)$. Since we have the resolvent estimate $\|\lambda R(\lambda, \bar{A} - \bar{B})\| \le \tilde{C}$ for $Re\, \lambda > \omega$, and $\bar{C}$ and $\bar{D}$ are bounded by the following lemma, it is clear that A generates an analytic semigroup.

Lemma 2.3. *The operators $\bar{D} : L_p(\Omega) \to \bar{X}$ and $\bar{C} : \bar{X} \to L_p(\Omega)$ are bounded.*

Proof. We have by $0 \le \varphi_l \le 1$, $\operatorname{supp}\varphi_l \subset Q_l$, and (11)

$$\sum_l \|J_l^{-1}\varphi_l f\|_p^p \le C_c \sum_l \|\varphi_l f\|_p^p \le C_c \sum_l \|f\|_{L_p(Q_l)}^p \le C_{c,K}\|f\|_p$$

for $f \in L_p(\Omega)$ which proves boundedness of $\bar{D}$. For any $(v_l) \in \bar{X}$, we get from (8) and (11)

$$\begin{aligned}
\|\sum_l J_l\tilde{\varphi}_l v_l\|_p^p &= \sum_{m\in\Gamma}\int_\Omega \varphi_m^2 |\sum_{l\sim m}\varphi_l J_l v_l|^p\,dx \\
&\le C_{K,p}\sum_m\sum_{l\sim m}\int_\Omega \varphi_m^2\varphi_l^p |J_l v_l|^p\,dx \\
&\le C_{K,p}K\sum_l \int_{Q_l}|J_l v_l|^p\,dx \\
&\le C_{K,p,c}\sum_l \int |v_l|^p\,dy,
\end{aligned}$$

which proves boundedness of $\bar{C}$. □

The abstract perturbation result Theorem 2.1 that we used in step III is a special case of a more general perturbation theorem for A-bounded operators B with sufficiently small A-bound (see [8]). Using the general result yields $A - \mu \in MR_p(L_p)$ for sufficiently large μ. By $\omega(A) < 0$ we thus have proved $A \in MR_p(L_p)$.

We now show optimality of the regularity angle, i.e., we prove $\theta_r(A) = \theta_\sigma(A)$. To this purpose we adapt arguments given in [12]. We fix $\theta \in (\theta_\sigma(A), \pi/2]$. We repeat the localization procedure by choosing η according to step III for the constants Mc^2, $c^{-2}\alpha$ and θ. Once again we know that, for $l \in \Gamma_0$, $R_\theta(A_l)$ is by the steps I and III bounded by a constant only depending on M, α and θ, and that, for $l \in \Gamma_1$, $R_\theta(\tilde{A}_l)$ is bounded by a constant only depending on M, α, θ and c. Recall that we had put $\tilde{A}_l := A_l$ for $l \in \Gamma_0$. For $0 \le \psi \le \pi - \theta$, $t \neq 0$ and $\lambda = \gamma_\psi(t)$, we rewrite (10) as

$$(\lambda - \tilde{A}_l)\tilde{u}_l = \tilde{f}_l - (J_l^{-1}B_l + (\tilde{b}^l)^t\nabla J_l^{-1}\varphi_l)u =: \tilde{f}_l - F_l u, \quad l \in \Gamma, \tag{13}$$

apply $R(\lambda, \tilde{A}_l)$ from the left, use the operator $\bar{C}$ defined above, and get

$$R(\lambda, A)f = \sum_l J_l\tilde{\varphi}_l R(\lambda, \tilde{A}_l)\tilde{f}_l - \sum_l J_l\tilde{\varphi}_l R(\lambda, \tilde{A}_l)F_l R(\lambda, A)f. \tag{14}$$

Using the arguments in the proof of Lemma 2.3 for $(\int_{\mathbb{R}} |f(t,\cdot)|^2\,dt)^{1/2}$ instead of f and for $(\int_{\mathbb{R}} |v_l(t,\cdot)|^2\,dt)^{1/2}$ instead of v_l we see that the induced operators $I \otimes \bar{D} : (L_2)L_p(\Omega) \to (L_2)\bar{X}$ and $I \otimes \bar{C} : (L_2)\bar{X} \to (L_2)L_p(\Omega)$ are continuous. Since we know that the $(L_2)L_p$-multiplier norms of $tR(\gamma_\psi(t), \tilde{A}_l)$ are uniformly bounded this gives

$$\|t\sum_l J_l\tilde{\varphi}_l R(\gamma_\psi(t), \tilde{A}_l)\tilde{f}_l(t,\cdot)\|_{(L_2)L_p} \le \text{const}\,\|f(t,\cdot)\|_{(L_2)L_p}.$$

Thus it rests to show boundedness of the second series in (14). The boundedness of $I \otimes \bar{C}$ gives

$$\left\{ \begin{array}{rl} & \|t \sum_l J_l \tilde{\varphi}_l R(\gamma_\psi(t), \tilde{A}_l) F_l R(\gamma_\psi(t), A) f(t,\cdot)\|^p_{(L_2)L_p} \\ \leq & \text{const} \sum_l \|F_l R(\gamma_\psi(t), A) f(t,\cdot)\|^p_{(L_2)L_p}. \end{array} \right. \tag{15}$$

We use the fact (see [15]) that $N(t) : L_p \to L_p$ is an $(L_2)L_p$-multiplier if $d/dt\, N(t)$ is integrable on $\mathbb{R}$. We have $d/dt\, F_l R(\gamma_\psi(t), A) = -e^{i\psi \operatorname{sgn} t} F_l R(\gamma_\psi(t), A)^2$. By $\omega(A) < 0$ we have $\|R(\gamma_\psi(t), A)\|_{p\to p} \leq \text{const}\,(1 + |t|)^{-1}$. Taking again a look at the definition of F_l in (13) we see that transforming the coordinates back to Q_l we have to estimate the following norms

$$\|B_l R(\gamma_\psi(t), A)^2 f\|^p_{L_p(Q_l)}, \quad \|J_l (\tilde{b}^l)^t \nabla J_l^{-1} \varphi_l R(\gamma_\psi(t), A)^2 f\|_{L_p(Q_l)}. \tag{16}$$

Now B_l respects the support and is bounded $W^s_p \to L_p$ as we have seen above. Hence by interpolation the 1st norm in (16) is $\leq \text{const}\,(1+|t|)^{-1-(2-s)/2}\|f\|_{L_p(Q_l)}$. A similar argument gives that the 2nd norm in (16) is $\leq \text{const}(1+|t|)^{-3/2}\|f\|_{L_p(Q_l)}$. Note that the constants are uniform in $l \in \Gamma$. Hence we can estimate the right hand side of (15) by

$$\text{const} \sum_l \|f(t,\cdot)\|^p_{(L_2)L_p(Q_l)},$$

and the $\sum_l$ is $\leq C_K \cdot \|f(t,\cdot)\|^p_{(L_2)L_p}$ by (11). This ends the proof of Theorem 1.1.

3. Non-autonomous Parabolic Equations

If we combine Theorem 1.1 with results in [11] we get maximal regularity for non-autonomous parabolic equations

$$\partial_t u(t,x) = \mathcal{A}(t,x) u(t,x) + f(t,x), \quad t \in [0,T], x \in \Omega \tag{17}$$

where

$$\mathcal{A}(t,x) := \sum_{j,k=1}^n a_{jk}(t,x) \partial_j \partial_k + \sum_{j=1}^n b_j(t,x) \partial_j + d(t,x).$$

Here we assume $a_{jk} \in BUC([0,T] \times \Omega, \mathbb{R})$ and, for each $t \in [0,T]$, that assumption (5) holds for $(a_{jk}(t,\cdot))$. Note that, by continuity in time and compactness of $[0,T]$, the constants M and α can be chosen independent of t. For the lower order coefficients we assume either

$$\left\{ \begin{array}{l} b \in C([0,T], L_{q_1,\text{unif}}(\Omega)^n) \text{ and } d \in C([0,T], L_{q_0,\text{unif}}(\Omega)) \\ \text{for some } q_1, q_0 > p \text{ with } q_1 > n,\ q_0 > \frac{n}{2}, \end{array} \right. \tag{18}$$

or we fix $s \in (1,\infty)$ and assume

$$\left\{ \begin{array}{l} b \in L_{r_1}([0,T], L_{q_1}(\Omega)^n) \text{ and } d \in L_{r_0}([0,T], L_{q_0}(\Omega)) \\ \text{for some } q_0, q_1 > p \text{ and } r_1, r_0 > s \text{ such that} \\ \frac{2s}{r_1} + \frac{n}{q_1} < 1 \text{ and } \frac{s}{r_0} + \frac{n}{2q_0} < 1. \end{array} \right. \tag{19}$$

Again, we consider Dirichlet boundary conditions and define for every $t \in [0,T]$ the operator $A(t)$ by $(A(t)u)(x) = \mathcal{A}(t,x)u(x)$ for u in the corresponding domain $W^2_{p,D}(\Omega)$. Then each $A(t)$ is the generator of an analytic semigroup in $L_p(\Omega)$ and we have the following result.

Theorem 3.1. *There is an $s_0 > 1$ such that for any $1 < s < s_0$ and any $f \in L_s([0,T], L_p(\Omega))$ there is a unique function $u \in W^1_s([0,T], L_p(\Omega)) \cap L_s([0,T], W^2_{p,D}(\Omega))$ satisfying*

$$u'(t) = A(t)u(t) + f(t), \text{ a.e. } t \in [0,T], u(0) = 0,$$

and we have the estimate

$$\|u\|_{L_s([0,T],L_p(\Omega))} + \|u\|_{L_s([0,T],W^2_p(\Omega))} + \|u'\|_{L_s([0,T],L_p(\Omega))} \leq c_p \|f\|_{L_s([0,T],L_p(\Omega))}.$$

This means that the non-autonomous equation (17) has maximal regularity of type L_s-L_p.

Observe that the s appearing in the theorem is connected to the assumed regularity of the lower order coefficients in (19) which means that in case of (19) the regularity in time of the solution depends on a combined space and time regularity of the lower order terms.

Proof. In the case of (18) we can directly apply [11, Theorem 2.5] to the operators $A(t)$. In this case we can take $s_0 = \infty$.

In the case of (19) we apply [11, Theorem 2.5] to the operators $A_0(t)$ obtained from the principal parts $\mathcal{A}_0(t,x)$ of $\mathcal{A}(t,x)$. Then we use [11, Theorem 3.1] as explained in [11], Section 4. There it is shown that for chosen r_k, q_k, $k = 0, 1$, one can take $s_0 := \min\{r_k(1 - \frac{n}{2q_k} - \frac{k}{2}) : k = 0, 1\}$ in (19). □

4. Other boundary conditions and higher order operators

In this section we comment on extension of Theorem 1.1.

Remark 4.1. The same method of proof may be used for (M,θ)-elliptic systems on $\mathbb{R}^n$ where no boundary conditions enter. For the notion of (M,θ)-elliptic systems we refer to [2], [5]. Then step II is not necessary and the arguments in steps I, III, IV apply and yield maximal regularity for (M,θ)-elliptic systems whose highest order coefficients are bounded and uniformly continuous. A sketch of proof may be found in [8]. This result is known, since [5] establishes even a bounded H^∞-calculus, but the proof in [5] is considerably more involved.

Remark 4.2. Consider a general elliptic boundary value problem $(A, B_1, \ldots, B_m)$ of order $2m$ whose boundary operators satisfy the Lopatinskij-Shapiro conditions (see, e.g., [17, II ch. 13]) on a smooth bounded domain Ω. If the coefficients of the principal part of A are uniformly continuous then we may argue as in steps I, III, and IV. The crucial part is then step II, i.e., A is a constant coefficient operator on the half space but $B_1, \ldots, B_n$ are variable coefficient operators. If the coefficients of the boundary operators are sufficiently smooth then maximal regularity follows from results due to Seeley ([13]).

Recently, general boundary value problems with infinte-dimensional state space have been studied in [4].

Acknowledgment

The author wants to thank Lutz Weis for the inspiration to this work and stimulating discussions on the subject.

References

[1] H. Amann, *Linear and quasilinear parabolic problems.* Vol. 1: Abstract linear theory, Birkhäuser, Basel, 1995.

[2] H. Amann, M. Hieber, G. Simonett, *Bounded H_∞-calculus for elliptic operators.* Differential and Integral Equations **7** (1994), 613–653.

[3] Ph. Clément, S. Li, *Abstract parabolic quasilinear equations and application to a groundwater flow problem.* Adv. Math. Sci. Appl. **3** (1994), 17–32.

[4] R. Denk, M. Hieber, J. Prüss, *$\mathcal{R}$-Boundedness, Fourier Multipliers and Problems of Elliptic and Parabolic Type.* Preprint 2001.

[5] X.T. Duong, G. Simonett, *H_∞-calculus for elliptic operators with non-smooth coefficients.* Differential and Integral Equations **10** (1997), 201–217.

[6] M. Hieber, J. Prüss, *Heat kernels and maximal L^p-L^q estimates for parabolic evolution equations.* Commun. Partial Differ. Equations **22** (1997), 1647–1669.

[7] N.J. Kalton, G. Lancien, *A solution to the problem of the L_p-maximal regularity.* Math. Z. **235** (2000), 559–568.

[8] P.C. Kunstmann, L. Weis, *Perturbation theorems for maximal L_p-regularity.* Ann Scuola Norm. Sup. Pisa **30** (2001), 415–435.

[9] A. Lunardi, *Analytic Semigroups and Optimal Regularity in Parabolic Problems.* Birkhäuser, Basel, 1995.

[10] A. Pazy, *Semigroups of linear operators and applications to partial differential equations.* Springer, New York, 1983.

[11] J. Prüss, R. Schnaubelt, *Solvability and maximal regularity of parabolic evolution equations with coefficients continuous in time.* J. Math. Anal. Appl. **256** (2001), 405–430.

[12] J. Prüss, H. Sohr, *Imaginary powers of elliptic second order differential operators in L^p-spaces.* Hiroshima Math. J. **23** (1993), 161–192.

[13] R. Seeley, *Complex powers of an elliptic operator.* In: Singular Integrals, Proc. Sympos. Pure Math., Vol. 10, 288–307, Amer. Math. Soc., 1967.

[14] H.B. Stewart, *Generation of analytic semigroups by strongly elliptic operators.* Trans. Amer. Math. Soc. **199** (1974), 141–162.

[15] L. Weis, *A new approach to L_p-maximal regularity.* In: Proceedings 6th Int. Conf. on Evolution Equations and their Applications, 1998, Bad Herrenalb, 195–214, Marcel Dekker, 2001.

[16] L. Weis, *Operator valued Fourier Multiplier Theorems and Maximal L_p-Regularity.* Math. Ann. **319** (2001), 735–758.

[17] J. Wloka, *Partial differential equations.* Cambridge University Press, 1987.

Peer C. Kunstmann
Mathematisches Institut I
Universität Karlsruhe
Englerstraße 2
D-76128 Karlsruhe, Germany
E-mail address: peer.kunstmann@math.uni-karlsruhe.de

Progress in Nonlinear Differential Equations
and Their Applications, Vol. 55, 307–321

An Age Structured S.I. Epidemic Problem in a Heterogeneous Environment

Michel Langlais and Jean-Marc Naulin

1. Introduction

In this work we analyze an S.I. epidemic model with age dependence and a heterogeneous spatial structure. Our main motivation rests on a tentative qualitative comparison in Naulin [18] between continuous models and matrix population models, i.e., discrete models, with age and space structures of the S.E.I.R. form, used in modelling the propagation of rabies in red fox populations (*Vulpes vulpes*) in Suppo et al. [19]; see also Naulin [17].

A derivation of continuous S.E.I.R.models from the corresponding matrix models is found in Naulin [18], along the lines of Cushing [7] and Gurtin [10]. Then, the first question to address is a global (in time) existence problem. While this does not pose any problem for matrix population models because they are set in an explicit form, $X(t+1) = P(t, X(t))X(t)$ where $P(t, X)$ is the reproduction matrix (Caswell [5]), this is far from being obvious for continuous models with age dependence because they are both non linear and non local. A further feature is that population fluxes are age dependent and may contain a transport term, as it is the case for territorial populations where juveniles must find their own territory to settle down and reproduce ([17]), [19]. This yields new age dependent mathematical problems much more structured than the one usullly considered (see the books of Webb [20], Iannelli [11] and Aniţa [3]).

The S.I. epidemic problem we shall consider here has a simpler form than the S.E.I.R. fox-rabies model quoted above, while preserving its main mathematical features. It contains as a submodel the domestic cat - F.I.V. model introduced in Courchamp et al. [6] through a system of O.D.Es, and analyzed in Ainseba et al. [1] through systems of reaction-diffusion equations with spatially dependent coefficients, but without any age-structure.

We first introduce a generic age and space population dynamic problem

$$(\partial_t + \partial_a)u - \operatorname{div}(\mathbf{D}(x,t,a)\nabla u - u\mathbf{C}(x,t,a))\,\mu(x,t,a,P(x,t))u = f(x,t,a), \quad (1.1)$$

in $\Omega \times (0,+\infty) \times (0, A_\dagger)$, where $u = u(x,t,a)$ is the space-time-age dependent density of individuals,

$$P(x,t) = \int_0^{A_\dagger} u(x,t,a)da, \qquad x \in \Omega, \quad a \in (0, A_\dagger), \quad (1.2)$$

is the total population at location $x \in \Omega$ and time $t > 0$, $\mathbf{D}(x,t,a)\nabla u - u\mathbf{C}(x,t,a)$ is the population flux, $\mu(x,t,a,P(x,t))$ is the mortality rate and $f(x,t,a)$ a possible external supply. An initial condition at $a = 0$ is given by the renewal equation

$$u(x,t,0) = \int_0^{A_\dagger} \beta(x,t,a,P(x,t))u(x,t,a)da, \qquad x \in \Omega, \quad a \in (0,A_\dagger), \qquad (1.3)$$

where $\beta(x,t,a,P(x,t))$ is the fertility function. An initial condition is prescribed at $t = 0$

$$u(x,0,a) = u_0(x,a) \geq 0, \qquad x \in \Omega, \quad a \in (0,A_\dagger), \qquad (1.4)$$

and a no flux boundary condition is assumed on $\partial\Omega \times (0,+\infty) \times (0,A_\dagger)$

$$(\mathbf{D}(x,t,a)\nabla u(x,t,a) - u(x,t,a)\mathbf{C}(x,t,a)) \cdot \eta(x) = 0. \qquad (1.5)$$

Concerning the S.I. problem, the state variables (s,i) are the space-time-age dependent densities $(s(x,t,a), i(x,t,a))$, $x \in \Omega$, $t > 0$, $a \in (0,A_\dagger)$ of susceptible and infected individuals; from these the population density is retrieved as $u(x,t,a) = s(x,t,a) + i(x,t,a)$, and the space-time density is still given by (1.2). Our S.I. model reads in $\Omega \times (0,+\infty) \times (0,A_\dagger)$

$$\begin{aligned}
(\partial_t + \partial_a)s - \ & \operatorname{div}(\mathbf{D}_s(x,t,a)\nabla s - s\mathbf{C}_s(x,t,a)) + \mu_s(x,t,a,P(x,t))s \\
&= -\sigma(x,t,a;i,s), \\
(\partial_t + \partial_a)i - \ & \operatorname{div}(\mathbf{D}_i(x,t,a)\nabla i - i\mathbf{C}_i(x,t,a)) + \mu_i(x,t,a,P(x,t))i \\
&= +\sigma(x,t,a;i,s),
\end{aligned} \qquad (1.6)$$

where P is given in (1.2), $\mathbf{D}_\ell(x,t,a)\nabla\ell - \ell\mathbf{C}_\ell(x,t,a)$ is the population flux and $\mu_\ell(x,t,a,P(x,t))$ is the density dependent mortality rate in population ℓ, $\ell = s,i$; note that one should have $\mu_s \leq \mu_i$ and actually $\mu_i = \mu_s + \alpha$ in [6] where $\alpha > 0$ is the additional mortality rate due to the virus. Last, $\sigma(x,t,a;i,s)$ is the incidence term, say the recruitment of infected individuals from the susceptible population. In this work, along the terminology of Busenberg and Cooke [4] and de Jong et al. [12] we shall assume it takes an inter-cohort proportionate mixing form

$$\sigma(x,t,a;i,s) = \int_0^{A_\dagger} k(x,t,a',a) \frac{i(x,t,a')}{s(x,t,a') + i(x,t,a')} da' \, s(x,t,a). \qquad (1.7)$$

The initial conditions at $a = 0$, for $x \in \Omega$ and $a \in (0,A_\dagger)$, are given by

$$\begin{aligned}
s(x,t,0) &= \int_0^{A_\dagger} (\beta_s(x,t,a,P(x,t))s(x,t,a) + q\beta_i(x,t,a,P(x,t))i(x,t,a))da, \\
i(x,t,0) &= (1-q)\int_0^{A_\dagger} \beta_i(x,t,a,P(x,t))i(x,t,a)da,
\end{aligned} \qquad (1.8)$$

where β_ℓ is the birth function in population ℓ, $\ell = s,i$, and q is the proportion of offspring born from infected individuals that are susceptible at birth, the proportion $1-q$ being infected; $q = 1$ means no vertical transmission as in the cat-F.I.V. model [6], where one also has $\beta_i = \beta_s$.

The initial conditions at $t = 0$ are

$$\ell(x,0,a) = \ell_0(x,a) \geq 0, \qquad x \in \Omega, \quad a \in (0,A_\dagger),\ \ell = s,i, \tag{1.9}$$

and we assumed no flux boundary conditions on $\partial\Omega \times (0,+\infty) \times (0,A_\dagger)$

$$(\mathbf{D}_\ell(x,t,a)\nabla\ell(x,t,a) - \ell(x,t,a)\mathbf{C}_\ell(x,t,a)) \cdot \eta(x) = 0,\ \ell = s,i. \tag{1.10}$$

2. Notations, definitions and preliminary results

2.1. Notations

Let Ω be a bounded open set in $I\!R^N$ with a smooth boundary $\partial\Omega$ so that locally Ω lies on one side of $\partial\Omega$. $H^1(\Omega)$ is the first order Sobolev space built on Ω, and $[H^1(\Omega)]'$ is its dual space; $< \cdot,\cdot >$ is the duality pairing between $H^1(\Omega)$ and $[H^1(\Omega)]'$.

Let T and $A_\dagger$ be a finite positive numbers; let $q \in [0,1]$.

One assumes the fertility rates β, β_s, β_i and the mortality rates μ, μ_s, μ_i are nonnegative and measurable functions $\Omega \times (0,+\infty) \times (0,A_\dagger) \times (0,+\infty) \to [0,+\infty)$, and satisfy

(A1.) $0 \leq \beta_{*n}(a) \leq \beta(x,t,a,P) \leq \beta_n(a) \leq \overline{\beta} < +\infty$, a.e. $\Omega \times (0,+\infty) \times (0,A_\dagger) \times (0,+\infty)$;
$P \to \beta(\cdot,\cdot,\cdot,P)$ locally Lipschitz continuous, uniformly on $\Omega \times (0,T) \times (0,A_\dagger)$, $\forall T > 0$;

(A2.) $0 \leq \mu_n(a) \leq \mu(x,t,a,P) \leq \mu_n^*(a)$, a.e. $\Omega \times (0,+\infty) \times (0,A_\dagger) \times (0,+\infty)$, with $\mu_n^* \in L^\infty(0,A)$, $\forall 0 < A < A_\dagger$ and $\int_0^{A_\dagger} \mu_n(a)da = +\infty$;
$P \to \mu(\cdot,\cdot,\cdot,P)$ locally Lipschitz continuous, uniformly on $\Omega \times (0,T) \times (0,A_\dagger)$, $\forall T > 0$;

with analoguous conditions for β_s, β_i and μ_s, μ_i; the support condition required for β_n, β, β_s, and β_i implies no reproduction at young ages.

(A3.) The initial data u_0, s_0 and i_0 are nonnegative elements of $L^\infty(\Omega \times (0,A_\dagger))$; the external supply f is a nonnegative element of $L^\infty(\Omega \times (0,\infty) \times (0,A_\dagger))$.

(A4.) Transport vectors $\mathbf{C}(x,t,a)$ and $\mathbf{C}_\ell(x,t,a)_\ell$, $\ell = s,i$, are bounded over $\Omega \times (0,+\infty) \times (0,A_\dagger)$.

(A5.) Diffusion matrices $\mathbf{D}(x,t,a)$ and $\mathbf{D}_\ell(x,t,a)$, $\ell = s,i$, have bounded coefficients over $\Omega \times (0,+\infty) \times (0,A_\dagger)$, and satisfy a uniform ellipticity condition, namely there exists $\alpha > 0$ such that $\forall \zeta \in I\!R^N$

$$\sum_{i,j=1}^N \mathbf{D}_{\ell,ij}(x,t,a)\zeta_i\zeta_j,\ \sum_{i,j=1}^N \mathbf{D}_{ij}(x,t,a)\zeta_i\zeta_j$$
$$\geq \alpha\|\zeta\|_{I\!R^N},\ \text{ a.e. } \Omega \times (0,+\infty) \times (0,A_\dagger). \tag{2.1}$$

(A6.) The contact number k is a nonnegative and measurable function $\Omega \times (0,+\infty) \times (0,A_\dagger)^2 \to [0,+\infty)$ with support in $\overline{\Omega} \times [0,+\infty) \times [0,A_k]^2$ for some $0 < A_k < A_\dagger$; it satisfies

$$0 \leq \int_0^{A_\dagger} k(x,t,a',a)\,da' \leq k_{max} < +\infty, \text{ a.e. } \Omega \times (0,+\infty) \times (0,A_\dagger). \tag{2.2}$$

2.2. Definitions

We shall refer to the set of equations (1.1)–(1.2)–(1.3)–(1.4)–(1.5) as problem $(\mathcal{P})$. A weak solution to problem $(\mathcal{P})$ on $\Omega \times (0,T) \times (0,A_\dagger)$ is a nonnegative function u, an element of $L^\infty(\Omega \times (0,T) \times (0,A_\dagger)) \cap L^2((0,T) \times (0,A_\dagger); H^1(\Omega))$, such that

$$\begin{aligned} &\mu(\cdot,\cdot,\cdot,P(\cdot,\cdot))u(\cdot,\cdot,\cdot) \in L^1(\Omega \times (0,T) \times (0,A_\dagger)), \\ &(\partial_t + \partial_a)u \in L^2((0,T) \times (0,A_\dagger); [H^1(\Omega)]'), \end{aligned} \tag{2.3}$$

satisfying (1.1) and (1.5) in a weak form: for all φ in $L^\infty(\Omega \times (0,T) \times (0,A_\dagger)) \cap L^2((0,T) \times (0,A_\dagger); H^1(\Omega))$,

$$\begin{aligned} \int_{(0,T)\times(0,A_\dagger)} < (\partial_t + \partial_a)u, \varphi > dadt + \int_{\Omega\times(0,T)\times(0,A_\dagger)} ([\mathbf{D}(x,t,a)\nabla u - u\mathbf{C}(x,t,a)] \\ \cdot \nabla\varphi + [\mu(x,t,a,P(x,t)u - f]\varphi)\,dxdadt = 0, \end{aligned} \tag{2.4}$$

and satisfying the initial conditions (1.3) in $L^2((0,T)\times\Omega)$ and (1.4) in $L^2((0,A_\dagger)\times\Omega)$. These latter make sense as soon as $u \in L^2((0,T) \times (0,A_\dagger); H^1(\Omega))$ and $(\partial_t + \partial_a)u \in L^2((0,T) \times (0,A_\dagger); [H^1(\Omega)]')$: see Garroni and Langlais [9].

We shall refer to the set of equations (1.6)–(1.2)–(1.7)–(1.8)–(1.9)–(1.10) as problem $(\mathcal{SI})$.
A weak solution to problem $(\mathcal{SI})$ on $\Omega \times (0,T) \times (0,A_\dagger)$ is defined in a straightforward fashion from the definition of a weak solution for problem $(\mathcal{P})$.

2.3. An auxiliary linear problem

We consider the linear problem

$$(\partial_t + \partial_a)w - \operatorname{div}(\mathbf{D}(x,t,a)\nabla w - w\mathbf{C}(x,t,a)) + m(x,t,a)w = f(x,t,a), \tag{2.5}$$

in $\Omega \times (0,+\infty) \times (0,A_\dagger)$, together with the initial conditions

$$w(x,0,a) = w_0(x,a), \qquad x \in \Omega, \quad a \in (0,A_\dagger); \tag{2.6}$$

$$w(x,t,0) = \int_0^{A_\dagger} b(x,t,a)w(x,t,a)da, \qquad x \in \Omega, \quad a \in (0,A_\dagger); \tag{2.7}$$

and the no flux boundary conditions on $\partial\Omega \times (0,+\infty) \times (0,A_\dagger)$

$$(\mathbf{D}(x,t,a)\nabla w(x,t,a) - w(x,t,a)\mathbf{C}(x,t,a)) \cdot \eta(x) = 0. \tag{2.8}$$

We still assume $\mathbf{D}$, $\mathbf{C}$, f and w_0 satisfy the generic conditions listed in Subsection 2.1.

We first state two results. They are quite similar to those given in Langlais [16] for the simpler cas $\mathbf{D}(x,t,a) = I_d$ and $\mathbf{C}(x,t,a) \equiv 0$; they are proved upon developing a fixed point argument on the initial condition at $a = 0$ (see [18]).

Proposition 2.1. *Let b and m be two given nonnegative and measurable functions on $\Omega \times (0,+\infty) \times (0, A_\dagger)$ such that*

$$\begin{aligned} &0 \le b(x,t,a) \le b_n(a) \le \overline{\beta} < +\infty, \qquad x \in \Omega, t > 0, a \in (0, A_\dagger);\\ &m \in L^\infty(\Omega \times (0,+\infty) \times (0,A)), \forall A \in (0, A_\dagger). \end{aligned} \tag{2.9}$$

Then, for each $T > 0$ there exists a unique weak solution of (2.5)–(2.6)–(2.7)–(2.8), this a unique function w such that w, $\sqrt{m}w \in L^2(\Omega \times (0,T) \times (0,A_\dagger))$ and $(\partial_t + \partial_a)u \in L^2((0,T) \times (0,A_\dagger); [H^1(\Omega)]')$, a solution of (2.5)–(2.8) in the weak form (2.4) for all φ in $L^2((0,T) \times (0,A_\dagger); H^1(\Omega))$ with $\sqrt{m}\varphi$ in $L^2(\Omega \times (0,T) \times (0,A_\dagger))$, and satisfying (2.6)–(2.7).

Proposition 2.2. *Let (2.9) hold. Then, the solution to (2.5)–(2.8) is nonnegative.*
Assume (b_i, m_i) satisfies (2.9), w_{0i} satisfies (A3.), and let w_i be the corresponding solution to (2.5)–(2.8), $i = 1,2$; then

$$\left.\begin{aligned} b_1 \le b_2, m_2 \le m_1 \ \text{a.e. in}\ \Omega \times (0,\infty) \times (0,A_\dagger),\\ f_1 \le f_2 \ \text{a.e. in}\ \Omega \times (0,\infty) \times (0,A_\dagger),\\ w_1 \le w_2, \text{a.e. in} \Omega \times (0,A_\dagger), \end{aligned}\right\} \Rightarrow u_1 \le u_2 \, \text{a.e.} \Omega \times (0,\infty) \times (0,A_\dagger).$$

The main result of this subsection is the following L^∞ estimate.

Theorem 2.1. *Let (2.9) hold with $b_n(a) \equiv 0$ for some $a_* > 0$ and $a \in (0, a_*)$. Assume $\mathbf{D}, \mathbf{C}, f$ and w_0 satisfy conditions listed in Subsection 2.1.*
Then, for each $T > 0$, there exists a constant $M_0(T) > 0$ depending on α, $\overline{\beta}$, $\|\mathbf{C}\|_{\infty,\Omega\times(0,T)\times(0,A_\dagger)}$, $\|w_0\|_{\infty,\Omega\times(0,A_\dagger)}$, $\|f\|_{\infty,\Omega\times(0,T)\times(0,A_\dagger)}$ and the spatial dimension N such that the weak solution w of (2.5)–(2.6)–(2.7)–(2.8) satisfies

$$\|w(\cdot,t,\cdot)\|_{\infty,\Omega\times(0,A_\dagger)} \le M_0(T), \quad t \in (0,T). \tag{2.10}$$

Proof. Using the comparison result stated in Proposition 2.2 it is sufficient to establish Theorem 2.1 for $m(x,t,a) \equiv 0$ and $b(x,t,a) \equiv b_n(a)$. The main difficulty lies in the dependence of $\mathbf{D}$ and $\mathbf{C}$ on the variables x, t and a. We proceed through three steps.

Step 1. Below the characteristic line $t = a$ one may change variable. For $a_0 > 0$ and $c \in (0, A_\dagger - a_0)$ set $t = c, a = a_0 + c$; let $v(x,c) = w(x,c,a_0+c)$. Then v is a solution to a linear parabolic problem,

$$\partial_c v - \operatorname{div}(\mathbf{D}(x,c,a_0+c)\nabla v - v\mathbf{C}(x,c,a_0+c)) = f(x,c,a_0+c), x \in \Omega, c \in (0, A_\dagger - a_0),$$

together with the initial and boundary conditions

$$v(x,0) = w(x,0,a_0) = w_0(x,a_0), \quad x \in \Omega;$$

$$(\mathbf{D}(x,c,a_0+c)\nabla v(x,c) - v(x,c)\mathbf{C}(x,c,a_0+c)) \cdot \eta(x) = 0, \quad x \in \partial\Omega, c \in (0, A_\dagger - a_0).$$

From the assumption on w_0 it follows

$$\|v(\cdot,0)\|_{\infty,\Omega} \leq \|w_0\|_{\infty,\Omega\times(0,A_\dagger)}.$$

Assume first that $f \equiv 0$. Then, a simple integration over $\Omega \times (0,c)$ yields

$$\|v(\cdot,c)\|_{1,\Omega} \leq \|v(\cdot,0)\|_{1,\Omega} \leq \|w_0(\cdot,a_0)\|_{1,\Omega} \leq |\Omega|\,\|w_0\|_{\infty,\Omega\times(0,A_\dagger)},\ 0<c<A_\dagger-a_0.$$

One may now reproduce a proof in Alikakos ([2], Theorem 3.1) to show there exists a constant M_1 depending on α, N, and the L^∞ norms of $\mathbf{C}$ and w_0 such that

$$\|v(\cdot,c)\|_{\infty,\Omega} \leq M_1 < +\infty, \quad 0<c<A_\dagger-a_0, 0<a_0<A_\dagger. \tag{2.11}$$

For non trivial f one may use results in Ladyzenskaya et al. [14], to show the existence of a constant M_1 now depending on α, N, and the L^∞ norms of $\mathbf{C}, w_0$ and f such that (2.11) holds.

Step 2. We now integrate (2.5)–(2.8), with $m=0$ and $b=b_n$, over $\Omega\times(0,A_\dagger)$ to get

$$\frac{d}{dt}\int_{\Omega\times(0,A_\dagger)} w(x,t,a)dadx \leq \int_{\Omega\times(0,A_\dagger)} b_n(a)w(x,t,a)dadx.$$

As a consequence

$$\|e^{-\overline{\beta}t}w(.,t,.)\|_{1,\Omega\times(0,A_\dagger)} \leq \|w_0\|_{1,\Omega\times(0,A_\dagger)}. \tag{2.12}$$

Step 3. Above the characteristic line $t=a$, one changes variables. For $t_0>0$ and $c\in(0,A_\dagger)$ set $a=c$, $t=t_0+c$; let $v(x,c)=e^{-\overline{\beta}t_0}w(x,t_0+c,c)$. Then, v is a solution of a parabolic problem,

$$\partial_c v - \operatorname{div}(\mathbf{D}(x,t_0+c,c)\nabla v - v\mathbf{C}(x,t_0+c,c)) = f(x,t_0+c,c), \quad x\in\Omega, c\in(0,A_\dagger),$$

together with the initial and boundary conditions

$$v(x,0)=e^{-\beta t_0}w(x,t_0,0)=e^{-\overline{\beta}t_0}\int_0^{A_\dagger} b_n(a)w(x,t_0,a)da, \quad x\in\Omega,$$

$$(\mathbf{D}(x,t_0+c,c)\nabla v(x,c) - v(x,c)\mathbf{C}(x,t_0+c,c))\cdot\eta(x)=0, \quad x\in\partial\Omega, c\in(0,A_\dagger).$$

Assume first that $f\equiv 0$. Then, a simple integration over $\Omega\times(0,c)$ together with (2.12), yields

$$\|v(\cdot,c)\|_{1,\Omega} \leq \|v(\cdot,0)\|_{1,\Omega} \leq \overline{\beta}\,\|w_0\|_{1,\Omega\times(0,A_\dagger)}, \quad c\in(0,A_\dagger). \tag{2.13}$$

To conclude as in *Step 1* the main problem now is to get L^∞ estimates on $v(\cdot,0)$; this is exactly where one uses the condition $b_n(a)\equiv 0$ for some $a_*>0$ and $a\in(0,a_*)$.

First, we work on the time interval $t_0\in(0,a_*)$; then, using (2.11), one gets

$$0\leq v(x,0)=e^{-\overline{\beta}t_0}\int_{a_*}^{A_\dagger} b_n(a)w(x,t_0,a)da \leq \overline{\beta}\,A_\dagger\,M_1 = m_{21}, \quad x\in\Omega;$$

thus one can invoke the same result in Alikakos [2] to get a constant M_{21} now depending on α, N, the L^∞ norm of $\mathbf{C}$ and m_{21} such that

$$\|v(\cdot,c)\|_{\infty,\Omega} \leq M_{21}, \qquad 0<c<A_\dagger, \quad 0\leq t_0\leq a_*. \tag{2.14}$$

Next, a recursion argument produces sequences $(m_{2n})_{n\geq 0}$ and $(M_{2n})_{n\geq 0}$ depending on α, N, the L^∞ norm of $\mathbf{C}, w_0$ and $\overline{\beta}$ such that for $t_0 \in (na_*, (n+1)a_*)$,

$$0 \leq v(x,0) \leq m_{2n}, \quad \|v(\cdot,c)\|_{\infty,\Omega} \leq M_{2n}, \quad 0 < c < A_\dagger, \quad na_* \leq t_0 \leq (n+1)a_*. \tag{2.15}$$

When $f \equiv 0$ this estimate together with estimate (2.11) complete the proof of Theorem 2.1.

For non trivial f one may use results in Ladyzenskaya et al. [14], to reach similar conclusions with constants M_{1n} now depending on α, N, $\overline{\beta}$, and the L^∞ norms of $\mathbf{C}, w_0$ and f.

3. The age-space dependent population dynamic problem

We come back to our original nonlinear population dynamics model $(\mathcal{P})$.

3.1. Main result

Theorem 3.2. *Let conditions (A1) to (A6) be satisfied. Assume*

(A7.) *for some $a_* > 0$, $\beta(x,t,a,P) = \beta_n(a) \equiv 0$ in $\Omega \times (0,+\infty) \times (0,a_*) \times (0,+\infty)$.*

Then, for each $T > 0$ problem $(\mathcal{P})$ has a unique weak solution globally defined on $\Omega \times (0,T) \times (0,A_\dagger)$; it satisfies

$$0 \leq u(x,t,a) \leq w^*(x,t,a), \qquad x \in \Omega, 0 < t < T, a \in (0,A_\dagger), \tag{3.1}$$

where w^* is the solution to (2.5)–(2.6)–(2.7)–(2.8) with $m = 0$, $b(x,t,a) \equiv \beta_n(a)$ and $w_0 = u_0$.

The proof given in Subsection 3.2 below goes through an iterative method. When $\mathbf{D}$ and $\mathbf{C}$ do not depend on the age variable, a similar result can be proved without condition (A7): see Langlais [15], Kubo and Langlais [13], and Naulin [18].

3.2. A contraction argument and existence result

To get existence of a solution on $\Omega \times (0,T) \times (0,A_\dagger)$ we use a standard fixed point argument for a contracting mapping on time intervals of small but constant amplitude τ^*.

Let $\mathcal{K}$, a convex and closed subset of $L^2(\Omega \times (0,T) \times (0,A_\dagger))$, be defined as

$$\mathcal{K} = \{v \in L^2(\Omega \times (0,T) \times (0,A_\dagger)), 0 \leq v \leq w^*, \text{ a.e. } \Omega \times (0,T) \times (0,A_\dagger)\}.$$

Given v in $\mathcal{K}$ set $Q(x,t) = \displaystyle\int_0^{A_\dagger} v(x,t,a)da$ and let u be a solution to the linear problem

$$(\partial_t + \partial_a)u - \operatorname{div}(\mathbf{D}(x,t,a)\nabla u - u\mathbf{C})(x,t,a) + \mu(x,t,a,Q(x,t))u = f, \tag{3.2}$$

on $\Omega\times(0,T)\times(0,A_\dagger)$ together with the initial and boundary conditions (1.4)–(1.5) and

$$u(x,t,0)=\int_0^{A_\dagger}\beta(x,t,a,Q(x,t))u(x,t,a)da,\qquad x\in\Omega.\quad t\in(0,T).\tag{3.3}$$

From the linear theory this problem (3.2)–(1.4)–(3.3)–(1.5) has a unique weak solution, which is nonnegative; it also follows from a comparison argument that

$$0\le u(x,t,a)\le w^*(x,t,a),x\in\Omega,t\in(0,T),a\in(0,A_\dagger),\tag{3.4}$$

Hence one may define a mapping $\phi:\mathcal{K}\to\mathcal{K}$, with $\phi(v)=u$ the solution to (3.2)–(1.4)–(3.3)–(1.5).

Lemma 3.1. *There exists two constants k_1 and k_2 depending solely on $R^*=\|w^*\|_{\infty,\Omega\times(0,A_\dagger)\times 0,T)}$, α, $\overline{\beta}$, $A_\dagger$, $\left\|\frac{\partial\chi}{\partial p}\right\|_{\infty,\Omega\times(0,T)\times(0,A_\dagger)\times(0,R^*)}$ for $\chi=\beta,\mu$, $\|\mathbf{C}\|_{\infty,\Omega\times(0,T)\times(0,A_\dagger)}$ such that for $t\in(0,T)$*

$$\begin{aligned}&\frac{d}{dt}\int_{\Omega\times(0,A_\dagger)}|u_1-u_2|^2(x,t,a)dxda\le\\&k_1\int_{\Omega\times(0,A_\dagger)}|u_1-u_2|^2(x,t,a)dxda+k_2\int_{\Omega\times(0,A_\dagger)}|v_1-v_2|^2(x,t,a)dxda.\end{aligned}\tag{3.5}$$

Proof. We denote by $k_i, i\ge 3$, various constants having properties identical to k_1 and k_2.
Taking the difference between two solutions u_1 and u_2 corresponding to two inputs v_1 and v_2, multiplying the resulting equation by (u_1-u_2) and integrating over Ω yields

$$\begin{aligned}\frac{1}{2}(\partial_t+\partial_a)\int_\Omega(u_1-u_2)^2dx+\int_\Omega\mathbf{D}\nabla(u_1-u_2)\cdot\nabla(u_1-u_2)dx\\-\int_\Omega(u_1-u_2)\mathbf{C}\cdot\nabla(u_1-u_2)dx+\int_\Omega\mu(Q_1)(u_1-u_2)^2dx\\=-\int_\Omega u_2(u_1-u_2)(\mu(Q_1)-\mu(Q_2))dx.\end{aligned}$$

Using the ellipticity condition on matrix $\mathbf{D}$, Cauchy-Schwarz inequality to handle the integral containing vector $\mathbf{C}$, the nonnegativity of the death-rate μ, the a priori bound (3.4) and the mean value theorem to handle $\mu(Q_1)-\mu(Q_2)$, one arrives to

$$\begin{aligned}\frac{1}{2}(\partial_t+\partial_a)\int_\Omega(u_1-u_2)^2dx+\frac{\alpha}{2}\int_\Omega|\nabla(u_1-u_2)|^2dx\le\\k_3\int_\Omega(u_1-u_2)^2dx+k_4\int_\Omega|u_1-u_2|.|Q_1-Q_2|dx.\end{aligned}$$

A further integration over age, from $a = 0$ to $a = A_\dagger$, gives

$$\frac{1}{2}\frac{d}{dt}\int_{\Omega\times(0,A_\dagger)}(u_1-u_2)^2(x,t,a)dxda + \frac{\alpha}{2}\int_{\Omega\times(0,A_\dagger)}|\nabla(u_1-u_2)|^2(x,t,a)dxda \le$$
$$k_3\int_{\Omega\times(0,A_\dagger)}(u_1-u_2)^2(x,t,a)dxda + I_1(t) + I_2(t). \tag{3.6}$$

Now,

$$I_1(t) = k_4\int_{\Omega\times(0,A_\dagger)}|u_1-u_2|(x,t,a).\left|\int_0^{A_\dagger}|v_1-v_2|(x,t,a)da\right|dxda \le$$
$$k_4\int_\Omega\left(\int_0^{A_\dagger}|v_1-v_2|(x,t,a)da\right).\left(\int_0^{A_\dagger}|u_1-u_2|(x,t,a)da\right)dx.$$

Hölder inequality implies $\int_0^{A_\dagger}|w_1-w_2|(a)da \le \sqrt{A_\dagger}.\left(\int_0^{A_\dagger}(w_1-w_2)^2(a)da\right)^{1/2}$; using this twice and applying again Cauchy-Schwarz inequality leads to

$$0 \le I_1(t) \le \frac{1}{2}A_\dagger k_4\int_{\Omega\times(0,A_\dagger)}\left[(u_1-u_2)^2+(v_1-v_2)^2\right](x,t,a)dxda. \tag{3.7}$$

Next,

$$I_2(t) = \frac{1}{2}\int_\Omega\left(\int_0^{A_\dagger}\left[\beta(Q_1)(u_1-u_2)+u_2(\beta(Q_1)-\beta(Q_2))\right](x,t,a)da\right)^2dx,$$

so that

$$I_2(t) \le \int_\Omega\left[\left(\int_0^{A_\dagger}\beta(Q_1)(u_1-u_2)da\right)^2+\left(\int_0^{A_\dagger}u_2(\beta(Q_1)-\beta(Q_2)da\right)^2\right]dx.$$

First, elementary algebraic calculations yields

$$\int_\Omega\left(\int_0^{A_\dagger}\beta(Q_1)(u_1-u_2)da\right)^2dx \le \overline{\beta}^2\int_\Omega\left(\int_0^{A_\dagger}|u_1-u_2|da\right)^2dx \le$$
$$\overline{\beta}^2A_\dagger\int_{\Omega\times(0,A_\dagger)}(u_1-u_2)^2dxda.$$

Second, the a priori bound (3.4) and the Mean Value Theorem give

$$\int_\Omega\left(\int_0^{A_\dagger}u_2(\beta(Q_1)-\beta(Q_2))da\right)^2dx \le k_5A_\dagger^2\int_\Omega|Q_1-Q_2|^2(x,t)dx$$
$$= k_5A_\dagger^2\int_\Omega\left|\int_0^{A_\dagger}(v_1-v_2)(x,t,a)da\right|^2dx \le k_5A_\dagger^2\int_{\Omega\times(0,A_\dagger)}(v_1-v_2)^2dadx.$$

As a consequence one may deduce that

$$0 \le I_2(t) \le k_6(1 + A_\dagger^2) \int_{\Omega\times(0,A_\dagger)} \left[(u_1 - u_2)^2 + (v_1 - v_2)^2\right](x,t,a)dxda. \quad (3.8)$$

substituting inequality (3.7) for $I_1(t)$ and (3.8) for $I_2(t)$ into inequality (3.6), one has the desired result, i.e., inequality (3.5).

End of the proof of Theorem 3.2. Take k_1, k_2 from Lemma 3.1, and choose τ^* such that

$$k_2 \frac{1 - e^{k_1\tau^*}}{k_1} = \rho(\tau^*) < 1.$$

Then, it follows from Lemma 3.1 that the mapping $\phi : \mathcal{K} \to \mathcal{K}$, is a strict contraction in $L^2(\Omega \times (0,\tau^*) \times (0, A_\dagger)) \cap \mathcal{K}$, with a rate $\rho(\tau^*)$. Let u be its unique fixed point. Routine calculations show it is also a solution to problem $(\mathcal{P})$ on $\Omega \times (0,\tau^*) \times (0, A_\dagger)$. To check this, one may construct u as a limit of a sequence $(u^n)_{n\ge0}$ defined as $u^0 = 0$ and $u^{n+1} = \phi(u^n)$, $n \ge 0$, converging in $L^2(\Omega \times (0,\tau^*) \times (0, A_\dagger))$ and satisfying an L^∞ bound: $0 \le u^n \le w^*$.
Using the Mean-Value Theorem, one gets that the sequence $(\mu(\cdot,\cdot,\cdot,Q^n(\cdot,\cdot))u^{n+1})_{n\ge0}$ is converging to $\mu(\cdot,\cdot,\cdot,P(\cdot,\cdot))u$ in $L^1(\Omega \times (0,\tau^*) \times (0,A))$, $0 < A < A_\dagger$; this sequence being also bounded in $L^2(\Omega \times (0,\tau^*) \times (0,A))$, $0 < A < A_\dagger$, convergence also holds weakly in $L^2(\Omega\times(0,\tau^*)\times(0,A))$, $0 < A < A_\dagger$.
Straighforward integrations by parts yields that the sequence $(\|\nabla u^n\|_{2,\Omega\times(0,\tau^*)\times(0,A_\dagger)})_{n\ge0}$ is bounded in $L^2(\Omega \times (0,\tau^*) \times (0, A_\dagger))$.
This is sufficient to show that the fixed point u is a solution of (2.4) for any $A \in (0, A_\dagger)$. A further integration by parts shows (2.3) holds, as well as (2.4).
The nonlocal condition at $a = 0$ is handled along the lines of the nonlocal death term.

To conclude one may reproduce the same proof on $(n\tau^*, (n+1)\tau^*)$ until T is reached.

4. The S.I. age-space dependent epidemic problem

We come back to our original epidemic problem $(\mathcal{SI})$.

4.1. Main result

Theorem 4.3. *Assume conditions (A1) to (A6) are satisfied. Moreover, assume that (A7) from Theorem 3.2 and the following condition hold*

(A8.) $s_0(x,a) \ge s_0 > 0$, *a.e.* $\Omega \times (0, a_{*n}))$, *for some* $a_{*n} \in supp\beta_{*n} \cap (0, A_k)$.

Then, for each $T > 0$ *problem* $(\mathcal{SI})$ *has a unique weak solution globally defined on* $\Omega \times (0,T) \times (0, A_\dagger)$.

This is proved in Subsection 3.2 below. We still use a standard fixed point argument for a contracting mapping on time intervals of small but constant amplitude τ^*. Namely, given a couple $(\bar{s}, \bar{\imath})$ of nonnegative elements in $L^\infty(\Omega \times (0,T) \times$

$(0, A_\dagger))$, let (s, i) be the solution on $\Omega \times (0, T) \times (0, A_\dagger)$ of the linear system

$$\begin{aligned}(\partial_t + \partial_a)s - \ &\mathrm{div}\,(\mathbf{D}_s(x,t,a)\nabla s - s\mathbf{C}_s(x,t,a)) + \beta_s(x,t,a,\overline{P}(x,t))s \\ &= -\gamma(x,t,a;\overline{s},\overline{\imath})\, s, \\ (\partial_t + \partial_a)i - \ &\mathrm{div}\,(\mathbf{D}_i(x,t,a)\nabla i - i\mathbf{C}_i(x,t,a)) + \beta_i(x,t,a,\overline{P}(x,t))i \\ &= +\gamma(x,t,a;\overline{s},\overline{\imath})\, s,\end{aligned} \tag{4.1}$$

where,

$$\gamma(x,t,a;\overline{s},\overline{\imath}) = \int_0^{A_\dagger} k(x,t,a',a)\frac{\overline{\imath}(x,t,a')}{\overline{s}(x,t,a') + \overline{\imath}(x,t,a')}\, da', \tag{4.2}$$

with the initial sconditions at $a = 0$, for $x \in \Omega$ and $a \in (0, A_\dagger)$, given by

$$\begin{aligned} s(x,t,0) &= \int_0^{A_\dagger} (\beta_s(x,t,a,\overline{P}(x,t))s(x,t,a) + q\beta_i(x,t,a,\overline{P}(x,t))i(x,t,a))da, \\ i(x,t,0) &= (1-q)\int_0^{A_\dagger} \beta_i(x,t,a,\overline{P}(x,t))i(x,t,a)da, \end{aligned} \tag{4.3}$$

where

$$\overline{P}(x,t) = \int_0^{A_\dagger} (\overline{s}(x,t,a) + \overline{\imath}(x,t,a))\, da, \tag{4.4}$$

the initial conditions at $t = 0$ in (1.9) and the no flux boundary conditions in (1.10).

4.2. An invariance result

The first point now is to find a suitable invariant closed and convex domain in $L^2(\Omega \times (0,T) \times (0, A_\dagger))$.

Lemma 4.2. *For each $T > 0$ there exits a constant $M(T)$ depending only on α, N, the L^∞ norm of $\mathbf{C}_s, \mathbf{C}_i, s_0, i_0$ and $\overline{\beta}$ such that given a couple $(\overline{s}, \overline{\imath})$ of nonnegative elements in $L^\infty(\Omega \times (0,T) \times (0, A_\dagger))$, then the solution (s, i) of (4.1)-(4.3)-(1.9)-(1.10) satisfies*

$$0 \le s(x,t,a),\, i(x,t,a) \le M(T), \quad 0 < t < T,\, x \in \Omega,\, a \in (0, A_\dagger). \tag{4.5}$$

Proof. We follow the steps in the proof of Theorem 2.1.

Stcp 1. Below the characteristic line $t = a$, for $a_0 > 0$ and $c \in (0, A_\dagger - a_0)$ set $t = c, a = a_0 + c$. Then $v(x,c) = s(x,c,a_0+c)$ is a solution to a linear parabolic inequality,

$$\partial_c v - \ \mathrm{div}(\mathbf{D}_s(x,c,a_0+c)\nabla v - v\mathbf{C}_s(x,c,a_0+c)) \le 0, \qquad x \in \Omega, c \in (0, A_\dagger - a_0),$$

together with the initial and boundary conditions

$$v(x,0) = s(x,0,a_0) = s_0(x,a_0), \quad x \in \Omega;$$

$$(\mathbf{D}_s(x,c,a_0+c)\nabla v(x,c) - v(x,c)\mathbf{C}_s(x,c,a_0+c))\cdot\eta(x) = 0,\ x \in \partial\Omega, c \in (0, A_\dagger - a_0).$$

From the assumption on s_0 it follows $\|v(\cdot,0)\|_{\infty,\Omega} \le \|s_0\|_{\infty,\Omega\times(0,A_\dagger)}$, and a simple integration over $\Omega \times (0,c)$ yields $\|v(\cdot,c)\|_{1,\Omega} \le \|v(\cdot,0)\|_{1,\Omega} \le |\Omega|\, \|s_0\|_{\infty,\Omega\times(0,A_\dagger)}$,

$0 < c < A_\dagger - a_0$. One may now reproduce a proof in Alikakos ([2], Theorem 3.1) to show that there exists a constant M_1 depending on α, N, and the L^∞ norms of $\mathbf{C}_s$ and s_0 such that

$$0 \leq s(x,c,a_0+c) \leq M_1 < +\infty, \quad x \in \Omega,\ 0 < c < A_\dagger - a_0,\ 0 < a_0 < A_\dagger. \quad (4.6)$$

In a similar fashion, $v(x,c) = i(x,c,a_0+c)$ is a solution to a linear parabolic inequality,

$$\partial_c v - \operatorname{div}(\mathbf{D}_i(x,c,a_0+c)\nabla v - v\mathbf{C}_i(x,c,a_0+c)) \leq k_{max} M_1, \quad x \in \Omega, c \in (0, A_\dagger - a_0),$$

together with the initial and boundary conditions

$$v(x,0) = i(x,0,a_0) = i_0(x,a_0), \quad x \in \Omega;$$

$$(\mathbf{D}_i(x,c,a_0+c)\nabla v(x,c) - v(x,c)\mathbf{C}_i(x,c,a_0+c))\cdot\eta(x) = 0, \quad x \in \partial\Omega, c \in (0, A_\dagger - a_0).$$

One may recall results in Ladyzenskaya et al [14], to show the existence of a constant M_2 now depending on α, N, M_1 and the L^∞ norms of $\mathbf{C}_i, i_0$ such that

$$0 \leq i(x,c,a_0+c) \leq M_2 < +\infty, \quad x \in \Omega,\ 0 < c < A_\dagger - a_0,\ 0 < a_0 < A_\dagger. \quad (4.7)$$

Step 2. We now integrate (4.1) over $\Omega \times (0, A_\dagger)$ to get for $p = s + i$

$$\frac{d}{dt}\int_{\Omega\times(0,A_\dagger)} p(x,t,a)\,da\,dx \leq \overline{\beta}\int_{\Omega\times(0,A_\dagger)} p(x,t,a)\,da\,dx.$$

As a consequence

$$\|e^{-\overline{\beta}t}p(.,t,.)\|_{1,\Omega\times(0,A_\dagger)} \leq \|p_0 = s_0 + i_0\|_{1,\Omega\times(0,A_\dagger)}. \quad (4.8)$$

Step 3. We proceed by time intervals of length a_*.
Let us begin with $[0, a_*]$. Above the characteristic line $t = a$, for $t_0 > 0$ and $c \in (0, A_\dagger)$ set $a = c$, $t = t_0 + c$. Then $v(x,c) = e^{-\overline{\beta}t_0}s(x, t_0+c, c)$ is a solution of a parabolic inequality,

$$\partial_c v - \operatorname{div}(\mathbf{D}_s(x,t_0+c,c)\nabla v - v\mathbf{C}_s(x,t_0+c,c)) \leq 0, \quad x \in \Omega, c \in (0, A_\dagger),$$

together with the initial and boundary conditions

$$v(x,0) = e^{-\overline{\beta}t_0}\int_{a_*}^{A_\dagger}(\beta_s(x,t_0,a,\overline{P}(x,t_0))s(x,t_0,a)$$
$$+ q\beta_i(x,t_0,a,\overline{P}(x,t_0))i(x,t_0,a))\,da, \quad x \in \Omega,$$

$$(\mathbf{D}_s(x,t_0+c,c)\nabla v(x,c) - v(x,c)\mathbf{C}_s(x,t_0+c,c))\cdot\eta(x) = 0, \quad x \in \partial\Omega, c \in (0, A_\dagger).$$

From (4.6)–(4.7) and $0 \leq t \leq a_*$ one gets

$$0 \leq v(x,0) \leq \overline{\beta}\,A_\dagger\,(M_1 + qM_2) = m_{31}, \quad x \in \Omega.$$

Another integration over $\Omega \times (0,c)$ yields $\|v(\cdot,c)\|_{1,\Omega} \leq \|v(\cdot,0)\|_{1,\Omega} \leq, |\Omega|\, m_{31}$, $c \in (0, A_\dagger)$. To conclude one can invoke the same result in Alikakos [2] to get a constant M_{31} now depending on α, N, the L^∞ norm of $\mathbf{C}_s$ and m_{31} such that

$$0 \leq s(x,t_0+c,c) \leq M_{31} < +\infty, \quad x \in \Omega,\ 0 < c < A_\dagger,\ 0 < t_0 < a_*. \quad (4.9)$$

In a similar fashion, $v(x,c) = e^{-\overline{\beta}t_0} i(x, t_0 + c, c)$ is a solution to a linear parabolic inequality,

$$\partial_c v - \operatorname{div}(\mathbf{D}_i(x, t_0+c, c)\nabla v - v\mathbf{C}_i(x, t_0+c, c)) \le k_{max} M_{31}, \qquad x \in \Omega, c \in (0, A_\dagger),$$

together with the initial and boundary conditions

$$v(x,0) = i(x,t_0,0) = (1-q)e^{-\overline{\beta}t_0} \int_{a_*}^{A_\dagger} \beta_i(x, t_0, a, \overline{P}(x,t_0)) i(x, t_0, a)da, \quad x \in \Omega;$$

$$(\mathbf{D}_i(x, t_0+c, c)\nabla v(x,c) - v(x,c)\mathbf{C}_i(x, t_0+c, c)) \cdot \eta(x) = 0, \qquad x \in \partial\Omega, c \in (0, A_\dagger).$$

One may refer to results in Ladyzenskaya et al. [14], to show the existence of a constant M_{41} now depending on α, N, M_{31} and the L^∞ norm of $\mathbf{C}_i$ such that

$$0 \le i(x, t_0 + c, c) \le M_{41} < +\infty, \quad x \in \Omega,\ 0 < c < A_\dagger,\ 0 < t_0 < a_*. \tag{4.10}$$

Last, a recursion argument produces sequences $(M_{3n}, M_{4n})_{n\ge 0}$ depending on α, N, the L^∞ norm of $\mathbf{C}_s, \mathbf{C}_i, s_0, i_0$ and $\overline{\beta}$ such that for $t \in (na_*, (n+1)a_*)$, $x \in \Omega$ and $0 \le a \le t$ $0 \le s(x,t,a) \le M_{3n}$, $0 \le i(x,t,a) \le M_{4n}$.

4.3. A positivity result

The second point now is to show a positivity property for solutions (s, i).

Lemma 4.3. *The solution s to (4.1)–(4.3)–(1.9)–(1.10) is bounded from below by the nonnegative solution w_* of the linear problem (2.5)–(2.6)–(2.7)–(2.8) with* $\mathbf{D} = \mathbf{D}_s$, $\mathbf{C} = \mathbf{C}_s$, $m(x,t,a) = \mu_n^*(a) + \int_0^{A_\dagger} k(x,t,a')da'$, $f = 0$, $w_0 = s_0$ *and* $b(x,t,a) = \beta_{*n}(a)$.
Moreover, there exists a positive constant $m(T)$ such that

$$0 < m(T) \le w_*(x,t,a), \quad x \in \Omega, 0 \le t \le T, 0 \le a \le A_k. \tag{4.11}$$

Proof. Rewriting

$$\gamma(x,t,a;\overline{s},\overline{\imath}) = \int_0^{A_\dagger} k(x,t,a',a)da' - \int_0^{A_\dagger} k(x,t,a',a)\frac{\overline{s}(x,t,a')}{\overline{s}(x,t,a') + \overline{\imath}(x,t,a')} da',$$

the first statement follows from a comparaison result for linear equations (Proposition 2.2).

The positive lower bound (4.11) is derived through integrations along the characteristic lines of $\partial_t + \partial_a$, as in Langlais [16], using condition $(A8)$ as in Fitzgibbon and Langlais [8].

End of the proof of Theorem 4.3. Starting with $s = 0$ and $i = 0$, say $\gamma \equiv 0$, let us introduce a sequence of nonnegative functions $(s^n, i^n)_{n\ge 0}$, where (s^{n+1}, i^{n+1}) is the solution to (4.1)–(4.3)–(1.9)–(1.10) with $\overline{s} = s^n$ and $\overline{\imath} = s^n$. One has $0 \le w_*(x,t,a) \le s^n(x,t,a)$.
In order to show that this sequence is a Cauchy sequence on $L^2(\Omega\times(0,\tau^*)\times(0,A_\dagger))$ for some positive τ^*, the main difference with respect to the proof of Theorem 3.2

lies in the incidence term $\gamma(x,t,a;\bar{s},\bar{\imath})\,s$. Note that from the support assumption in condition (A6) one has

$$\gamma(x,t,a;s^n,i^n)\,s^{n+1} - \gamma(x,t,a;s^{n-1},i^{n-1})\,s^n = \int_0^{A_k} k\frac{i^n}{p^n}(x,t,a')\,da'\,(s^{n+1}-s^n)$$
$$+\int_0^{A_k} k\frac{i^n - i^{n-1}}{p^n}(x,t,a')\,da'\,s^n + \int_0^{A_k} ki^n\frac{p^n - p^{n-1}}{p^n\,p^{n+1}}(x,t,a')\,da'\,s^n.$$

It follows from the uniform positivity property established in Lemma 4.11 that for some constatnt k independent of n, the $L^2(\Omega\times(0,T)\times(0,A_\dagger))$ norm of this term is bounded from above by

$$k\left(\int_{\Omega\times(0,A_\dagger)} |u^{n+1}-u^n|^2(x,t,a)dxda + \int_{\Omega\times(0,A_\dagger)} |u^n - u^{n-1}|^2(x,t,a)dxda\right).$$

References

[1] B.E. Ainseba, W.E. Fitzgibbon, M. Langlais and J.J. Morgan, *An application of homogenization techniques to population dynamics models.* Communications in Pure and Applied Analysis, in press.

[2] N.D. Alikakos, *An application of the invariance principle to reaction-diffusion equations.* J. Differential Equations **33** (1979), 201–225.

[3] S. Aniţa, *Analysis and Control of Age-Dependent Population Dynamics.* Kluwer Academic Publishers, Dordrecht, 2000.

[4] S. Busenberg, K.C. Cooke, *Vertically Transmitted Diseases.* Biomathematics Series **23**, Springer, Berlin 1993.

[5] H. Caswell, *Matrix Population Models.* Sinauer Associates, Inc. Publishers, 2nd edition, 2001.

[6] F. Courchamp, D. Pontier, M. Langlais, M. Artois, *Population dynamics of Feline Leukemia Virus within cat populations.* J. Theor. Biol. **175** (1995), 553–560.

[7] J. Cushing, *An introduction to structured population dynamics.* S.I.A.M. Philadelphia, 1998.

[8] W.E. Fitzgibbon and M. Langlais, *Weakly coupled hyperbolic systems modeling the circulation of infectious disease in structured populations.* Mathematical Biosciences, **165** (2000), 79–95.

[9] M.-G. Garroni, M. Langlais, *Age dependent population diffusion with external constraints.* J. Math. Biol. **14** (1982), 77–94.

[10] M-E. Gurtin, *A system of equations for age dependent population diffusion.* J. Theor. Biol. **40** (1972), 389–392.

[11] M. Iannelli, *Mathematical Theory of Age-Structured Population Dynamics.* Applied Mathematics Monographs No 7, C.N.R. Pisa, 1994.

[12] M.C.M. de Jong, O. Diekmann, A.A de Koeijer and P. Reinders, *The force of infection in populations of varying size: a modeling problem.* J. Biol. Systems **3** (1995), 519–529.

[13] M. Kubo, M. Langlais, *Periodic solutions for non-linear population dynamics models with age- dependence and spatial structure.* J. Differential Equations **109** (1994), 274–294.

[14] O.A. Ladyzhenskaya, V. Solonnikov and N. Ural'ceva, *Linear and quasilinear equations of parabolic type.* Transl. AMS **23**, Providence Rhode Island, 1968.

[15] M. Langlais, *A nonlinear problem in age dependent population diffusion.* S.I.A.M. J. Math. Analysis **16** (1985), 510–529.

[16] M. Langlais, *Large time behavior in a nonlinear age-dependent population dynamics problem with spatial diffusion.* J. Math. Biol. **26**, No.3 (1988), 319–346.

[17] J. M. Naulin, *A contribution of sparse matrices tools to matrix population model analysis.* Mathematical Biosciences, in press.

[18] J. M. Naulin, *Ph.D.*, U niversité Bordeaux 1, 2001.

[19] C. Suppo, J.-M. Naulin, M. Langlais, M. Artois, *A modelling approach of vaccination and sterilization programms for rabies control in fox populations.* Proc. R. Soc. London, série B, **267** (2000), 1575–1582.

[20] G.F. Webb, *Theory of nonlinear age-dependent population dynamics.* Monographs and Textbooks in Pure and Applied Mathematics, **89**. New York-Basel: Marcel Dekker, Inc. VI, 1985.

Michel Langlais
UMR CNRS 5466
Mathématiques Appliquées de Bordeaux,
case 26, Université Victor Segalen Bordeaux 2,
146, rue Léo Saignat
F-33076 Bordeaux Cedex, France
E-mail address: langlais@sm.u-bordeaux2.fr

Jean-Marc Naulin
UMR CNRS 5466
Mathématiques Appliquées de Bordeaux,
case 26, Université Victor Segalen Bordeaux 2,
146, rue Léo Saignat
F-33076 Bordeaux Cedex, France
E-mail address: jean-marc.naulin@sm.u-bordeaux2.fr

Progress in Nonlinear Differential Equations
and Their Applications, Vol. 55, 323–332

A General "Isotropic" Paley-Wiener Theorem and Some of its Applications

Günter Lumer

1. Introduction

In this paper, we give a very general "isotropic" ([1]) Paley-Wiener([2]) theorem, and a series of its applications. Among the latter: extensions of classical results like a well-known theorem on entire Laplace transforms([3]) of exponential type on L^2([4]), far beyond all L^p, $p \geq 1$([5]), up to non exponentially bounded continuous functions, distributions, and further generalized functions; extension (and also brief elegant proofs) of central results on nilpotent and superstable([6]) semigroups (as those of M. Krein([7]), A. Sinclair, see Section 5); extension (and also brief elegant proof) of a result of F. Neubrander and S. Flory on supp f in terms of growth of $\tilde{f}^{(n)}(\lambda_0)$ ($\tilde{}$ = L.T.) extended to the context of asymptotic Laplace transform.

In Section 2, we treat preliminary material (context, notations, recalling known facts); in Section 3 we give the general "isotropic" P-W result mentioned before; Sections 4 and 5 contain the mentioned applications; Section 6 contains a remark on (an implication on) the theory of superstable solutions.

A preprint version, where the basic lemma of Section 3 below is proved but no applications treated, appeared in [L2].

2. Background, Context, Notations. Recalling Known Facts

We shall treat the results of Section 3 in the context of hyperfunctions([8]), from which the rest follows. However for best understanding, and later for deriving consequences – since most readers are not necessarily familiar with hyperfunctions

[1] I.e., with estimate of the Laplace transform depending only on the modulus of the corresponding complex variable.
[2] Hereafter often abbreviated P-W.
[3] Hereafter often abbreviated L.T.
[4] Which says "For X any Banach space, $f \in L^2([0,\infty), X)$, $a \geq 0$, supp $f \subset [0,a]$ iff the L.T. of f, $\tilde{f}(z)$, is an entire function $O(e^{a|z|})$.
[5] For the L^p we give a "nearly instant" simultaneous proof of the P-W for all L^p, $1 \leq p \leq \infty$.
[6] See Section 6 below, or [L3].
[7] A beautiful old result that says: "a semigroup is 0 after a finite time iff its resolvent is entire of exponential type", for precision and details see Section 5 below.
[8] [K1], [L1].

but certainly quite familiar with distributions $\mathcal{D}'$ – we start with the $\mathcal{D}'(\mathbb{R}, X)$ context. This is all the more appropriate – indeed necessary – since : (A) many applications depend on understanding that (and how)

$$\mathcal{D}'(\mathbb{R}, X) \subset B(\mathbb{R}, X) \tag{1}$$

the latter denoting the set (linear space) of all hyperfunctions on $\mathbb{R}$ with values in X; (B) even if we look only at functions and distributions, our method of proof (with the advantages on top of generality of being short and direct) is "almost at the same price" as for B, and uses (1) with the advantages coming then from Lemma 1.1 below, the general L.T. theory, and the inversion formula (4) below([9]), as we shall see.

For (1), let us recall that from results going back to the sixties (see [B], [L1]), given any $T \in \mathcal{D}'(\mathbb{R}, X)$, $\exists F \in \mathcal{O}(\mathbb{C} \setminus \mathbb{R}, X)$([10]) such that T is the difference of the limits in the sense of $\mathcal{D}'$ of $F(\cdot + i\epsilon)$, $F(\cdot - i\epsilon)$, as $\epsilon \to 0_+$ (see [B] for details). Moreover, any two F_1, F_2, with the just mentioned property relative to one given T, differ by an entire function, i.e., $F_1 - F_2$ (extend so as to) $\in \mathcal{O}(\mathbb{C}, X)$, so that $\mathcal{D}'(\mathbb{R}, X)$ can be identified with a subspace of the hyperfunctions $B(\mathbb{R}, X)$

$$B(\mathbb{R}, X) = \frac{\mathcal{O}(\mathbb{C} \setminus \mathbb{R}, X)}{\mathcal{O}(\mathbb{C}, X)} \quad . \tag{2}$$

$\mathcal{D}'([\alpha, \infty), X)$ denotes the distributions with support in $[\alpha, \infty)$; $B([\alpha, \infty), X)$ is understood similarly. For any F in $\mathcal{O}(\mathbb{C} \setminus \mathbb{R}, X)$, $[F]$ will denote the corresponding element (coclass) it determines in $B(\mathbb{R}, X)$.

As already said, even for functions or distributions looking at them as described above, i.e., via (1), has advantages because of the flexibility of modifying conveniently an $F \in [F]$ modulo an appropriate entire function. In this respect we recall the following important result concerning $B([\alpha, \infty), X)$, (see [K3]):

2.1. Lemma. *(Sato-Komatsu-Kaneko). Let $b = [F] \in B([\alpha, \infty), X)$ and V be an open neighborhood in $\mathbb{C}$ of $[\alpha, \infty)$, $\epsilon > 0$ be given. Then $\exists F_0 \in [F]$ such that*

$$\|F_0(\cdot)\| < \epsilon \quad \text{in } \mathbb{C} \setminus V. \tag{3}$$

This makes it possible to construct a general L.T. via complex integration around. $[\alpha, \infty)$ as explained in [K1]. [K1] will be our basic working reference for hyperfunction L.T. matters. (We mention that the general L.T. also permits to define convolutions for appropriate pairs of type $B([\alpha, \infty), X)$, see [K1], via products of L.T.s, without ambiguity.)

We follow essentially the notations of [K1] with some small deviations/variants: here we use often $B([\alpha, \infty), X)$ instead of ${}^X B_{[\alpha,\infty)}$ (also, written briefly here $B_{[\alpha,\infty)}$), α here for a; here we use generally b (instead of f) for an element of B and usually $f, g, \ldots$, when dealing with a function (say when $b = f$).

[9] Which works precisely modulo entire functions.
[10] $\mathcal{O}(V, X)$ = set of all analytic (holomorphic) functions on V with values in X.

We recall from [K1] that for arbitrary b (f in [K1]) $\in B_{[\alpha,\infty)}$, the (general, Komatsu) L.T., $\mathcal{L}b$, is multiple-valued, a coclass, each element $\hat{g} \in \mathcal{L}b$ being an analytic function defined at least on some neighborhood Ω of $S = \{\infty e^{i\theta} : |\theta| < \pi/2\}$, and that given one such $\hat{g} \in \mathcal{L}b$, we can recapture b from $\hat{g}$ by obtaining one $F \in [F] = b$ via the (Komatsu) inversion formula ((23) p. 217 of [K1]) where here instead of "integration from Λ to ∞, $\int_\Lambda^\infty$" we write simply Λ for the integration path in the inversion formula mentioned, (and the same in other integrals). We have thus

$$F(z) = \frac{1}{2\pi i} \int_\Lambda e^{z\lambda} \hat{g}(\lambda) d\lambda. \tag{4}$$

(For more details see [K1], [K2]. One must be careful when using (4) to see exactly for which Λ, z, the integral in (4) actually converges.)

Notice that we can of course in general not continue analytically $F(z)$ freely around 0 across some a priori specified part of $[\alpha, \infty)$ "without going to another sheet of the Riemann surface", since supp b is in $[\alpha,\infty)$ – and because of what supp b is (see [Ka], [L1]): we cannot extend analytically F in $\mathbb{C}$ across supp b.

3. General "Isotropic" Paley-Wiener Theorem

Essential for the results in this section is the following

3.1. Lemma. *Let $b \in B([0,\infty), X) = B_{[0,\infty)}$. If $\mathcal{L}f$ (which is a coclass) contains a g entire satisfying*

$$\|g(\lambda)\| = O(e^{a|\lambda|}) \quad \text{on } \mathbb{C},\ a \geq 0, \tag{5}$$

then

$$\text{supp } b \subset [0, a]. \tag{6}$$

Proof. The idea is to obtain a defining function F of $b = [F]$, in $\mathcal{O}(\mathbb{C} \setminus [0,\infty), X)$ via (4) applied to $\hat{g} = g$, and show that it can be extended analytically across $[0,\infty)$ outside $D_a = \{z \in \mathbb{C} : |z| \leq a\}$. Because of what is said in Section 2, in the paragraph after (4), we need for the above a somewhat careful argument using appropriately (5).

We proceed as follows: (i) we first construct directly, independently of F, a function $F_1(z)$ well-defined and analytic in $\mathbb{C} \setminus D_a$; (ii) then we show that F_1 coincides with an appropriate defining F on some open subset of $(\mathbb{C} \setminus D_a) \setminus [0,\infty)$ so that $F = F_1$ on the latter set, and hence F_1 extends F across the positive real axis outside D_a, yielding (6).

To do the above, we introduce first the following notations: $\forall 0 \neq z \in \mathbb{C}$ we write ray (z) for $\{sz : s \text{ real } \geq 0\}$; $V(z, \alpha)$ given $0 < \alpha < \pi$ denotes the angular region (sector) with vertex 0, bisected by ray(z) and with opening α.

(i) $\forall |z| > a$, i.e., $z \in \mathbb{C} \setminus D_a$, we define,

$$\Lambda_z = \text{ray}(-\bar{z}) \quad \text{oriented from 0 to } \infty \tag{7}$$

($\overline{z}$ denotes as usual the complex conjugate of z), and we define $\forall z \in \mathbb{C} \setminus D_a$

$$F_1(z) = \frac{1}{2\pi i} \int_{\Lambda_z} e^{z\lambda} g(\lambda) d\lambda. \tag{8}$$

For $\lambda \in \Lambda_z$, $Re(\lambda z) = -|z|\,|\lambda|$ so (8) exists in view of (5). Moreover using again (5) and the Cauchy theorem it is not difficult to establish the following which we state as

Sublemma. *$\forall a_1 > a$ $\exists \alpha_1 = \alpha_1(a_1)$ such that $\forall z_1$ with $|z_1| > a_1$, $F_1(z_2)$ where $z_2 \in V(z_1, \alpha_1)$ and $|z_2| > a_1$ can be computed from (8) with Λ_{z_2} replaced at will by any ray in $V(-\overline{z}_1, \alpha_1)$.*

From the above sublemma we have in particular that for any z_1, $|z_1| > a_1$, $F_1(z)$ with $z \in V(z_1, \alpha_1)$ and $|z| > a_1$, can be written with fixed integration path Λ_{z_1} as

$$F_1(z) = \frac{1}{2\pi i} \int_{\Lambda_{z_1}} e^{z\lambda} g(\lambda) d\lambda. \tag{9}$$

From (9) follows then (derivation under the integral sign being now legitimate) that F_1 is analytic (holomorphic) for all z with $|z| > a_1$; but a_1 is arbitrary as long as it is $> a$. Hence F_1 is indeed well-defined and

$$F_1 \in \mathcal{O}(\mathbb{C} \setminus D_a, X). \tag{10}$$

(ii) For $0 \neq z = iy$, $y > 0$, on the positive imaginary axis $i\mathbb{R}_+$, we have $\Lambda_z = \Lambda_{iy} = \text{ray}(i)$, and using again the sublemma with $a_1 > a$, $V(i, \alpha_1(a_1))$, we can find a path Λ (ray contained in $V(i, \alpha_1)$) and z_0 with $|z_0|$ large enough, which together with an open neighborhood W of z_0 is in $V(i, \alpha_1)$, in such a way that the integral

$$\frac{1}{2\pi i} \int_{\Lambda} e^{z\lambda} g(\lambda) d\lambda \quad \text{for } z \in W, \tag{11}$$

gives by the inversion result (4) (Komatsu inversion formula) in W one defining function $F(z)$ of b, and by the sublemma also gives $F_1(z)$ in W. (We can also assume that z_0 with $|z_0|$ large enough has been chosen such that we get a W in $\mathbb{C} \setminus D_a$.) It follows now that

$$F = F_1 \text{ in } (\mathbb{C} \setminus D_a) \setminus [0, \infty). \tag{12}$$

As said at the start of the proof, (12) and (i) (first paragraph of (i)) yield (6). □

If instead of $[0, \infty)$ we consider any left bounded interval $[\alpha, \infty)$, one can show that

3.2. Lemma. *If in Lemma 2.1, we replace $[0, \infty)$ by $[\alpha, \infty)$, any α real, the conclusion reads*

$$\text{supp } b \subset [-a, a]. \tag{13}$$

Proof. $b = [F]$. It suffices to consider supp $b \subset [-\beta, \gamma]$ with $0 < \beta < \infty$, $0 \leq \gamma \leq \infty$. First, by translation, $b_\beta = [F_\beta] = [F(z - \beta)] = [G(z)]$, with clearly supp $b_\beta \subset [0, \gamma + \beta]$. Computing $\tilde{G}(\lambda)$ (see [K1]) gives $\tilde{G}(\lambda) = e^{-\beta\lambda} \tilde{F}(\lambda)$, so that (assuming, as we may, that our $\tilde{F}$ is the g of (5))

$$\left|\tilde{G}(\lambda)\right| = O\left(e^{(a+\beta)|\lambda|}\right),$$

so by 3.1, $\gamma + \beta \leq a + \beta$, i.e., already

$$\gamma \leq a. \tag{14}$$

Next, consider $\check{b} = [-\check{F}]$, where $\check{F}(z) = F(-z)$. (Actually $\check{}$ is indeed the usual $\check{}$ operator extended to hyperfunctions, but all we need to know is what follows.) Then supp $\check{b} \subset [-\gamma, \beta]$ and $(-\check{F})^{\sim}$, $(-\check{F})^{\sim}(\lambda)$, satisfies the same $O(e^{a|\lambda|})$ estimate as $\tilde{F}(\lambda)$. Hence as in (14), $\beta \leq a$. So finally, $[-\beta, \gamma] \subset [-a, a]$,

$$\text{supp } b \subset [-a, a] \tag{15}$$

as claimed. □

Lemma 3.2 (Lemma 3.1) has essentially a converse. We say "essentially" because beyond the case where b is limited to functions the converse will not hold in general without slightly modifying condition (5). This can be seen already for $b \in \mathcal{D}'$, $a = 0$, where $O(1)$ implies $\hat{g}$ = constant (which can only happen for b = constant $\times\delta$). We shall have to replace in general "$O\left(e^{a|\lambda|}\right)$" by "$O\left(e^{(a+\epsilon)|\lambda|}\right)$ $\forall\epsilon > 0$" or equivalent.

The general theorem is:

Theorem 1. *Let $b \in B([\alpha, \infty), X) = {}^X B_{[\alpha,\infty)}$, α real. Then given $a \geq 0$*

$$\textit{supp } b \subset [-a, a] \tag{16}$$

iff $\forall\epsilon > 0\, \exists g_\epsilon \in \mathcal{L}b$ with g_ϵ entire and $O\left(e^{(a+\epsilon)|\lambda|}\right)$ in $\mathbb{C}$.

Proof. The main part of the statement (sufficiency) follows at once from Lemma 3.2. The rest ("only if") follows by estimating a closed contour integral around, and sufficiently close to, $[-a, a]$ (representing the analytic functional associated to b), as in the proof of 4.1 of [L-N] ((24) and sentence thereafter in 4.1 just mentioned).

4. Extended and Classical P-W Theorems, for L^1_{loc}, L^p $1 \leq p \leq \infty$, Exponentially Unbounded $C(\mathbb{R}_+)$

We start with a general P-W theorem for L^1_{loc} (thus in particular applicable to all L^p $1 \leq p \leq \infty$ and continuous not exponentially bounded functions) derived from the general results 3.1, 3.3 for hyperfunctions, but now in terms of ordinary functions([11]) and asymptotic L.T. (again expressed by functions) close to the usual L.T.

Before stating that theorem, we recall some facts and notations. $L^1_{loc} = L^1_{loc}([0, \infty]$, $X) \subset \mathcal{D}'(\mathbb{R}, X) \subset B(\mathbb{R}, X)$ (see (1)), and for $f \in L^1_{loc}$, $f = b \in B$ has a L.T. $\mathcal{L}b$ as considered in Section 3 above, or [K1], which coincides with the (coclass of

[11]Either scalar, or in general X-valued.

analytic functions) Lf asymptotic L.T. associated to $f = b$ (precisions and proof of $Lf = \mathcal{L}f$ are in [L-N])([12]). We now have

Theorem 2. *Let $f \in L^1_{loc}([0,\infty], X)$, $a \geq 0$. Let L denote asymptotic L.T. Then*

$$supp\ f \subset [0,a] \tag{17}$$

iff $\forall \epsilon > 0\ \exists \hat{f}_\epsilon \in Lf$ with $\hat{f}_\epsilon(\lambda)$ entire and $O(e^{|\lambda|a+\epsilon})$ in $\mathbb{C}$ (i.e., as $|\lambda| \to \infty$ in $\mathbb{C}$). In particular (17) holds if $\exists$ at least one $\hat{f} \in Lf$ entire and $O(e^{a|\lambda|})$ in $\mathbb{C}$.

Proof. By results of [L-N] (Section 2, (15) of [L-N]) $L = \mathcal{L}$ on $L^1_{loc}([0,\infty], X)$ (the coclasses Lf, $\mathcal{L}f$, coincide $\forall f \in L^1_{loc}$), hence our statement follows from 3.1, 3.3, above. □

As a consequence we derive a very short proof of the following classical result, well-known for $p = 2$, probably in the literature for the other values of p as well, but anyway here as said with different very short proof. (Also the proof below, while not "easier" because of the hyperfunctions technique involved is not only shorter than the proofs we know (such as from [R], [H], for $p = 2$) but even more "direct" (with a very "direct line of thought"). So we have

Theorem 3. *Let $1 \leq p \leq \infty$, $f \in L^p([0,\infty), X)$, and the classical L.T. $\tilde{f}(\lambda)$ defined for $Re\lambda > 0$ by*

$$\tilde{f}(\lambda) = \int_0^\infty e^{-\lambda t} f(t)dt. \tag{18}$$

Then, for $a \geq 0$,

$$supp\ f \subset [0,a], \tag{19}$$

iff $\tilde{f}(\lambda)$ is entire and $O(e^{a|\lambda|})$ in $\mathbb{C}$.

Proof. (a) First take $p = 1$, then $\tilde{f}(\lambda)$ (as in (18)) $\in Lf$. Indeed the approximation condition needed ((3) of [L-N]) holds as $0 < T \to \infty$, since

$$\left\| \int_0^T e^{-\lambda t} f(t)dt - \tilde{f}(\lambda) \right\| = \left\| \int_T^\infty e^{-\lambda t} f(t)dt \right\| \leq e^{-\lambda T} \|f\|_{L^1}. \tag{20}$$

(b) For $1 < p \leq \infty$ consider $g(t) = e^{-(1/q)t} f(t) \in L^1$ where $\frac{1}{q} + \frac{1}{p} = 1$ (, use Hölder's inequality); $\tilde{g}(\lambda) = \tilde{f}(\lambda + 1/q)$.
(c) When the "if" condition holds for f, $\|\tilde{g}(\lambda)\| = O(e^{a|\lambda+1/q|}) = O(e^{a|\lambda|})$. So for $1 \leq p \leq \infty$ we get (19) from 4.1 at once. The "only if" is clear. □

The above proof of 4.2 in extremely short but relies in addition of our general P-W result 3.3, on asymptotic L.T. and thus also on the hyperfunction results of [L-N]. Therefore we give below another proof of 4.2 derived from 3.3 by more elementary means (integration, Fubini, $\mathcal{D}'$, ...) but clearly longer, requiring some more work.

[12] Lf coclass of analytic functions generically denoted $\hat{f}$ (see [L-N]) is obtained roughly speaking by "asymptotic limit" of $\int_0^T e^{-\lambda t} f(t)dt$ as $T \to \infty$ (see precise definition in [L-N]).

Another proof of 4.2. With the same context and notations as in 4.2 and proof above, $f \in L^p$, we only need to consider the "if" part, and to show directly that the classical L.T. $\tilde{f} = \tilde{f}(\lambda)$, (18), which always exists by Hölder's inequality, is an element of $\mathcal{L}f$, or even less: $\exists b \in B([0,\infty), X)$,

$$\tilde{f} \in \mathcal{L}b; \operatorname{supp} b \subset [0,a] \Rightarrow \operatorname{supp} f \subset [0,a]. \tag{21}$$

Again replacing as above $f(t)$ by $g(t) = e^{-(1/q)t} f(t)$ we need only consider the case $p = 1$ (, a big simplification). Define ($p = 1$):

$$b = \left[\frac{1}{2\pi i} \int_{\mathbb{R}_+} \frac{f(t)}{t - z} dt \right] \quad (Imz \neq 0). \tag{22}$$

Write $F(z)$ for the integral in (22), i.e., $b = [F(z)]$. Since $f \in L^1$, $\|F(z)\|$ is bounded when $Imz \geq$ some constant $c > 0$.
Hence F is of exponential type in $\mathbb{C} \setminus [0, \infty)$, so

$$\tilde{F}(\lambda) \in \mathcal{L}b, \tag{23}$$

$$\tilde{F}(\lambda) = \int_{\mathcal{C}} e^{-\lambda\xi} F(\xi) d\xi, \tag{24}$$

(see [K1] formula (21) of [K1]), $\mathcal{C}$ as in the figure on the left top of p. 217 of [K1]; here $\mathcal{C}$ going around $[0, \infty)$ at distance $d > 0$ from the latter with angles $\psi > 0$, $-\psi < 0$, above and below $[0,\infty)$, ψ small and later going to 0 so as to continue $\tilde{F}(\lambda)$ to all of $\mathbb{C} \setminus \mathbb{R}_+$. Then we have

$$\begin{aligned} \tilde{F}(\lambda) &= \int_{\mathcal{C}} e^{-\lambda\xi} d\xi \frac{1}{2\pi i} \int_{\mathbb{R}_+} \frac{f(t)}{t - \xi} dt \\ &= \int_0^\infty f(t) dt \frac{1}{2\pi i} \int_{\mathcal{C}} \frac{e^{-\lambda\xi}}{t - \xi} d\xi = \int_0^\infty e^{-\lambda t} f(t) dt = \tilde{f}(\lambda), \end{aligned} \tag{25}$$

using Fubini (easily justified because of $f \in L^1$, $d > 0$, for $|\arg \lambda| < \frac{\pi}{2} - \psi$) first, and then Cauchy's theorem afterwards.
Thus (23), (24), (25) give indeed $\tilde{f}(\lambda) = \tilde{F}(\lambda) \in \mathcal{L}b$, proving the first part in (21) for b as in (22) (via continuation as said to all of $\mathbb{C} \setminus \mathbb{R}_+$).
For the second part of (21), suppose that:

$$\operatorname{supp} b \subset [0,a] \text{ for } b \text{ as in } (22) \tag{26}$$

Consider then $a < \alpha < \beta < \infty$, and

$$\begin{aligned} F_1(z) &= \frac{1}{2\pi i} \int_0^\alpha \frac{f(t)}{t - z} dt + \frac{1}{2\pi i} \int_\beta^\infty \frac{f(t)}{t - z} dt, \\ F_2(z) &= \frac{1}{2\pi i} \int_\alpha^\beta \frac{f(t)}{t - z} dt, \end{aligned} \tag{27}$$

Then $F(z) = F_1(z) + F_2(z)$, $F_1(z)$ extends analytically across (α, β), and $F_2(z)$ is the analytic representation of the distribution with compact support corresponding to $f|(\alpha,\beta)$ (f restricted to α, β) so is well-known (see [B] 5.6, p. 48) to have distributional boundary value $f|(\alpha,\beta)$. Since by assumption $F(z)$ extends analytically across $\mathbb{R}\setminus[0,a]$, we see that $f=0$ on (α,β), α,β being arbitrary within $a<\alpha<\beta<\infty$, we have supp $f\subset[0,a]$, proving that also the second part of (21) holds for b as in (22).

Then finally, when the "if" condition of 4.2 holds, and since $\tilde{f}(\lambda)\in\mathcal{L}b$ direct application of 3.1, or 3.3, gives supp $b\subset[0,a]$ and hence supp $f\subset[0,a]$. The "only if" part is clear. □

Comparing the 2 proofs above of 4.2, the reader may consider the (sometimes) gain in brevity versus the cost of a bit more advanced technique or conversely a bit more lengthy....

4.3. Comment. In addition to the classical or close-to-classical facts above, in the general context and notations of 3.3 of Section 3, the latter applies directly to distributions and ultradistributions in view of the embedding results for the latter into $B(\mathbb{R},X)$.

5. Supports and Growth of nth Order Derivatives of L.T.s.; Nilpotent Semigroups and Growth of Resolvents

We start with an extension (and fast proof) of a recent (unpublished) result of Frank Neubrander and Simone Flory, basic for this section.

Theorem 4. *Let $f\in L^1_{loc}([0,\infty),X)$, Lf be its asymptotic L.T. Then*

$$\text{supp } f\subset[0,a], \tag{28}$$

iff for some $\hat{f}\in Lf$,

$$\left\|\hat{f}^{(n)}(\lambda_0)\right\|\le Ma^n, n=1,2,3,\ldots, \tag{29}$$

for some $\lambda_0>0$, and a constant $M>0$ ($^{(n)}$ denotes as usual "derivative of order n". Notice that any $\hat{f}$ is analytic for $0<\lambda$ large (in the postsector corresponding to Lf).

Proof. Suppose the "if" condition holds. Then locally near λ_0

$$\hat{f}(\lambda)=\sum_{n=0}^{\infty}\frac{\hat{f}^{(n)}(\lambda_0)}{n!}(\lambda-\lambda_0)^n, \tag{30}$$

but since here :

$$\frac{\|\hat{f}^{(n)}(\lambda_0)\|}{n!}\le M\frac{a^n}{n!}\quad n=1,2,3,\ldots, \tag{31}$$

$\hat{f}(\lambda)$ is entire and $O(e^{a|\lambda-\lambda_0|})$ hence $O(e^{a|\lambda|})$. Now 4.1 gives (28) at once. The rest is clear. □

As a consequence,

Theorem 5. *(A. Sinclair, [S]). Let $Q(t) = e^{tA}$ be any (C_0) semigroup on X, with generator A, and corresponding resolvent $R(\lambda, A)$. Then $Q(t)$ is nilpotent with $Q(t) = 0$ for $t \geq a > 0$, iff $R(\lambda, A)$ is entire and $\exists \lambda_0$, M, constants > 0 with:*

$$\|n!R^n(\lambda_0, A)\| \leq Ma^n, n = 1, 2, 3, \ldots \tag{32}$$

Proof. Since $R(\lambda, A)$ is the (strong) classical L.T. of $Q(t) \in L^1_{loc}$, and $\|R^{(n)}(\lambda_0, A)\| = \|n!R^{n+1}(\lambda_0, A)\|$, the present result follows from 4.1. □

The result 5.2 was proved in the 80 s (see [S]), however the following beautiful, a priori more general, result goes back to the sixties as far as we know to M. Krein, (see [Kr-G]), or earlier.

Theorem 6. *Let $Q(t) = e^{tA}$ be any (C_0) semigroup on X, $a > 0$. Then $Q(t) = 0$ for $t \geq a$ iff $R(\lambda, A)$ is entire and*

$$\|R(\lambda, A)\| = O(e^{a|\lambda|}). \tag{33}$$

Proof. Immediate from 4.2 since $R(\lambda, \cdot)$ is the strong classical L.T. of $Q(\cdot)$. □

6. A Remark on Superstable ("Small") Solutions

Finally we make a brief incursion into the general study of "small" (superstable) solutions [L3] and its relation with P-W type criteria. We recall that u continuous: $[0, \infty] \to X$ not identically 0 is called superstable (or small solution) iff

$$\overline{\lim}_{t\to\infty} \frac{\ln \|u(t)\|}{t} = -\infty \tag{34}$$

(Superstable semigroups are defined the same way, i.e., via (34) with $Q(\cdot)$ instead of $u(\cdot)$.) Superstable u have always entire classical L.T. $\tilde{u}$, and P-W tells us when supp u is compact and when not. Sometimes in a given general context the 2 latter possibilities can coexist depending on specifications, and one will look at the behavior of $\rho = \rho(u) =$ the (exponential) order of the entire function $\tilde{u}$, (see [L3]). It can be shown that with appropriate u, X, we can get all values of ρ between 1 and ∞ included; we now point out that as a simple consequence of our above P-W results we have always:

6.1. Proposition. *For u superstable ("small solution") whether of compact support or not, we have always*

$$\rho(u) \geq 1. \tag{35}$$

Proof. Suppose we had $\rho(u) < 1$. Then $\forall \epsilon > 0$ we would have $\|\tilde{u}(\lambda)\| = O\left(e^{\epsilon|\lambda|}\right)$ and hence by our P-W results, supp $u \subset [0, \epsilon]$. Thus u would be identically 0, a contradiction proving (35). □

Acknowledgment

For useful conversations, exchanges, concerning this work, we like to thank A.V. Balakrishnan, I. Cioranescu, H. Komatsu, P. Kunstmann, P. Malliavin, F. Neubrander, P. Shapira, L. Weis, L. Zsidó.

References

[B] H. Bremermann, *Distributions, complex variables, and Fourier transforms.* Addison-Wesley, 1965.

[H] K. Hoffman, *Banach spaces of analytic functions.* Prentice Hall, 1962.

[Ka] A. Kaneko, *An introduction to hyperfunctions.* KTK Scientific Publishers/Tokyo, Kluwer Academic Publishers/Dordrecht/Boston/London, 1988.

[K1] H. Komatsu, *Operational calculus and semigroups of operators.* Functional analysis and related topics 1991 (Kyoto), Lecture Notes in Math, vol. 1540, Springer (1993), 213–234.

[K2] H. Komatsu, *Laplace transforms of hyperfunctions - a new foundations of the Heaviside calculus.* J. Fac. Sci. Tokyo, Sect. IA, Math. **34** (1987), 805–820.

[K3] H. Komatsu, *Operational calculus, hyperfunctions and ultradistributions.* Algebraic analysis, vol.1, Academic Press, 1988, 357–372.

[Kr-G] M. Krein, and I. Gohberg, *Theory and applications of Volterra operators in Hilbert space.* AMS translations of Math Monographs, vol. 24, 1970.

[L1] G. Lumer, *An introduction to hyperfunctions and δ-expansions, in Generalized functions, operator theory, and dynamical systems.* Chapman & Hall/CRC Res. Notes in Math, vol. 399, 1999, 1–25.

[L2] G. Lumer, *Laplace transforms and supports : a general "isotropic" theorem of Paley-Wiener type.* Ulmer Seminare, Funktionalanalysis und Differentialgleichungen, Appl. Analysis, Univ. Ulm, vol. 5, 2001, 254–259.

[L3] G. Lumer, *On the growth orders of the resolvents for an explicit class of superstable semigroups.* Ulmer Seminare, Funktionalanalysis und Differentialgleichungen, Appl. Analysis, Univ. Ulm, vol. 6, 2002, 253–258.

[L-N] G. Lumer, and F. Neubrander, *The asymptotic Laplace transform: New results and relation to Komatsu's Laplace transform for hyperfunctions.* In Lect. Notes in Pure and Appl. Math., vol. 219, Marcel Dekker, New York, 2001, 147–162.

[R] W. Rudin, *Real and complex analysis.* Mc Graw-Hill, 1966.

[S] A. Sinclair, *Continuous semigroups in Banach algebras.* London Math. Soc. Lect. Notes, vol. 63, Cambridge Univ. Press, 1982.

Günter Lumer
Institut de Mathématique
Université de Mons-Hainaut
Mons, Belgium
and
Solvay Institutes for Physics and Chemistry
Brussels, Belgium
E-mail address: dufour@umh.ac.be

Progress in Nonlinear Differential Equations
and Their Applications, Vol. 55, 333–340

Asymptotic Stability for a Functional Differential Equation in Hilbert Space

Miklavž Mastinšek

1. Introduction

We consider the stability properties for the functional differential equation (FDE) of the form:

$$\begin{aligned} u'(t) &= Au(t) + bu(t-h) + \int_{-h}^{0} a(r)Au(t+r)dr \quad t > 0 \\ u(0) &= \phi^0 \\ u(r) &= \phi^1(r) \quad r \in [-h, 0) \end{aligned} \tag{1.1}$$

where A is the infinitesimal generator of an analytic semigroup of linear operators $S(t)$ on a Hilbert space X, b is a real number, $a(.)$ is measurable square integrable real function and the initial value $\phi = (\phi^0, \phi^1)$ belongs to the product space $F \times L^2(-h, 0; D(A))$ with F denoting a suitable intermediate space between $D(A)$ and X.

We note that for $\phi = (\phi^0, \phi^1) \in X \times L^2(-h, 0; X))$ the initial value problem is not well posed; see, e.g., [1]. The appropriate initial data is an element of the product space $Z := F \times L^2(-h, 0; D(A)))$, where F is the Lions real interpolation space between $D(A)$ and X given by:

$$F = \{x \in X; \int_0^\infty ||AS(t)x||^2 dt < \infty\} \text{ and norm } ||x||_F = (||x||^2 + \int_0^\infty ||AS(t)x||^2 dt)^{1/2} \tag{1.2}$$

The existence of solution of FDE (1.1) has been studied in [1]. The following result was obtained: For an arbitrary initial value $\phi = (\phi^0, \phi^1) \in Z$ there is a unique $u = u(t)$ of FDE (1.1) on the interval $[-h, T], T > 0$ such that $u \in L^2(0, T; D(A)) \cap W^{1,2}(0, T; X)$,

$$||u||_{L^2(0,T;D(A)) \cap W^{1,2}(0,T;X)} \leq c_1 ||\phi||_Z \tag{1.3}$$

and the constant c_1 depends on T, M_0, M_1, $||a||$, b. Moreover the solution semigroup $T(t)$ can be defined in the product space $Z = F \times L^2(-h, 0; D(A))$ in the following way:

$$T(t)\phi = (u(t), u_t(.)) \quad t \geq 0 \tag{1.4}$$

for every $\phi \in Z$ and u_t given by $u_t(s) = u(t+s), s \in [-h, 0]$. This semigroup is strongly continuous in Z, hence there exist real constants M and ω such that

$$||T(t)|| \leq Me^{\omega t} \quad t \geq 0 \tag{1.5}$$

Let Λ be the infinitesimal generator of the solution semigroup $T(t)$ in Z and $\sigma(\Lambda)$ its spectrum. It is known, when the semigroup $T(t)$ is norm continuous (i.e., it is continuous in the uniform operator topology) for $t \geq t_0 > 0$, then the following relation holds:

$$\sup \{\text{Re } \lambda : \lambda \in \sigma(\Lambda)\} = \inf \{\omega \in \mathbf{R}; ||T(t)|| \leq Me^{\omega t} \text{ for some } M > 0\} \tag{1.6}$$

and the study of the asymptotic behaviour of the semigroup $T(t)$ can be made through the location of the spectrum of Λ, see, e.g., [3], p. 281.
There are various conditions on the semigroup such that $T(t)$ is norm continuous for $t \geq t_0$ and (1.6) holds. For the case where $a(.)$ is continuous see, e.g., [2],[5]. In [7] equation (1.1) with distributed delays was considered. In this article we study equations where distributed and discrete delays appear. In Section 2 we show the following result: when $b \in R$ and $a(.)$ is a measurable square integrable real function ($a \in L^2(-h, 0; R)$), then the solution semigroup is norm continuous for $t > h$. In Section 3 we study solutions of the characteristic equation of (1.1) and prove stability of $u(.)$ for the case where the infinitesimal generator A has real point spectrum such that $\sigma(A) = \sigma_P(A) \subseteq (-\infty, -\alpha_0]$ for some $\alpha_0 > 0$. We give stability conditions in terms of α_0, b, $a(.)$ and thus generalize the results of [7] to equations where distributed and discrete delays appear.

Notation: Throughout X is a Hilbert space with norm $||.||$, R and C are the sets of real respectively complex numbers and $L(X)$ is the space of bounded linear operators in X. Thoughout A be the infinitesimal generator of a bounded analytic semigroup $\{S(t); t \geq 0\}$ on X with 0 in the resolvent set, so that there are constants M_0 and M_1 such that it holds:

$$||S(t)||_{L(X)} \leq M_0 \quad t \geq 0 \qquad ||tAS(t)||_{L(X)} \leq M_1 \quad t > 0 \tag{1.7}$$

For a closed linear injective operator A in X its domain $D(A)$ is regarded as a Banach space with the norm $||x||_{D(A)} = ||Ax||$. $C(t_0, T, Y)$ is the space of continuous functions from an interval $[t_0, T]$ to a Banach space Y and $L^2(t_0, T; Y)$ denotes the space of measurable square integrable functions from $[t_0, T]$ to Y. As usual for a reflexive space Y $W^{1,2}(t_0, T; Y)$ denotes the space of absolutely continuous functions on $[t_0, T]$ such that their derivative belongs to $L^2(t_0, T; Y)$. We note that there is a continuous embedding:

$$L^2(0, T; D(A)) \cap W^{1,2}(0, T; X) \rightarrow C(0, T; F) \tag{1.8}$$

so that there exists a constant c_0 such that

$$||u||_{C(0,T;F)} \leq c_0 ||u||_{L^2(0,T;D(A)) \cap W^{1,2}(0,T;X)} \tag{1.9}$$

for each $u \in L^2(0, T; D(A)) \cap W^{1,2}(0, T; X)$; see, e.g., [1], [6].

2. Norm Continuity

In this section we consider the continuity of the solution semigroup $T(t)$ for $t > h$ in the uniform operator topology. Actually we can prove that the solution semigroup of the FDE (1.1) is the norm continuous uniformly in t on any closed interval $[t_0, T]$ for $t_0 > h$. More precisely:

Theorem 1. *Let $\{T(t); t \geq 0\}$ be the solution semigroup of the functional differential equation (1.1) and suppose that $0 < h < t_0 \leq \tau \leq \tau' \leq T$, with t_0 and T fixed. Then for every $\epsilon > 0$ there is a $\triangle > 0$ such that*

$$||T(\tau')\phi - T(\tau)\phi||_Z < \epsilon||\phi||_Z \tag{2.1}$$

holds for every $\phi \in Z$ whenever $(\tau' - \tau) < \triangle$. The solution semigroup $\{T(t); t \geq 0\}$ of the functional differential equation (1.1) is continuous in the uniform operator topology at every $t > h$.

Proof. We give here the sketch of the proof. For the details we refer to [7], where the case with $b = 0$ was treated. Let $u \in L^2(-h, T; D(A))$ be the solution of equation (1.1) and $0 < h < \tau_0 \leq \tau \leq \tau' \leq T$. We have to show that

$$||T(\tau')\phi - T(\tau)\phi||_Z = ||u(\tau') - u(\tau)||_F + ||u_{\tau'} - u_\tau||_{L^2(-h,T;D(A))} \leq \epsilon||\phi||_Z \tag{2.2}$$

holds for every $\phi \in Z$, whenever $(\tau' - \tau) < \triangle$. First let us consider the L^2 norm of the second term. By definition we have:

$$||u_{\tau'} - u_\tau||_{L^2(-h,T;D(A))} = \int_{-h}^{0} ||Au(\tau' + r) - Au(\tau + r)||^2 dr = \tag{2.3}$$

Denoting $\tau' - \tau = \triangle$, $t = \tau + r$ and $t_0 = \tau_0 - h$ we get:

$$= \int_{\tau-h}^{\tau} ||Au(t + \triangle) - Au(t)||^2 \leq \int_{t_0}^{T-\triangle} ||Au(t + \triangle) - Au(t)||^2 dt$$

$$= ||u(. + \triangle) - u(.)||^2_{L^2(t_0,T-\triangle;D(A))}$$

In order to estimate the last integral we first rewrite FDE(1.1) as the integral equation and denote

$$u(t) = S(t)\phi^0 + b\int_0^t S(t-s)u(s-h)ds + \int_0^t S(t-s)\int_{-h}^{0} a(r)Au(s+r)dr\, ds \tag{2.4}$$

$$u(t) := u_1(t) + u_2(t) + u_3(t) \tag{2.5}$$

We will give the L_2 norm of the difference of each of the three terms in (2.5).

i) Let $t_0 > 0$ and $\epsilon_1 > 0$ be arbitrary fixed numbers. By assumption A is the infinitesimal generator of an analytic semigroup $S(t)$ and $\phi_0 \in F$. Hence we have:

$$\begin{aligned} &||u_1(. + \triangle) - u_1(.)||^2_{L^2(t_0,T;D(A))} = \int_{t_0}^{T} ||AS(t + \triangle)\phi^0 - AS(t)\phi^0||^2 dt \leq \\ &||S(t_0 + \triangle) - S(t_0)||^2 \int_{t_0}^{T} ||AS(t - t_0)\phi^0||^2 dt \leq \end{aligned} \tag{2.6}$$

By relation $S'(t) = AS(T)$ and by assumptions (1.2) and (1.7) then it follows:

$$\leq (M_1 log(t_0 + \triangle)/t_0||\phi^0||_F^2 \leq \epsilon_1^2||\phi||_Z^2$$

for $\triangle$ sufficiently small.

ii) For the second term in (2.5) we write

$$u_2(t+\triangle) - u_2(t) = b\int_t^{t+\triangle} S(s+\triangle - s)u(s-h)ds+$$

$$+b\int_0^t [S(t+\triangle - s) - S(t-s)]u(s-h)ds =: w_1(t) + w_2(t)$$

By assumption A is a closed operator, $u \in L^2(-h, T; D(A))$ and $S(t)$ commutes with A. Therefore for every $t \in [0, T-\triangle]$ we have:

$$||Aw_1(t)|| = M_0 b\int_t^{t+\triangle} ||Au(s-h)||ds \leq M_0 b\sqrt{\triangle}||u||_{L^2(-h,T;D(A))}$$

In order to obtain the estimate for $w_2(t)$ we note that from (1.7) it follows:

$$||S(t+\triangle) - S(t)||_{L(X)} \leq M_3 k(t)$$

for $t \in (0, T-\triangle)$, $M_3 = max(2M_0, M_1)$ and $k(t) = min(1, \frac{\triangle}{t})$ for $t > 0$. For the details see, e.g., [8], proof of Thm 3.1, p. 110. Hence by the same argument as above we get:

$$||Aw_2(t)|| = M_3 b\int_0^t k(t-s)||Au(s-h)ds$$

$$\leq M_3 b(\int_0^t (k(t-s))^2 ds)^{\frac{1}{2}}||u||_{L^2(-h,T;D(A))}$$

By integration we get for any $\epsilon_2 > 0$:

$$||u_2(.+\triangle) - u_2(.)||_{L^2(t_0,T-\triangle;D(A))} \leq M\sqrt{\triangle}\sqrt{T-t_0}||u||_{L^2(-h,T;D(A))} \leq \epsilon_2||\phi||_Z, \tag{2.7}$$

for $\triangle$ sufficiently small and $M > 0$.

iii) Let us define

$$f(s) = \int_{-h}^0 a(r)Au(s+r)dr \tag{2.8}$$

and write

$$u_3(t+\triangle) - u_3(t) = \int_t^{t+\triangle} S(s)f(t+\triangle - s)ds$$

$$+\int_0^t S(s)[f(t+\triangle - s) - f(t-s)]ds$$

$$= \int_{-\triangle}^0 S(t-s)f(s+\triangle)ds + \int_0^t S(t-s)[f(s+\triangle) - f(s)]ds$$

$$=: \nu_1(t) + \nu_2(t)$$

By the Hoelder inequality we get the estimate for $v_1(t)$:

$$||A\nu_1(t)|| \leq \frac{M_1}{t_0}\sqrt{\triangle}||f||_{L^2(0,\triangle;X)} \quad \forall t \geq t_0 \tag{2.9}$$

Integrating inequation (2.9) we get:

$$\begin{aligned}||\nu_1||_{L^2(t_0,T;D(A))} &\leq \sqrt{T-t_0}\frac{M_1}{t_0}\sqrt{\triangle}||f||_{L^2(0,\triangle;X)} \leq \\ &\leq \sqrt{T-t_0}\frac{M_1}{t_0}\sqrt{\triangle}||a||_{L^2(/h,0;R)}||u||_{L^2(-h,\triangle;D(A))} \\ &\leq \epsilon_0||\phi||_Z\end{aligned}$$

for any $\epsilon_0 > 0$ and $\triangle$ sufficiently small. Let us consider the term $\nu_2(t)$. By the maximal regularity result we have:

$$||\nu_2||_{L^2(0,T-\triangle;D(A))} \leq k_0||f(.+\triangle) - f(.)||_{L^2(0,T-\triangle;X)}$$

with the constant k_0 dependent of M_2 and T. By using the known estimates on the convolution of L^1 and L^2 functions, we obtain the following inequality:

$$||f(.+\triangle) - f(.)||_{L^2(0,T-\triangle;X)} \leq ||a(.-\triangle) - a(.)||_{L^1(R)}||u||_{L^2(-h,T;D(A))}$$

where $a(.) = 0$ outside $[-h, 0)$. By the continuity of the shift operator in L^1 we have:

$$||a(.-\triangle) - a(.)||_{L^1(R)} \leq \epsilon_a$$

for any $\epsilon_a > 0$ and $\triangle$ sufficiently small. Hence for any $\epsilon_3 > 0$

$$||u_3(.+\triangle) - u_3(.)||_{L^2(0,T-\triangle;D(A))} \leq \epsilon_3||\phi||_Z \tag{2.10}$$

and the third estimate is obtained. Therefore by estimates (2.3)–(2.10) we may conclude that for any $\epsilon > 0$ the inequalities:

$$||u_{\tau'} - u_\tau||_{L^2(-h,0;D(A))} \leq ||u(.+\triangle) - u(.)||_{L^2(0,T-\triangle;D(A))} \leq \epsilon||\phi||_Z$$

hold for every pair $\tau, \tau' \geq t_0$ such that $\tau' - \tau = \triangle$ is sufficiently small. By using the continuous embedding (1.8),(1.9) the estimate for $||u(\tau') - u(\tau)||_F$ in (2.2) can be obtained exactly in the same way as it is shown for the case $b = 0$ in [7], hence we omit the details. Therefore from the estimate (2.2) we may conclude that the solution semigroup is norm continuous uniformly over any closed interval $[t_0, T]$ and $t_0 > h$. This shows that it is norm continuous at every $t > h$. □

3. Asymptotic Stability

As we have noted in (1.6) in the introduction, the norm continuity of $T(t)$ for $t > h$ implies, that the spectrum of the solution semigroup $T(t)$ can be determined through the location of the spectrum of its infinitesimal generator Λ. It is known that for a functional differential equation the spectrum of Λ can be obtained by the so called characteristic equation of equation (1.1); see, e.g., [4],[9]. Therefore we define for every complex number λ the operator $\triangle(\lambda) : D(A) \to X$:

$$\triangle(\lambda)x = \lambda x - be^{-\lambda h}x - Ax - m(\lambda)Ax \quad x \in D(A), \quad \text{where}$$

$$m(\lambda) = 1 + \int_{-h}^{0} a(s)e^{\lambda s}ds$$

The equation $\triangle(\lambda)x = 0$ is the characteristic equation of equation (1.1). The spectrum of Λ can be determined from solutions of the characteristic equation. This was done in detail in [2]. We will apply some of those results to the case where the infinitesimal generator A has only real spectrum satisfying the assumption:

Assumption 1: $\sigma(A) = \sigma_P(A) \subseteq (-\infty, -\alpha_0]$, for some $\alpha_0 > 0$ (3.1)

In that case the following characterization of the spectrum of Λ can be obtained by the same argument as that given for the case $b = 0$ in Proposition 5.2 in [2]: If the function $a(.)$ is such that $m(0) \neq 0$, then the spectrum of Λ satisfies the relation: $\sigma(\Lambda) \subseteq \Gamma_1 \cup \Gamma_0$, where

$$\Gamma_1 = \{\lambda \in C; m(\lambda) \neq 0 \text{ and } \frac{\lambda - be^{-\lambda h}}{m(\lambda)} \in \sigma(A)\} \quad \text{and} \tag{3.2}$$

$$\Gamma_0 = \{\lambda \in C; \lambda \neq 0 \text{ and } m(\lambda) = 0\} \tag{3.3}$$

The condition $m(0) \neq 0$ means that 0 is not in the spectrum of Λ. For instance this occurs when $|a|_{L^1(-h,0;R)} < 1$. From the characterizations (3.2) and (3.3) of the spectrum of Λ it follows that the stability properties of the solution can be given in terms of parameters $a(.)$, b and α_0. In that direction we can prove the following result.

Theorem 2. *Suppose A is the infinitesimal generator of a bounded analytic semigroup on X satisfying assumption (3.1). If b and $a(.)$ satisfy the condition*

$$\frac{|b|}{\alpha_0} + |a|_{L^1(-h,0;R)} < 1 \quad , \tag{3.4}$$

then the solution u of the functional differential equation (1.1) is asymptotically stable and the following estimate holds

$$||u(t)||_F + ||u_t||_{L^2(-h,0;D(A))} \leq Me^{\omega t}||\phi||_Z \,, \tag{3.5}$$

for some $M > 0$, $\omega > 0$ and all $t \geq 0$.

Proof. Let $\omega_0 \in (0, \alpha_0)$ be arbitrary, so that $\epsilon_0 = \omega_0/\alpha_0 < 1$. Furthermore let us define

$$a_0 = (1 - \epsilon_0)e^{-\omega_0 h} \tag{3.6}$$

We will show that under the assumption

$$\frac{|b|}{\alpha_0} + |a|_{L^1(-h,0;R)} < a_0$$

it follows that $\text{Re}\,\lambda \leq -\omega_0$ for $\lambda \in \Gamma_1 \cup \Gamma_0$ and hence every $\lambda \in \sigma(\Lambda)$.

i) Suppose first that $\lambda \in \Gamma_0$. Then by (3.3) λ satisfies the equation

$$1 + \int_{-h}^{0} a(s)e^{\lambda s} ds = 0$$

Put $\lambda = x + yi$ and assume that $\operatorname{Re}\lambda > -\omega_0$. In that case we get:

$$\begin{aligned} 1 &= |\int_{-h}^{0} a(s)e^{\lambda s} ds| \le \\ &\le \int_{-h}^{0} |a(s)|e^{xs} ds \le e^{\omega_0 h}(\int_{-h}^{0} |a(s)|ds) = \\ &= e^{\omega_0 h}(|a|_{L^1(-h,0;R)}) \le e^{\omega_0 h} a_0 = 1 - \epsilon_0 \end{aligned}$$

which contradicts the assumption.

ii) Suppose now that $\lambda \in \Gamma_1$. Hence by (3.2) λ satisfies the equation $\frac{\lambda - be^{-\lambda h}}{m(\lambda)} = -\alpha$ for some $\alpha \ge \alpha_0$ and $-\alpha \in \sigma(A)$. The equation can be rewritten as

$$1 + \frac{\lambda}{\alpha} = \frac{be^{\lambda s}}{\alpha} - \int_{-h}^{0} a(s)e^{\lambda s} ds$$

For the real part we get

$$1 + \frac{x}{\alpha} = \frac{be^{-xh}\cos y}{\alpha} - \int_{-h}^{0} a(s)\cos ys\, ds$$

Let us assume that $-\omega_0 < x$. In that case we get

$$\begin{aligned} 1 - \epsilon_0 = 1 - \frac{\omega_0}{\alpha_0} &< 1 + \frac{x}{\alpha_0} \le \\ &\le 1 + \frac{x}{\alpha} = |\frac{be^{xh}\cos y}{\alpha} - \int_{-h}^{0} a(s)e^{xs}\cos ys\, ds| \le \\ &\le e^{\omega_0 h}(\frac{|b|}{\alpha_0} + \int_{-h}^{0} |a(s)|ds = e^{\omega_0 h}(\frac{|b|}{\alpha_0} + |a|_{L^1(-h,0;R)}) \le e^{\omega_0 h} a_0 = 1 - \epsilon_0 \end{aligned}$$

which is a contradiction. Thus we have obtained the following estimate:

$$\operatorname{Re}\lambda = x \le -\omega_0 ,$$

where ω_0 is an arbitrary number from the interval $(0, \alpha_0)$. By (3.6) for sufficiently small ω_0 the upper bound a_0 will attain any value close to and less than 1. This means whenever

$$\frac{|b|}{\alpha_0} + |a|_{L^1(-h,0;R)} \le a_0 < 1$$

there is an ω_0 such that $\operatorname{Re}\lambda \le -\omega_0 < 0$ for every $\lambda \in \sigma(\Lambda)$. Therefore by (1.6) there exist $\omega > 0$ and $M > 0$ such that

$$||T(t)|| \le Me^{-\omega t} \quad \textit{for all} \quad t \ge 0 ,$$

which completes the proof of Theorem 2. □

References

[1] G. Di Blasio, K. Kunisch and E. Sinestrari, *L^2-regularity for parabolic integrodifferential equations with delay in the highest order derivatives.* J. Math. Anal. Appl. **102** (1984), 38–57.

[2] G. Di Blasio, K. Kunisch and E. Sinestrari, *Stability for abstract linear functional differential equations.* Israel J. Math. **50** (1985), 231–263.

[3] K.-J. Engel and R. Nagel, *One-Parameter Semigroups for Linear Evolution Equations.* Graduate Texts in Mathematics 194, Springer-Verlag, 1999.

[4] J.K. Hale, *Theory of Functional Differential Equations.* Springer-Verlag, New York, 1977.

[5] J.M. Jeong, *Stabilizability of retarded functional differential equation in Hilbert space.* Osaka J. Math. **28** (1991), 347–365.

[6] J.L. Lions and E. Magenes, *Problemes aux limites non homogenes et applications.* Vol. 1, Dunod, Paris, 1968.

[7] M. Mastinšek, *Norm continuity and stability for a functional differential equation in Hilbert space.* To appear in J. Math. Anal. Appl.

[8] A. Pazy, *Semigroups of Linear Operators and Applications to Partial Differential Equations.* Springer, New York, 1983.

[9] C. Travis and G. Webb, *Existence and stability for partial functional differential equations.* Trans. Amer. Math. Soc. **200** (1974), 395–418.

Miklavž Mastinšek
EPF-University of Maribor
Razlagova 14
2000 Maribor, Slovenia
E-mail address: `mastinsek@uni-mb.si`

Progress in Nonlinear Differential Equations
and Their Applications, Vol. 55, 341–348

Self-Intersections for Willmore Flow

Uwe F. Mayer and Gieri Simonett

1. Introduction

In this paper we consider the Willmore flow in three space dimensions. We prove that embedded surfaces can be driven to a self-intersection in finite time. This situation is in strict contrast to the behavior of hypersurfaces under the mean curvature flow, where the maximum principle prevents self-intersections, but very much analogous to the surface diffusion flow.

The Willmore flow is a geometric evolution law in which the normal velocity of a moving surface equals the Laplace-Beltrami of the mean curvature plus some lower order terms. More precisely, we assume in the following that Γ_0 is a closed compact immersed and orientable surface in $\mathbb{R}^3$. Then the Willmore flow is governed by the law

$$V(t) = \Delta_{\Gamma(t)} H_{\Gamma(t)} + 2H_{\Gamma(t)}(H^2_{\Gamma(t)} - K_{\Gamma(t)}), \qquad \Gamma(0) = \Gamma_0 . \tag{1.1}$$

Here $\Gamma = \{\Gamma(t)\,;\, t \geq 0\}$ is a family of smooth immersed orientable surfaces, $V(t)$ denotes the velocity of Γ in the normal direction at time t, while $\Delta_{\Gamma(t)}$, $H_{\Gamma(t)}$, and $K_{\Gamma(t)}$ stand for the Laplace-Beltrami operator, the mean curvature, and the Gauß curvature of $\Gamma(t)$, respectively.

The evolution law (1.1) does not depend on the local choice of the orientation. However, if $\Gamma(t)$ is embedded and encloses a region $\Omega(t)$ we always choose the outer normal, so that $V(t)$ is positive if $\Omega(t)$ grows, and so that $H_{\Gamma(t)}$ is positive if $\Gamma(t)$ is convex with respect to $\Omega(t)$.

Any equilibrium of (1.1), that is, any closed smooth surface that satisfies the equation

$$\Delta H + 2H(H^2 - K) = 0 \tag{1.2}$$

is called a Willmore surface [18, p. 282]. There has been much interest over the last years in characterizing Willmore surfaces, see for instance [15, 18] and the references cited therein. Willmore surfaces arise as the critical points of the functional

$$W(f) := \int_{f(M)} H^2 \, dS, \tag{1.3}$$

see [18, Section 7.4]. Here, M denotes a smooth closed orientable surface and $f : M \to \mathbb{R}^3$ is a smooth immersion of M into $\mathbb{R}^3$. Associated with this functional is a variational problem: Given a smooth closed orientable surface M_g of genus g determine the infimum $W(M_g)$ of $W(f)$ over all immersions $f : M_g \to \mathbb{R}^3$ and classify all manifolds $f(M_g)$ which minimize W. We refer to [4, 8, 14, 15, 17, 18] and the references therein for more details and interesting results.

The Willmore flow is the L^2-gradient flow for the functional (1.3) on the moving boundary, see for example [7], and also [10] for related work on gradient flows. Thus the Willmore flow has the distinctive property that it evolves surfaces in such a way as to reduce the total quadratic curvature. To be more precise, we show that the flow decreases the total quadratic curvature for any $C^{2+\beta}$ initial surface Γ_0.

Proposition 1. *Let $0<\beta<1$ and let Γ_0 be a closed compact immersed orientable surface that is $C^{2+\beta}$-smooth. Then*

$$\int_{\Gamma(t)} H^2(t)\, d\mu \leq \int_{\Gamma_0} H^2(0)\, d\mu, \qquad 0 \leq t \leq T,$$

where $[0,T]$ denotes the interval of existence guaranteed in the existence theorem of [16], and where $H(t)$ denotes the mean curvature of $\Gamma(t)$.

To the best of our knowledge, the result of Proposition 1 is new (under the given assumptions).

Next we show that the flow can force $\Gamma(t)$ to lose embeddedness in order to decrease the total quadratic curvature.

Theorem 2. *Let $0<\beta<1$ be fixed.*

There exist a closed embedded surface $\Sigma_0 \in C^{2+\beta}$, a constant $T_0 > 0$, numbers $t_0, t_1 \in (0, T_0]$ with $t_0 < t_1$, and a $C^{2+\beta}$–neighborhood U_0 of Σ_0 such that

- **(a)** *the Willmore flow (1.1) has a unique classical solution $\Gamma = \{\Gamma(t)\,;\, t \in [0,T_0]\}$ for all $\Gamma_0 \in U_0$,*
- **(b)** *$\Gamma(t)$ ceases to be embedded for every $t \in (t_0, t_1)$ and every $\Gamma_0 \in U_0$.*
- **(c)** *each surface $\Gamma(t)$ is of class C^∞ for $t \in (0, T_0]$ and smooth in $t \in (0, T_0)$.*

It should be noted that the neighborhood U_0 of Theorem 2 also contains C^∞-surfaces that will be driven to a self-intersection in finite time. Our approach relies on results and techniques in [6, 12, 16], and we follow closely the original argument in [12].

Lastly we mention that numerical simulations [13] seem to indicate that the Willmore flow can drive immersed surfaces to topological changes in finite time.

2. The mathematical setting

We first introduce some notations. Given an open set $U \subset \mathbb{R}^3$, let $h^s(U)$ denote the little Hölder spaces of order $s > 0$, that is, the closure of $BUC^\infty(U)$ in $BUC^s(U)$, the latter space being the Banach space of all bounded and uniformly Hölder continuous functions of order s. If Σ is a (sufficiently) smooth submanifold of $\mathbb{R}^3$ then the spaces $h^s(\Sigma)$ are defined by means of a smooth atlas for Σ. It is known that $BUC^t(\Sigma)$ is continuously embedded in $h^s(\Sigma)$ whenever $t > s$. In the following, we assume that Σ is a smooth compact closed immersed oriented surface in $\mathbb{R}^3$. Let ν be the unit normal field on Σ commensurable with the chosen orientation. Then we can find $a > 0$ and an open covering $\{U_l\,;\, l = 1, \dots, m\}$ of Σ such that

$$X_l : U_l \times (-a, a) \to \mathbb{R}^3\,, \qquad X_l(s, r) := s + r\nu(s)\,,$$

is a smooth diffeomorphism onto its image $\mathcal{R}_l := \operatorname{im}(X_l)$, that is,

$$X_l \in \mathit{Diff}^\infty(U_l \times (-a, a), \mathcal{R}_l)\,, \qquad 1 \le l \le m\,.$$

This can be done by selecting the open sets $U_l \subset \Sigma$ in such a way that they are embedded in $\mathbb{R}^3$ instead of only immersed, and then taking $a > 0$ sufficiently small so that each of the U_l has a tubular neighborhood of radius a. It follows that $\mathcal{R} := \cup \mathcal{R}_l$ consists of those points in $\mathbb{R}^3$ with distance less than a to Σ. Let $\beta \in (0, 1)$ be fixed. Then we choose numbers α, $\beta_1 \in (0, 1)$ with $\alpha < \beta_1 < \beta$. Let

$$W := \{\rho \in h^{2+\beta_1}(\Sigma)\,;\, \|\rho\|_\infty < a\}\,. \tag{2.1}$$

Given any $\rho \in W$ we obtain a compact oriented immersed manifold Γ_ρ of class $h^{2+\beta_1}$ by means of the following construction:

$$\Gamma_\rho := \bigcup_{l=1}^{m} \operatorname{Im}\left(X_l : U_l \to \mathbb{R}^3\,,\ [s \mapsto X_l(s, \rho(s))]\right). \tag{2.2}$$

Thus Γ_ρ is a graph in normal direction over Σ and ρ is the signed distance between Σ and Γ_ρ. On the other hand, every compact immersed oriented manifold Γ that is a smooth graph over Σ and that is contained in $\mathcal{R}$ can be obtained in this way. For convenience we introduce the mapping

$$\theta_\rho : \Sigma \to \Gamma_\rho, \qquad \theta_\rho(s) := X_l(s, \rho(s)) \text{ for } s \in U_l\,, \quad \rho \in W\,.$$

It follows that θ_ρ is a well-defined global $(2+\beta_1)$–diffeomorphism from Σ onto Γ_ρ. The Willmore flow (1.1) can now be expressed as an evolution equation for the distance function ρ over the fixed reference manifold Σ,

$$\partial_t \rho = G(\rho)\,, \qquad \rho(0) = \rho_0\,. \tag{2.3}$$

Here $G(\rho) := L_\rho \theta_\rho^*(\Delta_{\Gamma_\rho} H_{\Gamma_\rho} + 2H_{\Gamma_\rho}(H_{\Gamma_\rho}^2 - K_{\Gamma_\rho}))$ for $\rho \in h^{4+\alpha}(\Sigma) \cap W$, while Δ_{Γ_ρ}, H_{Γ_ρ}, and K_{Γ_ρ} are the Laplace-Beltrami operator, the mean curvature, and the Gauss curvature of Γ_ρ, respectively, and $L(\rho)$ is a factor that comes in by calculating the normal velocity in terms of ρ, see [6] for more details. We are now ready to state the following existence result for solutions of (2.3).

Proposition 2.1. *Let $\sigma \in W$ be given.*

(a) *There exist a positive constant $T_0 > 0$ and a neighborhood $W_0 \subset W$ of σ in $h^{2+\beta_1}(\Sigma)$ such that (2.3) has a unique solution*

$$\rho(\cdot, \rho_0) \in C([0, T_0], W) \cap C^\infty((0, T_0) \times \Sigma) \text{ for every } \rho_0 \in W_0.$$

(b) *The map $[(t, \rho_0) \mapsto \rho(t, \rho_0)]$ defines a smooth local semiflow on W_0.*

(c) *$\rho(\cdot, \rho_0) \in C([0, T_0], h^{4+\alpha}(\Sigma)) \cap C^1([0, T_0], h^\alpha(\Sigma))$ for all $\rho_0 \in h^{4+\alpha}(\Sigma) \cap W_0$.*

Proof. (a) and (b) follow from [16, Proposition 2.2]. Moreover, [16, Lemma 2.1] shows that the mapping $[\rho \mapsto G(\rho)] : h^{4+\alpha}(\Sigma) \cap W \to h^\alpha(\Sigma)$ is smooth and that the derivative is given by $G'(\rho) = P(\rho) + B(\rho)$, where

$$P(\rho) \in L(h^{4+\alpha}(\Sigma), h^\alpha(\Sigma)), \quad B(\rho) \in L(h^{2+\alpha}(\Sigma), h^\alpha(\Sigma)), \quad \rho \in h^{4+\alpha}(\Sigma) \cap W.$$

In the following we fix $\rho \in h^{4+\alpha}(\Sigma) \cap W$. [16, Lemma 2.1] also shows that $P(\rho)$ generates a strongly continuous analytic semigroup on $h^\alpha(\Sigma)$. A well-known perturbation result, see [1, Theorem I.1.3.1], then implies $G'(\rho) \in L(h^{4+\alpha}(\Sigma), h^\alpha(\Sigma))$ also generates a strongly continuous analytic semigroup on $h^\alpha(\Sigma)$. It is known that the little Hölder spaces are stable under the continuous interpolation method [1, 2, 5, 9]. Therefore, the spaces $(h^{4+\alpha}(\Sigma), h^\alpha(\Sigma))$ form a pair of maximal regularity for $G'(\rho)$, see [1, Theorem III.3.4.1] or [2, 5, 9]. Part (c) follows now from maximal regularity results, for instance [2, Theorem 2.7]. □

3. The proof of Proposition 1

We first note that any function in $C^{2+\beta}$ is also in $h^{2+\beta_1}$ for $\beta_1 \in (0, \beta)$. Let Γ_0 be a given surface in $\mathbb{R}^3$ that satisfies the assumptions of Proposition 1. We can find a smooth surface Σ as in Section 2 and a function $\rho_0 \in W$ such that $\Gamma_0 = \Gamma_{\rho_0}$, where Γ_{ρ_0} is defined in (2.2). According to Proposition 2.1(a) there exists a number $T = T(\rho_0) > 0$ such that equation (2.3) has a unique solution $\rho(\cdot, \rho_0)$ with the smoothness properties stated in the proposition. It follows from the construction in Section 2 that the family $\Gamma := \{\Gamma(t)\ ;\ 0 \le t \le T\}$, where $\Gamma(t) := \Gamma_{\rho(t)}$ for $0 \le t \le T$, is the unique classical solution for the Willmore flow (1.1). In particular, we conclude that

$$[t \mapsto \int_{\Gamma(t)} H^2(t)\, d\mu] \in C^\infty((0,T), \mathbb{R}).$$

Given $x \in \Gamma(t)$, let $\{z(\tau, x) \in \mathbb{R}^3\ ;\ \tau \in (-\varepsilon, \varepsilon)\}$ be an orthogonal flow line through x, that is, $z(\cdot, x)$ satisfies

$$\begin{aligned} &z(\tau, x) \in \Gamma(t+\tau) \text{ for } \tau \in (-\varepsilon, \varepsilon), \\ &\dot z(\tau) = (VN)(t+\tau, z(\tau)) \text{ for } \tau \in (-\varepsilon, \varepsilon), \quad z(0) = x, \end{aligned}$$

where $N(t, \cdot)$ denotes the unit normal field on $\Gamma(t)$, and $V(t, \cdot)$ is the normal velocity of $\Gamma(t)$. A proof for the existence of a unique trajectory $\{z(\tau, x) \in \mathbb{R}^3; \tau \in (-\varepsilon, \varepsilon)\}$ with the above properties can for instance be found in [11, Lemma 2.1]. For further use we introduce the manifold

$$\mathcal{M} := \bigcup_{t \in (0,T)} \{t\} \times \Gamma(t).$$

Given any smooth function u on $\mathcal{M}$ we define

$$\frac{d}{dt} u(t,x) := \frac{d}{d\tau} u(t+\tau, z(\tau, x))\Big|_{\tau=0}, \qquad (t,x) \in \mathcal{M}.$$

The following differentiation rule is well-known in differential geometry,

$$\frac{d}{dt}\int_{\Gamma(t)} u(t,x)\,d\mu(x) = \int_{\Gamma(t)} \frac{d}{dt}u(t,x)\,d\mu(x) + 2\int_{\Gamma(t)} (uHV)(t,x)\,d\mu(x). \tag{3.1}$$

Let $(t,x) \in \mathcal{M}$ be fixed and let $\{z(\tau,x)\,;\,\tau \in (-\varepsilon,\varepsilon)\}$ be a flow line trough x. Then one can show that

$$\frac{d}{d\tau}H^2(t+\tau, z(\tau,x))\Big|_{\tau=0} = -H[\Delta_{\Gamma(t)}V + (4H^2 - 2K)V](t,x), \tag{3.2}$$

see for instance [18, Section 7.4]. If follows from (3.1)–(3.2), from the divergence theorem, and from (1.1) that

$$\frac{d}{dt}\int_{\Gamma(t)} H^2(t)\,d\mu = -\int_{\Gamma(t)} [\Delta H + 2H(H^2 - K)]V\,d\mu \le 0. \tag{3.3}$$

This is true for any $t \in (0,T)$. The mean value theorem now implies that

$$\int_{\Gamma(t)} H^2(t)\,d\mu - \int_{\Gamma(\tau)} H^2(\tau)\,d\mu \le 0 \qquad \text{for } \ 0 < \tau \le t < T.$$

Taking the limit as $\tau \to 0$ and using that $[\tau \mapsto \int_{\Gamma(\tau)} H^2(\tau)\,d\mu] \in C([0,T],\mathbb{R})$, see Proposition 2.1(b), yields the assertion of Proposition 1. □

4. The proof of Theorem 2

In order to provide a proof of Theorem 2 we now choose Σ to be any smooth compact closed immersed orientable surface in $\mathbb{R}^3$ such that its image contains the flat two-dimensional disk $U := \{(s,0) \in \mathbb{R}^2 \times \mathbb{R}\,;\,|s| \le 1\}$ twice, and with opposite orientations. To be precise, let $i : \Sigma \to \mathbb{R}^3$ be the immersion under consideration, then we ask that

$$i^{-1}(U) = U^+ \cup U^-$$

with $U^+ \cap U^- = \emptyset$ and both U^+ and U^- are flat two-dimensional disks of radius 1. Additionally we ask that $\Sigma \setminus (U^+ \cup U^-)$ is embedded in $\mathbb{R}^3$. Identifying U^+ for the moment with its image U we ask that the normal on U^+ points upwards, that is, $\nu(\cdot)|_{U^+} = e_3$, the 3rd basis vector of $\mathbb{R}^3$. It follows that $\nu(\cdot)|_{U^-} = -e_3$.

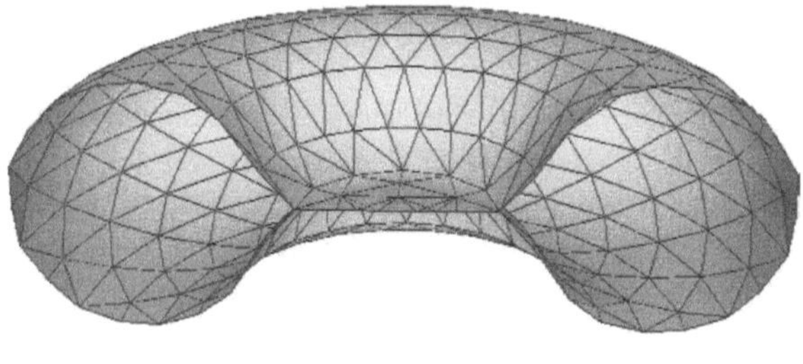

FIGURE 1. This is a possible choice of Σ, cut in halves.

Let W be as in (2.1) and let $\sigma \in h^{4+\alpha} \cap W$ locally be radially symmetric with regards to the centers of $U^{\pm}$. This implies $\partial_j \sigma(0) = 0$ for $j = 1, 2$. Observe that $\theta_\sigma(s) = (s, \pm\sigma(s))$ (these are coordinates in $\mathbb{R}^3$) for $s \in U^{\pm}$ and that $\theta_\sigma : U^{\pm} \to \theta_\sigma(U^{\pm})$ is an $h^{4+\alpha}$–diffeomorphism. It is straightforward to compute

$$G(\sigma)|_{U^\pm} := L(\sigma)\theta_\sigma^*(\Delta_{\Gamma_\sigma} H_{\Gamma_\sigma} + 2H_{\Gamma_\sigma}(H^2_{\Gamma_\sigma} - K_{\Gamma_\sigma}))|_{U^\pm}$$

in local coordinates, yielding

$$\begin{aligned} 2G(\sigma)|_{U^\pm}(0) &= -\Delta^2\sigma(0) + \sum_{j,k=1}^{2} (\partial_j\partial_k\sigma(0))^2 \Delta\sigma(0) \\ &\quad +2 \sum_{j,k,l=1}^{2} \partial_j\partial_k\sigma(0)\partial_j\partial_l\sigma(0)\partial_k\partial_l\sigma(0)\,, \end{aligned}$$

where Δ is the Laplacian in Euclidean coordinates of $\mathbb{R}^2$ (see [6, Section 2] for more details). Because of the radial symmetry of σ we have $H^2_{\Gamma_\sigma} = K_{\Gamma_\sigma}$ at the center of the disks $U^{\pm}$, so that lower order term $\theta_\sigma^*(2H_{\Gamma_\sigma}(H^2_{\Gamma_\sigma} - K_{\Gamma_\sigma}))$ vanishes at the center of $U^{\pm}$. We will now specify one more property of σ. We choose $r > 0$ small and we require that $\sigma(s) = |s|^4$ for $s \in U_r^{\pm} = \{s \in U^{\pm}\, ;\, |s| < r\}$. If r is small enough then this is compatible with $\sigma \in h^{4+\alpha}(\Sigma) \cap W$. We conclude that

$$G(\sigma)|_{U^\pm}(0) = -16 < 0\,. \tag{4.1}$$

It follows from Proposition 2.1 that the evolution equation (2.3) with initial value $\rho(0) = \sigma$ has a unique solution

$$\rho(\cdot, \sigma) \in C([0, T_0], h^{4+\alpha}(\Sigma)) \cap C^1([0, T_0], h^\alpha(\Sigma))\,. \tag{4.2}$$

Next we consider the restriction $\rho^{\pm}(t, \sigma)$ on $U^{\pm}$ of the function $\rho(t, \sigma)$, that is, $\rho^{\pm}(t, \sigma) := \rho(t, \sigma)|_{U^\pm}$ for $0 \le t \le T_0$, and we set $d^{\pm}(t) := \rho^{\pm}(t, \sigma)(0)$, to track the position of the center. It follows from (4.2) that $d^{\pm} \in C^1([0, T_0])$. Moreover, using the local character of G, we conclude that $d^{\pm}$ satisfies the equation

$$(d^{\pm})'(t) = G(\rho(t, \sigma))|_{U^\pm}(0) \quad \text{for} \quad 0 \le t \le T_0\,, \qquad d^{\pm}(0) = 0\,. \tag{4.3}$$

Equations (4.1)–(4.3) and the mean value theorem yield

$$d^{\pm}(t) = -Mt + \Big(\int_0^1 ((d^{\pm})'(\tau t) - (d^{\pm})'(0))\, d\tau\Big)t\,, \tag{4.4}$$

where $M := 16$. It follows from (4.4) that there exists a positive constant $\mu > 0$ and an interval $(t_0, t_1) \subset (0, T_0]$ such that $\rho^{\pm}(t, \sigma)(0) = d^{\pm}(t) \le -\mu$ for $t \in (t_0, t_1)$. By Proposition 2.1(b) we can find a function $\sigma_0 \in W_0$ such that $\Sigma_0 := \Gamma_{\sigma_0}$ is embedded and such that $\Gamma(t) := \Gamma_{\rho(t,\sigma_0)}$ is immersed for at least $t \in (t_0, t_1)$. By employing Proposition 2.1(b) once more we conclude there is a neighborhood $W(\sigma_0) \subset W_0$ of σ_0 in $h^{2+\beta_1}(\Sigma)$ such that Γ_{ρ_0} is still embedded, whereas $\Gamma_{\rho(t,\rho_0)}$ is immersed for $t \in (t_0, t_1)$ and all $\rho_0 \in W(\sigma_0)$. We note that $C^{2+\beta}(\Sigma)$ is contained in $h^{2+\beta_1}(\Sigma)$ with continuous injection $j : C^{2+\beta}(\Sigma) \to h^{2+\beta_1}(\Sigma)$. Hence $U_0 := j^{-1}(W(\sigma_0))$ is a $C^{2+\beta}$–neighborhood of σ_0 and Theorem 2 follows by choosing $\Sigma_0 := \Gamma_{\sigma_0}$ and $\Gamma_0 := \Gamma_{\rho_0}$ for $\rho_0 \in U_0$. □

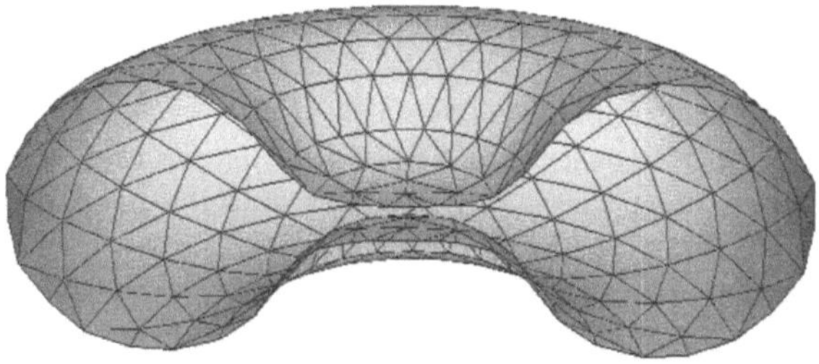

FIGURE 2. This is half of Γ_0, a surface that loses embeddedness and becomes immersed. The gap might have to be much smaller than depicted.

Remark 4.2. The following is the essence of the construction: Γ_σ is an immersed surface such that its image contains two opposing fourth-order paraboloids touching only at the vertex. The global symmetry of Γ_σ is irrelevant, we only need the local symmetry at the center. Locally we can compute the initial velocity of Γ_σ, and it is such as to create an overlapping of the fourth-order paraboloids. A continuity argument then guarantees the same behavior for nearby embedded surfaces, which do exist by construction of Γ_σ. We have chosen a fourth-order paraboloid in order to facilitate the computation of $G(\sigma)|_{U^\pm}$. Any other configuration that produces the same sign as in (4.1) will work as well.

References

[1] H. Amann, *Linear and quasilinear parabolic problems.* Vol. I, Birkhäuser, Basel, 1995.

[2] S.B. Angenent, *Nonlinear analytic semiflows.* Proc. Roy. Soc. Edinburgh Sect. A **115** (1990), 91–107.

[3] R. Bryant, *A duality theorem for Willmore surfaces.* J. Diff. Geom. **20** (1984), 23–53.

[4] B.-Y. Chen, *On a variational problem on hypersurfaces.* J. London Math. Soc. **2** (1973), 321–325.

[5] G. Da Prato and P. Grisvard, *Equations d'évolution abstraites nonlinéaires de type parabolique.* Ann. Mat. Pura Appl. (4) **120** (1979), 329–396.

[6] J. Escher, U.F. Mayer, and G. Simonett, *The surface diffusion flow for immersed hypersurfaces.* SIAM J. Math. Anal. **29** (1998), 1419–1433.

[7] E. Kuwert and R. Schätzle, *Gradient flow for the Willmore functional.* Comm. Anal. Geom. **10** (2002), 307–339.

[8] R. Kusner, *Estimates for the biharmonic energy on unbounded planar domains, and the existence of surfaces of every genus that minimize the squared-mean-curvature integral.* In Elliptic and parabolic methods in geometry, (Minneapolis, MN, 1994), A K Peters, Wellesley, MA, 1996, 67–72.

[9] A. Lunardi, *Analytic semigroups and optimal regularity in parabolic problems.* Birkhäuser, Basel, 1995.

[10] U.F. Mayer, *A numerical scheme for free boundary problems that are gradient flows for the area functional.* Europ. J. Appl. Math. **11** (2000), 61–80.

[11] U.F. Mayer and G. Simonett, *Classical solutions for diffusion-induced grain-boundary motion.* J.Math. Anal. Appl. **234** (1999), 660–674.

[12] U.F. Mayer and G. Simonett, *Self-intersections for the surface diffusion and the volume preserving mean curvature flow.* Differential Integral Equations **13** (2000), 1189–1199.

[13] U.F. Mayer and G. Simonett, *A numerical scheme for axisymmetric solutions of curvature driven free boundary problems, with applications to the Willmore flow.* Interfaces and Free Boundaries **4** (2002), 1–22.

[14] U. Pinkall, *Hopf tori in* S^3. Invent. Math. **81** (1985), 379–386.

[15] U. Pinkall and I. Sterling, *Willmore surfaces.* Math. Intelligencer **9** (1987), 38–43.

[16] G. Simonett, *The Willmore flow near spheres.* Differential Integral Equations **14** (2001), 1005–1014.

[17] L. Simon, *Existence of surfaces minimizing the Willmore functional.* Comm. Anal. Geom. **1** (1993), 281–326.

[18] T.J. Willmore, *Riemannian Geometry.* Claredon Press, Oxford, 1993.

Uwe F. Mayer
387 Trailview Road
Encinitas
CA 92024, USA
E-mail address: mayer@math.utah.edu

Gieri Simonett
Department of Mathematics
Vanderbilt University
Nashville
TN 37240, USA
E-mail address: simonett@math.vanderbilt.edu

Progress in Nonlinear Differential Equations
and Their Applications, Vol. 55, 349–358

Chernoff's Theorem and the Construction of Semigroups

O.G. Smolyanov, H. v. Weizsäcker, and O. Wittich

1. Introduction

The classical Chernoff Theorem is a statement about the convergence of discrete-time approximations of semigroups using a family of contraction operators strongly differentiable at $t = 0$. By Chernoff's construction, different families may yield the same semigroup. Such families are called *Chernoff equivalent*, i.e., their derivatives at $t = 0$ coincide. This is outlined in Section 2. Moreover Chernoff's Theorem still holds if the families used as input of the construction are only asymptotically contractive in a sense made precise in the notion of *properness* (see Definition 2).

In order to show that these simple observations can indeed have nontrivial consequences, we use well known results about the short time asymptotic of Gaussian integrals – summarized in Section 3 – to construct in Section 4 basically three slightly different families which are Chernoff equivalent to the heat semigroup on a closed and connected Riemannian manifold. The first of these is a contraction family, the second is a proper family and the third is a proper family with the additional property that its integral kernel with respect to the Riemannian volume on M is symmetric.

The paper provides a systematic basis for the semigroup part of previously known approximation results for the Wiener measure on manifolds in [1] and [8].

2. Chernoff's Theorem

In this first section we summarize some facts about families of contractions in a Banach space V. Let $K(V)$ denote the contractions on V. A strongly continuous family $S : [0, \infty) \to K(V)$ with $S(0) = 1$ is called *proper* if it is strongly differentiable at $t = 0$ and its differential DS given by $\mathcal{D}_S := \{f \in V : \lim_{t\to 0}(S(t)f - f)/t \text{ exists}\}$ and

$$G_S f := \lim_{t\to 0}(S(t)f - f)/t$$

for $f \in \mathcal{D}_S$ is a (generally unbounded) operator $(G_S, \mathcal{D}_S)$ generating a strongly continuous contraction semigroup on V. Under these assumptions Chernoff's Theorem reads as follows:

Theorem 1. *Let S be a proper contraction family with $DS = (G_S, \mathcal{D}_S)$. Then*

$$\mathrm{s} - \lim_{n\to\infty} S(t/n)^n = e^{tG_S}$$

compactly on $[0, \infty)$.

Proof. See [5], Theorem 11.3, p. 316. □

Remark. The same statement holds for arbitrary partitions of the time interval. If $\mathcal{P} = \{0 = t_0 < t_1 < \ldots < t_n = t\}$ is a finite partition with *mesh* $|\mathcal{P}| := \max |t_{k+1} - t_k|$ we also have

$$\mathrm{s} - \lim_{|\mathcal{P}|\to 0} S(t_n - t_{n-1}) \ldots S(t_1 - t_0) = e^{tG_S}$$

compactly on $[0, \infty)$ due to the cofinality of equipartitions. For a proof see [8], Proposition 3.

Thus the contraction family S serves as a tool to construct discrete semigroups $U^{(n)}$, $n = 1, 2, \ldots$, approximating e^{tG_S} in the sense of [4], (3.9), p. 511. That means, letting $U^{(n)}(k) := S(1/n)^k$, $k = 0, 1, \ldots$, we obtain strong convergence

$$U^{(n)}(k_n) \to e^{tG_S}$$

as long as k_n is chosen such that $k_n/n \to t$ as $n \to \infty$ (cf. [6], III.5). However, the observation is that the limit semigroup only depends on the derivative of the contraction family at $t = 0$.

2.1. Chernoff Equivalence

Let Π denote the set of all proper families. We now reformulate Chernoff's Theorem focussing on the map which assigns to each proper family its corresponding contraction semigroup. The map $P : \Pi \to \Pi$ given by

$$P(S)_t := e^{tDS}$$

we call *Chernoff map.* P maps proper families onto the subset $\Sigma \subset \Pi$ of strongly continuous contraction semigroups. Σ remains pointwise fixed under P. We are interested in the attracting domains for each fixed point, i.e., the set of preimages of a given semigroup.

Definition 1. *Two proper families $S, T \in \Pi$ are called Chernoff equivalent if one of the two equivalent conditions holds.*

(i) $P(S) = P(T)$,
(ii) $DS = DT$.

The following simple criterion for Chernoff equivalence will be applied in the sequel.

Lemma 1. *Let $T, S \in \Pi$ with $DT = (G_T, \mathcal{D}_T)$. If there is some determining subset $\mathcal{D}' \subset \mathcal{D}_T$ for DT such that*

$$\|T_t f - S_t f\| = o(t) \tag{1}$$

for all $f \in \mathcal{D}'$, then S and T are Chernoff equivalent.

Proof. Let $f \in \mathcal{D}'$. Then by our assumption

$$\frac{\|T_t f - S_t f\|}{t} \to 0$$

as $t \to 0$. Therefore

$$\lim_{t\to 0}\left\|\frac{T_t f - f}{t} - \frac{S_t f - f}{t}\right\| = \left\|G_T f - \lim_{t\to 0}\frac{S_t f - f}{t}\right\| = 0.$$

This calculation implies $(G_T, \mathcal{D}_T) \subset (G_S, \mathcal{D}_S)$ and since generators of C_0-semigroups are maximal in the sense that they have no *proper extension* ([2], Proposition 3.1.15, p. 177), we finally obtain $DS = DT$. □

2.2. Relaxing Contractivity

In order to include some further examples we relax our assumptions to include non-contractions as well.

Definition 2. *A strongly continuous family $B : [0,\infty) \to B(V)$ of bounded operators with $B(0) = 1$ is called proper if*

(i) $c(t) := \|B(t)\|$ *is continuously differentiable at* $t = 0$.

(ii) $K_B(t) := \frac{B(t)}{c(t)} \in \Pi$ *is a proper family of contractions.*

Since for all f in the domain of DK_B

$$\frac{B(t)f - f}{t} = \frac{c(t) - 1}{t} K_B(t) f + \frac{K_B(t)f - f}{t},$$

we obtain that B is differentiable at zero with $DB = (DK_B + \dot{c}(0))$. On the other hand

$$B(t/n)^n f = c(t/n)^n K_B(t/n)^n f,$$

and we finally obtain the following result:

Lemma 2. *We may extend the Chernoff map to proper families of bounded operators. If B is such a family with $c(t) := \|B(t)\|$ and corresponding contraction family K_B, then*

$$P(B)(t) = e^{t(DK_B + \dot{c}(0))}.$$

Corollary 1. *If $c(t) = 1 + o(t)$, then $P(B) = P(K_B)$.*

Remark. Note that $c(t)$ is not continuous in general even if $B(t)$ is a contraction semigroup.

3. Gaussian Integrals

In the sequel contraction families as described in Section 1 will be constructed by families of integral operators with probabilistic kernel. In order to compute the derivative at zero – and therefore the Chernoff equivalence class – of such a family, we first collect some facts about the short time asymptotic of Gaussian integrals. Since the computations in Section 4 involve the calculation of those integrals for products of polynomials, we find it most convenient to determine the relevant terms in advance. Therefore we introduce some degree d on the space of polynomials such that the short-time contribution of a monomial p either vanishes or is of order $t^{d(p)}$. Using this notion, we reformulate Wick's formula in an algebraic way and conclude Corollary 2. It states that in our situation the only relevant terms are of homogeneous degree one.

3.1. Wick's Formula

Let $x \in \mathbb{R}^n$ and $t > 0$. By Fubini's theorem and using the fact that the Gaussian integral solves the heat equation, we obtain the following result also known as *Wick's formula.*

Lemma 3. *Let $k \in \mathbb{Z} \times \mathbb{N}^n$ and*

$$d(k) := k_0 + \frac{1}{2}(k_1 + \ldots + k_n).$$

Let

$$p_k(t,y) = t^{k_0} y_1^{k_1} \ldots y_n^{k_n}$$

be a monomial in (t,y) such that $d(k) \geq 0$. Then for

$$\mathcal{G}_t(p_k) := \int_{\mathbb{R}^n} \frac{e^{-\frac{|y|^2}{2t}}}{\sqrt{2\pi t}^n} p_k(t,y)dy$$

we obtain

$$\mathcal{G}_t(p_k) = \begin{cases} 0 & (k_1, \ldots, k_n) \notin (2\mathbb{N})^n \\ t^{d(k)} \prod_{i=1}^n (k_i - 1)!! & \text{else} \end{cases}.$$

Here we use the standard notation $(2n-1)!! = 1 \cdot 3 \cdot \ldots \cdot (2n-1)$.

Let now $\mathcal{L}$ denote the real algebra generated by all monomials of the form $p_k(t,x)$, $k \in \mathbb{Z} \times \mathbb{N}^n$. The map

$$D : \mathcal{L} \to \frac{1}{2}\mathbb{Z}$$

given by $D(p_k) := d(k)$ induces a *grading* $\mathcal{L} = \oplus_{s \in \frac{1}{2}\mathbb{Z}} \mathcal{L}^s$, where $\mathcal{L}^s := \langle p_k : d(k) = s \rangle$ denotes the subspace of homogeneous elements of degree s. We consider the *associated filtration* by ideals $\mathcal{I}^r := \oplus_{s \geq r} \mathcal{L}^s$. Let furthermore $f \mapsto [f]$ denote the quotient map

$$q : \mathcal{I}^0 \to \mathcal{I}^0 / \mathcal{I}^{3/2}$$

and Q the projection onto the subalgebra generated by monomials p_k with $(k_1, \ldots, k_n) \in (2\mathbb{N})^n$.

That means, $Q[p_k]$ for the monomial p_k with $k_0 = 0$ either equals zero or p_k in case that $p_k \in \{1, x_1^2, \ldots, x_n^2\}$. Thus the following proposition about the short time asymptotic of Gaussian integrals is just a rewriting of Lemma 3.

Proposition 1. *Let $f \in \mathcal{I}^0$. Then $\lim_{t\to 0} \mathcal{G}_t(f)$ exists and we have asymptotically*

$$\mathcal{G}_t(f) = \mathcal{G}_t(Q[f]) + o(t) \tag{2}$$

as $t \to 0$. In other words, if $f = f_0 + f_{1/2} + f_1 + \ldots$ is the decomposition of f into homogeneous elements, then

$$\mathcal{G}_t(f) = \mathcal{G}_t(Qf_0) + \mathcal{G}_t(Qf_1) + o(t). \tag{3}$$

Remark. q is a ring homomorphism, whereas Q is not.

In the sequel we will use the following fact concerning quotients of Gaussian integrals.

Corollary 2. *Let $f, h \in \mathcal{L}$ such that*

(i) $f = f_0 + f_{1/2} + f_1$ *and* $f_0 \in \mathbb{R}$ *is constant.*
(ii) $h = 1 + h_1$.

Then

$$\frac{\mathcal{G}_t(fh)}{\mathcal{G}_t(h)} - \frac{\mathcal{G}_0(fh)}{\mathcal{G}_0(h)} = \mathcal{G}_t(f_1) + o(t). \tag{4}$$

Proof. First we compute

$$\begin{aligned} Q[hf] &= Q[[h]\,[f]] = Q[(1+h_1)(f_0 + f_{1/2} + f_1)] \\ &= Q(f_0 + f_{1/2} + f_1 + h_1 f_0) = f_0 + Qf_1 + f_0 Q h_1 \end{aligned}$$

and $Q[h] = 1 + Q(h_1)$. Hence, by formula (3), $\mathcal{G}_t(fh) = f_0(1+\mathcal{G}_t(Qh_1)) + \mathcal{G}_t(Qf_1) + o(t) = f_0\mathcal{G}_t(h) + \mathcal{G}_t(Qf_1) + o(t)$. That implies, since $\mathcal{G}_t(h_1) = O(t)$

$$\begin{aligned} \frac{\mathcal{G}_t(fh)}{\mathcal{G}_t(h)} - \frac{\mathcal{G}_0(fh)}{\mathcal{G}_0(h)} &= \frac{f_0\mathcal{G}_t(h) + \mathcal{G}_t(Qf_1)}{\mathcal{G}_t(h)} - \frac{f_0}{1} + o(t) \\ &= \frac{\mathcal{G}_t(Qf_1)}{1 + O(t)} + o(t) = \mathcal{G}_t(Qf_1) + o(t). \end{aligned}$$ □

That means in particular that in the situation above, f_1 is the only relevant term in order to determine the short-time asymptotic.

3.2. Short Time Asymptotic

In order to make use of the discussion in the preceding subsection, we have to show how it applies to more general situations. We observe first that the polynomial short time asymptotic of a Gaussian integral is in some sense independent of the domain of integration.

Lemma 4. *Let $U \subset \mathbb{R}^n$ be a subset containing a ball $B := \{x \in \mathbb{R}^n : |x| \le \rho\}$ around zero for some $\rho > 0$. Let $f \in C_b(\mathbb{R}^n)$ be bounded and continuous. Then as $t \to 0$*

$$\int_U \frac{e^{-\frac{|y|^2}{2t}}}{\sqrt{2\pi t}^n} f(y)dy - \mathcal{G}_t(f) = o(t^N)$$

for every $N \in \mathbb{N}$.

Proof. We use the abbreviation

$$\gamma_t(dy) := \frac{1}{\sqrt{2\pi t}^n} \exp(-|y|^2/2t)dy$$

and denote the sup-norm by $\|-\|$. Then

$$\begin{aligned} \left|\int_U \gamma_t(dy)f(y) - \mathcal{G}_t(f)\right| &= \left|\int_{\mathbb{R}^n-U} \gamma_t(dy)f(y)\right| \\ &\leq \|f\| \int_{\mathbb{R}^n-U} \gamma_t(dy) \leq \|f\| \int_{\mathbb{R}^n-B} \gamma_t(dy) \\ &\leq \sqrt{2}^n \|f\| \int_{\mathbb{R}^n-B} \gamma_{2t}(dy) e^{-\frac{|y|^2}{4t}} \leq \sqrt{2}^n \|f\| e^{-\frac{\rho^2}{4t}}. \end{aligned}$$

This expression tends to zero divided by any power of t. □

From this we draw the following conclusion for the polynomial short time asymptotic of more general functions:

Corollary 3. *Let $U \subset \mathbb{R}^n$ be a bounded open neighbourhood of the origin. Let $f \in C^r(U) \cap C(\overline{U})$ be continuous on the closure of U and m times continuously differentiable in the interior. Denote the Taylor polynomial of f up to order $r-1$ ((r-1)-jet) around 0 by $\hat{f}$. For every $k \in \mathbb{Z}$ such that $t^k \hat{f} \in \mathcal{I}^0$ we have asymptotically as $t \to 0$*

$$\int_U \frac{e^{-\frac{|y|^2}{2t}}}{\sqrt{2\pi t}^n} t^k f(y)dy = \mathcal{G}_t(t^k \hat{f}) + o(t^{k+\frac{r-1}{2}}).$$

Proof. First of all we may extend f to a bounded continuous function in $C_b(\mathbb{R}^n)$ which coincides with f on U. By Lemma 4 the actual neighbourhood of 0 does not play any role for the polynomial short time asymptotic. Therefore we may choose a suitably small ball B of radius $\rho > 0$ around 0 such that by Taylor's formula ([3], (8.14.3), p. 186) we have $|f(x) - \hat{f}(x)| \leq C_r(f)|x|^r$ for all $x \in B$. Here

$$C_r(f) := \sup_{x\in B\ |\nu|=r} \frac{|D^\nu f(x)|}{\nu_1! \dots \nu_n!}, \tag{5}$$

where $\nu = (\nu_1, \dots, \nu_n)$ denotes a *multi-index*, $|\nu| = \nu_1 + \dots + \nu_n$, and we use D^ν as shorthand for $\partial^{|\nu|}/\partial x_1^{\nu_1} \dots \partial x_n^{\nu_n}$. That implies

$$\begin{aligned} &\left|\int_B \gamma_t(dy)t^k f(y) - \int_B \gamma_t(dy)t^k \hat{f}(y)\right| \leq t^k \int_B \gamma_t(dy)|f(y) - \hat{f}(y)| \\ \leq\ & C_r(f)t^k \int_B \gamma_t(dy)|y|^r \leq C_r(f)t^{k+\frac{m}{2}}, \end{aligned}$$

which is $o(t^{k+\frac{m-1}{2}})$ and tends to zero as $t \to 0$ by our assumption about k. □

4. Proper Families Equivalent to the Heat Semigroup on Manifolds

Using the results from the preceding sections, we will now construct basically two examples of families which are equivalent to the heat semigroup on a closed and connected riemannian manifold M of dimension m. We consider the semigroup $e^{-t\Delta_L/2} : C(M) \to C(M)$, on the space of continuous functions and start with the following elementary lemma which closes the gap between the pointwise estimates of the preceding section and the global estimates necessary now.

Lemma 5. *Let $k \in C(M \times M)$, $k \geq 0$ and consider the integral operator*

$$I_k f(x) := \int_M k(x,y) f(y) \mathrm{vol}_M(y).$$

Then $I_k : C(M) \to C(M)$ is a bounded operator with norm

$$\| I_k \| = \sup_{x \in M} \left| \int_M k(x,y) \mathrm{vol}_M(y) \right| = \| I_k 1 \|.$$

Let $d_M(-,-)$ denote the distance function on M. Then by the preceding lemma

$$T_t f(q) := \frac{\int_M e^{-\frac{d_M(p,q)^2}{2t}} f(p) \mathrm{vol}_M(p)}{\int_M e^{-\frac{d_M(p,q)^2}{2t}} \mathrm{vol}_M(p)} \tag{6}$$

is a family of contractions on the Banach space $V = C(M)$.

The subspace $C^3(M) \subset C(M)$ is a determining set for $G = -\frac{1}{2}\Delta_M$, where Δ_M denotes the scalar Laplacian on M. By Lemma 1 we can prove Chernoff equivalence of the family above and the semigroup generated by G, if we can show that

$$\lim_{t \to 0} \frac{T_t f - f}{t} = G f \tag{7}$$

for all $f \in C^3(M)$. By the very same estimate as in Lemma 4 for the polynomial asymptotics we may restrict ourselves to integration over an arbitrary open neighbourhood $U(q)$ instead of over all of M. We choose $U(q)$ so small that we can use the exponential map to construct a *normal coordinate neighbourhood*

$$\phi_q : V(0) \to U(q).$$

Then we have

$$d_M(q,p) = d_M(\phi_q(0), \phi_q(x)) = |x|$$

in these coordinates. Furthermore

$$\mathrm{vol}_M(\phi_q(x)) = \sqrt{\det g}(x)\, dx,$$

where g is the metric tensor. Therefore, using again Lemma 4, we obtain

$$(T_t f - f)(q) = \frac{\mathcal{G}_t(\sqrt{\det g} f)}{\mathcal{G}_t(\sqrt{\det g})}(0) - \frac{\mathcal{G}_0(\sqrt{\det g} f)}{\mathcal{G}_0(\sqrt{\det g})}(0).$$

But now $\sqrt{\det g}$ is infinitely differentiable due to our assumptions on the manifold and f is in $C^3(M)$. We may therefore apply Corollary 3 with $k = 0$, $r = 3$ and in the sequel we just have to consider the Taylor expansion of these functions up to second order. The expansion of the metric tensor is given by

Lemma 6. *In normal coordinates the Taylor expansion of g is given by*

$$g_{ab}(x) = \delta_{ab} + \frac{1}{3} R^M_{auvb}(0) x^u x^v + O(|x|^3),$$

where R^M denotes the curvature tensor of M.

Proof. See [7], (1.14) Proposition, p. 8. □

That implies for the volume form:

Corollary 4. *In normal coordinates the Taylor expansion of $\sqrt{\det g}$ is given by*

$$\sqrt{\det g}(x) = 1 + \frac{1}{6} R^M_{auva} x^u x^v + O(|x|^3),$$

where R^M denotes the curvature tensor of M.

Proof. By Lemma 6 we get

$$\sqrt{\det g}(x) = \det\left(e^{\frac{1}{3} R^M_{auvb}(0) x^u x^v} + O(|x^3|) \right)^{1/2} = e^{\frac{1}{6} \operatorname{tr}(R^M_{auvb}(0) x^u x^v)} + O(|x^3|)$$
$$= 1 + \frac{1}{6} R^M_{auva} x^u x^v + O(|x|^3).$$ □

Therefore the 2-jet of $\sqrt{\det g}$ has exactly the properties required for the function h in Corollary 2. So we apply Corollary 2 and get

$$(T_t f - f)(q) = \mathcal{G}_t(f_1)(0) + o(t) = \frac{t}{2} (\Delta f)(0) + o(t).$$

But in normal coordinates, the Laplacian on M coincides with $-\Delta$ since we always assume our Laplace-Beltrami operator to be positive. Therefore we may write invariantly

$$(T_t f - f)(q) = -\frac{t}{2} \Delta_M f(q) + o(t)$$

for all functions $f \in C^3(M)$. This is a pointwise statement. But inspecting the proof of Corollary 3 above shows that

$$\left| T_t f - f + \frac{t}{2} \Delta_M f \right| (q) \leq C_3(f_q) t^{3/2} \tag{8}$$

with $f_q := f \circ \phi_q$. Due to the compactness of M the numbers $C_3(f_q), q \in M$ are even *uniformly* bounded. This translates (8) into the global statement

$$\left\| T_t f - f + \frac{t}{2} \Delta_M f \right\| \leq C t^{3/2}.$$

Since the functions $f \in C^3(M)$ are known to provide a determining subset for the Laplacian, this finally implies

Proposition 2. *The family T_t defined above is Chernoff equivalent to the heat semigroup on M.*

Next, we want to prove a corresponding statement in the spirit of Subsection 1.3. We consider the short time asymptotic of the denominator, namely

$$a(t,q) := \int_M e^{-\frac{d_M(p,q)^2}{2t}} \mathrm{vol}_M(p)$$

and calculate its asymptotic:

Lemma 7. *Let* Scal_M *be the scalar curvature of* M. *Then*

$$a(t,q) = \sqrt{2\pi t}^m \left(e^{-\frac{t\mathrm{Scal}_M(q)}{6}} + o(t)\right).$$

Proof. Using again the Taylor expansion of the volume form, we get

$$\begin{aligned} a(t,q) &= \sqrt{2\pi t}^m \left(\mathcal{G}_t([\sqrt{\det g}]) + o(t)\right) \\ &= \sqrt{2\pi t}^m \left(1 + \frac{t}{2}\Delta\left(\frac{1}{6}R^M_{auva}x^u x^v\right)_{x=0} + o(t)\right) \\ &= \sqrt{2\pi t}^m \left(1 + \frac{t}{6}R^M_{auua}(0) + o(t)\right) \\ &= \sqrt{2\pi t}^m \left(1 - \frac{t}{6}\mathrm{Scal}_M(q) + o(t)\right). \end{aligned}$$

□

Now we consider the proper family of bounded operators

$$B_t f(q) := \frac{1}{\sqrt{2\pi t}^m} \int_M e^{-\frac{d_M(p,q)^2}{2t} + \frac{t\mathrm{Scal}_M(q)}{6}} f(p)\mathrm{vol}_M(p). \tag{9}$$

By Lemma 7 the norm of the operators is given by

$$c(t) = \sup_{q\in M}\left|\frac{1}{\sqrt{2\pi t}^m}\int_M e^{-\frac{d_M(p,q)^2}{2t} + \frac{t\mathrm{Scal}_M(q)}{6}} \mathrm{vol}_M(p)\right| = 1 + o(t).$$

For the corresponding contraction family K_B we obtain by Corollary 1, that its derivative DK_B equals the derivative DB. But on the other hand

$$\begin{aligned} \|K_B(t)f - T_t f\| &= \left\|\frac{B_t f}{c(t)} - T_t f\right\| \\ &= \left\|\frac{B_t f}{c(t)} - \sqrt{2\pi t}^m e^{-\frac{t}{6}\mathrm{Scal}_M(q)} \frac{B_t f}{a(t,q)}\right\| \\ &= o(t). \end{aligned}$$

Therefore Lemma 1 implies that the families K_B and T are Chernoff equivalent. Since Chernoff equivalence is clearly transitive, we obtain

Corollary 5. *The family* B_t *is Chernoff equivalent to the heat semigroup on* M.

Finally, it should be noted that there is also a symmetric version of the approximating kernel. By the same arguments as in Lemma 7 with

$$\hat{a}(t,q) := \int_M e^{-\frac{d_M(p,q)^2}{2t} + \frac{t\mathrm{Scal}_M(p)}{12}} \mathrm{vol}_M(p)$$

instead of $a(t,q)$ we obtain

Corollary 6. *The family*

$$\hat{B}_t f(q) := \frac{1}{\sqrt{2\pi t}^m} \int_M e^{-\frac{d_M(p,q)^2}{2t} + \frac{t\mathrm{Scal}_M(q)+t\mathrm{Scal}_M(p)}{12}} f(p)\mathrm{vol}_M(p)$$

is Chernoff equivalent to the heat semigroup on M.

Remark. (i) The approximations B_t and $\hat{B}_t$ were already considered in [1]. (ii) The preceding results hold as well if we consider the heat equation on the Banach space $V = L^\infty(M, \mathrm{vol}_M)$.

References

[1] L. Andersson, B.K. Driver, *Finite dimensional approximations to Wiener measure and path integral formulas on manifolds.* Journ. Funct. Anal. **165** (1999), 430–498.

[2] O. Bratteli, D. W. Robinson, *Operator Algebras and Quantum Statistical Mechanics 1.* 2nd. Ed., Springer, New York, 1987.

[3] J. Dieudonne, *Foundations of Modern Analysis.* Academic Press, New York, 1960.

[4] T. Kato, *Perturbation Theory for Linear Operators.* 2nd Ed., Springer, Berlin, 1980.

[5] R.A. Kunze, I.E. Segal, *Integrals and Operators.* Springer, Berlin, 1978.

[6] K.J. Engel, R. Nagel, *One-Parameter Semigroups for Linear Evolution Equations.* Graduate Texts in Mathematics 194, Springer-Verlag, Berlin, 1999.

[7] J. Roe, *Elliptic Operators, Topology and Asymptotic Methods.* Longman, Harlow, 1988.

[8] O. G. Smolyanov, H. v. Weizsäcker, O. Wittich, *Brownian motion on a manifold as a limit of stepwise conditioned standard Brownian motions.* In: J. Jost (Ed.), Infinite Dimensional (Stochastic) Analysis, CMS Conference Proceedings **29**, 2000.

O.G. Smolyanov
Faculty of Mechanics and Mathematics
Moscow State University
119899 Moscow, Russia
E-mail address: `smolyan@mail.ru`

H. v. Weizsäcker
Fachbereich Mathematik
Universität Kaiserslautern
D-67663 Kaiserslautern, Germany
E-mail address: `weizsaecker@mathematik.uni-kl.de`

O. Wittich
GSF – National Research Center for Environment and Health
D-85764 Neuherberg, Germany
E-mail address: `wittich@gsf.de`

Progress in Nonlinear Differential Equations
and Their Applications, Vol. 55, 359–373

Harmonic Map Heat Flow generated by Markovian Semigroups

Karl-Theodor Sturm

1. The Domain Space

Our domain space will be a measurable space $(M, \mathcal{M})$ with a given Markov semigroup $p = (p_t)_{t>0}$ on it. That is, M is an arbitrary set, $\mathcal{M}$ is a σ-field on M and $p :]0, \infty[\times M \times \mathcal{M} \to [0, \infty]$ satisfies

- $\forall t > 0, \forall A \subset \mathcal{M} : x \mapsto p_t(x, A)$ is a $\mathcal{M}$-measurable function on M;
- $\forall t > 0, \forall x \in M : A \mapsto p_t(x, A)$ is a probability measure on $(M, \mathcal{M})$;
- $\forall s, t > 0, \forall x \in M, \forall A \in \mathcal{M} : \; p_{s+t}(x, A) = \int_M p_t(y, A) p_s(x, dy)$.

Occasionally, we require $(M, \mathcal{M})$ to be a *Radon measurable space.* All locally compact spaces with countable bases as well as all Polish spaces (= complete separable metric spaces) – equipped with their Borel σ-fields – are Radon measurable spaces.

Example 1.1. Let M be a Riemannian manifold, $\mathcal{M}$ its Borel σ-field, m the Riemannian volume measure and $k :]0, \infty[\times M \times M \to [0, \infty]$ be the minimal heat kernel on M (= fundamental solution of $\frac{1}{2}\Delta - \frac{\partial}{\partial t}$). Then $p_t(x, dy) := k_t(x, y)m(dy)$ defines a Markov semigroup provided M is *stochastically complete*, i.e., provided $p_t(x, M) = 1$ for all $x \in M$ and some (hence all) $t > 0$.
The latter is always satisfied if M is connected and complete and if the Ricci curvature of M is bounded from below or, more generally, if $\mathrm{Ric}|_{B(r,x_0)} \geq C(r^2 + 1)$ or, even more generally, if $m(B_r(x_0)) \leq \exp[C(r^2 + 1)]$ (for some $x_0 \in M, C \in \mathbb{R}$ and all $r > 0$), cf. Grigoryan (2000).

Example 1.2. Let $M = \mathbb{R}^{d+1}$ equipped with its Borel σ-field $\mathcal{M}$ and let $k(s, x, t, dy)$ be the transition kernel for the parabolic partial differential equation

$$\frac{\partial}{\partial s}u(x, s) = \sum_{i,j=1}^{d} a_{ij}(x, s)\frac{\partial^2}{\partial x_i \partial x_j}u(x, s) + \sum_{i=1}^{d} b_i(x, s)\frac{\partial}{\partial x_i}u(x, s)$$

on $\mathbb{R}^d$, where a_{ij} and b_i are bounded measurable functions on $\mathbb{R}^{d+1}$ and (a_{ij}) is locally uniformly elliptic, symmetric and continuous. Then

$$p_t((x, s), A) := \int 1_A((y, s + t))k(s, x, s + t, dy)$$

defines a Markov semigroup on $\mathbb{R}^{d+1}$. If the coefficients a_{ij} and b_i do not depend on time then $p_t(x,B) := k(0,x,t,B)$ defines a Markov semigroup on $\mathbb{R}^d$. See STROOCK, VARADHAN (1981).

Similar results hold true for hypo- and subelliptic operators (cf. FEFFERMAN, PHONG (1983), JERISON, SANCHEZ-CALLE (1986)) as well as for certain pseudo-differential operators, for instance for $(-\Delta)^{\alpha/2}$ with $\alpha < 2$ which is covered by the next result.

Example 1.3. Given a symmetric matrix $a \in \mathbb{R}^{d\times d}$, a vector $b \in \mathbb{R}^d$ and a measure μ on $\mathbb{R}^d$ satisfying $\int_{\mathbb{R}^d} \|y\|^2/(1+\|y\|^2)\mu(dy) < \infty$ there exists a unique convolution semigroup $(q_t)_{t>0}$ of probability measures on $\mathbb{R}^d$ such that $p_t(x,B) := q_t(B-x)$ defines the Markov semigroup for the Lévy operator

$$\sum_{i,j=1}^{d} a_{ij}\frac{\partial^2}{\partial x_i \partial x_j}u(x) + \sum_{i=1}^{d} b_i \frac{\partial}{\partial x_i}u(x) + \int \left(u(x+y) - u(x) - \frac{y\cdot\nabla u(x)}{1+\|y\|^2}\right)\mu(dy).$$

See, e.g., ETHIER, KURTZ (1986), JACOB (1996, 2001), TAIRA (1991).

Lemma 1.4. *Each quasi-regular conservative Dirichlet form $(\mathcal{E},\mathcal{D}(\mathcal{E}))$ on a σ-finite measure space $(M,\mathcal{M},m)$ defines a Markov semigroup $(p_t)_{t>0}$ such that $\forall u \in L_2(M)\cap L_\infty(M)$ and for m-a.e. $x \in M$*

$$e^{ta}u(x) = \int_M u(y)p_t(x,dy). \tag{1}$$

Here a denotes the generator of $(\mathcal{E},\mathcal{D}(\mathcal{E}))$. See MA, RÖCKNER (1992).

Standard examples here are Dirichlet forms associated with elliptic differential operators in divergence form (with bounded measurable coefficients) on $\mathbb{R}^d$. Let us mention some non-classical examples of quasi-regular conservative Dirichlet forms:

- Dirichlet form on the Wiener space $C(\mathbb{R}_+,\mathbb{R}^n)$ and Ornstein-Uhlenbeck semi-group;
- Dirichlet forms on path or loop spaces $C(R_+,M)$ or $C(S^1,M)$, resp., over Riemannian manifolds.

The "quasi-regularity" of the Dirichlet form is not really essential here since in the sequel we only use p_t for $t \in \mathbb{T} := \{k\cdot 2^{-n} : k,n \in \mathbb{N}\}$ and for each conservative Dirichlet form on a Radon space $(M,\mathcal{M})$ there exists a Markov semigroup $(p_t)_{t\in\mathbb{T}}$ satisfying (1).

Lemma 1.5. *Each Markov process $(\Omega,\mathcal{A},\mathbb{P},X_t^x)$ with values in some measurable space $(M,\mathcal{M})$ defines a Markov semigroup on that space by*

$$p_t(x,A) := \mathbb{P}(X_t^x \in A). \tag{2}$$

If $(M,\mathcal{M})$ is a Radon measurable space then vice versa: each Markov semigroup on $(M,\mathcal{M})$ defines via (2) a Markov process (unique up to equivalence). See, e.g., BAUER (1996).

One of the main examples for such Markov processes are solutions of stochastic differential equations

$$dX_t^x = x + b(X_t^x)dt + \sigma(X_t^x)dW_t$$

on $\mathbb{R}^d$ where (W_t) denotes Brownian motion and $b : \mathbb{R}^d \to \mathbb{R}$ and $\sigma : \mathbb{R}^d \to \mathbb{R}^d$ are locally Lipschitz and bounded. Other remarkable examples are

- Super-Brownian motion on the space of measures on $\mathbb{R}^n$;
- Fleming-Viot processes on the space of probability measures on $\mathbb{R}^n$;
- Interacting particle systems as processes on the configuration space over $\mathbb{R}^n$.

In all the above-mentioned examples, the Markov processes can be chosen to be *right Markov processes* which means that they have some additional minimal regularity properties.

Definition 1.6. *Let* $(\Omega,\mathcal{A},\mathbb{P},X_t^x)$ *be a right Markov process associated with a Markov semigroup* $(p_t)_{t>0}$ *on a Radon measurable space* $(M,\mathcal{M})$. *Then for each measurable subset* $D \subset M$ *the stopped semigroup* $(p_{D,t})_{t>0}$ *is the Markov semigroup on* $(M,\mathcal{M})$ *defined by*

$$p_{D,t}(x,A) := \mathbb{P}(X^x_{t\wedge\tau(D,x)} \in A)$$

where $\tau(D,x) := \inf\{t > 0 : X_t^x \notin D\}$ *denotes the first exit time of* D

Definition 1.7. *Given a Markov semigroup* $(p_t)_{t>0}$ *on a measurable space* $(M,\mathcal{M})$ *we define for each* $t > 0$ *the matching operator* p_t^∇ *acting on symmetric functions* $\rho : M \times M \to \mathbb{R}_+$ *by*

$$p_t^\nabla \rho(x_1,x_2) := \sup_u |p_t u(x_1) - p_t u(x_2)|$$

where the supremum is over all bounded measurable $u : M \to \mathbb{R}$ *satisfying* $|u(y_1) - u(y_2)| \leq \rho(y_1,y_2)$ *for all* $y_1, y_2 \in M$. *The coupling semigroup* $(p_t^\diamond)_{t>0}$ *acting on symmetric functions* $\rho : M \times M \to \mathbb{R}_+$ *is defined by*

$$p_t^\diamond \rho(x_1,x_2) := \sup\left\{ p_{t_n}^\nabla \circ \ldots \circ p_{t_1}^\nabla \rho(x_1,x_2) : n \in \mathbb{N}, t_i > 0, \sum_{i=1}^n t_i = t \right\}.$$

Remark 1.8. For each Markov semigroup $(p_t)_{t>0}$ on a measurable space $(M,\mathcal{M})$

$$p_t^\diamond \rho(x_1,x_2) \leq \inf \mathbb{E}\rho(X_t^{x_1}, X_t^{x_2})$$

where the infimum is over all Markov processes $(\Omega,\mathcal{A},\mathbb{P},X_t^x)$ associated with $(p_t)_{t>0}$. Cf. RACHEV, RÜSCHENDORF (1998), KENDALL (1990).

2. The Target Space

Our target space will be a complete metric space (N,d) with a given barycenter contraction b on it.

We denote by $\mathcal{N}$ the Borel σ-field of N, and for each $\theta > 0$, by $\mathcal{P}^\theta(N)$ the set of all probability measures p on $(N,\mathcal{N})$ with separable support and with

$\int_N d^\theta(z,x)p(dx) < \infty$ for some/all $z \in N$. Given two measures $p,q \in \mathcal{P}^1(N)$, a measure $\mu \in \mathcal{P}^1(N \times N)$ is called *coupling* of p and q iff

$$\mu(A \times N) = p(A), \; \mu(N \times A) = q(A). \qquad (\forall A \in \mathcal{N})$$

The L_1-*Wasserstein distance* or *Kantorovich-Rubinstein distance* of $p,q \in \mathcal{P}^1(N)$ is defined as

$$d^W(p,q) = \inf\left\{\int_{N^2} d(x_1,x_2)\mu(dx) : \; \mu \in \mathcal{P}^1(N^2) \text{ is a coupling of } p \text{ and } q\right\}.$$

Definition 2.1. *A barycenter contraction is a map* $b : \mathcal{P}^1(N) \to N$ *such that*

- $b(\delta_x) = x$ *for all* $x \in N$;
- $d(b(p),b(q)) \le d^W(p,q)$ *for all* $p,q \in \mathcal{P}^1(N)$.

Remark 2.2. If there exists a barycenter contraction on (N,d) then (N,d) is a geodesic space: For each pair of points $x_0, x_1 \in N$ we can define one geodesic $t \mapsto x_t$ connecting x_0 and x_1 by $x_t := b((1-t)\delta_{x_0} + t\delta_{x_1})$.

Given any for points $x_0, x_1, y_0, y_1 \in N$, the function $t \mapsto d(x_t, y_t)$ is convex. In particular, the geodesic $t \mapsto x_t$ depends continuously on x_0 and x_1. However, it is not necessarily the only geodesic connecting x_0 and x_1.

If geodesics in N are unique then the existence of a barycenter contraction implies that $d : N \times N \to \mathbb{R}$ is convex. Thus N has (globally) "nonpositive curvature" in the sense of Busemann.

Example 2.3. Let (N,d) be a complete metric space with (globally) "nonpositive curvature" in the sense of A.D. Alexandrov. Then for each $p \in \mathcal{P}^2(N)$ there exists a unique $b(p) \in N$ which minimizes the uniformly convex function

$$z \mapsto \int_N d^2(z,x)p(dx)$$

on N. The map $b : \mathcal{P}^2(N) \to N$ extends to a barycenter contraction $\mathcal{P}^1(N) \to N$. See STURM (2001).
Equivalently, $b(p)$ can be defined via the law of large numbers as the unique accumulation point of the sequence

$$\frac{1}{n}\sum_{i=1}^{n} X_i(w)$$

for a.e. ω where $(X_i)_i$ is a sequence of independent random variables with distribution p. The point $\frac{1}{n}\sum_{i=1}^n X_i(w)$ is defined by induction on n as the point $\gamma_{1/n}$ on the geodesic from $\gamma_0 := \frac{1}{n-1}\sum_{i=1}^{n-1} X_i(w)$ to $\gamma_1 := X_n(w)$. See STURM (2002).
Examples of spaces with nonpositive curvature in the sense of A.D. Alexandrov are

- complete, simply connected Riemannian manifolds with nonpositive sectional curvature;
- trees and, more generally, Euclidean Bruhat-Tits buildings;
- Hilbert spaces;

- L_2-spaces of maps into such spaces;
- Finite or infinite (weighted) products of such spaces;
- Gromov-Hausdorff limits of such spaces.

See, e.g., BALLMANN (1995), EELLS, FUGLEDE (2001), JOST (1994, 1997), KOREVAAR, SCHOEN (1993, 1997).

Example 2.4. Let N be a complete, simply connected Riemannian manifold and let d be a Riemannian distance. *Then (N, d) admits a barycenter contraction b if and only if N has nonpositive sectional curvature.*
Indeed, if (N, d) admits a barycenter contraction then so does (N_0, d) for each closed convex $N_0 \subset N$. Hence, geodesics in N_0 are unique and thus $t \mapsto d(\gamma_t, \eta_t)$ is convex for any pair of geodesics γ and η in N_0. This implies that N has nonpositive curvature (JOST (1997)).
Conversely, if N has nonpositive curvature then it admits a barycenter contraction by the previous Example 2.3. It is worthwhile to mention that also the *martingale mean* defines a barycenter contraction (different from the previous one, cf. THALMAIER (1996, 1996a)). Here for each $p \in \mathcal{P}^1(N)$ the martingale mean $b(p)$ is defined to be the starting point Y_0 of a N-valued martingale $(Y_t)_{t\in[0,\infty]}$ whose terminal variable Y_∞ is distributed according to p.

Example 2.5. Let (N, d) be a locally compact separable complete metric space with *negative curvature in the sense of Busemann.* Then ES-SAHIB, HEINICH (1999) have constructed a barycenter contraction. For Riemannian manifolds, this is different from those in Examples 2.3 and 2.4, and also for trees, it is different from that in Example 2.3.

Example 2.6. Let $(N, \|.\|)$ be a (real or complex) Banach space and put $d(x, y) := \|x-y\|$. Then $\mathcal{P}^1(N)$ is the set of Radon measures p on N satisfying $\int_N \|x\|\, p(dx) < \infty$. For each $p \in \mathcal{P}^1(N)$, the identity $x \mapsto x$ on N is *Bochner integrable* and

$$b(p) := \int_N x\, p(dx)$$

defines a barycenter contraction on (N, d). Cf. LEDOUX, TALAGRAND (1991), for instance.

Lemma 2.7. *Let I be a countable set and for each $i \in I$, let (N_i, d_i) be a complete metric space with barycenter contraction b_i and "base" point $o_i \in N_i$. Given $\theta \in [1, \infty]$, define a complete metric space (N, d) with base point $o = (o_i)_{i\in I}$ by*

$$N := \left\{ x = (x_i)_{i\in I} \in \bigotimes_{i\in I} N_i : \ d(x, o) < \infty \right\}, \quad d(x, y) := \left[\sum_{i\in I} d_i^\theta(x_i, y_i) \right]^{\frac{1}{\theta}}$$

provided $\theta < \infty$ or by $d(x, y) = \sup_{i\in I} d_i(x_i, y_i)$ if $\theta = \infty$. One can define a barycenter contraction b on $\mathcal{P}^1(N)$ by

$$b(p) := (b_i(p_i))_{i\in I}$$

where $p_i \in \mathcal{P}^1(N_i)$ *with* $p_i : A \mapsto p(\{x = (x_j)_{j\in I} \in N : x_i \in A\})$ *denotes the projection of* $p \in \mathcal{P}^1(N)$ *onto the i-th factor of* N.

For instance, this applies to $N = \mathbb{R}^n, n \geq 2$ with the usual notion of barycenter but with "unusual" metric $d(x,y) = \sup\{|x_i - y_i| : i = 1,..,n\}$. In this case, geodesics are not unique, e.g., each curve $t \mapsto (t, \varphi_2(t), \ldots, \varphi_n(t))$ with $\varphi \in C^1(\mathbb{R}), \varphi_i(0) = \varphi_i(1) = 0$ and let $|\varphi_i'| \leq 1$ is a geodesic connecting $(0,0,\ldots,0)$ and $(1,0,\ldots,0)$.

3. The Nonlinear Heat Semigroup

Let $(M, \mathcal{M})$, $p = (p_t)_{t>0}$ and (N,d,b) be as in Chapters 1, 2 and let $\mathcal{L}(M,N,p)$ denote the set of all measurable maps $f : M \to N$ with separable ranges and with

$$\partial_t f(x) := \int_M d(f(x), f(y)) p_t(x, dy) < \infty$$

for all $t > 0$ and all $x \in M$. For each such f, t and x, the probability measure $p_t(x, f^{-1}(\bullet))$ lies in $\mathcal{P}^1(N)$ and thus

$$P_t f(x) := b(p_t(x, f^{-1}(\bullet)))$$

is well-defined.

Lemma 3.1. *For all* $f, g \in \mathcal{L}(M,N,p)$, *all* $s, t > 0$, *and all* $x, y \in M$

(i) $d(P_t f(x), P_t g(x)) \leq \int d(f(y), g(y)) p_t(x, dy)$;
(ii) $d(P_t f(x), f(x)) \leq \partial_t f(x)$;
(iii) $\partial_s (P_t f)(x) \leq \partial_{s+t} f(x) + \partial_t f(x)$;
(iv) $d(P_t f(x), P_t f(y)) \leq p_t^\nabla d_f(x,y)$ with p_t^∇ from Definition 1.7 and d_f denoting the function $(x,y) \mapsto d(f(x), f(y))$ on $M \times M$;
(v) $P_t f \in \mathcal{L}(M,N,p)$.

The map $P_t : \mathcal{L}(M,N,p) \to \mathcal{L}(M,N,p)$ *is called nonlinear Markov operator associated with the kernel* p_t.

For the sequel, fix once for all a subsequence $(\delta_n)_{n\in\mathbb{N}}$ of $(2^{-n})_{n\in\mathbb{N}}$ and put $\mathbb{T} = \{k \cdot 2^{-n} : k, n \in \mathbb{N}\}$. Let $\mathcal{L}^*(M,N,p)$ denote the set of all $f \in \mathcal{L}(M,N,p)$ for which

$$P_t^* f(x) := \lim_{n\to\infty} (P_{\delta_n})^{t/\delta_n} f(x)$$

exists for all $t \in \mathbb{T}$ and all $x \in M$. Here $(P_\delta)^k$ denotes the k-th iteration of the nonlinear Markov operator P_δ.

Note that if $N = \mathbb{R}$ (equipped with the usual d and b) then $\mathcal{L}^*(M,\mathbb{R},p) = \mathcal{L}(M,\mathbb{R},p)$ is the set of all measurable $f : M \to \mathbb{R}$ with $\int |f(y)| p_t(x,dy) < \infty$ $(\forall t > 0, x \in M)$, and

$$P_t^* f(x) = P_t f(x) = \int_M f(y) p_t(x, dy).$$

Lemma 3.2. *For all $f, g \in \mathcal{L}^*(M,N,p)$, all $s,t \in \mathbb{T}$ and all $x,y \in M$*

(i) $d(P_t^* f(x), P_t^* g(x)) \leq \int d(f(y), g(y)) p_t(x, dy)$;

(ii) $d(P_t^* f(x), f(x) \leq \partial_t f(x))$ and $d(P_t^* f(x), P_{t+s}^* f(x)) \leq p_t(\partial_s f)(x)$;

(iii) $\partial_s(P_t^*)(x) \leq \partial_{s+t} f(x) + \partial_t f(x)$;

(iv) $d(P_t f(x), P_t f(y)) \leq p_t^\diamond d_f(x,y)$ with $p_t^\diamond$ from Definition 1.7 and d_f denoting the function $(x,y) \mapsto d(f(x), f(y))$ on $M \times M$;

(v) $P_t^* f \in \mathcal{L}^*(M,N,p)$ and $P_s^*(P_t^* f)(x) = P_{s+t}^* f(x)$.

The operator P_t^* on $\mathcal{L}^*(M,N,p)$ is called nonlinear heat operator associated with the "linear heat semigroup" $(p_t)_{t>0}$. The semigroup $(P_t^*)_{t\in\mathbb{T}}$ of operators on $\mathcal{L}^*(M,N,p)$ is called nonlinear heat semigroup.

4. Lipschitz Continuity

Let $(M, \mathcal{M})$, $p = (p_t)_{t>0}$ and (N, d, b) be as before. In addition, throughout this Chapter we fix a nonnegative symmetric function ρ on $M \times M$. (Typically, ρ will be a metric on M. But $\mathcal{M}$ will not necessarily be the Borel σ-field of ρ.) For $f, g : M \to N$ put $d_\infty(f,g) := \sup_{x \in M} d(f(x), g(x))$,

$$\mathrm{dil}_\rho f := \sup_{x,y \in M} \frac{d(f(x), f(y))}{\rho(x,y)}$$

(with $\frac{0}{0} := 0$) and let $\mathcal{C}_\rho^{Lip}(M,N)$ denote the set of measurable $f : M \to N$ with separable range and $\mathrm{dil}_\rho f < \infty$. Moreover, let $\mathcal{L}_\infty(M,N)$ denote the set of bounded measurable $f : M \to N$ with separable range.

Theorem 4.1. *Assume that (N,d) has nonpositive and lower bounded curvature (in the sense of Alexandrov). Moreover, assume $\exists C, \beta > 0$ and $\forall t > 0 : \exists c_t$ such that* $\sup_{s \leq t} c_s < \infty$ *and $\forall x, y \in M$:*

$$p_t^\diamond \rho(x,y) \leq c_t \cdot \rho(x,y) \tag{3}$$

$$\int \rho^4(x,z) p_t(x,dz) \leq C \cdot t^{1+\beta} \tag{4}$$

Then $\mathcal{C}_\rho^{Lip}(M,N) \subset \mathcal{L}^*(M,N,p)$ and $(P_t^*)_{t\in\mathbb{R}_+}$ is a strongly continuous semigroup on $\mathcal{C}_\rho^{Lip}(M,N)$. More precisely, for all $x \in M, t \in \mathbb{R}_+$ and $f \in \mathcal{C}_\rho^{Lip}(M,N)$

$$P_t^* f(x) = \lim_{s \to 0} P_s^{\lfloor t/s \rfloor} f(x)$$

exists and the limit is continuous in each variable:

$$\mathrm{dil}_\rho P_t^* f \leq c_t \cdot \mathrm{dil}_\rho f \tag{5}$$

$$d_\infty(P_t^* f, P_t^* g) \leq d_\infty(f,g) \tag{6}$$

$$d_\infty(P_s^* f, P_t^* f) \leq C^{1/4} \cdot \mathrm{dil}_\rho f \cdot |t-s|^{\frac{1+\beta}{4}}. \tag{7}$$

Remark 4.2. (i) If $c_t = e^{-\kappa t}$ for some $\kappa \in \mathbb{R}$ and $N = \mathbb{R}$ then condition (3) is already *necessary* for (5). Indeed, for any Markov semigroup $(p_t)_t$ on a metric space (M, ρ) and any $\kappa \in \mathbb{R}$ the following are equivalent:

- $\mathrm{dil}_\rho p_t u \le e^{-\kappa t} \cdot \mathrm{dil}_\rho u \qquad (\forall t, \forall u \in \mathcal{C}^{Lip}_\rho(M,\mathbb{R}))$
- $p_t^\diamond \rho(x,y) \le e^{-\kappa t} \rho(x,y). \qquad (\forall t)$

(ii) Condition (4) excludes jump-diffusion processes. However, for diffusions it is a very weak assumption. E.g., for solutions of SDEs with bounded measurable coefficients on $\mathbb{R}^d$ or for the Markov semigroups from Example 1.2, $\int |x-y|^4 p_t(x,dy) \le C \cdot t^2$ for all $x, y \in \mathbb{R}^d$ and all t.

(iii) The assumption on the lower bounded curvature of (N,d) can be weakened in order to include also "spaces with a reverse variance inequality of some order > 2", e.g., the result of gluing together two copies of the set $\{z = (x,t) \in \mathbb{R}^k : t \le \psi(x)\}$ along their boundary $\{z = (x,t) \in \mathbb{R}^k : t = \psi(x)\}$ where $\psi : \mathbb{R}^{k-1} \to \mathbb{R}$ is any smooth convex function.

For typical examples satisfying (3), see Examples 4.5–4.9 below.

The main point in Theorem 4.1 is that it yields convergence independent of the choice of the sequence $(\delta_n)_n$. Our next result will give convergence for suitable choices of sequences $(\delta_n)_n$. Here the advantages will be:

- it does not require any kind of lower curvature bound for (N,d);
- it applies also to jump processes and to nonlocal equations;
- it also yields smoothing from $\mathcal{L}_\infty(M,N)$ to $\mathcal{C}^{Lip}_\rho(M,N)$.

In order to formulate the latter, let ρ_0 be another nonnegative symmetric function on $M \times M$ (besides ρ) and define dil_{ρ_0} and $\mathcal{C}^{Lip}_{\rho_0}(M,N)$ in an analogous way.

Theorem 4.3. *Assume that $(M,\mathcal{M})$ is a Radon measurable space, (M,ρ) is separable and (N,d) is locally compact. Moreover, assume that* $\forall t \in \mathbb{T} : \exists C_t : \forall x, y \in M$:

$$p_t^\diamond \rho_0(x,y) \le C_t \cdot \rho(x,y) \tag{8}$$

$$\int \rho_0(x,z) p_t(x,dz) < \infty. \tag{9}$$

Then for each sequence $(s_k)_k \subset (2^{-k})_k$ there exists a subsequence $(\delta_k)_k = (s_{n_k})_k$ such that $\mathcal{C}^{Lip}_{\rho_0}(M,N) \subset \mathcal{L}^*(M,N,p)$ and for each $t \in \mathbb{T}$

$$P_t^* : \mathcal{C}^{Lip}_{\rho_0}(M,N) \to \mathcal{C}^{Lip}_\rho(M,N).$$

More precisely, $\forall f \in \mathcal{C}^{Lip}_{\rho_0}(M,N), \forall x \in M, \forall t \in \mathbb{T}$ the limit

$$P_t^* f(x) = \lim_{k\to\infty} P_{\delta_k}^{t/\delta_k} f(x)$$

exists and

$$\mathrm{dil}_\rho P_t^* f \le C_t \cdot \mathrm{dil}_{\rho_0} f. \tag{10}$$

Remark 4.4. (i) The most important choices for ρ_0 are either $\rho_0 \equiv \rho$ or $\rho_0 \equiv 1$. In the latter case, $\mathrm{dil}_{\rho_0} f = \mathrm{osc} f := \sup_{x,y\in M} d(f(x), f(y))$ and thus $\mathcal{C}^{Lip}_{\rho_0}(M,N) = \mathcal{L}_\infty(M,N)$. Theorem 4.3 then proves existence of the limit $P_t^* f$ for all $f \in \mathcal{L}_\infty(M,N)$ and smoothing $P_t^* f : \mathcal{L}_\infty(M,N) \to \mathcal{C}^{Lip}_\rho(M,N)$.

(ii) If in addition to the assumptions of the previous Theorem $\gamma_s(x) := \int \rho(x,z)p_s(x,dz) \to 0$ for $s \to 0$ then the limit

$$P_t^* f(x) = \lim_{k\to\infty} P_{\delta_k}^{\lfloor t/\delta_k \rfloor} f(x)$$

exists for all $t \in \mathbb{R}_+$ and for $s \to 0$

$$d(P_t^* f(x), P_{t+s}^* f(x) \leq C_t \cdot \mathrm{dil}_{\rho_0} f \cdot \gamma_s(x) \to 0. \tag{11}$$

(iii) Assume that (8) only holds true for all x, y in a ρ-open set $M_1 \subset M$. Then the limit $P_t^* f(x) = \lim_{k\to\infty} P_{\delta_k}^{t/\delta_k} f(x)$ exists $\forall f \in \mathcal{C}_{\rho_0}^{Lip}(M,N), \forall x \in M_1, \forall t \in \mathbb{T}$ and is ρ-Lipschitz continuous on M_1.

Example 4.5. Let $(q_t)_{t>0}$ be a convolution semigroup of probability measures on an Abelian group M and define a translation invariant Markov semigroup by $p_t(x,A) := q_t(x^{-1}A)$. Then for each symmetric $\rho : M \times M \to \mathbb{R}_+$

$$p_t^\diamond \rho(x,y) \leq \int \rho(x+z, y+z)\, q_t(dz).$$

In particular, if ρ is translation invariant then

$$p_t^\diamond \rho(x,y) \leq \rho(x,y).$$

For instance, this applies to all Lévy semigroups on $\mathbb{R}^n$ as introduced in Example 1.3. For various other examples, see BENDIKOV (1995) and BENDIKOV, SALOFF-COSTE (2001).

Example 4.6. Let $(p_t)_{t>0}$ be a Markov semigroup on $M = \mathbb{R}^n$ such that

$$p_t(x,B) = \int_{\mathbb{R}^n} 1_B(y) k_t(\|x-y\|) dy$$

for all $t > 0$ with some nonincreasing function $r \mapsto k_t(r)$ on $\mathbb{R}_+$ with

$$C_t := \int_{\mathbb{R}^{n-1}} k_t(\|z\|) dz < \infty.$$

Put $\rho(x,y) = \|x-y\|$. Then

$$p_t^\diamond 1(x,y) \leq C_t \cdot \rho(x,y).$$

For instance, if $(p_t)_{t>0}$ is the classical heat semigroup on $\mathbb{R}^n$, i.e., $k_t(r) = (2\pi t)^{-n/2} \exp(-r^2/(2t))$, this yields

$$C_t = \frac{1}{\sqrt{2\pi t}}.$$

More generally, for $\alpha < 2$ let $(p_t)_{t>0}$ be the *symmetric α-stable semigroup* on $\mathbb{R}^n$, i.e., the Markov semigroup associated with the Lévy operator $(-\frac{1}{2}\Delta)^{\alpha/2}$. Then $\int_{\mathbb{R}^n} \exp(ix\xi) k_t(\|x\|) dx = \exp(-\|\xi\|^\alpha \cdot t/2)$ and thus

$$C_t = \frac{1}{2\pi} \int_{\mathbb{R}} \exp(-s^\alpha \cdot t/2) ds = C'_\alpha \cdot t^{-1/\alpha}.$$

Example 4.7. Let $(p_t)_{t>0}$ be the heat semigroup on a complete Riemannian manifold M of Ricci curvature $\geq \kappa$ and dimension n, and let ρ be the Riemannian distance ρ on M. Then $\exists C = C(\kappa, n) : \forall x, y, t :$

- $p_t^\diamond \rho(x,y) \leq e^{-\kappa t} \cdot \rho(x,y)$
- $p_t^\diamond 1(x,y) \leq C \cdot t^{-1/2} \cdot e^{-\kappa t} \cdot \rho(x,y)$
- $\int \rho^4(x,y) p_t(x,dy) \leq C \cdot t^2$.

For examples of Alexandrov spaces M (= complete metric spaces of lower bounded "sectional" curvature) where the same estimates hold true, see VON RENESSE (2002).

Example 4.8. Let M and $(p_t)_{t>0}$ as in the previous Example 4.7 and fix an open subset $D \subset M$. Let $(p_{D,t})_{t>0}$ be the stopped semigroup as introduced in Definition 1.6. Then for this semigroup, (8) with $\rho_0 = 1$ is satisfied on each relatively compact open subset $M_1 \subset D$.

Example 4.9. Let $(p_t)_{t>0}$ be a strongly continuous, symmetric semigroup on a σ-finite measure space $(M, \mathcal{M}, m)$ and assume a "curvature-dimension condition" in the sense of Bakry-Emery (see, e.g., LEDOUX (2000) holds true with curvature bound κ and dimension bound n. Moreover, let ρ be a symmetric nonnegative function on $M \times M$ with the "Rademacher property"

$$\text{dil}_\rho u \leq 1 \quad \Longleftrightarrow \quad u \in \mathcal{D}(\Gamma), \Gamma(u) \leq 1 \ m-\text{a.e.}$$

where Γ denotes the square field operator associated with $(p_t)_{t>0}$. Then

$$p_t^\diamond \rho(x,y) \leq e^{-\kappa t} \cdot \rho(x,y).$$

For instance, this applies to the Ornstein-Uhlenbeck semigroup on the Wiener space $M = \mathcal{C}(\mathbb{R}_+, \mathbb{R}^n)$. Here m=Wiener measure, ρ=Cameron-Martin distance, $\kappa = 1$ and $n = \infty$.

5. L_θ-Contraction Properties

In the previous Chapters, we have presented the pointwise approach to nonlinear Markov operators and nonlinear heat semigroups. In this Chapter, we present the L_θ-approach. As before (N, d, b) will be a complete metric space with barycenter contraction and $(M, \mathcal{M})$ will be a measurable space with a Markov semigroup $(p_t)_{t>0}$ on it. In addition, we now fix a measure m on $(M, \mathcal{M})$ and a number $\theta \in [1, \infty]$ and we assume that $\exists C, \alpha \in \mathbb{R}$:

$$\|p_t u\|_\theta \leq C \cdot e^{\alpha t} \cdot \|u\|_\theta \tag{12}$$

for all $u \in L_\theta(M)$ and $t > 0$. Here $L_\theta(M) = L_\theta(M, \mathcal{M}, m)$ denotes the Lebesgue space of (m-equivalence classes of) measurable functions $u : M \to \mathbb{R}$ with norm $\|u\|_\theta < \infty$.

Example 5.1. (i) Each semigroup $(p_t)_{t>0}$ derived from a symmetric Dirichlet form on $L_2(M, \mathcal{M}, m)$ satisfies $\|p_t u\|_\theta \le \|u\|_\theta$ for each $\theta \in [1, \infty]$. For $\theta = 2$ we even obtain

$$\|p_t u\|_2 \le e^{\alpha t} \cdot \|u\|_2$$

with $\alpha = \sup \operatorname{spec}(A) = -\inf_{u \in L_2(M)} \frac{\mathcal{E}(u)}{\|u\|_2^2} \le 0$ being the top of the L_2-spectrum of the generator $A = \lim_{t \to 0} \frac{1}{t}(p_t - 1)$ of the Dirichlet form.

(ii) Let $(p_t)_{t>0}$ be any Markov semigroup on a measurable space $(M, \mathcal{M})$. Choose $m := \sum_{x \in M} \delta_x$ to be the counting measure. Then for all measurable $u : M \to \mathbb{R}$ and all $t > 0$

$$\|p_t u\|_\infty \le \|u\|_\infty.$$

For measurable $f, g : M \to N$ put $d_\theta(f, g) := \|d(f, g)\|_\theta$ where $d(f, g)$ denotes the function $x \mapsto d(f(x), g(x))$ on M. In particular, for $\theta < \infty$

$$d_\theta(f, g) = \left(\int_M d^\theta(f(x), g(x))\, m(dx) \right)^{1/\theta}.$$

Let $L_\theta(M, N, p)$ denote the set of (equivalence classes of) measurable $f : M \to N$ with separable ranges and with $\partial_t f \in L_\theta(M)$ for all $t > 0$. One easily verifies that $(L_\theta(M, N, p), d_\theta)$ is a complete metric space and each constant map lies in $L_\theta(M, N, p)$. If $f \in L_\theta(M, N)$ and $t > 0$ then $\partial_t f(x) < \infty$ for m-a.e. $x \in M$. Hence, $P_t f(x)$ is well-defined for m-a.e. $x \in M$ and $P_t f$ defines an m-equivalence class of measurable maps with separable ranges. Moreover,

- for each measurable $g : M \to N$ with separable range $f \in L_\theta(M, N, p),\ d_\theta(f, g) < \infty \Longrightarrow g \in L_\theta(M, N, p)$;
- $f \in L_\theta(M, N, p) \Longrightarrow P_t f \in L_\theta(M, N, p)$ and $d_\theta(f, P_t f) \le \|\partial_t f\|_\theta$.

Let $L_\theta^*(M, N, p)$ denote the set of $f \in L_\theta(M, N, p)$ for which the d_θ-limit

$$P_t^* f := \lim_{n \to \infty} (P_{\delta_n})^{t/\delta_n} f$$

exists for all $t \in \mathbb{T}$.

Example 5.2. Let $N = \mathbb{R}$ (with the usual d and b) and let $(p_t)_{t>0}$ be the heat semigroup on $M = \mathbb{R}^1$ (with m being Lebesgue measure). Then $L_\theta(M, N, p) = L_\theta^*(M, N, p) \supset L_\theta(M)$ with strict inclusion. Indeed, consider the function $f(x) = (1 + |x|)^\alpha$. Then for $\theta < \infty$

$$f \in L_\theta(M) \Longleftrightarrow \alpha < -1/\theta \quad \text{and} \quad f \in L_\theta(M, N, p) \Longleftrightarrow \alpha < 1 - 1/\theta$$

(since $\partial_t f(x) \approx C \cdot \alpha \cdot \sqrt{t} \cdot |x|^{\alpha - 1}$ for large x). Similarly,

$$f \in L_\infty(M) \Longleftrightarrow \alpha \le 0 \quad \text{and} \quad f \in L_\infty(M, N, p) \Longleftrightarrow \alpha \le 1.$$

With exactly the same arguments as for Lemma 3.2 we deduce

Proposition 5.3. *For all $f, g \in L_\theta^*(M, N, p)$ and all $s, t \in \mathbb{T}$:*

(i) $P_t^* f \in L_\theta^*(M, N, p)$ and $P_s^*(P_t^* f) = P_{s+t}^* f$;
(ii) $d_\theta(P_t^* f, P_t^* g) \le C \cdot e^{\alpha t} \cdot d_\theta(f, g)$;
(iii) $d_\theta(P_t^* f, P_{t+s}^* f) \le C \cdot e^{\alpha t} \cdot \|\partial_s f\|_\theta$.

Remark 5.4. (i) The set $L^*_\theta(M,N,p)$ is closed w.r.t. d_θ.
(ii) For all $f \in L^*_\theta(M,N,p)$ with $\lim_{t\to 0} \|\partial_t f\|_\theta = 0$ the map $t \mapsto P^*_t f$ is continuous in $t \in \mathbb{T}$ (according to 5.3(iii)) and thus

$$P^*_t f = \lim_{\mathbb{T}\ni s\to t} P^*_s f = \lim_{n\to\infty} (P_{\delta_n})^{\lfloor t/\delta_n \rfloor} f \ \in L^*_\theta(M,N,p)$$

is well-defined for all $t > 0$. Here $\lfloor s \rfloor$ denotes the integer part of $s \in \mathbb{R}$.
(iii) For each $\theta < \infty$

$$L^*_\theta(M,N,p) \supset L_\theta(M,N,p) \cap \mathcal{L}^*(M,N,p).$$

Indeed, $f \in L_\theta(M,N,p) \cap \mathcal{L}^*(M,N,p)$ implies $u_n := d(P^{t/\delta_n}_{\delta_n} f, P^*_t f) \longrightarrow 0$ pointwise on M for $n \to \infty$ and $u_n \le 2\partial_t f \in L_\theta(M)$ for all $\delta_n \le t$. Hence, by Lebesgue's dominated convergence theorem $u_n \to 0$ in $L_\theta(M)$ and thus $P^{t/\delta_n}_{\delta_n} f \to P^*_t f$ in $L_\theta(M,N,p)$.
(iv) Assume that $p_t(x,.) \ll m$ for all $x \in M$ and all $t \in \mathbb{T}$ ("absolute continuity of p_t"). Then $f(x) = g(x)$ for m-a.e. $x \in M$ implies $P^*_t f(x) = P^*_t g(x)$ for all $x \in M$ and $t \in \mathbb{T}$. In particular, for each $f \in L^*_\theta(M,N,p)$ and each $t \in \mathbb{T}$ the map $P^*_t f$ is pointwise well-defined on M.

Theorem 5.5. *For all $t \in \mathbb{T}$ and all $\theta' \in [1,\infty]$ with $\|p_t\|_{\theta,\theta'} := \sup_{u\in L_\theta(M)} \frac{\|p_t u\|_{\theta'}}{\|u\|_\theta} < \infty$*

$$d_{\theta'}(P^*_t f, P^*_t g) \le \|p_t\|_{\theta,\theta'} \cdot d_\theta(f,g). \qquad (\forall f,g \in L^*_\theta(M,N,p))$$

As an immediate corollary we deduce that if the linear semigroup $(p_t)_{t>0}$ acting on $L_\theta(M), 1 \le \theta \le \infty$, is hyper-,ultra- or supercontractive then so is the nonlinear semigroup $(P^*_t)_{t\in\mathbb{T}}$ acting on $L_\theta(M,N,p), 1 \le \theta \le \infty$. We quote the following main examples.

Corollary 5.6. *Given $t \in \mathbb{T}$, assume that the Markov kernel p_t has a bounded density $k_t(x,y) := \frac{p_t(x,dy)}{m(dy)} \le C_t$. Then for all $1 \le \theta \le \theta' \le \infty$*

$$d_{\theta'}(P^*_t f, P^*_t g) \le C_t^{1/\theta - 1/\theta'} \cdot d_\theta(f,g). \qquad (\forall f,g \in L^*_\theta(M,N,p))$$

Corollary 5.7. *Let the Markov semigroup $(p_t)_t$ be associated with a symmetric Dirichlet form $(\mathcal{E},\mathcal{D}(\mathcal{E}))$ on $L_2(M)$.*

(i) Assume either that a "Nash inequality" holds true for $\mu > 0$ (with constants C_0, C_1):

$$\|u\|_2^{2+4/\mu} \le \left[C_0 \cdot \mathcal{E}(u) + C_1 \cdot \|u\|^2\right] \cdot \|u\|_1^{4/\mu} \qquad (\forall u \in \mathcal{D}(\mathcal{E}))$$

or that a "Sobolev inequality" holds true for $\mu > 2$ (with constants C_0, C_1):

$$\|u\|^2_{2\mu/(\mu-2)} \le C_0 \cdot \mathcal{E}(u) + C_1 \cdot \|u\|^2. \qquad (\forall u \in \mathcal{D}(\mathcal{E}))$$

Then for some constant C, all $t \in \mathbb{T}, t \le 1$ (or even all $t \in \mathbb{T}$ if $C_1 = 0$) and all $1 \le \theta \le \theta' \le \infty$

$$d_{\theta'}(P^*_t f, P^*_t g) \le C \cdot t^{-\frac{\mu}{2}(\frac{1}{\theta} - \frac{1}{\theta'})} \cdot d_\theta(f,g). \qquad (\forall f,g \in L^*_\theta(M,N,p))$$

(ii) Assume that m is a probability measure and that a "logarithmic Sobolev inequality" with constant $\nu > 0$ holds true for all $u \in \mathcal{D}(\mathcal{E})$ with $\|u\|_2 = 1$

$$\int_M u^2 \log u^2 \, dm \leq \frac{2}{\nu} \cdot \mathcal{E}(u).$$

Then for all $t \in \mathbb{T}$ and all $1 < \theta < \theta' < \infty$ with $\frac{\theta'-1}{\theta-1} < e^{2\nu t}$

$$d_{\theta'}(P_t^* f, P_t^* g) \leq d_\theta(f, g). \qquad (\forall f, g \in L_\theta^*(M, N, p))$$

Cf. DAVIES (1989), LEDOUX (1998).

Theorem 5.8. *Assume that $\alpha < 0$ in (12). Then for each $f \in L_\theta^*(M, N, p)$ there exists a unique $h \in L_\theta^*(M, N, p)$ with $d_\theta(h, f) < \infty$ and*

$$P_t^* h = h \tag{13}$$

for all $t \in \mathbb{T}$. Indeed, $h = \lim_{\mathbb{T} \ni t \to \infty} P_t^* f$ and for $t \to \infty$

$$d_\theta(h, P_t^* f) \leq C \cdot e^{\alpha t} \cdot d_\theta(h, f) \longrightarrow 0.$$

The above result may be used to deduce existence and uniqueness of the solution to the Dirichlet problem. Namely, given a Markov semigroup $(p_t)_{t>0}$ on a complete separable metric space M and a bounded open subset $D \subset M$, let $(p_{D,t})_{t>0}$ be the stopped semigroup as introduced in Chapter 1 and replace the measure $m(dx)$ by $(1_D(x) + \infty \cdot 1_{M \setminus D}(x)) m(dx)$. Then in most examples (due to the boundedness of D)

$$\|p_{D,t} u\|_\theta \leq C \cdot e^{\alpha_D t} \cdot \|u\|_\theta \tag{14}$$

with some $\alpha_D < 0$. Hence, for each $f \in L_\theta^*(M, N, p)$ there exists a unique $h \in L_\theta^*(M, N, p)$ with $d_\theta(h, f) < \infty$ and $h = f$ m-a.e. on $M \setminus D$ ("boundary condition" or more precisely "complement condition") and $P_{D,t}^* h = h$ for all $t \in \mathbb{T}$, in particular,

$$\lim_{\mathbb{T} \ni t \to 0} \frac{1}{t} d(P_{D,t}^* h, h) = 0$$

("harmonicity in D").

For further details we refer to the forthcoming paper STURM (2002B).

References

W. BALLMANN (1995), *Lectures on spaces of nonpositive curvature.* DMV Seminar Band **25**, Birkhäuser.

H. BAUER (1996), *Probability theory.* de Gruyter Studies in Mathematics, **23**, Walter de Gruyter & Co., Berlin.

A. BENDIKOV (1995), *Potential theory on infinite-dimensional Abelian groups.* de Gruyter Studies in Mathematics, **21.**, Walter de Gruyter,Berlin.

A. BENDIKOV, L. SALOFF-COSTE (2001), *On the absolute continuity of Gaussian measures on locally compact groups.* J. Theor. Probab. **14**, No.3, 887–898.

J. EELLS, B. FUGLEDE (2001), *Harmonic maps between Riemannian polyhedra.* Cambridge Tracts in Mathematics, **142.**, Cambridge University Press, Cambridge.

M. EMERY (1989), *Stochastic calculus in manifolds.* Universitext. Springer, Berlin.

A. ES-SAHIB, H. HEINICH (1999), *Barycentre canonique pour un espace métrique à courbure négative.* Séminaire de Probabilités, XXXIII, 355–370, Lecture Notes in Math., **1709**, Springer, Berlin.

S.N. ETHIER, T.G. KURTZ (1986), *Markov processes. Characterization and convergence.* Wiley Series in Probability and Mathematical Statistics: Probability and Mathematical Statistics, John Wiley, New York.

C. FEFFERMAN, D.H. PHONG (1983), *Subelliptic eigenvalue problems.* Conference on harmonic analysis, Chicago, Wadsworth Math. Ser., 590–606.

A. GRIGORYAN (1999), *Estimates of heat kernels on Riemannian manifolds.* Spectral theory and geometry, 140–225, London Math. Soc. Lecture Note Ser., **273**, Cambridge Univ. Press.

N. JACOB (1996), *Pseudo-differential operators and Markov processes.* Mathematical Research, **94**, Akademie Verlag, Berlin.

N. JACOB (2001), *Pseudo differential operators and Markov processes. Vol. I. Fourier analysis and semigroups.* Imperial College Press, London.

D.S. JERISON, A. SANCHEZ-CALLE (1986), *Estimates for the heat kernel for a sum of squares of vector fields.* Indiana Univ. Math. J. **35** no. 4, 835–854.

J. JOST (1994), *Equilibrium maps between metric spaces.* Calc. Var. Partial Differential Equations 2, **2**, 173–204.

J. JOST (1997), *Nonpositive curvature: geometric and analytic aspects.* Lectures in Mathematics, ETH Zürich, Birkhäuser, Basel.

J. JOST (1997), *Generalized Dirichlet forms and harmonic maps.* Calc. Var. Partial Differential Equations **5**, no. 1, 1–19.

W.S. KENDALL (1990), *Probability, convexity, and harmonic maps with small images. I. Uniqueness and fine existence.* Proc. London Math. Soc. **61**, 371–406.

N. KOREVAAR, R. SCHOEN (1993), *Sobolev spaces and harmonic maps for metric space targets.* Comm. Anal. Geom. **1**, 561–569.

N. KOREVAAR, R. SCHOEN (1997), *Global existence theorems for harmonic maps to non-locally compact spaces.* Comm. Anal. Geom. **5**, no. 2, 333–387.

M. LEDOUX (2000), *The geometry of Markov diffusion generators.* Probability theory. Ann. Fac. Sci. Toulouse Math. (6) 9 (2000), no. 2, 305–366

M. LEDOUX, TALAGRAND (1991), *Probability in Banach spaces. Isoperimetry and processes.* Ergebnisse der Mathematik und ihrer Grenzgebiete (3), **23**, Springer, Berlin.

Z.M. MA, M. RÖCKNER (1992), *Introduction to the theory of (nonsymmetric) Dirichlet forms.* Universitext, Springer, Berlin.

S.T. RACHEV, L. RÜSCHENDORF (1998), *Mass transportation problems. Vol. I. Theory.* Probability and its Applications, Springer, New York.

M.-K. VON RENESSE (2002), *Comparison properties of diffusion semigroups on spaces with lower curvature bounds.* PhD Thesis, Bonn.

D.W. STROOCK, S.R.S. VARADHAN (1979), *Multidimensional diffusion processes.* Grundlehren der Mathematischen Wissenschaften **233**, Springer-Verlag, Berlin-New York.

K.T. STURM (2001), *Nonlinear Markov operators associated with symmetric Markov kernels and energy minimizing maps between singular spaces.* Calc. Var. **12**, 317–357

K.T. STURM (2002), *Nonlinear martingale theory for processes with values in metric spaces of nonpositive curvature.* To appear in Ann. Prob.

K.T. STURM (2002A), *Nonlinear Markov operators, discrete heat flow, and harmonic maps between singular spaces.* To appear in Potential Analysis.

K.T. STURM (2002B), *A semigroup approach to harmonic maps.* Preprint, Bonn.

K. TAIRA (1991), *Boundary value problems and Markov processes.* Lecture Notes in Mathematics, 1499, Springer-Verlag, Berlin.

A. THALMAIER (1996), *Martingales on Riemannian manifolds and the nonlinear heat equation.* Stochastic analysis and applications, 429–440, World Sci. Publishing, River Edge, NJ.

A. THALMAIER (1996A), *Brownian motion and the formation of singularities in the heat flow for harmonic maps.* Probab. Theory Related Fields 105, no. 3, 335–367.

Karl-Theodor Sturm
Institut für Angewandte Mathematik
Universität Bonn
Wegelerstrasse 6
D-53115 Bonn, Germany
E-mail address: sturm@uni.bonn.de

Progress in Nonlinear Differential Equations
and Their Applications, Vol. 55, 375–401

The Hamilton-Jacobi-Bellman Equation and the Stochastic Noether Theorem

Jan A. van Casteren

1. Introduction

It is pointed out that Zambrini and coworkers [24, 1, 19, 20, 5] have kind of a transition scheme to go from classical stochastic calculus (with non-reversible processes) to physical real time (reversible) quantum mechanics and vice versa. An important tool in this connection is the so-called Noether theorem. In fact, in Zambrini's words, reference [24] contains the first concrete application of this theorem. In [24] the author formulates a theorem like Theorem 1.1 below, he also uses so-called "Bernstein diffusions" (see, e.g., [7]) for the "Euclidean Born interpretation" of quantum mechanics. The Bernstein diffusions are related to solutions of $\left(\frac{\partial}{\partial t} - (K_0 \dot{+} V)\right) \eta = 0$, and of $\left(\frac{\partial}{\partial t} + K_0 \dot{+} V\right) \eta^* = 0$. It is noted that Aubin in [2] employs epigraph methods rather than viscosity solutions as discussed in Crandall, Ishii, and Lions in [6]. Aubin treats HJB set-valued inclusions instead of HJB equations. In his work set-valued analysis is quite relevant. Indeed, Aubin's solutions are more general than viscosity solutions. In the present paper we prove a version of the stochastic Noether theorem in terms of the carré du champ operator and ideas from stochastic control: see Theorem 3.1, which should be compared with Theorem 2.4 in [24]. It is a substantial extension of the results in [21]. The operator K_0 generates a *diffusion* in the following sense: for every C^∞-function $\Phi : \mathbb{R}^n \to \mathbb{R}$, with $\Phi(0, \dots, 0) = 0$, the following identity is valid:

$$\begin{aligned} &K_0 \left(\Phi(f_1, \dots, f_n)\right) \qquad (1.1) \\ &= \sum_{j=1}^{n} \frac{\partial \Phi}{\partial x_j} (f_1, \dots, f_n) \, K_0 f_j - \frac{1}{2} \sum_{j,k=1}^{n} \frac{\partial^2 \Phi}{\partial x_j \partial x_k} (f_1, \dots, f_n) \, \Gamma_1(f_j, f_k) \end{aligned}$$

for all functions $f_1, \dots, f_n$ in a rich enough algebra of functions $\mathcal{A}$, contained in the domain of the generator K_0, as described below. The algebra $\mathcal{A}$ has to be "large" enough. To be specific, we assume that there exists a reference measure m on the Borel field $\mathcal{E}$ of E, and that $\mathcal{A}$ possesses the following properties (cfr. Bakry [3]). Its domain is dense in $L^p(E, m)$ for all $1 \le p < \infty$ and it is a core for K_0 considered as an operator in these spaces. The same is true for the space

$C_0(E)$. In addition, it is assumed that $\mathcal{A}$ is stable under composition with C^∞-functions of several variables, that vanish at the origin. Moreover, in order to obtain some nice results a rather technical condition is required: whenever $(f_n : n \in \mathbb{N})$ is a sequence in $\mathcal{A}$ that converges to f with respect to the graph norm of K_0 (in $L^2(E, m)$) and whenever $\Phi : \mathbb{R} \to \mathbb{R}$ is a C^∞-function, vanishing at 0, with bounded derivatives of all orders (including the order 0), then one may extract a subsequence $(\Phi(f_{n_k}) : k \in \mathbb{N})$ that converges to $\Phi(f)$ in $C_0(E)$, whereas the sequence $(K_0\Phi(f_{n_k}) : k \in \mathbb{N})$ converges in $C_0(E)$ and also in $L^1(E, m)$ to $K_0\Phi(f)$. Notice that all functions of the form $e^\psi f$, ψ, $f \in \mathcal{A}$, belong to $\mathcal{A}$. Also notice that the required properties of $\mathcal{A}$ depend on the generator K_0. In fact we will assume that the algebra $\mathcal{A}$ is also large enough for all operators of the form $f \mapsto e^{-\psi}K_0(e^\psi f)$, where ψ belongs to $\mathcal{A}$. Sometimes it is quite natural to assume that $K_0 1 = 0$. If this is is the case, then it is not necessary to assume $\Phi(0, \ldots, 0) = 0$ in the definition of diffusion. For the notion of the squared gradient operator (carré du champ opérateur) see equality (1.4). The operator K_0 acts on the first variable and so does the squared gradient operator Γ_1. The symbol D_1 stands for the operator $D_1 = \dfrac{\partial}{\partial t}$. Fix $u > t_0 \geq 0$. We work in the continuous function space like $C_0(E \times (t_0, u])$ and also in the space $C(E \times (t_0, u])$. If we write $D(D_1 - K_0)$ for the domain of the operator $D_1 - K_0$, then the corresponding space should be specified. In fact the space $C_0(E \times (t_0, u])$ is endowed with the topology of uniform convergence and the operator $D_1 - K_0$ is considered as the generator of the semigroup $S(s)\exp(-sK_0) f(x,t) = \exp(-sK_0) f(\cdot, t+s)(x)$. The operator $S(s)$ is uniform motion to the right. The process

$$\{(\Omega, \mathcal{F}, \mathbb{P}_x), (X(t) : t \geq 0), (\vartheta_t : t \geq 0), (E, \mathcal{E})\}$$

is the strong Markov process generated by $-K_0$; it is supposed to have continuous paths. In the space $C(E)$ the operator K_0 is considered as a local operator in the sense that a function $f \in C(E)$ belongs to its domain if there exists a function $g \in C(E)$ such that for every open subset U of E together with every compact subset K of U we have

$$\lim_{t \downarrow 0} \sup_{x \in K} \left| g(x) - \frac{f(x) - \mathbb{E}_x\left[f(X(t)) : \tau_U > t\right]}{t} \right| = 0.$$

Here τ_U is the first exit time from U: $\tau_U = \inf\{t > 0 : X(t) \in E \setminus U\}$. We write $g = -K_0 f$. From Proposition 1.6 in [9] page 9 it follows that the constant function 1 belongs to the domain of K_0 and that $K_0 1 = 0$. In a more classical context in, e.g., L^p-spaces the operator K_0 can often be considered as a differential operator in "distributional" sense. It is noticed that there exists a close relationship between the viscous Burgers' equation (in an open subset of $\mathbb{R}^d$)

$$-\frac{\partial U}{\partial t} + U \cdot \nabla U - \frac{1}{2}\Delta U = \nabla V,$$

and the Hamilton-Jacobi-Bellman equation. If we write the vector field U in the form $U = \nabla\varphi$, then the function φ satisfies

$$-\frac{\partial\varphi}{\partial t} + \frac{1}{2}\nabla\varphi\cdot\nabla\varphi - \frac{1}{2}\Delta\varphi = V + \text{ constant.}$$

In the sequel we will mainly be concerned with the Hamilton-Jacobi-Bellman equation as exhibited in (1.2). We have the following result for generators of diffusions: it refines Theorem 2.4 in Zambrini [24].

Theorem 1.1. *Let $\chi : E \times (t_0, u] \to [0, \infty]$ be a function such that*

$$\mathbb{E}_{x,t}^{M_{v,t}}\left[|\log\chi\left(X(u), u\right)|\right], \quad v \in D\left(D_1 - K_0\right)$$

is finite for $t_0 < t \le u$. Here $u > t_0 \ge 0$ are fixed times. Let S_L be a (classical) solution to the following Riccati type equation. This equation is called the Hamilton-Jacobi-Bellman equation. For $t_0 < s \le u$ and $x \in E$ the following identity is true:

$$\begin{cases} -\dfrac{\partial S_L}{\partial s}(x,s) + \dfrac{1}{2}\Gamma_1\left(S_L, S_L\right)(x,s) + K_0 S_L(x,s) - V(x,s) = 0; \\ S_L(x,u) = -\log\chi(x,u), \quad x \in E. \end{cases} \tag{1.2}$$

Then for any real valued $v \in D\left(D_1 - K_0\right)$ the following inequality is valid:

$$S_L(x,t) \le \mathbb{E}_{x,t}^{M_{v,t}}\left[\int_t^u \left(\frac{1}{2}\Gamma_1\left(v,v\right) + V\right)\left(X(\tau),\tau\right)d\tau\right] - \mathbb{E}_{x,t}^{M_{v,t}}\left[\log\chi\left(X(u),u\right)\right], \tag{1.3}$$

and equality is attained for the "Lagrangian action" $v = S_L$.

By definition $E_{x,t}\left[Y\right]$ is the expectation, conditioned at $X(t) = x$, of the random variable Y which is measurable with respect to the information from the future: i.e., with respect to $\sigma\left\{X(s) : s \ge t\right\}$. The measure $\mathbb{P}_{x,t}^{M_{v,t}}$ is defined in equality (1.5) below. Put $\eta_\chi = \exp\left(-S_L\right)$, where S_L satisfies (1.2). From (1.1) it follows that $\left(\dfrac{\partial}{\partial t} - \left(K_0 \dot{+} V\right)\right)\eta_\chi = 0$, provided that $K_0 1$ is interpreted as 0, i.e., $\int K_0 f dm = 0$ for all functions $f \in D\left(K_0\right)$, $K_0 f \in L^1\left(E, m\right)$.

Fix a function $v : E \times (t_0, u] \to \mathbb{R}$ in $D\left(D_1 - K_0\right)$, where, as above, $D_1 = \dfrac{\partial}{\partial t}$ is differentiation with respect to t. Let the process

$$\left\{\left(\Omega, \mathcal{F}, \mathbb{P}_{x,t}\right), \left(\left(q_v(t), t\right) : t \ge 0\right), \left(\vartheta_t : t \ge 0\right), \left(E \times \mathbb{R}, \mathcal{E} \otimes \mathcal{B}\right)\right\}$$

be the Markov process generated by the operator $-K_v + D_1$, where K_v is defined by $K_v(f)(x,t) = K_0 f(x,t) + \Gamma_1\left(v, f\right)(x,t)$. Here, $\mathcal{B}$ denotes the Borel field of $\mathbb{R}$, and by $\Gamma_1(v,f)(x,t)$ we mean (notice that the operator Γ_1 only acts on the space (or state) variable, not on the time variable)

$$\Gamma_1(v,f)(x,t) = \lim_{s\downarrow 0}\frac{1}{s}\mathbb{E}_x\left[\left(v(X(s),t) - v(X(0),t)\right)\left(f(X(s),t) - f(X(0),t)\right)\right]. \tag{1.4}$$

We also believe that the following version of the Cameron-Martin formula is true. For all finite n-tuples $t_1, \ldots, t_n$ in $(0, \infty)$ the identity (1.6) is valid:

$$\mathbb{E}_{x,t}^{M_{v,t}}\left[\prod_{j=1}^{n} f_j\left(X\left(t_j+t\right), t_j+t\right)\right] \tag{1.5}$$

$$= \mathbb{E}_{x,t}\left[\exp\left(-\frac{1}{2}\int_t^u \Gamma_1(v,v)(X(\tau),\tau)\,d\tau - M_{v,t}(u)\right)\prod_{j=1}^{n} f_j\left(X\left(t_j+t\right), t_j+t\right)\right]$$

$$= \mathbb{E}_{x,t}\left[\prod_{j=1}^{n} f_j\left(q_v\left(t_j+t\right), t_j+t\right)\right] \tag{1.6}$$

where the $\mathbb{E}_{x,t}$-martingale $M_{v,t}(s)$, $s \geq t$, is given by

$$M_{v,t}(s) = v(X(s),s) - v(X(t),t) + \int_t^s \left(-\frac{\partial}{\partial\tau} + K_0\right) v(X(\tau),\tau)\,d\tau. \tag{1.7}$$

Its quadratic variation part $\langle M_{v,t}\rangle(s) := \langle M_{v,t}, M_{v,t}\rangle(s)$ is given by

$$\langle M_{v,t}\rangle(s) = \int_t^s \Gamma_1(v,v)(X(\tau),\tau)\,d\tau. \tag{1.8}$$

The equality in (1.5) serves as a definition of the measure $\mathbb{P}_{x,t}^{M_{v,t}}(\cdot)$, and the equality in (1.6) is a statement. The proof of Theorem 1.1 can be found in [22].

Remark. The equation in (1.2) can be phrased in a semi-linear setting as follows. Find a function $S_L(x,t)$ which satisfies the following partial differential equation:

$$\begin{cases} -\dfrac{\partial S_L}{\partial s}(x,s) + \dfrac{1}{2} g\left(x, s, S_L(x,s), \nabla S_L(\cdot,s)(x)\right)\Gamma_1\left(S_L, S_L\right)(x,s) \\ +K_0 S_L(x,s) - V\left(x, s, S_L(x,s), \nabla S_L(\cdot,s)(x)\right) = 0; \\ S_L(x,u) = -\log\chi(x,u), \quad x \in E. \end{cases} \tag{1.9}$$

Here ∇f_2 is the mapping $f_1 \mapsto \Gamma_1(f_1, f_2)$ for smooth enough functions f_1 and f_2, and $\nabla f_2(\cdot)(x)$ is the linear functional $f_1 \mapsto \Gamma_1(f_1, f_2)(x)$. One way to treat this kind of equation is considering the following backward problem. Find adapted processes $Y(s)$ and $Z(s)$, with $Z(s)(Y_1(\cdot), Y_2(\cdot)) = \dfrac{d}{ds}\langle Y_1(\cdot), Y_2(\cdot)\rangle(s)$, satisfying:

$$Y(t) - Y(u) - \int_t^u V(X(s), s, Y(s), Z(s)(\cdot, Y))\,ds$$

$$+\frac{1}{2}\int_t^u g(X(s), s, Y(s), Z(s)(\cdot, Y))\,Z(s)(Y,Y)\,ds = M(t) - M(u), \tag{1.10}$$

where $M(s)$, $t_0 < t \leq s \leq u$, is a forward local $\mathbb{P}_{x,t}$-martingale (for every $u > t > t_0$). If $Y(s) = S_L(X(s), s)$, S_L is "smooth", and $S_L(x,t)$ satisfies (1.9), then $Y(s)$

satisfies (1.10), and vice versa. If $g(x,s,y,z)$ only depends on $y \in \mathbb{R}$, then, by the occupation formula,

$$\int_t^u g(Y(s))\, Z(s)\, (Y,Y)\, ds = \int_t^u g(Y(s))\, d\left(Y(\cdot), Y(\cdot)\right)(s)$$
$$= \int_{\mathbb{R}} \left(L_u^y(Y) - L_t^y(Y)\right) g(y) dy,$$

where dy is Lebesgue measure, and $L_t^y(Y)$ is the (density of) local time of the process $Y(t)$. If $g \equiv 1$ and $Y(s) = S_L(X(s), s)$, then (1.10) is also equivalent to the following assertion: the process $s \mapsto \exp\left(Y(s) - Y(u) - \int_s^u V(\tau)d\tau\right)$, $t_0 < t \leq s \leq u$, is a local backward (exponential) $\mathbb{P}_{x,t}$-martingale (for every $u > t > t_0$). Here we wrote $V(s) = V\left(X(s), s, Y(s), Z(s)(\cdot, Y)\right)$. The functions V and g depend on $x \in E$, $s \in (t_0, u]$, $y \in \mathbb{R}$, and on the squared gradient operator $(f_1, f_2) \mapsto \Gamma_1(f_1, f_2)$, or, more generally, on the covariance mapping $(Y_1, Y_2) \mapsto \langle Y_1, Y_2\rangle(s)$ of the local semi-martingales $Y_1(s)$ and $Y_2(s)$. In order to introduce boundary conditions it is required to insert in equation (1.9) a term of the form

$$\int_t^u h\left(X(s), s, Y(s), Z(s)(\cdot, Y)\right) dA(s),$$

where $A(s)$ is a process which is locally of bounded variation, and which only increases when, e.g., $X(s)$ hits the boundary. We hope to come back on this and similar problems in a later paper. In order to be sure about uniqueness and existence of solutions we probably will need some Lipschitz and linear growth conditions on the function V and some boundedness condition on g. For more details on backward stochastic differential equations see, e.g., Pardoux [15] and [16].

2. The Hamilton-Jacobi-Bellman equation and viscosity solutions

Let $\chi : E \times [t,u] \to [0,\infty]$ be as in Theorem 1.1. In what follows we write $D_1 = \dfrac{\partial}{\partial t}$. We also write $D_1\varphi = \dot{\varphi}$. What is the relationship between the following expressions?

$$\sup_{\Phi \in D(D_1 - K_0)} \left\{\Phi(x,t) :\ -\dot{\Phi} + K_0\Phi + \frac{1}{2}\Gamma_1(\Phi,\Phi) \leq V,\ \Phi(\cdot, u) \leq -\log\chi(\cdot, u)\right\}; \tag{2.1}$$

$$-\log \mathbb{E}_{x,t}\left[\exp\left(-\int_t^u V(X(\sigma),\sigma)\, d\sigma\right) \chi(X(u), u)\right]; \tag{2.2}$$

$$\inf_{v \in D(D_1 - K_0)} \left\{\mathbb{E}_{x,t}^{M_v,t}\left[\int_t^u \left(\frac{1}{2}\Gamma_1(v,v) + V\right)(X(\tau),\tau) d\tau\right] - \mathbb{E}_{x,t}^{M_v,t}\left[\log\chi(X(u),u)\right]\right\}; \tag{2.3}$$

$$\inf_{\Phi\in D(D_1-K_0)}\left\{\Phi(x,t):\ -\dot{\Phi}+K_0\Phi+\frac{1}{2}\Gamma_1\left(\Phi,\Phi\right)\geq V,\ \Phi(\cdot,u)\geq-\log\chi\left(\cdot,u\right)\right\}. \tag{2.4}$$

A somewhat more general result than Theorem 1.1 reads as follows. As above, we work in the space $C_0\left(E\times(t_0,u]\right)$, where $u>t_0$ is fixed.

Theorem 2.1. (i) *The following inequalities are valid:*

$$\begin{aligned}&\sup_{\Phi\in D(D_1-K_0)}\left\{\Phi(x,t):\ -\dot{\Phi}+K_0\Phi+\frac{1}{2}\Gamma_1\left(\Phi,\Phi\right)\leq V,\ \Phi(\cdot,u)\leq-\log\chi\left(\cdot,u\right)\right\}\\ &\leq-\log\mathbb{E}_{x,t}\left[\exp\left(-\int_t^u V\left(X(\sigma),\sigma\right)d\sigma\right)\chi\left(X(u),u\right)\right]\\ &\leq\inf_{v\in D(D_1-K_0)}\left\{\mathbb{E}_{x,t}^{M_{v,t}}\left[\int_t^u\left(\frac{1}{2}\Gamma_1\left(v,v\right)+V\right)\left(X(\tau),\tau\right)d\tau\right]\right.\\ &\qquad\left.-\mathbb{E}_{x,t}^{M_{v,t}}\left[\log\chi\left(X(u),u\right)\right]\right\}\\ &\leq\inf_{\Phi\in D(D_1-K_0)}\left\{\Phi(x,t):\ -\dot{\Phi}+K_0\Phi+\frac{1}{2}\Gamma_1\left(\Phi,\Phi\right)\geq V,\ \Phi(\cdot,u)\geq-\log\chi\left(\cdot,u\right)\right\}.\end{aligned}$$

(ii) *If the function S_L defined by the non-linear Feynman-Kac formula*

$$S_L(x,t)=-\log\mathbb{E}_{x,t}\left[\exp\left(-\int_t^u V\left(X(\sigma),\sigma\right)d\sigma\right)\chi\left(X(u),u\right)\right] \tag{2.5}$$

belongs to $D\left(D_1-K_0\right)$, then the above 4 quantities are equal. Moreover the function S_L satisfies the Hamilton-Jacobi-Bellman equation (1.2). *The same is true if the expressions in* (2.1) *and in* (2.4) *are equal.*

(iii) *In general the function in* (2.5) *is a viscosity solution of the Hamilton-Jacobi-Bellman equation* (1.2). *This means that if (x,t) belongs to $E\times(t_0,u]$ and if $\varphi\in D\left(D_1-K_0\right)$ has the property that*

$$\left[S_L-\varphi\right](x,t)=\sup\left\{\left[S_L-\varphi\right](y,s):(y,s)\in E\times[t,u]\right\},$$

then

$$\left[-\dot{\varphi}+K_0\varphi\right](x,t)+\frac{1}{2}\Gamma_1\left(\varphi,\varphi\right)(x,t)\leq V(x,t). \tag{2.6}$$

It also means that if (x,t) belongs to $E\times(t_0,u]$ and if $\varphi\in D\left(D_1-K_0\right)$ has the property that

$$\left[S_L-\varphi\right](x,t)=\inf\left\{\left[S_L-\varphi\right](y,s):(y,s)\in E\times[t,u]\right\},$$

then

$$\left[-\dot{\varphi}+K_0\varphi\right](x,t)+\frac{1}{2}\Gamma_1\left(\varphi,\varphi\right)(x,t)\geq V(x,t). \tag{2.7}$$

(iv) *If for all $(x,t)\in E\times(t_0,u]$ the expression*

$$\mathbb{E}_{x,t}\left[\exp\left(-\int_t^u V\left(X(\sigma),\sigma\right)d\sigma\right)\chi\left(X(u),u\right)\right],$$

is strictly positive, then the following equality is valid:

$$-\log \mathbb{E}_{x,t}\left[\exp\left(-\int_t^u V(X(\sigma),\sigma)\,d\sigma\right)\chi(X(u),u)\right] \tag{2.8}$$

$$= \inf_{v\in D(D_1-K_0)} \mathbb{E}_{x,t}^{M_v,t}\left[\int_t^u \left(\frac{1}{2}\Gamma_1(v,v)+V\right)(X(\tau),\tau)d\tau - \log\chi(X(u),u)\right]. \tag{2.9}$$

(v) *Let S be a viscosity solution to* (1.2). *Suppose that for every $(x,t)\in E\times(t_0,u]$ there exists a functions φ_1 and $\varphi_2\in D(K_0)$ such that*

$$(S-\varphi_1)(x,t) = \sup_{y\in E, u>s>t}(S-\varphi_1)(y,s), \text{ and} \tag{2.10}$$

$$(S-\varphi_2)(x,t) = \inf_{y\in E, u>s>t}(S-\varphi_2)(y,s). \tag{2.11}$$

Then $S = S_L$. More precisely, in the presence of (2.11) *and* (2.10) *the 4 quantities in assertion* (i) *are equal.*

Remark. It would be nice to have explicit, and easy to check, conditions on the function V which guarantee the strict positivity of the expression

$$\mathbb{E}_{x,t}\left[\exp\left(-\int_t^u V(X(\sigma),\sigma)\,d\sigma\right),\ X(u)\in B\right],$$

where B is any compact subset of E. Another problem which poses itself is the following. What can be done if in equation 1.2 the expression $\Gamma_1(S_L,S_L)$ is replaced with $(\Gamma_1(S_L,S_L))^p$, $p>0$. If $0<p<1$, then the equation probably can be treated by the use of *branching processes*: see, e.g., Etheridge [12] or Dawson and Perkins [8]. It seems that the conclusion in (iv) is also valid if we assume that $\mathbb{E}_{x,t}\left[\int_t^u V_-(X(\tau),\tau)d\tau\right]<\infty$ for all $(x,t)\in E\times(t_0,u]$. If

$$-\log\mathbb{E}_{x,t}\left[\exp\left(-\int_t^u V\big(X(\tau),\tau\big)d\tau\right)\right]=\infty,$$

then we may choose $v=0$, and apply Jensen inequality.

Let $t\mapsto M(t)$ be a martingale. Another point of concern is the Novikov condition which is required to be sure that processes of the form

$$t\mapsto \exp\left(M(t)-\frac{1}{2}\langle M,M\rangle(t)\right) \tag{2.12}$$

are martingales. The Novikov condition reads as follows. Let $M(t)$ be a martingale, and suppose that $\mathbb{E}\left[\exp\left(\frac{1}{2}\langle M,M\rangle(t)\right)\right]$ is finite for all $t\geq 0$. Then the process in (2.12) is a martingale. So, strictly speaking, we have to assume in the sequel that the Novikov condition is satisfied: i.e., all the expectations ($x\in E$, $t_0\leq t<s\leq u$)

$$\mathbb{E}_{x,t}\left[\exp\left(\frac{1}{2}\int_t^s \Gamma_1(\varphi,\varphi)\big(X(\tau),\tau\big)d\tau\right)\right]$$

are supposed to be finite; otherwise we will only get local martingales. For more details on the Novikov condition see, e.g., Revuz and Yor [17], Corollary 1.16, page 309.

Another problem is about the uniqueness of the viscosity solution of equation (1.2). In order to address this problem we use a technique, which is related to the methods used in Dynkin and Kuznetsov [10] p 26 ff, and [11], p 1969 ff. Among other things we tried the method of "doubling the number of variables" as advertised in ([13]) Page 547, but it did not work out so far. We also tried (without success) the jet bundle technique in Crandall, Ishii and Lions [6]. To be precise we use a martingale technique combined with sub- and supersolutions: see assertion (v) of Theorem 2.1.

First we insert the following proposition.

Proposition 2.2. (i) *The operator $D_1 - K_0 - V$ extends to a generator of a semigroup* $\exp(s(D_1 - K_0 - V))$, $s \geq 0$, *given by*

$$\begin{aligned} &\exp(s(D_1 - K_0 - V))\,\Phi(x,t) \\ &= \mathbb{E}_x\left[\exp\left(-\int_0^s V(X(\tau),\tau+t)\,d\tau\right)\Phi(X(s),s+t)\right] \qquad (2.13) \\ &= \mathbb{E}_{x,t}\left[\exp\left(-\int_t^{s+t} V(X(\tau),\tau)\,d\tau\right)\Phi(X(s+t),s+t)\right]. \qquad (2.14) \end{aligned}$$

(ii) *Let the function $S_L(x,t)$ be given by*

$$\begin{aligned} S_L(x,t) &= -\log \mathbb{E}_x\left[\exp\left(-\int_0^{u-t} V(X(\tau),\tau+t)\,d\tau\right)\chi(X(u-t),u)\right] \\ &= -\log \mathbb{E}_{x,t}\left[\exp\left(-\int_t^u V(X(\tau),\tau)\,d\tau\right)\chi(X(u),u)\right]. \end{aligned}$$

Then the following identity is valid ($s \geq 0$, $t \geq 0$, $s+t \leq u$)

$$[\exp(s(D_1 - K_0 - V))\exp(-S_L)](x,t) = \exp(-S_L(x,t)). \qquad (2.15)$$

(iii) *Let $\Psi : E \times (t_0,t] \to \mathbb{R}$ be a function belonging to $D(D_1 - K_0)$, and let $V_0 : E \times (t_0,t] \to \mathbb{R}$ be a function for which ($(y,s) \in E \times (t_0,t]$)*

$$\Psi(y,s) = -\log \mathbb{E}_{y,s}\left[\exp\left(-\int_s^u V_0(X(\tau),\tau)d\tau - \Psi(X(t),t)\right)\right]. \qquad (2.16)$$

Then $-D_1\Psi + K_0\Psi + \frac{1}{2}\Gamma_1(\Psi,\Psi) = V_0$ on $(t_0,t]$.

Remark. Suppose that the semigroup $\exp(-tK_0)$, $t \geq 0$, has an integral kernel $p_0(t,x,y)$, which is continuous on $(0,\infty) \times E \times E$, and hence, for $f : E \to [0,\infty)$ Borel measurable we have $\exp(-tK_0) f(x) = \int_E p_0(t,x,y)f(y)dm(y)$, where m is a non-negative Radon measure on E. Define the measure $\mu_{x,t}^{y,u}$ on the σ-field generated by $X(\tau)$, $t \leq \tau < u$ by $\mu_{x,t}^{y,u}(A) = \mathbb{E}_{x,t}[p_0(u-s,X(s),y)\,1_A]$, where A belongs to the σ-field generated by $X(\tau)$, $t \leq \tau \leq s$, with $t \leq s < u$. By the $\mathbb{E}_{x,t}$-martingale property of the process $s \mapsto p_0(u-s,X(s),y)$, $t \leq s < u$, the measure is well defined and can be extended to the σ-field generated by $X(\tau)$, $t \le \tau < u$. The latter can be done via the classical Kolmogorov extension theorem. The integral kernel of the operator $\exp(s(D_1 - K_0 - V))$ is given by the Feynman-Kac formula:

$$\exp(s(D_1 - K_0 - V))(x,y) = \int \exp\left(-\int_t^{s+t} V(X(\tau),\tau)d\tau\right) d\mu_{x,t}^{y,t+s}.$$

Under appropriate conditions on V, the integral kernel of $\exp(s(D_1 - K_0 - V))$ is again continuous on $(0,u] \times E \times E$. Details for time-independent functions V can be found in, e.g., [9].

Proof of Proposition 2.2. (i) Let s_1 and s_2 be positive real numbers, and let Φ be a non-negative Borel measurable function defined on $E \times [0,\infty)$. Then we have:

$$\begin{aligned}
&[\exp(s_1(D_1 - K_0 - V))\exp(s_2(D_1 - K_0 - V))\Phi](x,t) \\
&= \mathbb{E}_x\left[\exp\left(-\int_0^{s_1} V(X(\tau),\tau+t)\,d\tau\right)\{\exp(s_2(D_1 - K_0 - V))\Phi(X(s_1),s_1+t)\}\right] \\
&= \mathbb{E}_x\Bigg[\exp\left(-\int_0^{s_1} V(X(\tau),\tau+t)\,d\tau\right) \\
&\qquad \mathbb{E}_{X(s_1)}\left\{\exp\left(-\int_0^{s_2} V(X(\tau),\tau+s_1+t)\,d\tau\right)\Phi(X(s_2),s_1+s_2+t)\right\}\Bigg]
\end{aligned}$$

(Markov property)

$$\begin{aligned}
&= \mathbb{E}_x\Bigg[\exp\left(-\int_0^{s_1} V(X(\tau),\tau+t)\,d\tau\right) \\
&\qquad \exp\left(-\int_0^{s_2} V(X(\tau+s_1),\tau+s_1+t)\,d\tau\right)\Phi(X(s_1+s_2),s_1+s_2+t)\Bigg] \\
&= \mathbb{E}_x\left[\exp\left(-\int_0^{s_1+s_2} V(X(\tau),\tau+t)\,d\tau\right)\Phi(X(s_1+s_2),s_1+s_2+t)\right] \\
&= [\exp((s_1+s_2)(D_1 - K_0 - V))\Phi](x,t).
\end{aligned} \tag{2.17}$$

Next we prove assertion (ii):

$$\begin{aligned}
&[\exp(s(D_1 - K_0 - V))\exp(-S_L)](x,t) \\
&= \mathbb{E}_x\left[\exp\left(-\int_0^{s} V(X(\tau),\tau+t)\,d\tau\right)\exp(-S_L(X(s),s+t))\right] \\
&= \mathbb{E}_x\Bigg[\exp\left(-\int_0^{s} V(X(\tau),\tau+t)\,d\tau\right) \\
&\qquad \mathbb{E}_{X(s)}\left\{\exp\left(-\int_0^{u-t-s} V(X(\tau),\tau+s+t)\,d\tau\right)\chi(X(u-s-t),u)\right\}\Bigg]
\end{aligned}$$

(Markov property)

$$= \mathbb{E}_x \left[\exp\left(- \int_0^s V(X(\tau), \tau + t)\, d\tau \right) \right.$$
$$\left. \exp\left(- \int_0^{u-t-s} V(X(\tau + s), \tau + s + t)\, d\tau \right) \chi(X(u-t), u) \right]$$
$$= \mathbb{E}_x \left[\exp\left(- \int_0^{u-t} V(X(\tau), \tau + t)\, d\tau \right) \chi(X(u-t), u) \right]$$
$$= \exp(-S_L(x,t)). \tag{2.18}$$

This proves assertion (ii).

(iii) From (1.1) and the proof of assertion (ii) of Proposition 2.2 it follows that

$$- D_1\Psi + K_0\Psi + \frac{1}{2}\Gamma_1(\Psi, \Psi) = e^{\Psi}(D_1 - K_0)\, e^{-\Psi}$$
$$= e^{\Psi}(D_1 - K_0 - V_0)\, e^{-\Psi} + V_0 = \lim_{s\downarrow 0} e^{\Psi} \frac{1}{s} \left(\exp(s(D_1 - K_0 - V_0)) - I\right) e^{-\Psi} + V_0$$
$$= V_0, \tag{2.19}$$

where we used the invariance $\exp(s(D_1 - K_0 - V_0))\, e^{-\Psi} = e^{-\Psi}$, $0 < s < t$. This proves assertion (iii). □

Proof. (Proof of Theorem 2.1.) (i) The first inequality in (i) follows from the non-linear Feynman-Kac formula. For $\Phi \in D(D_1 - K_0)$ we have:

$$\Phi(x,t)$$
$$= -\log \mathbb{E}_{x,t} \left[\exp\left(- \int_t^u \left[-\dot{\Phi} + K_0\Phi + \frac{1}{2}\Gamma_1(\Phi, \Phi) \right] (X(\tau), \tau)\, d\tau - \Phi(X(u), u) \right) \right].$$

The second inequality of (i) is a consequence of Jensen inequality, and should be compared with the arguments in Zambrini [24], who used ideas from Fleming and Soner: see Chapter VI in [14]. The reader is also referred to Sheu [18] and to [22]. The inequality we have in mind is the following one:

$$-\log \mathbb{E}_{x,t}^{M_{v,t}} \left[\exp(-\varphi)\right] \leq \mathbb{E}_{x,t}^{M_{v,t}} [\varphi], \tag{2.20}$$

with equality only if φ is constant $\mathbb{P}_{x,t}$-almost surely. We apply (2.20) for the stochastic variable $\varphi = \varphi_v$, given by

$$\varphi_v = - \int_t^u \left[\frac{1}{2}\Gamma_1(v, v) + V \right] (X(\tau), \tau)\, d\tau - M_{v,t}(u) - \log \chi(X(u), u). \tag{2.21}$$

We also notice that the following processes are $\mathbb{P}_{x,t}$-martingales on the interval $[t, u]$:

$$\exp\left(-\frac{1}{2} \langle M_{v,t} \rangle (s) - M_{v,t}(s) \right) \quad \text{and} \tag{2.22}$$

$$\exp\left(-\frac{1}{2} \langle M_{v,t} \rangle (s) - M_{v,t}(s) \right) \left(\langle M_{v,t} \rangle (s) + M_{v,t}(s) \right). \tag{2.23}$$

Since the process in (2.23) is a $\mathbb{P}_{x,t}^{M_{v,t}}$-martingale we have

$$\mathbb{E}_{x,t}^{M_{v,t}}\left[\frac{1}{2}\langle M_{v,t}\rangle(u)+\int_t^u V(X(\tau),\tau)\,d\tau-\log\chi(X(u),u)\right]$$
$$=\mathbb{E}_{x,t}^{M_{v,t}}\left[-\frac{1}{2}\langle M_{v,t}\rangle(u)-M_{v,t}(u)+\int_t^u V(X(\tau),\tau)\,d\tau-\log\chi(X(u),u)\right]$$

(here we apply Jensen inequality)

$$\geq-\log\mathbb{E}_{x,t}^{M_{v,t}}\left[\exp\left(\frac{1}{2}\langle M_{v,t}\rangle(u)+M_{v,t}(u)-\int_t^u V(X(\tau),\tau)\,d\tau+\log\chi(X(u),u)\right)\right]$$

(definition of the probability measure $\mathbb{E}_{x,t}^{M_{v,t}}$)

$$=-\log\mathbb{E}_{x,t}\left[\exp\left(-\int_t^u V(X(\tau),\tau)\,d\tau+\log\chi(X(u),u)\right)\right].$$

The final inequality is obtained by selecting $v=\Phi$, and using the appropriate properties as indicated in the fourth expression of (i). We also use the martingale property of the process in (2.23).

(ii) The assertion in (ii) immediately follows from (i).

(iii) Let (x,t) belong to $E\times(t_0,u]$, and let φ be as in (2.6). Then we have

$$[-\dot{\varphi}+K_0\varphi-V](x,t)+\frac{1}{2}\Gamma_1(\varphi,\varphi)(x,t)$$
$$=\exp(\varphi(x,t))\left[(D_1-K_0-V)\exp(-\varphi)\right](x,t)$$
$$=\exp(\varphi(x,t))\lim_{s\downarrow 0}\frac{1}{s}\left[(\exp(sD_1-sK_0-sV)-I)\exp(-\varphi)\right](x,t)$$
$$=\exp(\varphi(x,t))\liminf_{s\downarrow 0}\frac{1}{s}\Big(\left[\exp(sD_1-sK_0-sV)\{\exp(S_L-\varphi)\}\exp(-S_L)\right](x,t)$$
$$-\exp(S_L(x,t)-\varphi(x,t))\exp(-S_L(x,t))\Big)$$
$$=\exp(\varphi(x,t))\liminf_{s\downarrow 0}\frac{1}{s}\Big(\left[\exp(sD_1-sK_0-sV)\{\exp(S_L-\varphi)\}\exp(-S_L)\right](x,t)$$
$$-\left\{\sup_{(y,\sigma)\in E\times[t,u]}\exp(S_L(y,s)-\varphi(y,s))\right\}\exp(-S_L(x,t))\Big)$$
$$\leq\exp(\varphi(x,t))\liminf_{s\downarrow 0}\frac{1}{s}$$
$$\left(\left[\exp(sD_1-sK_0-sV)\left\{\sup_{(y,\sigma)\in E\times[t,u]}\{\exp(S_L-\varphi)\}(y,\sigma)\right\}\exp(-S_L)\right](x,t)\right.$$

$$- \left\{ \sup_{(y,\sigma)\in E\times[t,u]} \exp\left(S_L(y,\sigma) - \varphi(y,\sigma)\right) \right\} \exp\left(-S_L(x,t)\right) \Big)$$
$$\leq \sup\left\{ \left\{\exp\left(S_L - \varphi\right)\right\}(y,\sigma) : (y,\sigma) \in E\times[t,u] \right\} \exp\left(\varphi(x,t)\right)$$
$$\liminf_{s\downarrow 0} \frac{1}{s}\left(\left[\exp\left(sD_1 - sK_0 - sV\right)\exp\left(-S_L\right)\right](x,t) - \exp\left(-S_L(x,t)\right)\right)$$
$$= \exp\left(S_L(x,t) - \varphi(x,t)\right)\exp\left(\varphi(x,t)\right)\cdot 0 = 0. \tag{2.24}$$

The latter equality follows because for $u \geq s+t \geq s > 0$ the equality

$$\left[\exp\left(sD_1 - sK_0 - sV\right)\exp\left(-S_L\right)\right](x,t) = \exp\left(-S_L(x,t)\right)$$

is valid: see Proposition 2.2, assertion (ii).

The reverse inequality (2.7) follows in a similar manner.

(iv) In view of assertion (i) we only have to prove that the expression in (2.9) is less than or equal to the one in (2.8). To this end we consider ($v \in D(D_1 - K_0)$)

$$\mathbb{E}_{x,t}^{M_{v,t}}\left[\frac{1}{2}\int_t^u \Gamma_1(v,v)(X(\tau),\tau)\,d\tau + \int_t^u V(X(\tau),\tau)\,d\tau - \log\chi(X(u),u)\right]$$
$$= \mathbb{E}_{x,t}^{M_{v,t}}\left[\frac{1}{2}\langle M_{v,t}\rangle(u) + \int_t^u V(X(\tau),\tau)\,d\tau - \log\chi(X(u),u)\right]$$

(the process in (2.23) is a martingale)

$$= \mathbb{E}_{x,t}^{M_{v,t}}\left[-\frac{1}{2}\langle M_{v,t}\rangle(u) - M_{v,t}(u) + \int_t^u V(X(\tau),\tau)\,d\tau - \log\chi(X(u),u)\right]$$

(definition of the martingale $M_{v,t}()$)

$$= \mathbb{E}_{x,t}^{M_{v,t}}\left[-\frac{1}{2}\int_t^u \Gamma_1(v,v)(X(\tau),\tau)\,d\tau - v(X(u),u) + v(X(t),t)\right.$$
$$\left. - \int_t^u (-D_1 + K_0)v(X(\tau),\tau)\,d\tau + \int_t^u V(X(\tau),\tau)\,d\tau - \log\chi(X(u),u)\right]$$
$$= \mathbb{E}_{x,t}^{M_{v,t}}\left[\int_t^u \left\{(D_1 - K_0)v - \frac{1}{2}\Gamma_1(v,v)(X(\tau),\tau)\right\}d\tau + \int_t^u V(X(\tau),\tau)\,d\tau\right.$$
$$\left. + v(X(t),t) - v(X(u),u) - \log\chi(X(u),u)\right]$$

($(X(t),t) = (x,t)$, $\mathbb{P}_{x,t}$-almost surely)

$$= v(x,t) + \mathbb{E}_{x,t}^{M_{v,t}}\left[\int_t^u \left\{(D_1 - K_0)v - \frac{1}{2}\Gamma_1(v,v)(X(\tau),\tau)\right\}d\tau\right.$$
$$\left. + \int_t^u V(X(\tau),\tau)\,d\tau - v(X(u),u) - \log\chi(X(u),u)\right]$$

$$
\begin{aligned}
&= v(x,t) + \mathbb{E}_{x,t}^{M_{v,t}}\Bigg[\int_t^u \exp\left(v\left(X(\tau),\tau\right)\right)\left[\left(-D_1+K_0+V\right)\exp(-v)\right]\left(X(\tau),\tau\right)d\tau \\
&\qquad - v\left(X(u),u\right) - \log\chi\left(X(u),u\right)\Bigg] \\
&= v(x,t) + \mathbb{E}_{x,t}\Bigg[\exp\left(-\frac{1}{2}\left\langle M_{v,t}\right\rangle(u) - M_{v,t}(u) + \int_t^u V\left(X(\tau),\tau\right)d\tau\right) \\
&\qquad \exp\left(-\int_t^u V\left(X(\tau),\tau\right)d\tau\right) \\
&\qquad \int_t^u \exp\left(v\left(X(\tau),\tau\right)\right)\left[\left(-D_1+K_0+V\right)\exp(-v)\right]\left(X(\tau),\tau\right)d\tau \\
&\qquad - v\left(X(u),u\right) - \log\chi\left(X(u),u\right)\Bigg] \\
&= v(x,t) + \mathbb{E}_{x,t}\Bigg[\exp\left(\int_t \exp\left(v\left(X(\tau),\tau\right)\right)\left[\left(-D_1+K_0+V\right)\exp\left(-v\right)\right]\left(X(\tau),\tau\right)d\tau\right) \\
&\qquad \exp\left(v\left(X(t),t\right) - v\left(X(u),u\right)\right)\exp\left(-\int_t^u V\left(X(\tau),\tau\right)d\tau\right) \\
&\qquad \int_t^u \exp\left(v\left(X(\tau),\tau\right)\right)\left[\left(-D_1+K_0+V\right)\exp(-v)\right]\left(X(\tau),\tau\right)d\tau \\
&\qquad - v\left(X(u),u\right) - \log\chi\left(X(u),u\right)\Bigg].
\end{aligned}
\tag{2.25}
$$

Choose $w \in D\left(D_1 - K_0\right)$ and define for $s > 0$ the function v_s by

$$
\exp\left(-v_s\right) = \frac{1}{s}\int_0^s \exp\left(\sigma\left(D_1 - K_0 - V\right)\right)\exp(-w)\,d\sigma.
$$

Then

$$
\exp\left(v_s\right)\left(-D_1+K_0+V\right)\exp\left(-v_s\right) = \frac{\left(I - \exp\left(s\left(D_1 - K_0 - V\right)\right)\right)\exp\left(-w\right)}{\int_0^s \exp\left(\sigma\left(D_1 - K_0 - V\right)\right)\exp(-w)d\sigma}.
$$

So from (2.25) we obtain for w in the domain of $D_1 - K_0$ and $s > 0$ the inequality

$$
\begin{aligned}
&\mathbb{E}_{x,t}^{M_{v_s,t}}\left[\frac{1}{2}\int_t^u \Gamma_1\left(v_s,v_s\right)\left(X(\tau),\tau\right)d\tau + \int_t^u V\left(X(\tau),\tau\right)d\tau - \log\chi\left(X(u),u\right)\right] \\
&\le v_s(x,t) + \mathbb{E}_{x,t}\Bigg[\exp\left(\int_t^u \frac{\left[\left(I - \exp\left(s\left(D_1 - K_0 - V\right)\right)\right)\exp(-w)\right]\left(X(\tau),\tau\right)}{\int_0^s \left[\exp\left(\sigma\left(D_1 - K_0 - V\right)\right)\exp(-w)\right]\left(X(\tau),\tau\right)d\sigma}d\tau\right) \\
&\qquad \exp\left(v_s\left(X(t),t\right) - v_s\left(X(u),u\right)\right)\exp\left(-\int_t^u V\left(X(\tau),\tau\right)d\tau\right)
\end{aligned}
$$

$$\int_t^u \frac{\left[\left(I-\exp\left(s\left(D_1-K_0-V\right)\right)\right)\exp(-w)\right]\left(X(\tau),\tau\right)}{\int_0^s \left[\exp\left(\sigma\left(D_1-K_0-V\right)\right)\exp(-w)\right]\left(X(\tau),\tau\right)d\sigma}\,d\tau$$

$$\left.-\,v_s\left(X(u),u\right)-\log\chi\left(X(u),u\right)\right]. \tag{2.28}$$

Upon letting $w \in D\left(D_1 - K_0\right)$ tend to the function S_L in an appropriate manner, we obtain by invoking Proposition 2.2 the inequality

$$\inf_{v\in D(D_1-K_0)} \mathbb{E}_{x,t}^{M_{v_s,t}}\left[\frac{1}{2}\int_t^u \Gamma_1\left(X(\tau),\tau\right)d\tau + \int_t^u V\left(X(\tau),\tau\right)d\tau - \log\chi\left(X(u),u\right)\right]$$
$$\le S_L(x,t).$$

This proves assertion (iv).

(v). Here we use a martingale approach together with the idea of germs of a function. We will prove the following inequalities:

$$S(x,t) \le \sup\left\{\varphi_1(x,t) : V_{\varphi_1} \le V,\ \ \varphi_1(\cdot,u) \le S_L(\cdot,u)\right\} \tag{2.29}$$
$$\le \inf\left\{\varphi_2(x,t) : V_{\varphi_2} \ge V,\ \ \varphi_2(\cdot,u) \ge S_L(\cdot,u)\right\} \le S(x,t), \tag{2.30}$$

where $V_\varphi = -\dot{\varphi} + K_0\varphi + \dfrac{1}{2}\Gamma_1\left(\varphi,\varphi\right)$. In view of assertion (i) in Theorem 2.1 we then infer $S = S_L$. Fix $(x,t) \in E \times (t_0,u]$. Let $\varphi_1 \in D\left(D_1 - K_0\right)$ be such that

$$(S-\varphi_1)(x,t) = \sup\left\{(S-\varphi_1)(y,s) : y\in E,\ u\ge s\ge t\right\}.$$

We notice that the processes $M_{\varphi,t}$ and $M_{S_L,t}$, defined by respectively

$$M_{\varphi,t}(s) = \exp\left(-\int_t^s V_\varphi\left(X(\tau),\tau\right)d\tau - \varphi\left(X(s),s\right) + \varphi\left(X(t),t\right)\right),\ \text{and}$$
$$M_{S_L,t}(s) = \exp\left(-\int_t^s V\left(X(\tau),\tau\right) - S_L\left(X(s),s\right) + S_L\left(X(t),t\right)\right),$$

$t \le s \le u$, are $\mathbb{P}_{x,t}$-martingales. The latter assertion follows from the Markov property together with the Feynman-Kac formula: see (2.15), which is also true for V_φ instead of V and φ replacing S_L. Let $\mathbb{P}_{x,t}^{M_\varphi}$ denote the probability measure defined by $\mathbb{P}_{x,t}^{M_\varphi}(A) = \mathbb{E}_{x,t}\left[M_\varphi(s_2)1_A\right]$, $s_2 \ge s_1$, where A is $\mathcal{F}_{s_1}$-measurable. Since S is a viscosity subsolution we see that $V_{\varphi_1}(x,t) \le V(x,t)$. Fix $\varepsilon > 0$ and choose $\delta > 0$ in such a way that, for some neighborhood U of x in E, the inequality $V_{\varphi_1}(y,s) \le V(y,s) + \frac{1}{2}\varepsilon$ is valid for $(y,s) \in U \times [t, t+\delta]$. Here we use the continuity of $V(y,s)$ in $y = x$ and its right continuity in $s = t$. Then we choose a family of germs of "smooth" functions $(U_\alpha, \varphi_\alpha)$, $\alpha \in \mathcal{A}$, with the following properties:

(a) $\bigcup U_\alpha \supseteq E \times [t,u]$, i.e., the family U_α, $\alpha \in \mathcal{A}$, forms an open cover of the set $E \times [t,u]$;
(b) For every α, $\beta \in \mathcal{A}$, $\varphi_\alpha = \varphi_\beta$ on $U_\alpha \cap U_\beta$;
(c) For every $\alpha \in \mathcal{A}$ there exists $(x_\alpha, t_\alpha) \in U_\alpha$ such that $(S-\varphi_\alpha)(y,s) \le (S-\varphi_\alpha)(x_\alpha,t_\alpha)$, for $(y,s) \in U_\alpha$ and $s_\alpha \le s$;
(d) For every $\alpha \in \mathcal{A}$, the inequality $V_{\varphi_\alpha} \le V + \frac{1}{2}\varepsilon$ is valid on U_α;

(e) If (x,t) belongs to U_α, then $(S-\varphi_\alpha)(x,t) \le 0$;
(f) If (y,u) belongs to U_α, then $\varphi_\alpha(y,u) \le S(y,u) + \frac{1}{2}\varepsilon(u-t) = S_L(y,u) + \frac{1}{2}\varepsilon(u-t)$.

Since S is a viscosity subsolution property (d) is in fact a consequence of (c); we will need (d). Then we define the function $\psi_1 : E \times [t,u] \to \mathbb{R}$ as follows $\psi_1(y,s) = \varphi_\alpha(y,s)$, for $(y,s) \in U_\alpha$. Then, on U_α, $V_{\psi_1} = V_{\varphi_\alpha} \le V + \frac{1}{2}\varepsilon$. We write $V_1 = V_{\psi_1}$. By assertion (iii) of Proposition 2.2

$$\begin{aligned}\Psi_1^\varepsilon(y,s) &:= \psi_1(y,s) - \frac{1}{2}\varepsilon(u-s) - \frac{1}{2}\varepsilon(u-t) \qquad (2.31)\\ &= -\log \mathbb{E}_{y,s}\left[\exp\left(-\int_s^u \left(V_1(X(\tau),\tau) - \frac{1}{2}\varepsilon\right)d\tau - \left(\psi_1(X(u),u) - \frac{1}{2}\varepsilon(u-t)\right)\right)\right]\\ &= -\log \mathbb{E}_{y,s}\left[\exp\left(-\int_t^u V_{\Psi_1^\varepsilon}(X(\tau),\tau)d\tau - \Psi_1^\varepsilon(X(u),u)\right)\right].\end{aligned}$$

Then

$$\Psi_1^\varepsilon(y,s) \le -\log \mathbb{E}_{y,s}\left[\exp\left(-\int_s^u V(X(\tau),\tau)d\tau - S_L(X(u),u)\right)\right] = S_L(y,s),$$

and hence $\psi_1(x,t) \le S_L(x,t) + \varepsilon(u-t)$. By construction we also have $S(x,t) \le \psi_1(x,t)$. Consequently $S(x,t) \le S_L(x,t) + \varepsilon(u-t)$. Since $\varepsilon > 0$ is arbitrary we see $S(x,t) \le S_L(x,t)$. In fact, since $V_{\Psi_1^\varepsilon} \le V$, and since $\Psi_1^\varepsilon(y,u) \le S_L(y,u)$, we see that

$$S(x,t) \le \sup\left\{\varphi_1(x,t) : V_{\varphi_1} \le V, \quad \varphi_1(\cdot,u) \le S_L(\cdot,u)\right\}.$$

A similar argument shows the inequality

$$S(x,t) \ge \inf\left\{\varphi_2(x,t) : V_{\varphi_2} \ge V, \quad \varphi_2(\cdot,u) \ge S_L(\cdot,u)\right\}.$$

To be precise, again we fix $\varepsilon > 0$, and let $\varphi_2 \in D(D_1 - K_0)$ be a function such that $(S-\varphi_2)(x,t) = \inf\{(S-\varphi_2)(y,s) : (y,s) \in E \times [t,u]\}$. We choose $\delta > 0$ and a neighborhood U of x in such a way that $V_{\varphi_2}(y,s) \ge V(y,s) - \frac{1}{2}\varepsilon$ for $(y,s) \in U \times [t,t+\delta]$. Then we choose a family of germs of "smooth" functions $(U_\alpha, \varphi_\alpha)$, $\alpha \in \mathcal{A}$, with the following properties:

(a) $\bigcup U_\alpha \supseteq E \times [t,u]$, i.e., the family U_α forms an open cover of the set $E \times [t,u]$;
(b) α, $\beta \in \mathcal{A}$ implies $\varphi_\alpha = \varphi_\beta$ on $U_\alpha \cap U_\beta$;
(c) For every $\alpha \in \mathcal{A}$ there exists $(x_\alpha, t_\alpha) \in U_\alpha$ such that $(S-\varphi_\alpha)(y,s) \ge (S-\varphi_\alpha)(x_\alpha,t_\alpha)$, for $(y,s) \in U_\alpha$ and $s_\alpha \le s$;
(d) For every $\alpha \in \mathcal{A}$, the inequality $V_{\varphi_\alpha} \ge V - \frac{1}{2}\varepsilon$ is valid on U_α;
(e) If (x,t) belongs to U_α, then $(S-\varphi_\alpha)(x,t) \ge 0$;
(f) If (y,u) belongs to U_α, then $\varphi_\alpha(y,u) \ge S(y,u) - \frac{1}{2}\varepsilon(u-t) = S_L(y,u) - \frac{1}{2}\varepsilon(u-t)$.

Since S is a viscosity supersolution property (d) is in fact a consequence of (c). Then we define the function $\psi_2 : E \times [t,u] \to \mathbb{R}$ as follows $\psi_2(y,s) = \varphi_\alpha(y,s)$, for $(y,s) \in U_\alpha$. Then, on U_α, $V_{\psi_2} = V_{\varphi_\alpha} \leq V + \frac{1}{2}\varepsilon$. We write $V_2 = V_{\psi_2}$. As above, assertion (iii) of Proposition 2.2 implies

$$\Psi_2^\varepsilon(y,s) := \psi_2(y,s) + \frac{1}{2}\varepsilon(u-s) + \frac{1}{2}\varepsilon(u-t) \geq S_L(y,s). \tag{2.33}$$

By construction we have $S(x,t) \geq \psi_2(x,t)$, and hence $S_L(x,t) \leq \Psi_2^\varepsilon(x,t) \leq \psi_2(x,t) + \varepsilon(u-t) \leq S(x,t) + \varepsilon(u-t)$. Since $\varepsilon > 0$ is arbitrary we infer $S_L(x,t) \leq S(x,t)$. In fact, since $V_{\Psi_2^\varepsilon} \geq V$, and since $\Psi_2^\varepsilon(y,u) \geq S_L(y,u)$, we see that

$$S(x,t) \geq \sup\left\{\varphi_1(x,t) : V_{\varphi_1} \leq V, \quad \varphi_1(\cdot,u) \leq S_L(\cdot,u)\right\}.$$

In the mean time we also proved that the 4 quantities in assertion (i) are equal. □

3. A stochastic Noether theorem

The following theorem may be called the *stochastic Noether theorem*: cfr. Zambrini [24] Proposition 2.3 and Theorem 2.4. For a discussion and formulation of the classical (deterministic) Noether theorem, which in fact can be considered as the second constant of motion for a mechanical system, the reader is referred to Thieullen and Zambrini [19] page 300 – 302, and [20] page 423. A more extensive account can be found in Christodoulou [4]. It would be interesting to have a notion of *stochastic divergence invariance* in the present situation: cfr. Zambrini [24]. In the future we hope to return to this issue.

Theorem 3.1. *Let T be a differentiable function which only depends on time. As above the operator D_1 stands for $D_1 = \frac{\partial}{\partial t}$. Suppose that the functions φ, w, and T satisfy the following identities.*

(a) $K_0 f \frac{dT}{dt} = K_0\Gamma_1(f,w) - \Gamma_1(K_0 f, w) - \Gamma_1\left(f, \frac{\partial w}{\partial t} - \varphi\right)$ *for all functions* $f \in D(K_0 - D_1)$ *for which* $\Gamma_1(K_0 f, w)$ *exists as well.*

(b) $\frac{\partial \varphi}{\partial t} - K_0\varphi = \Gamma_1(V,w) + \frac{\partial(TV)}{\partial t}$.

Put

$$\mathcal{H}(f) = \frac{\partial f}{\partial t} - (K_0 \dot{+} V) f, \quad \mathcal{N}(f) = \Gamma_1(f,w) + T\frac{\partial f}{\partial t} - \varphi f, \text{ and}$$
$$\mathcal{D}(f) = \frac{\partial f}{\partial t} - \Gamma_1(\sigma_L, f) - K_0 f. \tag{3.1}$$

Suppose that the function σ_L satisfies:

(c) $\left(\mathcal{D}\varepsilon - \frac{\partial V}{\partial t}\right) T = \Gamma_1\left(\frac{\partial \sigma_L}{\partial t} + \varepsilon, w\right)$, *where* $\varepsilon = -K_0\sigma_L - \frac{1}{2}\Gamma_1(\sigma_L,\sigma_L) + V$.

Write $n := -\Gamma_1(\sigma_L, w) + \varepsilon T - \varphi$. The following assertions hold true.

(i) *If $\mathcal{H}(f) = 0$, then $\mathcal{H}(\mathcal{N}(f)) = 0$ as well.*
(ii) $\mathcal{D}n = 0$.

(iii) *The process* $t \mapsto n(X(t),t)$ *is a martingale with respect to the probability measures*

$$A \mapsto \mathbb{E}_{x,t_0}\left[\exp\left(-M_{\sigma_L}(t) - \frac{1}{2}\int_{t_0}^{t}\Gamma_1(\sigma_L,\sigma_L)(X(s),s)ds\right)1_A\right],$$

where

$$M_{f,t_0}(t) = f(X(t),t)) - f(X(t_0),t_0) + \int_{t_0}^{t}(K_0 - D_1)f(X(s),s)ds.$$

Remark. Define the operator $\mathcal{N}_0$ by

$$\mathcal{N}_0(f) = \Gamma_1(f,w) + T(K_0 \dot{+} V)f - \varphi f;$$

compare with Theorem 3.4. The proof of assertion (i) shows that the operators $\mathcal{H}$ and $\mathcal{N}_0$ commute: $\mathcal{H}(\mathcal{N}_0 f) = \mathcal{N}_0(\mathcal{H}f)$, $f \in D(\mathcal{H}) \cap D(\mathcal{N}_0)$, $\mathcal{H}(f) \in D(\mathcal{N}_0)$, and $\mathcal{N}_0(f) \in D(\mathcal{H})$.

The following proposition shows a situation where (c) is satisfied.

Proposition 3.2. *Suppose* S_L*, the minimal Lagrangian action, belongs to the domain of* $D_1 - K_0$*. Here* $D_1 = \frac{\partial}{\partial t}$*. Set* $\sigma_L = S_L$ *in Theorem* 3.1*. Then* (c) *is satisfied; more precisely,* $\mathcal{D}\varepsilon = D_1 V$ *and* $D_1\sigma_L + \varepsilon = 0$.

Proof of Proposition 3.2. Notice that

$$\varepsilon = -K_0 S_L - \frac{1}{2}\Gamma_1(S_L,S_L) + V = -D_1 S_L, \tag{3.2}$$

and hence

$$\begin{aligned}\mathcal{D}\varepsilon = -\mathcal{D}(D_1 S_L) &= -\frac{\partial^2 S_L}{\partial t^2} + \Gamma_1\left(S_L,\frac{\partial S_L}{\partial t}\right) + K_0\left(\frac{\partial S_L}{\partial t}\right)\\ &= -\frac{\partial^2 S_L}{\partial t^2} + \frac{1}{2}\frac{\partial}{\partial t}\Gamma_1(S_L,S_L) + \frac{\partial}{\partial t}K_0(S_L)\\ &= \frac{\partial}{\partial t}\left(-\frac{\partial S_L}{\partial t} + \frac{1}{2}\Gamma_1(S_L,S_L) + K_0 S_L\right) = \frac{\partial V}{\partial t}.\end{aligned} \tag{3.3}$$

Proposition 3.2 easily follows from (3.2) and (3.3). □

The equality in (3.4) below will be used in the proof Theorem 3.1.

Lemma 3.3. *For all appropriate functions* f*,* w*,* T*, and* φ *the following identity is true:*

$$\begin{aligned}&-\frac{1}{2}\Gamma_1(f,f)\frac{\partial T}{\partial t} - \frac{1}{2}\Gamma_1(\Gamma_1(f,f),w) + \Gamma_1(f,\Gamma_1(f,w))\\ &= \frac{1}{2}\left(K_0(f^2)\frac{\partial T}{\partial t} + \Gamma_1(K_0(f^2),w) - K_0\Gamma_1(f^2,w) + \Gamma_1\left(f^2,\frac{\partial w}{\partial t} - \varphi\right)\right)\\ &\quad - f\left(K_0(f)\frac{\partial T}{\partial t} + \Gamma_1(K_0(f),w) - K_0\Gamma_1(f,w) + \Gamma_1\left(f,\frac{\partial w}{\partial t} - \varphi\right)\right).\end{aligned} \tag{3.4}$$

Proof of Lemma 3.3. The equality $-\frac{1}{2}\Gamma_1(f,f) = \frac{1}{2}K_0(f^2) - fK_0 f$ together with $\Gamma_1(fK_0f, w) = f\Gamma_1(K_0 f, w) + K_0 f\Gamma_1(f,w)$ yields

$$
\begin{aligned}
&-\frac{1}{2}\Gamma_1(f,f)\frac{\partial T}{\partial t} - \frac{1}{2}\Gamma_1(\Gamma_1(f,f),w) + \Gamma_1(f,\Gamma_1(f,w))\\
&=\frac{1}{2}\left\{K_0(f^2)\frac{\partial T}{\partial t} + \Gamma_1(K_0(f^2),w) - K_0\Gamma_1(f^2,w)\right\}\\
&\quad - f\left\{K_0 f\frac{\partial T}{\partial t} + \Gamma_1(K_0 f,w) - K_0\Gamma_1(f,w)\right\}\\
&\quad + \frac{1}{2}K_0\Gamma_1(f^2,w) - fK_0\Gamma_1(f,w) - (K_0 f)\Gamma_1(f,w) + \Gamma_1(f,\Gamma_1(f,w))
\end{aligned}
$$

$$\left(\tfrac{1}{2}K_0\Gamma_1(f^2,w) = K_0(f\Gamma_1(f,w)) = (K_0 f)\Gamma_1(f,w) - \Gamma_1(f,\Gamma_1(f,w)) + fK_0\Gamma_1(f,w)\right)$$

$$
\begin{aligned}
&=\frac{1}{2}\left\{K_0(f^2)\frac{\partial T}{\partial t} + \Gamma_1(K_0(f^2),w) - K_0\Gamma_1(f^2,w)\right\}\\
&\quad - f\left\{K_0 f\frac{\partial T}{\partial t} + \Gamma_1(K_0 f,w) - K_0\Gamma_1(f,w)\right\}.
\end{aligned}
\tag{3.5}
$$

Since $\Gamma_1(f^2,\psi) = 2f\Gamma_1(f,\psi)$, equality (3.4) in Lemma 3.3 follows from (3.5). □

Proof of Theorem 3.1. (i) We calculate:

$$
\begin{aligned}
&\mathcal{H}(\mathcal{N}f) - \mathcal{N}(\mathcal{H}f)\\
&= \mathcal{H}(\Gamma_1(\omega,f)) + \mathcal{H}\left(T\frac{\partial f}{\partial t}\right) - \mathcal{H}(\varphi f) - \Gamma_1(\mathcal{H}f,w) - T\frac{\partial}{\partial t}(\mathcal{H}f) + \varphi\mathcal{H}f\\
&= \frac{\partial}{\partial t}(\Gamma_1(f,w)) - (K_0\dot{+}V)\Gamma_1(f,w)\\
&\quad \frac{\partial}{\partial t}\left(T\frac{\partial f}{\partial t} - \varphi f\right) - (K_0\dot{+}V)\left(T\frac{\partial f}{\partial t}\right) + (K_0\dot{+}V)(\varphi f)\\
&\quad - \Gamma_1\left(\frac{\partial f}{\partial t},w\right) + \Gamma_1(K_0 f,w) + \Gamma_1(Vf,w)\\
&\quad - T\left(\frac{\partial^2 f}{\partial t^2} - \frac{\partial}{\partial t}((K_0\dot{+}V)f)\right) + \varphi\frac{\partial f}{\partial t} - \varphi(K_0\dot{+}V)f\\
&= \Gamma_1\left(f,\frac{\partial w}{\partial t} - \varphi\right) - K_0\Gamma_1(f,w) + \Gamma_1(K_0 f,w)\\
&\quad + \left(\frac{\partial T}{\partial t} - K_0 T\right)\frac{\partial f}{\partial t} + \Gamma_1\left(T,\frac{\partial f}{\partial t}\right) + \left(-\frac{\partial\varphi}{\partial t} + K_0\varphi + \Gamma_1(V,w) + T\frac{\partial V}{\partial t}\right)f
\end{aligned}
$$

$$\left(\frac{\partial f}{\partial t} = (K_0\dot{+}V)f\right)$$

$$= \Gamma_1\left(\frac{\partial w}{\partial t} - \varphi, f\right) - K_0\Gamma_1(f,w) + \Gamma_1(K_0 f,w)$$

$$+ (K_0 f) \frac{\partial T}{\partial t} - (K_0 T) K_0 f + \Gamma_1 (T, K_0 f) + \Gamma_1 (T, V f)$$
$$+ \left(-\frac{\partial \varphi}{\partial t} + K_0 \varphi + \Gamma_1 (V, w) + T \frac{\partial V}{\partial t} + \frac{\partial T}{\partial t} V \right) f$$

(T only depends on t)

$$= \Gamma_1 \left(\frac{\partial w}{\partial t} - \varphi, f \right) - K_0 \Gamma_1 (f, w) + \Gamma_1 (K_0 f, w) + (K_0 f) \frac{\partial T}{\partial t}$$
$$+ \left(-\frac{\partial \varphi}{\partial t} + K_0 \varphi + \Gamma_1 (V, w) + T \frac{\partial V}{\partial t} + \frac{\partial T}{\partial t} V \right) f = 0. \tag{3.6}$$

The equality in (3.6) follows by our assumptions (a) and (b).

(ii) As an auxiliary function we introduce

$$g := \Gamma_1 (\sigma_L, \Gamma_1 (\sigma_L, w)) - \Gamma_1 (\sigma_L, \varepsilon) T + K_0 (\Gamma_1 (\sigma_L, w)) - K_0 (\varepsilon) T + K_0 \varphi.$$

We compute

$$\begin{aligned}
\mathcal{D}(n) &= \frac{\partial n}{\partial t} - \Gamma_1 (\sigma_L, n) - K_0 n \\
&= -\frac{\partial}{\partial t} (\Gamma_1 (\sigma_L, w)) + \frac{\partial}{\partial t} (\varepsilon T) - \frac{\partial}{\partial t} \varphi \\
&\quad + \Gamma_1 (\sigma_L, \Gamma_1 (\sigma_L, w)) - \Gamma_1 (\sigma_L, \varepsilon T) + \Gamma_1 (\sigma_L, \varphi) \\
&\quad + K_0 (\Gamma_1 (\sigma_L, w)) - K_0 (\varepsilon T) + K_0 \varphi \\
&= -\Gamma_1 \left(\frac{\partial \sigma_L}{\partial t}, w \right) - \Gamma_1 \left(\sigma_L, \frac{\partial w}{\partial t} \right) + \varepsilon \frac{\partial T}{\partial t} + \frac{\partial \varepsilon}{\partial t} T - \frac{\partial \varphi}{\partial t} + \Gamma_1 (\sigma_L, \varphi) + g \\
&= -\Gamma_1 \left(\frac{\partial \sigma_L}{\partial t} + V, w \right) - \Gamma_1 \left(\sigma_L, \frac{\partial w}{\partial t} - \varphi \right) + \varepsilon \frac{\partial T}{\partial t} + \frac{\partial \varepsilon}{\partial t} T - \frac{\partial \varphi}{\partial t} + g + \Gamma_1 (V, w) \\
&= -\Gamma_1 \left(\frac{\partial \sigma_L}{\partial t} + V, w \right) - \Gamma_1 \left(\sigma_L, \frac{\partial w}{\partial t} - \varphi \right) + V \frac{\partial T}{\partial t} \\
&\quad - \left(K_0 \sigma_L + \frac{1}{2} \Gamma_1 (\sigma_L, \sigma_L) \right) \frac{\partial T}{\partial t} + \frac{\partial \varepsilon}{\partial t} T - \frac{\partial \varphi}{\partial t} + g + \Gamma_1 (V, w) \\
&= -\Gamma_1 \left(\frac{\partial \sigma_L}{\partial t} + V, w \right) - \Gamma_1 \left(\sigma_L, \frac{\partial}{\partial t} w - \varphi \right) + \frac{\partial (VT)}{\partial t} \\
&\quad - \left(K_0 \sigma_L + \frac{1}{2} \Gamma_1 (\sigma_L, \sigma_L) \right) \frac{\partial T}{\partial t} + \frac{\partial (\varepsilon - V)}{\partial t} T - \frac{\partial \varphi}{\partial t} + g + \Gamma_1 (V, w) \\
&= -\Gamma_1 \left(\frac{\partial \sigma_L}{\partial t} + \varepsilon, w \right) + \left(\mathcal{D}(\varepsilon) - \frac{\partial V}{\partial t} \right) T - \Gamma_1 \left(\sigma_L, \frac{\partial w}{\partial t} - \varphi \right) \\
&\quad - \frac{\partial \varphi}{\partial t} + K_0 \varphi + \Gamma_1 (V, w) + \frac{\partial (VT)}{\partial t}
\end{aligned}$$

$$
\begin{aligned}
&- \left(K_0\sigma_L + \frac{1}{2}\Gamma_1\left(\sigma_L,\sigma_L\right)\right)\frac{\partial T}{\partial t} - \Gamma_1\left(K_0\sigma_L + \frac{1}{2}\Gamma_1\left(\sigma_L,\sigma_L\right), w\right) \\
&+ \Gamma_1\left(\sigma_L,\Gamma_1\left(\sigma_L,w\right)\right) + K_0\left(\Gamma_1\left(\sigma_L,w\right)\right) \\
= &-\Gamma_1\left(\frac{\partial \sigma_L}{\partial t} + \varepsilon, w\right) + \left(\mathcal{D}(\varepsilon) - \frac{\partial V}{\partial t}\right) T \\
&- \left(K_0\sigma_L\right)\frac{\partial T}{\partial t} - \Gamma_1\left(K_0\sigma_L, w\right) + K_0\Gamma_1\left(\sigma_L, w\right) - \Gamma_1\left(\sigma_L, \frac{\partial w}{\partial t} - \varphi\right) \\
&- \frac{\partial \varphi}{\partial t} + K_0\varphi + \Gamma_1\left(V,w\right) + \frac{\partial (VT)}{\partial t} \\
&- \frac{1}{2}\Gamma_1\left(\sigma_L,\sigma_L\right)\frac{\partial T}{\partial t} - \frac{1}{2}\Gamma_1\left(\Gamma_1\left(\sigma_L,\sigma_L\right), w\right) + \Gamma_1\left(\sigma_L,\Gamma_1\left(\sigma_L,w\right)\right) \\
= &-\Gamma_1\left(\frac{\partial \sigma_L}{\partial t} + \varepsilon, w\right) + \left(\mathcal{D}(\varepsilon) - \frac{\partial V}{\partial t}\right) T \qquad (3.7) \\
&- \left(K_0\sigma_L\right)\frac{\partial T}{\partial t} - \Gamma_1\left(K_0\sigma_L, w\right) + K_0\Gamma_1\left(\sigma_L, w\right) - \Gamma_1\left(\sigma_L, \frac{\partial w}{\partial t} - \varphi\right) \\
&- \frac{\partial \varphi}{\partial t} + K_0\varphi + \Gamma_1\left(V,w\right) + \frac{\partial (VT)}{\partial t} \\
&+ \frac{1}{2}\left(K_0\left(\sigma_L^2\right)\frac{\partial T}{\partial t} + \Gamma_1\left(K_0\left(\sigma_L^2\right), w\right) - K_0\Gamma_1\left(\sigma_L^2, w\right) + \Gamma_1\left(\sigma_L^2, \frac{\partial w}{\partial t} - \varphi\right)\right) \\
&- \sigma_L\left(K_0\left(\sigma_L\right)\frac{\partial T}{\partial t} + \Gamma_1\left(K_0\left(\sigma_L\right), w\right) - K_0\Gamma_1\left(\sigma_L, w\right) + \Gamma_1\left(\sigma_L, \frac{\partial w}{\partial t} - \varphi\right)\right).
\end{aligned}
$$

In order to obtain equality (3.7) we employed Lemma 3.3 with $f = \sigma_L$. Substituting the equalities (a), (b) and (c) in (3.7) shows (ii), i.e., $\mathcal{D}(n) = 0$.

(iii) Let f be a function in the domain of $K_0 - D_1$. As in equation (1.7) of §1 the $\mathbb{E}_{x,t_0}$-martingale $M_{f,t_0}(t)$, $t \geq t_0$, is given by

$$M_{f,t_0}(s) = f\left(X(t),t\right) - f\left(X(t_0),t_0\right) + \int_{t_0}^{t} \left(-D_1 + K_0\right) v\left(X(\tau),\tau\right) d\tau. \qquad (3.8)$$

Let f and g be two functions in $D\left(K_0 - D_1\right)$. Then the quadratic co-variation $\left\langle M_{f,t_0}, M_{g,t_0}\right\rangle(t)$ of $M_{f,t_0}(t)$ and $M_{g,t_0}(t)$ is given by

$$\left\langle M_{f,t_0}, M_{g,t_0}\right\rangle(t) = \int_{t_0}^{t} \Gamma_1\left(f,g\right)\left(X(\tau),\tau\right) d\tau. \qquad (3.9)$$

We notice that

$$n\left(X(t),t\right) - n\left(X(t_0),t_0\right) = M_{n,t_0}(t) - \int_{t_0}^{t} \left(K_0 - D_1\right) n\left(X(\tau),\tau\right) d\tau$$

(by (ii) $\mathcal{D}(n) = 0$ and hence $\left(K_0 - D_1\right) n = -\Gamma_1\left(\sigma_L, n\right)$)

$$= M_{n,t_0}(t) + \int_{t_0}^{t} \Gamma_1\left(\sigma_L, n\right)\left(X(\tau),\tau\right) d\tau. \qquad (3.10)$$

Let f be a function in $D(K_0 - D_1)$. From Itô's formula we obtain:

$$
\begin{aligned}
&\exp\left(M_{f,t_0}(t) - \frac{1}{2}\langle M_{f,t_0}, M_{f,t_0}\rangle(t)\right) n(X(t),t) - n(X(t_0),t_0) \\
&= \int_{t_0}^{t} \exp\left(M_{f,t_0}(s) - \frac{1}{2}\langle M_{f,t_0}, M_{f,t_0}\rangle(s)\right) n(X(s),s)\, dM_{f,t_0}(s) \\
&\quad - \frac{1}{2}\int_{t_0}^{t} \exp\left(M_{f,t_0}(s) - \frac{1}{2}\langle M_{f,t_0}, M_{f,t_0}\rangle(s)\right) n(X(s),s)\, d\langle M_{f,t_0}, M_{f,t_0}\rangle(s) \\
&\quad + \frac{1}{2}\int_{t_0}^{t} \exp\left(M_{f,t_0}(s) - \frac{1}{2}\langle M_{f,t_0}, M_{f,t_0}\rangle(s)\right) n(X(s),s)\, d\langle M_{f,t_0}, M_{f,t_0}\rangle(s) \\
&\quad + \int_{t_0}^{t} \exp\left(M_{f,t_0}(s) - \frac{1}{2}\langle M_{f,t_0}, M_{f,t_0}\rangle(s)\right) dM_{n,t_0}(s) \\
&\quad - \int_{t_0}^{t} \exp\left(M_{f,t_0}(s) - \frac{1}{2}\langle M_{f,t_0}, M_{f,t_0}\rangle(s)\right) (K_0 - D_1)\, n(X(s),s)\, ds \\
&\quad + \int_{t_0}^{t} \exp\left(M_{f,t_0}(s) - \frac{1}{2}\langle M_{f,t_0}, M_{f,t_0}\rangle(s)\right) \Gamma_1(f,n)(X(s),s)\, ds \\
&= \mathbb{E}_{x,t_0}\text{-martingale} \qquad\qquad (3.11) \\
&\quad + \int_{t_0}^{t} \exp\left(M_{f,t_0}(s) - \frac{1}{2}\langle M_{f,t_0}, M_{f,t_0}\rangle(s)\right) \left(\Gamma_1(f,n) + \Gamma_1(\sigma_L,n)\right)(X(s),s)\, ds,
\end{aligned}
$$

where we used (3.10). Suppose $\Gamma_1(f + \sigma_L, w) = 0$. From (3.11) it follows that the process

$$
\begin{aligned}
t \mapsto &\exp\left(M_{f,t_0}(t) - \frac{1}{2}\langle M_{f,t_0}, M_{f,t_0}\rangle(t)\right) \\
&= \exp\left(M_{f,t_0}(t) - \frac{1}{2}\int_{t_0}^{t} \Gamma_1(f,f)(X(s),s)\, ds\right) n(X(t),t)
\end{aligned}
$$

is a $\mathbb{E}_{x,t_0}$-martingale. So, with $f = -\sigma_L$, assertion (iii) of Theorem 3.1 follows. □

The following theorem can be considered as a complex version of the Noether theorem: see Theorem 3.1 above and Theorem 3.1 in Zambrini [24]. It has a physical interpretation: $\mathcal{N}(t)$ is called a *Noether observable.*

Theorem 3.4. *Let the functions T, w, and φ be related as in* (a′) *and* (b′) *below:*

(a′) $K_0 f \dfrac{dT}{dt} = K_0\Gamma_1(f,w) - \Gamma_1(K_0 f, w) + i\Gamma_1\left(f, \dfrac{\partial w}{\partial t} - \varphi\right)$ *for all functions f belonging to $D(K_0 - D_1)$ for which $\Gamma_1(K_0 f, w)$ makes sense as well.*

(b′) $\dfrac{\partial \varphi}{\partial t} - iK_0\varphi = -\Gamma_1(V,w) - \dfrac{\partial (TV)}{\partial t}$.

Set

$$\mathcal{N}(t)f = i\Gamma_1(f,w) - T(t)\left(K_0\dot{+}V\right)f - \varphi f.$$

Then the operators $\mathcal{N}(t)$ *and* $\frac{\partial}{i\partial t} - \left(K_0\dot{+}V\right)$ *commute.*

Suppose $\int K_0 f dm = 0$, $f \in D\left(K_0\right) \cap L^1(E,m)$. Then the adjoint $\mathcal{N}(t)^*$ is given by

$$\mathcal{N}(t)^* f = i\Gamma_1\left(f,\overline{w}\right) - 2i\left(K_0\overline{w}\right)f - \overline{T(t)}\left(K_0\dot{+}V\right)f - \overline{\varphi}f.$$

Hence the self-adjoint operator $\frac{\partial}{i\partial t} - \left(K_0\dot{+}V\right)$ also commutes with the operators $\mathcal{N}(t) + \mathcal{N}(t)^*$ and $\mathcal{N}(t) - \mathcal{N}(t)^*$.

Proof. Let f be a "smooth enough" function. Then a calculation yields:

$$\begin{aligned}
&\mathcal{N}(t)\left(\frac{\partial}{i\partial t} - \left(K_0\dot{+}V\right)\right)f - \left(\frac{\partial}{i\partial t} - \left(K_0\dot{+}V\right)\right)\mathcal{N}(t)f \\
&= i\Gamma_1\left(\frac{\partial f}{i\partial t} - \left(K_0\dot{+}V\right)f, w\right) - T(t)\left(K_0\dot{+}V\right)\left(\frac{\partial f}{i\partial t} - \left(K_0\dot{+}V\right)f\right) \\
&\quad - \varphi\left(\frac{\partial f}{i\partial t} - \left(K_0\dot{+}V\right)f\right) - \left(\frac{\partial}{i\partial t} - \left(K_0\dot{+}V\right)\right)\left(i\Gamma_1\left(f,w\right) - T(t)\left(K_0\dot{+}V\right)f - \varphi f\right) \\
&= \Gamma_1\left(\frac{\partial f}{\partial t}, w\right) - i\Gamma_1\left(K_0 f, w\right) - i\Gamma_1\left(Vf,w\right) - \frac{1}{i}T(t)K_0\left(\frac{\partial f}{\partial t}\right) - \frac{1}{i}T(t)V\frac{\partial f}{\partial t} \\
&\quad + T(t)\left(K_0\dot{+}V\right)^2 f - \frac{1}{i}\varphi\frac{\partial f}{\partial t} + \varphi K_0 f + \varphi V f \\
&\quad - \frac{\partial}{\partial t}\Gamma_1\left(f,w\right) + \frac{1}{i}\frac{\partial}{\partial t}\left(T(t)\left(K_0\dot{+}V\right)f\right) + \frac{1}{i}\frac{\partial\left(\varphi f\right)}{\partial t} \\
&\quad + i\left(K_0\dot{+}V\right)\Gamma_1\left(f,w\right) - T(t)\left(K_0\dot{+}V\right)^2 f - K_0\left(\varphi f\right) - \varphi V f \\
&= \Gamma_1\left(\frac{\partial f}{\partial t}, w\right) - i\Gamma_1\left(K_0 f, w\right) - iV\Gamma_1\left(f,w\right) - if\Gamma_1\left(V,w\right) \\
&\quad - \frac{1}{i}T(t)K_0\left(\frac{\partial f}{\partial t}\right) - \frac{1}{i}T(t)V\frac{\partial f}{\partial t} - \frac{1}{i}\varphi\frac{\partial f}{\partial t} + \varphi K_0 f \\
&\quad - \Gamma_1\left(\frac{\partial f}{\partial t}, w\right) - \Gamma_1\left(f, \frac{\partial w}{\partial t}\right) + \frac{1}{i}\frac{\partial T(t)}{\partial t}K_0 f + \frac{1}{i}\frac{\partial T(t)}{\partial t}Vf \\
&\quad + \frac{1}{i}T(t)\frac{\partial V}{\partial t}f + \frac{1}{i}T(t)K_0\left(\frac{\partial f}{\partial t}\right) + \frac{1}{i}T(t)V\frac{\partial f}{\partial t} \\
&\quad + \frac{1}{i}\frac{\partial\varphi}{\partial t}f + \frac{1}{i}\varphi\frac{\partial f}{\partial t} + iK_0\Gamma_1\left(f,w\right) + iV\Gamma_1\left(f,w\right) \\
&\quad - \left(K_0\varphi\right)f + \Gamma_1\left(f,\varphi\right) - \varphi K_0 f
\end{aligned}$$

$$= -i\Gamma_1(K_0f,w) - if\Gamma_1(V,w) - \Gamma_1\left(f,\frac{\partial w}{\partial t}-\varphi\right) + \frac{1}{i}\frac{\partial T(t)}{\partial t}K_0f$$
$$+\frac{1}{i}\frac{\partial(T(t)V)}{\partial t}f + \frac{1}{i}\frac{\partial\varphi}{\partial t}f + iK_0\Gamma_1(f,w) - (K_0\varphi)f$$
$$= \frac{1}{i}\frac{\partial T(t)}{\partial t}(K_0f) - \frac{1}{i}K_0\Gamma_1(f,w) + \frac{1}{i}\Gamma_1(K_0f,w) - \Gamma_1\left(f,\frac{\partial w}{\partial t}-\varphi\right)$$
$$+\frac{1}{i}f\left\{\Gamma_1(V,w) + \frac{\partial(T(t)V)}{\partial t} + \frac{\partial\varphi}{\partial t} + \frac{1}{i}K_0\varphi\right\}. \tag{3.12}$$

The result in Theorem 3.4 follows from the assumptions (a′) and (b′). □

Corollary 3.5. *Suppose that the functions w, T (which only depends on t), and ψ possess the following properties:*

(a′) *The set of functions f for which the equality*

$$K_0f\frac{dT}{dt} = K_0\Gamma_1(f,w) - \Gamma_1(K_0f,w) + \Gamma_1(f,K_0w+\psi)$$

makes sense and is valid is dense in the space $L^2(E\times[t_0,u],dm\times dt)$.

(b′) *The following equality is valid:*

$$\left(\frac{\partial}{\partial t}\right)^2 w + K_0^2w + i\frac{\partial\psi}{\partial t} + K_0\psi = -\Gamma_1(V,w) - \frac{\partial(TV)}{\partial t}.$$

Put

$$\mathcal{N}(t)f = i\Gamma_1(f,w) - T(t)\left(K_0\dot{+}V\right)f - \left(\left(\frac{\partial}{\partial t}-iK_0\right)w + i\psi\right)f,\quad f\in D\left(K_0\dot{+}V\right).$$

Then $\mathcal{N}(t)$ commutes with $\dfrac{\partial}{i\partial t} - \left(K_0\dot{+}V\right)$.

Proof. Set $\varphi = \dfrac{\partial w}{\partial t} + iK_0w + i\psi$ in Theorem 3.4. Then

$$K_0\Gamma_1(f,w) - \Gamma_1(K_0f,w) + i\Gamma_1\left(f,\frac{\partial w}{\partial t}-\varphi\right)$$
$$= K_0\Gamma_1(f,w) - \Gamma_1(K_0f,w) + \Gamma_1(f,K_0w+\psi) = K_0f\frac{dT}{dt}. \tag{3.13}$$

This shows (a′) of Theorem 3.4. It is clear that (b′) of Theorem 3.4 is satisfied as well. This proves the corollary. □

Let w and χ be related as in (3.15) of the proof of Proposition 3.6. This proposition isolates the properties of the function w. Its proof shows that in the operator

$$f\mapsto K_0(f)\frac{dT}{dt} - K_0\Gamma_1(f,w) + \Gamma_1(K_0(f),w) + \Gamma_1(f,\chi)$$

is a derivation if and only (3.14) is true.

Proposition 3.6. *Suppose that the function w has property* (a) *of Theorem* 3.1, *or* (a′) *of Theorem* 3.4, *or* (a′) *of its Corollary* 3.5. *Then*

$$\Gamma_1(f,g)\frac{dT}{dt}+\Gamma_1\left(\Gamma_1(f,g),w\right)=\Gamma_1\left(\Gamma_1(f,w),g\right)+\Gamma_1\left(f,\Gamma_1(g,w)\right), \tag{3.14}$$

for all functions f and g in $D\left(D_1-K_0\right)$.

Proof. Let f and g be functions in $D\left(D_1-K_0\right)$ with the property that its product fg also belongs to $D\left(D_1-K_0\right)$. We write

$$\chi=\frac{\partial w}{\partial t}-\varphi,\quad \chi=-i\frac{\partial w}{\partial t}-\varphi,\quad \text{or}\quad \chi=-K_0w-\psi, \tag{3.15}$$

as the case may be. Then

$$\begin{aligned}
&K_0(fg)\frac{dT}{dt}-K_0\Gamma_1(fg,w)+\Gamma_1(K_0(fg),w)+\Gamma_1(fg,\chi)\\
&=((K_0f)g-\Gamma_1(f,g)+fK_0g)\frac{dT}{dt}-K_0(\Gamma_1(f,w)g+f\Gamma_1(g,w))\\
&\quad+\Gamma_1((K_0f)g-\Gamma_1(f,g)+fK_0g,w)+f\Gamma_1(g,\chi)+\Gamma_1(f,\chi)g\\
&=((K_0f)g-\Gamma_1(f,g)+fK_0g)\frac{dT}{dt}-(K_0f)\Gamma_1(f,w)+\Gamma(f,\Gamma_1(g,w))-fK_0\Gamma_1(g,w)\\
&\quad+\Gamma_1(K_0f,w)g+(K_0f)\Gamma_1(g,w)-\Gamma_1(\Gamma_1(f,g),w)+\Gamma_1(f,w)K_0g+f\Gamma_1(K_0g,w)\\
&\quad+f\Gamma_1(g,\chi)+\Gamma_1(f,\chi)g\\
&=\left((K_0f)\frac{dT}{dt}-K_0\Gamma_1(f,w)+\Gamma_1(K_0f,w)+\Gamma_1(f,\chi)\right)g\\
&\quad+f\left((K_0g)\frac{dT}{dt}-K_0\Gamma_1(g,w)+\Gamma_1(K_0g,w)+\Gamma_1(g,\chi)\right)\\
&\quad-\Gamma_1(f,g)\frac{dT}{dt}+\Gamma_1(\Gamma_1(f,w),g)+\Gamma_1(f,\Gamma_1(g,w))-\Gamma_1(\Gamma_1(f,g),w)
\end{aligned}\tag{3.16}$$

An application of either (a) of Theorem 3.1 or (a′) of Theorem 3.4 or of Corollary 3.5 then yields equation (3.14) in Proposition 3.6. □

Remark. Let the functions T, w and ψ satisfy (a′) and (b′) of Corollary 3.5. Put $\chi=\psi+K_0w$. Then the triple (T,w,χ) satisfies:

(a) $K_0f\dfrac{dT}{dt}=K_0\Gamma_1\left(f,w\right)-\Gamma_1\left(K_0f,w\right)+\Gamma_1\left(f,\chi\right)$ (for f in a dense subspace of $L^2\left(E,m\right)$);

(b) $\dfrac{\partial^2w}{\partial t^2}+K_0\chi+i\dfrac{\partial\left(\chi-K_0w\right)}{\partial t}=-\Gamma_1\left(V,w\right)-\dfrac{\partial\left(TV\right)}{\partial t}$;

(c) $\dfrac{\partial\left(\chi-K_0w\right)}{\partial t}=0$ if V is real valued.

In order to find Noether observables the equations (a), and (b) have to be integrated simultaneously. Proposition 3.6 simplifies this somewhat in the sense that one first tries to find w, then χ. The couple (w,χ) also has to satisfy (b).

Note that in case

$$E = \mathbb{R}^d \quad \text{and} \quad K_0 f = -\frac{1}{2} \sum_{j,k=1}^{d} \frac{\partial}{\partial x_j} a_{j,k} \frac{\partial}{\partial x_k} f,$$

then

$$\Gamma_1(f,g) = \sum_{j,k=1}^{d} a_{j,k} \frac{\partial f}{\partial x_j} \frac{\partial g}{\partial x_k}.$$

Upon choosing linear functions f and g we see that w has to satisfy:

$$\begin{aligned} & a_{j,k}\frac{dT}{dt} + \sum_{\ell,m=1}^{d} a_{\ell,m} \frac{\partial a_{j,k}}{\partial x_m} \frac{\partial w}{\partial x_\ell} \\ &= 2\sum_{\ell,m=1}^{d} a_{j,\ell} a_{k,m} \frac{\partial^2 w}{\partial x_\ell \partial x_m} + \sum_{\ell,m=1}^{d} a_{j,m} \frac{\partial a_{k,\ell}}{\partial x_m} \frac{\partial w}{\partial x_\ell} + \sum_{\ell,m=1}^{d} \frac{\partial a_{j,\ell}}{\partial x_m} a_{k,m} \frac{\partial w}{\partial x_\ell}. \end{aligned}$$

It follows that the matrix with entries $\dfrac{\partial^2 w}{\partial x_l \partial x_m}$ is, up to a first order perturbation, $\dfrac{1}{2}\dfrac{dT}{dt} \times$ the inverse of the matrix $\left(a_{\ell,m}\right)_{\ell,m=1}^{d}$.

Conclusion

One of the main results of the present paper is contained in Theorem 2.1. The method of proof is based on martingale methods. For more information on viscosity solutions the reader is referred to [6]. Another feature of the present paper is the statement and proof of a generalized Noether theorem (Theorem 3.1) and its complex companion (Theorem 3.4). The proofs are of a computational character; they only depend on the properties of the generator of the diffusion and the corresponding carré du champ operator. They imitate and improve results obtained by Zambrini in [24]. Moreover the results solve problems posed in [22] (Problem 4, Theorem 16, pp. 257-258) and in §2 of [23]. In particular see Problem 4 and the question prior and related to the suggested Theorem 6 on pp. 48–50 of [23]. The present paper extends the results in [21] considerably.

Acknowledgment

Thanks are due to the University of Antwerp (UIA) for its material support. The author is indebted to J.-C. Zambrini (Lisbon) for interesting discussions on the subject and for some references. The author is also obliged to K.D. Elworthy for his hospitality while he was visiting the University of Warwick during the Workshop on SPDE's (July 16–27, 2001), where the present paper was finished. He also thanks the referee for pointing out some errors in a previous version.

References

[1] S. Albeverio, J. Rezende and J.C. Zambrini, *Probability and quantum symmetries 2.* 1999, (in preparation).

[2] J.-P. Aubin, *Systems of Hamilton-Jacobi-Bellman equations: A viability approach.* Istituto Nazionale di Alta Matematica, Università degli Studi di Roma "Tor Vergata", Notes, June–July 2001.

[3] D. Bakry, *L'hypercontractivité et son utilisation en théorie de semigroupes.* Lectures on probability theory **1581**, École d'été de Saint-Flour XXII-1992, P. Bernard (editor), Springer Verlag, Berlin, 1994, 1–114.

[4] D. Christodoulou, *The action principle and partial differential equations.* Annals of Mathematics Studies **146**, 2000, Princeton University Press, Princeton, NJ, 08540 USA.

[5] K.L. Chung and J.-C. Zambrini, *Introduction to random time and quantum randomness.* McGraw-Hill, Lisbon, 2001.

[6] M.G. Crandall and H. Ishii and P.L. Lions, *User's guide to viscosity solutions of second order partial differential equations.* Bull. Amer. Math. Soc. (New Series) **27** (1992), 1–67.

[7] A.B. Cruzeiro and J.C. Zambrini, *Malliavin Calculus and Euclidean quantum mechanics I; Functional Calculus.* J. Functional Analysis **96** (1991), 62–95.

[8] D.A. Dawson and E.A. Perkins, *Measure valued processes and renormalization of branching particle systems.* Stochastic partial differential equations: six perspectives, 45–106, Math. Surveys Monogr., 64, Amer. Math. Soc., Providence, RI, 1999.

[9] M. Demuth and J.A. van Casteren, *Stochastic spectral theory of selfadjoint Feller operators: a functional integration approach.* Birkhäuser Verlag, Basel, 2000.

[10] E.B. Dynkin and S.E. Kuznetsov, *Solutions of $Lu = u^\alpha$ dominated by L-harmonic functions.* J. Anal. Math. **68** (1996), 15–37.

[11] E.B. Dynkin and S.E. Kuznetsov, *Linear additive superdiffusions and related nonlinear P.D.E.* Trans. Amer. Math. Soc. **348** (1996), no. 5, 1959–1987.

[12] A.M. Etheridge, *An introduction to superprocesses.* University Lecture Series, 20, American Math. Soc., Providence, RI, 2000.

[13] L.C. Evans, *Partial differential equations.* Amer. Math. Soc. Providence R.I. 1998, Graduate Texts in Mathematics **19**.

[14] W.H. Fleming and H. M. Soner, *Controlled Markov processes and viscosity solutions.* Applications of Mathematics, **25**, Springer Verlag, Berlin, 1993.

[15] É. Pardoux, *Backward stochastic differential equations and viscosity solutions of systems of semilinear parabolic and elliptic PDEs of second order.* Stochastic Analysis and related topics VI (Geilo, 1996), 79–127, Progr. Probab., **42**, Birkhäuser, Boston, MA, 1998.

[16] É. Pardoux, *BSDEs, weak convergence and homogenization of semilinear PDEs.* Nonlinear analysis, differential equations and control (Montreal, QC, 1998), 503–549, NATO Sci. Ser. C Math. Phys. Sci., 528, Kluwer Acad. Publ., Dordrecht, 1999.

[17] D. Revuz and M. Yor, *Continuous martingales and Brownian motion.* Grundlehren der Mathematischen Wissenschaften **293**, 1991.

[18] S.J. Sheu, *Stochastic control and principal eigenvalue.* Stochastics **11** (1984) no 3–4, 191–211.

[19] M. Thieullen and J.C. Zambrini, *Probability and quantum symmetries 1. The theorem of Noether in Schrodinger's Euclidean quantum mechanics.* Ann. Inst. H. **67** no. 3 (1997), 297–338.

[20] M. Thieullen and J.C. Zambrini, *Symmetries in the stochastic calculus of variations.* Prob. Theory and related fields, **107**, (1997), 401–427.

[21] J.A. Van Casteren, *Viscosity solutions and the Hamilton-Jacobi-Bellman equation.* Proceedings of CIMASI'2000, Casablanca, October 23, 24 and 25 (on CD-rom).

[22] J. A. Van Casteren, *Feynman-Kac semigroups, martingales and wave operators.* J. Korean Math. Soc. **38** no. 2 (2001), 227–274.

[23] J.A. Van Casteren, *Some problems in stochastic analysis and semigroup theory.* Semigroups of operators: theory and applications (Newport Beach, CA, 1998), 43–60, Progr. Nonlinear Differential Equations Appl. **42**, Birkhäuser, Basel, 2000.

[24] J.-C. Zambrini, *A special time-dependent quantum invariant and a general theorem on quantum symmetries.* Proceedings of the second International Workshop Stochastic Analysis and Mathematical Physics: ANESTOC'96, R. Rebolledo ed., Workshop: Vina del Mar Chile 16–20 December 1996, World Scientific, Singapore, 1998, 197–210.

Jan A. van Casteren
Department of Mathematics and Computer Science
University of Antwerp (UIA)
2610 Wilrijk/Antwerp, Belgium
E-mail address: `vcaster@uia.ua.ac.be`

Progress in Nonlinear Differential Equations
and Their Applications, Vol. 55, 403–413

Marcinkiewicz and Mihlin Multiplier Theorems, and R-Boundedness

Alberto Venni

1. Introduction

Let $\mathcal{T}$ be a set of bounded linear operators on a Banach space X to a Banach space Y. $\mathcal{T}$ is said to be R-bounded if $\exists C > 0$ such that for any positive integer N and for arbitrary choices of $T_1, \dots, T_N \in \mathcal{T}$ and $x_1, \dots, x_N \in X$ one has

$$\sum_{\varepsilon \in \{-1,1\}^N} \Big\| \sum_{k=1}^{N} \varepsilon_k T_k x_k \Big\|_Y \le C \sum_{\varepsilon \in \{-1,1\}^N} \Big\| \sum_{k=1}^{N} \varepsilon_k x_k \Big\|_X. \tag{1}$$

If $j_1, \dots, j_N$ are positive integers such that $j_h \neq j_k$ for $h \neq k$, then the Rademacher functions $r_{j_1}, \dots, r_{j_N}$ are equidistributed, so that for any function f on $\{-1,1\}^N$ we have $\int_0^1 f(r_{j_1}(t), \dots, r_{j_N}(t))\, dt = 2^{-N} \sum_{\varepsilon \in \{-1,1\}^N} f(\varepsilon)$. Therefore (1) becomes

$$\int_0^1 \Big\| \sum_{k=1}^{N} r_{j_k}(t)\, T_k x_k \Big\| \, dt \le C \int_0^1 \Big\| \sum_{k=1}^{N} r_{j_k}(t)\, x_k \Big\| \, dt,$$

and now, because of the Khintchine-Kahane inequality (see [3, 11.1]), formula (1) can be written in an equivalent form as

$$\Big(\sum_{\varepsilon \in \{-1,1\}^N} \Big\| \sum_{k=1}^{N} \varepsilon_k T_k x_k \Big\|_Y^q \Big)^{1/q} \le C_q \Big(\sum_{\varepsilon \in \{-1,1\}^N} \Big\| \sum_{k=1}^{N} \varepsilon_k x_k \Big\|_X^q \Big)^{1/q}$$

where $q \in \mathbb{R}^+$. The best C_q will be called the R_q-bound of $\mathcal{T}$.

In the last years the notion of R-boundedness has proven to be very useful in connection with such problems as the maximal regularity for linear differential equations of parabolic type and the boundedness of the H^∞ functional calculus for sectorial operators. We refer to [2, 5] for further information on the subject.

In this note it is proved that if a family of Fourier multipliers on the real line satisfies a uniform bound of Marcinkiewicz type, then the corresponding set of bounded operators on $L^p(\mathbb{T}, X)$ to itself (where X is a UMD Banach space with the property (α) and $1 < p < \infty$) is R-bounded. An analogous result for a family of Mihlin multipliers in $L^p(\mathbb{R}, X)$ is proved.

2. The discrete case

Let X be a complex Banach space, and let $\mathbb{T} = \mathbb{R}/\mathbb{Z}$ be the one-dimensional torus. Let $(\lambda_k)_{k\in\mathbb{Z}}$ be a family of complex numbers, and let Λ be the operator defined on the space of the X-valued trigonometric polynomials in the following way: $(\Lambda f)\hat{}\,(k) = \lambda_k \hat{f}(k)$, where for an arbitrary function $g \in L^1(\mathbb{T}, X)$ $\hat{g}(k) := \int_0^1 e^{-2\pi ikt} g(t)\,dt$ is the k-th Fourier coefficient of g. One says that $(\lambda_k)_{k\in\mathbb{Z}}$ is a family of multipliers in $L^p(\mathbb{T}, X)$ (where $p \in [1, \infty[$) if the operator Λ has a continuous extension on $L^p(\mathbb{T}, X)$ to $L^p(\mathbb{T}, X)$. In this case, as the operator $g \mapsto \hat{g}(k)$ is continuous on $L^p(\mathbb{T}, X)$ to X, the equality $(\Lambda f)\hat{}\,(k) = \lambda_k \hat{f}(k)$ also holds when Λ denotes the continuous extension to $L^p(\mathbb{T}, X)$ of the homonymous operator originally defined on the space of the trigonometric polynomials.

In theorem 2 we state an extension of the Marcinkiewicz multiplier theorem to the case of vector valued functions that was proved in [1, Theorem 4]. We agree that $(\Delta_k)_{k\in\mathbb{Z}}$ is the usual dyadic decomposition of $\mathbb{Z}$: precisely we set $\Delta_0 = \{0\}$, and for any positive integer k, $\Delta_k = \{n \in \mathbb{Z};\ 2^{k-1} \leq n < 2^k\}$, $\Delta_{-k} = -\Delta_k$.

Theorem 2. *Let $(\lambda_k)_{k\in\mathbb{Z}}$ be a family of complex numbers. Assume that*

$$\max\Big\{\sup_{r\in\mathbb{Z}} |\lambda_r|,\ \sup_{k\in\mathbb{Z}} \sum_{r, r-1\in\Delta_k} |\lambda_r - \lambda_{r-1}|\Big\} = \gamma \in \mathbb{R}^+.$$

Then for each Banach space X of type UMD and $\forall p \in]1, \infty[$ $(\lambda_k)_{k\in\mathbb{Z}}$ is a family of multipliers in $L^p(\mathbb{T}, X)$, and the norm of the corresponding operator $\Lambda \in \mathcal{L}(L^p(\mathbb{T}, X))$ satisfies the inequality $\|\Lambda\|_{\mathcal{L}(L^p(\mathbb{T},X))} \leq C(p, X)\,\gamma$.

Our first aim is to prove the following R-version of theorem 2.

Theorem 3. *Let Ξ be any set of indices, and $\forall \xi \in \Xi$ let $(\lambda_{k,\xi})_{k\in\mathbb{Z}}$ be a family of complex numbers. Assume that*

$$\sup_{\xi\in\Xi}\Big(\max\Big\{\sup_{r\in\mathbb{Z}} |\lambda_{r,\xi}|,\ \sup_{k\in\mathbb{Z}}\ ;\ \sum_{r, r-1\in\Delta_k} |\lambda_{r,\xi} - \lambda_{r-1,\xi}|\Big\}\Big) = \gamma \in \mathbb{R}^+.$$

Let X be a UMD Banach space with the property (α), $p \in]1, \infty[$, and let Λ_ξ be the operator in $L^p(\mathbb{T}, X)$ associated to the family of multipliers $(\lambda_{k,\xi})_{k\in\mathbb{Z}}$. Then $\{\Lambda_\xi;\ \xi \in \Xi\}$ is a R-bounded set in $\mathcal{L}(L^p(\mathbb{T}, X))$, and its R_p-bound is $\leq C(p, X)\,\gamma$.

Let us establish some notations and recall some results.

If X is a UMD Banach space, then $(-i\,\mathrm{sgn}\,k)_{k\in\mathbb{Z}}$ is a family of multipliers in $L^p(\mathbb{T}, X)$ for every $p \in]1, \infty[$: in fact that could be taken as a definition of the class UMD. The corresponding operator is the discrete Hilbert transform H. The Riesz projection operator $R = R_0$ is the operator defined by $Rf = \frac{1}{2}(f + iHf) + \frac{1}{2}\hat{f}(0)$. This is obviously a bounded operator on $L^p(\mathbb{T}, X)$ to itself ($p \in]1, \infty[$), whose norm only depends on the norm of H in $\mathcal{L}(L^p(\mathbb{T}, X))$, and it is the operator associated to the family $(\lambda_k)_{k\in\mathbb{Z}}$ of multipliers, where $\lambda_k = 1$ if $k \geq 0$ and $\lambda_k = 0$ if $k < 0$.

Let $m \in \mathbb{Z}$. The shifted Riesz operator R_m is defined by $R_m f = \varphi_m R(\varphi_{-m} f)$, where, $\forall k \in \mathbb{Z}$, $\varphi_k(t) = e^{2\pi i k t}$. This is obviously a bounded operator on $L^p(\mathbb{T}, X)$ to itself (always, as it is understood from now on, for $p \in]1, \infty[$), with $\|R_m\|_{\mathcal{L}(L^p(\mathbb{T},X))} = \|R\|_{\mathcal{L}(L^p(\mathbb{T},X))}$, and it is in fact the operator associated to the family of multipliers $(\lambda_k)_{k\in\mathbb{Z}}$, where $\lambda_k = 1$ if $k \geq m$ and $\lambda_k = 0$ if $k < m$. Therefore, if Δ is a "segment" of $\mathbb{Z}$, say $\Delta = \{k \in \mathbb{Z};\ m \leq k \leq n\}$, then the characteristic function of Δ is a family of multipliers, and the associated operator is $S_\Delta = R_m - R_{n+1}$, so that the operator norm of S_Δ is $\leq 2\,\|R\|_{\mathcal{L}(L^p(\mathbb{T},X))}$.

The property (α) for a Banach space X can be expressed as follows:

$\exists C > 0$ such that for any positive integer n and for arbitrary families $(\alpha_{j,k})_{1\leq j,k\leq n} \in \mathbb{C}^{n^2}$ and $(x_{j,k})_{1\leq j,k\leq n} \in X^{n^2}$ the following inequality holds

$$\Big(\sum_{\varepsilon\in\{-1,1\}^n} \sum_{\varepsilon'\in\{-1,1\}^n} \Big\| \sum_{j,k=1}^{n} \varepsilon_j\, \varepsilon'_k\, \alpha_{j,k}\, x_{j,k} \Big\|^2 \Big)^{1/2} \leq$$

$$C \max_{j,k} |\alpha_{j,k}| \Big(\sum_{\varepsilon\in\{-1,1\}^n} \sum_{\varepsilon'\in\{-1,1\}^n} \Big\| \sum_{j,k=1}^{n} \varepsilon_j\, \varepsilon'_k\, x_{j,k} \Big\|^2 \Big)^{1/2}.$$

In order to prove theorem 3, we need some lemmas.

Lemma 4. *Let X be a UMD Banach space, $1 < p < \infty$. Suppose that $T_1, \ldots, T_n$ belong to the linear subspace of $\mathcal{L}(L^p(\mathbb{T}, X))$ generated by $\{R_k; k \in \mathbb{Z}\}$, $T_k = \sum_{j\in\mathbb{Z}} \lambda_{kj} R_j$. If $\mu = \sup_{1\leq k\leq n} \sum_{j\in\mathbb{Z}} |\lambda_{kj}|$, then*

$$\Big(\sum_{\varepsilon\in\{-1,1\}^n} \| \sum_{k=1}^{n} \varepsilon_k\, T_k\, f_k \|^p_{L^p(\mathbb{T},X)} \Big)^{1/p}$$

$$\leq \mu\, C(p, X) \Big(\sum_{\varepsilon\in\{-1,1\}^n} \| \sum_{k=1}^{n} \varepsilon_k\, f_k \|^p_{L^p(\mathbb{T},X)} \Big)^{1/p}.$$

Lemma 5. *Let X be a UMD Banach space, $1 < p < \infty$. Then for $f \in L^p(\mathbb{T}, X)$ and $m, n \in \mathbb{Z}$ with $m \leq n$,*

$$\Big(2^{-(n-m+1)} \sum_{\varepsilon} \| \sum_{k=m}^{n} \varepsilon_k\, S_{\Delta_k}\, f \|^p_{L^p(\mathbb{T},X)} \Big)^{1/p} \leq C(p, X) \|f\|_{L^p(\mathbb{T},X)}.$$

Lemma 6. *Let X be a UMD Banach space, $1 < p < \infty$, $m, n \in \mathbb{Z}$ with $m \leq n$, $f_m, \ldots, f_n \in L^p(\mathbb{T}, X)$. Then*

$$\| \sum_{k=m}^{n} S_{\Delta_k}\, f_k \|_{L^p(\mathbb{T},X)} \leq C(p, X) \Big(2^{-(n-m+1)} \sum_{\varepsilon} \| \sum_{k=m}^{n} \varepsilon_k\, S_{\Delta_k}\, f_k \|^p_{L^p(\mathbb{T},X)} \Big)^{1/p}.$$

Lemma 7. *Let X be a Banach space with the property (α), and let $\mathcal{T}$ be a R-bounded subset of $\mathcal{L}(X)$. Then if we call C the constant that appears in the definition of the property (α) and M the R_2-bound of $\mathcal{T}$ we have*

$$\Big(\sum_{\varepsilon\in\{-1,1\}^n}\sum_{\varepsilon'\in\{-1,1\}^n}\Big\|\sum_{j,k=1}^{n}\varepsilon_j\,\varepsilon'_k\,T_{j,k}\,x_{j,k}\Big\|^2\Big)^{1/2}$$
$$\leq C^2 M\Big(\sum_{\varepsilon\in\{-1,1\}^n}\sum_{\varepsilon'\in\{-1,1\}^n}\Big\|\sum_{j,k=1}^{n}\varepsilon_j\,\varepsilon'_k\,x_{j,k}\Big\|^2\Big)^{1/2}$$

whenever n is a positive integer, $(x_{j,k})_{1\leq j,k\leq n}\in X^{n^2}$ and $(T_{j,k})_{1\leq j,k\leq n}\in\mathcal{T}^{n^2}$.

A proof of lemma 4 can be found in [1, Lemma 7] (see also [6, Lemma 1.2]) and relies on the so-called "contraction principle" (that is the R-boundedness of $\{\lambda I_X; |\lambda|\leq M\}$, see [3, 12.2]) plus a convexity argument. The Littlewood-Paley type lemma 5 above is Theorem 3 of [1]. In order to obtain lemma 6 it is enough to apply to $f := \sum_{k=m}^{n} S_{\Delta_k} f_k$ the inequality

$$\|f\|_{L^p(\mathbb{T},X)}\leq C(p,X)\Big(2^{-(n-m+1)}\sum_{\varepsilon}\|\sum_{j=m}^{n}\varepsilon_j\,S_{\Delta_j}\,f\|^p_{L^p(\mathbb{T},X)}\Big)^{1/p}$$

that can be obtained from lemma 5 by means of a duality argument (see [6, (1.9)]). Lemma 7 is proved in [2, Lemma 3.13].

Proof of Theorem 3.
We have to prove that for a positive integer n, $(\xi_1,\ldots,\xi_n)\in\Xi^n$ and $f_1,\ldots,f_n\in L^p(\mathbb{T},X)$ the following inequality holds:

$$\Big(\sum_{\varepsilon\in\{-1,1\}^n}\Big\|\sum_{i=1}^{n}\varepsilon_i\,\Lambda_{\xi_i}f_i\Big\|^2_{L^2(\mathbb{T},X)}\Big)^{1/2}\leq C(X)\gamma\Big(\sum_{\varepsilon\in\{-1,1\}^n}\Big\|\sum_{i=1}^{n}\varepsilon_i f_i\Big\|^2_{L^2(\mathbb{T},X)}\Big)^{1/2}.$$

It is enough to prove this inequality when $f_1,\ldots,f_n$ are X-valued trigonometric polynomials, since the space of the trigonometric polynomials is dense in $L^2(\mathbb{T},X)$. In this case there exist q, $r\in\mathbb{Z}$, with $q<r$, such that the support of every f_i is contained in $\bigcup_{k=q}^{r}\Delta_k$. Now we call $\Lambda_{i,k}$ the operator associated to the family of multipliers $(\chi_{\Delta_k}(m)\,\lambda_{m,\xi_i})_{m\in\mathbb{Z}}$ which vanishes outside of Δ_k and equals λ_{m,ξ_i} if $m\in\Delta_k$. We have obviously $\Lambda_{i,k}=\Lambda_{\xi_i}\,S_{\Delta_k}=S_{\Delta_k}\,\Lambda_{\xi_i}$, and we have also $\Lambda_{i,k}=\Lambda_{i,k}\,S_{\Delta_k}=S_{\Delta_k}\,\Lambda_{i,k}$. Moreover the support of $\Lambda_{\xi_i}f_i$ is contained in $\bigcup_{k=q}^{r}\Delta_k$,

as well as the support of f_i; therefore

$$\Lambda_{\xi_i} f_i = \sum_{k=q}^{r} S_{\Delta_k} \Lambda_{\xi_i} f_i = \sum_{k=q}^{r} \Lambda_{i,k} f_i = \sum_{k=q}^{r} S_{\Delta_k} \Lambda_{i,k} f_i.$$

Hence we have

$$\sum_{\varepsilon\in\{-1,1\}^n} \Big\| \sum_{i=1}^{n} \varepsilon_i \Lambda_{\xi_i} f_i \Big\|^2_{L^2(\mathbb{T},X)} = \sum_{\varepsilon\in\{-1,1\}^n} \Big\| \sum_{k=q}^{r} S_{\Delta_k} \sum_{i=1}^{n} \varepsilon_i \Lambda_{i,k} f_i \Big\|^2_{L^2(\mathbb{T},X)} \leq$$

(by lemma 6)

$$\leq \frac{C(X)^2}{2^{r-q+1}} \sum_{\varepsilon} \sum_{\varepsilon'} \Big\| \sum_{k=q}^{r} \varepsilon'_k S_{\Delta_k} \sum_{i=1}^{n} \varepsilon_i \Lambda_{i,k} f_i \Big\|^2_{L^2(\mathbb{T},X)}$$

$$= \frac{C(X)^2}{2^{r-q+1}} \sum_{(\varepsilon,\, \varepsilon')} \Big\| \sum_{i,k} \varepsilon_i \varepsilon'_k \Lambda_{i,k} S_{\Delta_k} f_i \Big\|^2_{L^2(\mathbb{T},X)}.$$

Now we remark that the operators $\Lambda_{i,k}$ satisfy the assumptions of lemma 4, with $\mu \leq 3\gamma$. Indeed, if we set $\Delta_k = \{\alpha_k, \dots, \beta_k\}$, then

$$\Lambda_{i,k} = \sum_{j=\alpha_k}^{\beta_k} \lambda_{i,j} (R_j - R_{j+1}) = \lambda_{i,\alpha_k} R_{\alpha_k} - \lambda_{i,\beta_k} R_{\beta_k+1} + \sum_{j,j-1\in\Delta_k} (\lambda_{i,j} - \lambda_{i,j-1}) R_j.$$

Therefore, by lemmas 4 and 7 we obtain

$$\sum_{\varepsilon\in\{-1,1\}^n} \Big\| \sum_{i=1}^{n} \varepsilon_i \Lambda_{\xi_i} f_i \Big\|^2_{L^2(\mathbb{T},X)} \leq \frac{(3\gamma C(X))^2}{2^{r-q+1}} \sum_{(\varepsilon,\, \varepsilon')} \Big\| \sum_{i,k} \varepsilon_i \varepsilon'_k S_{\Delta_k} f_i \Big\|^2_{L^2(\mathbb{T},X)}$$

$$= \frac{(3\gamma C(X))^2}{2^{r-q+1}} \sum_{\varepsilon} \sum_{\varepsilon'} \Big\| \sum_{k} \varepsilon'_k S_{\Delta_k} \Big(\sum_{i} \varepsilon_i f_i \Big) \Big\|^2_{L^2(\mathbb{T},X)} \leq$$

(by lemma 5)

$$\leq (3\gamma C(X))^2 \sum_{\varepsilon} \Big\| \sum_{i} \varepsilon_i f_i \Big\|^2_{L^2(\mathbb{T},X)}. \qquad \square$$

3. The continuous case

Now we want to obtain a result analogous to theorem 3 for multipliers of Mihlin type in $L^p(\mathbb{R}, X)$. This can be made by a transference trick based on lemmas 8 and 9 (see [6]).

$\forall f \in L^1(\mathbb{R}, X)$, the Fourier transform of f is the function $\mathcal{F}f : \mathbb{R} \to X$ defined by $(\mathcal{F}f)(\tau) = \int_{\mathbb{R}} e^{-2\pi i t\tau} f(t)\, dt$. We denote by J the operator acting on functions on $\mathbb{R}$ in the following way: $(Jf)(t) = f(-t)$: thus $(J\mathcal{F}f)(\tau) = \int_{\mathbb{R}} e^{2\pi i t\tau} f(t)\,dt = (\mathcal{F}Jf)(\tau)$.

The following results hold.

Lemma 8 (Poisson summation formula). *Let X be a Banach space and let $f : \mathbb{R} \to X$ be a continuous function. Suppose that for some $\varepsilon > 0$ we have, as $|t| \to +\infty$, $\|f(t)\| = \mathrm{O}(|t|^{-1-\varepsilon})$ and $\|(\mathcal{F}f)(t)\| = \mathrm{O}(|t|^{-1-\varepsilon})$. Then $\forall t \in \mathbb{R}$ and $\forall a > 0$*

$$\sum_{n\in\mathbb{Z}} f(t+na) = \frac{1}{a}\sum_{n\in\mathbb{Z}} e^{2\pi i n t/a}(\mathcal{F}f)(n/a).$$

Lemma 9. (see [6, Lemma 2.1]). *Let X be a Banach space, $f \in \mathcal{S}(\mathbb{R}, X)$, $p \in]1, \infty[$. $\forall k \in \mathbb{N}$ we define $g_k : \mathbb{R} \to X$, $g_k(t) = 2^{-k/p'}\sum_{n\in\mathbb{Z}} e^{2\pi i n t}(\mathcal{F}f)(2^{-k}n)$, where p' is the exponent conjugate to p. Then $g_k \in L^p(\mathbb{T}, X)$ and $\|f\|_{L^p(\mathbb{R},X)} = \lim_{k\to+\infty} \|g_k\|_{L^p(\mathbb{T},X)}$.*

A proof of lemma 8 can be given by comparing the Fourier coefficients of the 1-periodic functions $t \mapsto \sum_{n\in\mathbb{Z}} f(t+n)$ and $t \mapsto \sum_{n\in\mathbb{Z}} e^{2\pi i n t}(\mathcal{F}f)(n)$. A sketch of proof of lemma 9 works as follows: g_k is a C^∞ 1-periodic function on $\mathbb{R}$, and therefore can be considered as an element of $L^p(\mathbb{T}, X)$. Taking into account the Poisson summation formula, one obtains

$$\begin{aligned}\|g_k\|_{L^p(\mathbb{T},X)} &= \Big(\int_{-2^{k-1}}^{2^{k-1}} \Big\|\sum_{n\in\mathbb{Z}} f(s+2^k n)\Big\|_X^p\, ds\Big)^{1/p}\\ &\le \Big(\int_{-2^{k-1}}^{2^{k-1}} \|f(s)\|_X^p\, ds\Big)^{1/p} + \sum_{0\neq n\in\mathbb{Z}} 2^{k/p} \sup_{|s|\ge 2^k(|n|-(1/2))} \|f(s)\|_X.\end{aligned}$$

Here, as $k \to +\infty$, the first addendum tends to $\|f\|_{L^p(\mathbb{R},X)}$, while the following sum tends to 0 since f is rapidly decreasing at infinity.

The R-version of the Mihlin multipliers theorem for Banach valued functions is the following.

Theorem 10. *Let $(\varphi_i)_{i\in I}$ be a family of functions that belong to $L^\infty(\mathbb{R}) \cap W^{1,1}_{loc}(\mathbb{R}\setminus\{0\})$, and assume that the functions $t \mapsto t\varphi_i'(t)$ (shortly denoted $t\varphi_i'$) belong to $L^\infty(\mathbb{R})$. Assume moreover that*

$$\sup_{i\in I} \max\Big\{\|\varphi_i\|_{L^\infty(\mathbb{R})}, \|t\varphi_i'\|_{L^\infty(\mathbb{R})}\Big\} =: \eta < +\infty.$$

Let X be a UMD Banach space with the property (α), and $p \in]1, \infty[$. Then $\exists\, C(p, X) \in \mathbb{R}^+$ such that for any positive integer n, $(i_1,\ldots,i_n) \in I^n$ and $(f_1,\ldots,f_n)$

$\in (\mathcal{S}(\mathbb{R},X))^n$ *we have*

$$\Big(\sum_{\varepsilon\in\{-1,1\}^n}\Big\|\sum_{r=1}^{n}\varepsilon_r\,\mathcal{F}(\varphi_{i_r}\,\mathcal{F}f_r)\Big\|^p_{L^p(\mathbb{R},X)}\Big)^{1/p}$$
$$\le C(p,X)\,\eta\Big(\sum_{\varepsilon\in\{-1,1\}^n}\Big\|\sum_{r=1}^{n}\varepsilon_r\,f_r\Big\|^p_{L^p(\mathbb{R},X)}\Big)^{1/p}.$$

Proof. We perform the proof in several steps.

(I) Assume that $\forall i\in I$ there is a sequence $(\psi_{i,k})_{k\in\mathbb{N}}$ of functions satisfying the following conditions:

(a) $\varphi_i(t)=\lim_{k\to+\infty}\psi_{i,k}(t)$ almost everywhere on $\mathbb{R}$

(b) $\forall(i,k)\in I\times\mathbb{N}\ \psi_{i,k}\in L^\infty(\mathbb{R})$ and $\forall i\in I\ \exists\, M_i\in\mathbb{R}^+$ such that $\sup_{k\in\mathbb{N}}\|\psi_{i,k}\|_{L^\infty(\mathbb{R})}$ $\le M_i$

(c) $\exists\,\tilde{M}\in\mathbb{R}^+$ such that for any positive integer n, $\forall\, k\in\mathbb{N}$ and for arbitrary $(i_1,\dots,i_n)\in I^n$, $(f_1,\dots,f_n)\in(\mathcal{S}(\mathbb{R},X))^n$ it holds

$$\Big(\sum_{\varepsilon\in\{-1,1\}^n}\Big\|\sum_{r=1}^{n}\varepsilon_r\,\mathcal{F}(\psi_{i_r,k}\,\mathcal{F}f_r)\Big\|^p_{L^p(\mathbb{R},X)}\Big)^{1/p}$$
$$\le\tilde{M}\Big(\sum_{\varepsilon\in\{-1,1\}^n}\Big\|\sum_{r=1}^{n}\varepsilon_r\,f_r\Big\|^p_{L^p(\mathbb{R},X)}\Big)^{1/p}.$$

If we now choose n, $(i_1,\dots,i_n)$ and $(f_1,\dots,f_n)$ as above, since $\mathcal{F}f_r\in\mathcal{S}(\mathbb{R},X)\subseteq L^1(\mathbb{R},X)$, from (a), (b) and the dominated convergence theorem we get

$$\|\psi_{i_r,k}\,\mathcal{F}f_r-\varphi_{i_r}\,\mathcal{F}f_r\|_{L^1(\mathbb{R},X)}\underset{k\to+\infty}{\to}0,$$

so that

$$\|\mathcal{F}(\psi_{i_r,k}\,\mathcal{F}f_r)-\mathcal{F}(\varphi_{i_r}\,\mathcal{F}f_r)\|_{L^\infty(\mathbb{R},X)}\underset{k\to+\infty}{\to}0$$

and also

$$\Big\|\sum_{r=1}^{n}\varepsilon_r\,\mathcal{F}(\psi_{i_r,k}\,\mathcal{F}f_r)-\sum_{r=1}^{n}\varepsilon_r\,\mathcal{F}(\varphi_{i_r}\,\mathcal{F}f_r)\Big\|_{L^\infty(\mathbb{R},X)}\underset{k\to+\infty}{\to}0.$$

Therefore, by Fatou's lemma and assumption (c) we get

$$\Big(\sum_{\varepsilon\in\{-1,1\}^n}\Big\|\sum_{r=1}^{n}\varepsilon_r\mathcal{F}(\varphi_{i_r}\mathcal{F}f_r)\Big\|^p_{L^p(\mathbb{R},X)}\Big)^{1/p}\le\tilde{M}\Big(\sum_{\varepsilon\in\{-1,1\}^n}\Big\|\sum_{r=1}^{n}\varepsilon_r f_r\Big\|^p_{L^p(\mathbb{R},X)}\Big)^{1/p}.$$

(II) Assume that $\forall i\in I\ \varphi_i\in C^\infty(\mathbb{R})$. Then $\forall f\in\mathcal{S}(\mathbb{R},X)$ we have $\varphi_i\,\mathcal{F}f\in\mathcal{S}(\mathbb{R},X)$, and so $\varphi_i\,\mathcal{F}f=\mathcal{F}J\mathcal{F}(\varphi_i\,\mathcal{F}f)$. Now let us fix n, $(i_1,\dots,i_n)$ and $(f_1,\dots,f_n)$ as above.

If $\varepsilon \in \{-1,1\}^n$, then $\sum_{r=1}^{n} \varepsilon_r \, \mathcal{F}(\varphi_{i_r} \, \mathcal{F} f_r) \in \mathcal{S}(\mathbb{R}, X)$, and hence, by lemma 9,

$$\Big\| \sum_{r=1}^{n} \varepsilon_r \, \mathcal{F}(\varphi_{i_r} \, \mathcal{F} f_r) \Big\|_{L^p(\mathbb{R},X)} = \Big\| \sum_{r=1}^{n} \varepsilon_r \, J\mathcal{F}(\varphi_{i_r} \, \mathcal{F} f_r) \Big\|_{L^p(\mathbb{R},X)}$$
$$= \lim_{k \to +\infty} \|h_{k,\varepsilon}\|_{L^p(\mathbb{T},X)}$$

where $h_{k,\varepsilon}(t) = 2^{-k/p'} \sum_{m\in\mathbb{Z}} e^{2\pi i m t} \sum_{r=1}^{n} \varepsilon_r \, \varphi_{i_r}(2^{-k}m) \, (\mathcal{F} f_r)(2^{-k}m)$, so that $h_{k,\varepsilon}$ is the 1-periodic function whose m-th Fourier coefficient is

$$2^{-k/p'} \sum_{r=1}^{n} \varepsilon_r \, \varphi_{i_r}(2^{-k}m) \, (\mathcal{F} f_r)(2^{-k}m).$$

Likewise

$$\Big\| \sum_{r=1}^{n} \varepsilon_r f_r \Big\|_{L^p(\mathbb{R},X)} = \lim_{k \to +\infty} \|g_{k,\varepsilon}\|_{L^p(\mathbb{T},X)}$$

where $g_{k,\varepsilon}(t) = 2^{-k/p'} \sum_{m\in\mathbb{Z}} e^{2\pi i m t} \sum_{r=1}^{n} \varepsilon_r \, (\mathcal{F} f_r)(2^{-k}m)$, so that $g_{k,\varepsilon}$ is the 1-periodic function whose m-th Fourier coefficient is $2^{-k/p'} \sum_{r=1}^{n} \varepsilon_r \, (\mathcal{F} f_r)(2^{-k}m)$. Now we remark that $\sup_{m\in\mathbb{Z}} |\varphi_i(2^{-k}m)| \le \|\varphi_i\|_{L^\infty(\mathbb{R})} \le \eta$. Moreover, if the dyadic interval Δ_j contains two different elements, then $|j| \ge 2$; if $j \ge 2$, then

$$\sum_{m,m-1\in\Delta_j} |\varphi_i(2^{-k}m) - \varphi_i(2^{-k}(m-1))| = \sum_{m=2^{j-1}+1}^{2^j-1} \Big| \int_{2^{-k}(m-1)}^{2^{-k}m} \varphi_i'(s) \, ds \Big|$$
$$\le \sum_{m=2^{j-1}+1}^{2^j-1} \int_{m-1}^{m} |\varphi_i'(2^{-k}\sigma)| \, 2^{-k} \, d\sigma = \int_{2^{j-1}}^{2^j-1} |\varphi_i'(t)| \, dt$$
$$\le \|t\varphi_i'\|_{L^\infty(\mathbb{R})} \int_{2^{j-1}}^{2^j-1} t^{-1} \, dt \le \eta \log 2.$$

The same estimate is obtained when $j \le -2$, by means of a similar argument.

Therefore if we apply theorem 3, we obtain

$$\Big(\sum_{\varepsilon\in\{-1,1\}^n} \|h_{k,\varepsilon}\|^p_{L^p(\mathbb{T},X)} \Big)^{1/p} \le C(p,X) \, (\log 2) \, \eta \Big(\sum_{\varepsilon\in\{-1,1\}^n} \|g_{k,\varepsilon}\|^p_{L^p(\mathbb{T},X)} \Big)^{1/p},$$

and as $k \to +\infty$ we get

$$\Big(\sum_{\varepsilon \in \{-1,1\}^n} \Big\| \sum_{r=1}^{n} \varepsilon_r \mathcal{F}(\varphi_{i_r} \mathcal{F} f_r) \Big\|_{L^p(\mathbb{R},X)}^p \Big)^{1/p}$$
$$\leq C(p,X)\,(\log 2)\,\eta \Big(\sum_{\varepsilon \in \{-1,1\}^n} \Big\| \sum_{r=1}^{n} \varepsilon_r f_r \Big\|_{L^p(\mathbb{R},X)}^p \Big)^{1/p}.$$

(III) Finally we drop the additional assumption that $\varphi_i \in C^\infty$ $\forall i$. We fix $\omega \in C_0^\infty(\mathbb{R})$, with $\omega \geq 0$, $\displaystyle\int_{\mathbb{R}} \omega = 1$ and $\operatorname{supp}\omega \subseteq [-1,1]$. $\forall \rho \in \mathbb{R}^+$ we set $\omega_\rho(t) = \rho^{-1}\omega(\rho^{-1}t)$ and $\psi_{i,\rho} = \omega_\rho * \varphi_i$. Then $\psi_{i,\rho} \in C^\infty(\mathbb{R}) \cap L^\infty(\mathbb{R})$ and (by Young's inequality) $\|\psi_{i,\rho}\|_{L^\infty(\mathbb{R})} \leq \|\varphi_i\|_{L^\infty(\mathbb{R})}$. Moreover

$$t\,\psi_{i,\rho}'(t) = t\,(\omega_\rho' * \varphi_i)(t) = t \int_{t-\rho}^{t+\rho} \omega_\rho'(t-s)\varphi_i(s)\,ds,$$

and if $0 < \rho < |t|$ partial integration also gives

$$t\,\psi_{i,\rho}'(t) = t \int_{t-\rho}^{t+\rho} \omega_\rho(t-s)\varphi_i'(s)\,ds.$$

Let $t \neq 0$ be fixed. If $\rho \geq \dfrac{|t|}{2}$, then

$$|t\,\psi_{i,\rho}'(t)| = |t \int_{\mathbb{R}} \omega_\rho'(t-s)\varphi_i(s)\,ds|$$
$$\leq 2\rho\,\|\varphi_i\|_\infty \int_{\mathbb{R}} \rho^{-2}\,|\omega'(\rho^{-1}(t-s))|\,ds = 2\|\omega'\|_{L^1(\mathbb{R})}\,\|\varphi_i\|_{L^\infty(\mathbb{R})}.$$

If, on the contrary, $\rho \leq \dfrac{|t|}{2}$, then

$$|t\,\psi_{i,\rho}'(t)| = |t \int_{t-\rho}^{t+\rho} \omega_\rho(t-s)\varphi_i'(s)\,ds|$$
$$\leq |t| \int_{\mathbb{R}} |\omega_\rho(\sigma)|\,d\sigma \sup_{|t-s|\leq\rho} |\varphi_i'(s)| \leq \|s\,\varphi_i'\|_\infty \sup_{|t-s|\leq|t|/2} \frac{|t|}{|s|} = 2\,\|s\,\varphi_i'\|_{L^\infty(\mathbb{R})}.$$

Therefore

$$\max\{\|\psi_{i,\rho}\|_{L^\infty(\mathbb{R})}, \|t\,\psi_{i,\rho}'\|_{L^\infty(\mathbb{R})}\}$$
$$\leq \max\{\|\varphi_i\|_{L^\infty(\mathbb{R})}, 2\|\omega'\|_{L^1(\mathbb{R})}\,\|\varphi_i\|_{L^\infty(\mathbb{R})}, 2\|t\,\varphi_i'\|_{L^\infty(\mathbb{R})}\} \leq 2\eta(1 + \|\omega'\|_{L^1(\mathbb{R})}).$$

Next we remark that on $\mathbb{R}\setminus\{0\}$ φ_i is almost everywhere equal to a continuous function. Therefore for almost all $t \neq 0$ we have

$$\begin{aligned}|\psi_{i,\rho}(t) - \varphi_i(t)| &= \left| \int_{\mathbb{R}} \omega_\rho(s)\,(\varphi_i(t-s) - \varphi_i(t))\,ds \right| \\ &\leq \sup_{|s|\leq\rho} |\varphi_i(t-s) - \varphi_i(t)| \underset{\rho\to 0}{\longrightarrow} 0.\end{aligned}$$

Thus, combining steps (I) and (II), we get the desired inequality. □

Corollary 11. *Let $(\psi_i)_{i\in I}$ be a family of elements of $L^1(\mathbb{R})$ and assume that $\forall i \in I$ $\mathcal{F}f_i \in W^{1,1}_{loc}(\mathbb{R}\setminus\{0\})$, with $t\,(\mathcal{F}f_i)' \in L^\infty(\mathbb{R})$. Assume moreover that*

$$\sup_{i\in I} \max\left\{ \|\mathcal{F}f_i\|_{L^\infty(\mathbb{R})},\, \|t\,(\mathcal{F}f_i)'\|_{L^\infty(\mathbb{R})} \right\} =: \eta < +\infty.$$

Let $p \in]1,\infty[$ and let X be a UMD Banach space with the property (α). $\forall i \in I$ let $T_i : L^p(\mathbb{R},X) \to L^p(\mathbb{R},X)$ be the bounded linear operator defined by $T_i f = \psi_i * f$. Then $\{T_i;\ i\in I\}$ is a R-bounded set of operators. More precisely, for any positive integer n and for arbitrary $(i_1,\ldots,i_n) \in I^n$, $(f_1,\ldots,f_n) \in (L^p(\mathbb{R},X))^n$ the following inequality holds:

$$\begin{aligned}&\Big(\sum_{\varepsilon\in\{-1,1\}^n} \Big\| \sum_{r=1}^n \varepsilon_r\, T_{i_r} f_r \Big\|^p_{L^p(\mathbb{R},X)} \Big)^{1/p} \\ &\leq C(p,X)\,\eta \Big(\sum_{\varepsilon\in\{-1,1\}^n} \Big\| \sum_{r=1}^n \varepsilon_r\, f_r \Big\|^p_{L^p(\mathbb{R},X)} \Big)^{1/p}.\end{aligned}$$

Proof. It is sufficient to deal with the particular case of $f_1,\ldots,f_n \in \mathcal{S}(\mathbb{R},X)$. In this case $\psi_{i_r} * f_r \in L^1(\mathbb{R},X)$ and $\mathcal{F}(\psi_{i_r} * f_r) = (\mathcal{F}\psi_{i_r})\,(\mathcal{F}f_r) \in L^1(\mathbb{R},X)$ (since $\mathcal{F}\psi_{i_r} \in L^\infty(\mathbb{R})$ and $\mathcal{F}f_r \in \mathcal{S}(\mathbb{R},X)$). Therefore $\psi_{i_r} * f_r = J\,\mathcal{F}((\mathcal{F}\psi_{i_r})\,(\mathcal{F}f_r))$, and so the result follows at once from theorem 10. □

Corollary 11 has been used in [4].

Acknowledgment

I want to thank Shangquan Bu and Jan Prüss for pointing out to me a mistake in a previous version of this paper.

References

[1] J. Bourgain, *Vector-valued singular integrals and the H_1-BMO duality.* In: Probability Theory and Harmonic Analysis, Dekker, New York, 1986, 1–19.

[2] P. Clément, B. de Pagter, F.A. Sukochev, H. Witvliet, *Schauder decompositions and multiplier theorems.* Studia Math. **138** (2000), 135–163.

[3] J. Diestel, H. Jarchow, and A. Tonge, *Absolutely Summing Operators.* Cambridge Studies in Advanced Mathematics **43**, Cambridge University Press, Cambridge, 1995.

[4] G. Dore, and A. Venni, H^∞ *functional calculus for an elliptic operator on a half-space with general boundary conditions.* Preprint.

[5] L. Weis, *Operator-valued Fourier multiplier theorems and maximal* L^p*-regularity.* Preprint.

[6] F. Zimmermann, *On vector-valued Fourier multiplier theorems.* Studia Math. **93** (1989), 201–222.

Alberto Venni
Dipartimento di Matematica
Università di Bologna
Piazza di Porta S. Donato 5
I-40127 Bologna, Italy
E-mail address: venni@dm.unibo.it

Progress in Nonlinear Differential Equations
and Their Applications, Vol. 55, 415–423

Quasilinear P.D.E.s with Memory Operators

Augusto Visintin

1. Memory Operators

In this note we deal with some models of memory, and in particular of hysteresis. Let us consider a system characterized by two scalar variables, u and w, and assume that at any instant $t \in [0,T]$, $w(t)$ depends on the previous evolution of u (memory effect) and on the initial state, that is,

$$w(t) = \left[\mathcal{F}\left(u, w^0\right)\right](t) \qquad \forall t \in [0,T]. \tag{1.1}$$

For any fixed w^0, $\mathcal{F}\left(\cdot, w^0\right)$ represents an operator acting in some space of time-dependent functions. It is natural to assume that $\mathcal{F}$ is *causal:*

$$\begin{aligned}&\forall \left(u_1, w^0\right), \left(u_2, w^0\right) \in \mathrm{Dom}\,(\mathcal{F}), \forall t \in [0,T],\\ &\text{if } u_1 = u_2 \text{ in } [0,t], \text{then } \left[\mathcal{F}\left(u_1, w^0\right)\right](t) = \left[\mathcal{F}\left(u_2, w^0\right)\right](t).\end{aligned} \tag{1.2}$$

$\mathcal{F}$ is then named a *memory operator*, or an (abstract) *Volterra operator*. We do not exclude the degenerate case in which $\mathcal{F}$ is a superposition operator. In several cases the state (u,w) is confined to a set $\mathcal{L} \subset \mathbf{R}^2$; we then assume that

$$\text{if } \left(u(0), w^0\right) \in \mathcal{L}, \text{then } \left[\mathcal{F}\left(u, w^0\right)\right](0) = w^0. \tag{1.3}$$

Rate-Independence

An important class of memory operators is characterized by the property that the path of the pair $(u(t), w(t))$ is invariant w.r.t. any increasing homeomorphism $\varphi : [0,T] \to [0,T]$, that is,

$$\mathcal{F}\left(u \circ \varphi, w^0\right) = \mathcal{F}\left(u, w^0\right) \circ \varphi \qquad \text{in } [0,T]. \tag{1.4}$$

In other terms, if $\mathcal{F}\left(\cdot, w^0\right)$ maps u to w, then it maps $u\circ\varphi$ to $w\circ\varphi$. This means that at any instant t, $w(t)$ only depends on the range of the restriction $u([0,t])$, and on the order in which values have been attained. In particular, if the input function u is periodic, the $\vec{M}$ vs. $\vec{H}$ relation does not depend on the frequency. This property is regarded as characteristic of hysteresis, and accordingly rate-independent memory operators are named *hysteresis operators*. A typical hysteresis loop is outlined in Fig. 1; however, the above definition of hysteresis operator is independent from occurrence of such loops, and in the next section we shall see an example where no hysteresis loop appears.

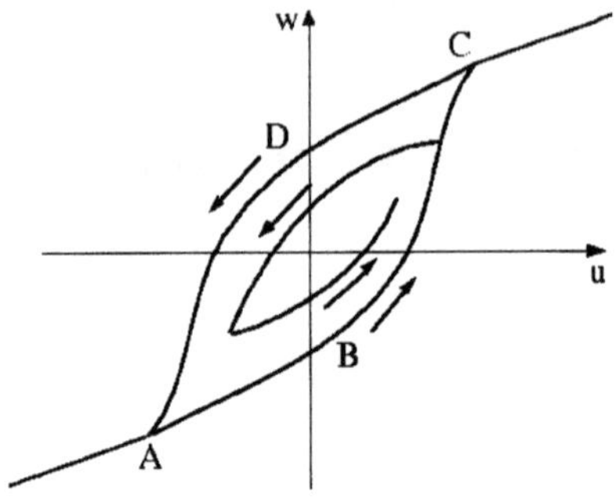

FIGURE 1

Rate-independence excludes any viscous-type memory, such as that represented by (nonlinear) *time-convolution operators* of the form

$$[\alpha(u) * k](t) := \int_0^t \alpha(u(t-\tau))\, k(\tau)\, d\tau \qquad \forall t \in [0,T], \tag{1.5}$$

for given functions k, α, the latter being a Caratheodory function. In reality, even in most typical hysteresis phenomena, memory is not purely rate-independent, as rate-dependent effects are superposed to hysteresis. In this note we then extend some results known for hysteresis operators to more general memory operators.

Dealing with rate-dependent components of memory, one can distinguish between long-time-scale and short-time-scale contributions. The former are typically represented by means of a time-convolution operator, as in (1.5). On the other hand, short-time memory is usually accounted for via a time derivative. There is a strict correlation between these representations; this can easily be grasped, considering that the solution of the equation $dw/dt + w = u$ can be written as a time-convolution.

Operators like (1.5) have been known and studied for a long time, whereas hysteresis operators have been systematically explored just in the last 30 years or so; see, e.g., [2],[4],[5],[6],[7]. In particular results have been derived for quasilinear parabolic equations of the form

$$\frac{\partial}{\partial t}\left[u + \mathcal{F}\left(u, w^0\right)\right] + Au = f, \tag{1.6}$$

and for hyperbolic equations like

$$\frac{\partial^2}{\partial t^2}\left[u + \mathcal{F}\left(u, w^0\right)\right] + Au = f, \qquad \frac{\partial}{\partial t}\left[u + \mathcal{F}\left(u, w^0\right)\right] + \frac{\partial u}{\partial x} = f. \tag{1.7}$$

Here $\mathcal{F}$ is a (possibly discontinuous) hysteresis operator, A is an elliptic operator, and f is a datum; the last of these equations is set in a one-dimensional domain.

In this note we deal with (1.6), for the case in which $\mathcal{F}$ is the sum of a hysteresis operator and a (nonlinear) time-convolution operator. In Sect. 2 we discuss monotonicity properties for hysteresis operators, and in Sect. 3 we prove an existence result. Finally, in Sect. 4 we discuss some other problems with hysteresis.

2. Monotonicity

Hysteresis and Monotonicity

Dealing with quasilinear equations with hysteresis, it is often useful to dispose of some monotonicity-type property for $\mathcal{F}$. If this is a hysteresis operator, the standard L^2-monotonicity property,

$$\int_0^T \left[\mathcal{F}\left(u_1, w^0\right) - \mathcal{F}\left(u_2, w^0\right)\right](u_1 - u_2)dt \geq 0 \qquad \forall\left(u_1, w^0\right), \left(u_2, w^0\right) \in \mathrm{Dom}\,(\mathcal{F}), \tag{2.1}$$

is fulfilled only if $\mathcal{F}$ is reduced to a superposition operator, that is, it has no memory. This is illustrated by the following counterample, which can easily be extended to any nondegenerate hysteresis operator. Let $\mathcal{F} : W^{1,1}(0,T) \times \mathbf{R} : \left(u, w^0\right) \mapsto w$ be defined by the following Cauchy problem

$$\frac{dw}{dt} = \left(\frac{du}{dt}\right)^+ \quad \text{in }]0,T[, \qquad w(0) = w^0.$$

This is a hysteresis operator: causality and rate-independence are straightforward. Let us take $u_1(t) := \sin t$ in $[0, \frac{3}{2}\pi]$, $u_1(t) := -1$ in $[\frac{3}{2}\pi, T]$ $(T > \frac{3}{2}\pi)$, $u_2 := 0$ in $[0,T]$. Setting $w_i := \mathcal{F}(u_i, 0)$ for $i = 1, 2$, we have $(w_1 - w_2)(u_1 - u_2) = -1$ for $t \geq \frac{3}{2}\pi$. Hence

$$\int_0^T (w_1 - w_2)(u_1 - u_2)dt < 0 \quad \text{if } T \text{ is large enough.}$$

Several hysteresis operators are *order preserving:*

$$\begin{aligned}&\forall(u_1, w_1^0), (u_2, w_2^0) \in \mathrm{Dom}\,(\mathcal{F}), \text{ if } u_1 \leq u_2 \text{ in } [0,T] \text{and } w_1^0 \leq w_2^0,\\ &\text{then } \mathcal{F}\left(u_1, w_1^0\right) \leq \mathcal{F}\left(u_2, w_2^0\right) \text{ in } [0,T].\end{aligned}$$

But it is the following property of *piecewise monotonicity* that seems especially appropriate for hysteresis operators in connection with quasilinear differential equations:

$$\begin{aligned}&\forall\left(u, w^0\right) \in \mathrm{Dom}\,(\mathcal{F}), \forall[t_1, t_2] \subset [0,T],\\ &\text{if } u \text{ is nondecreasing (nonincreasing, resp.) in } [t_1, t_2],\\ &\text{then } \mathcal{F}\left(u, w^0\right) \text{ is also nondecreasing (nonincreasing, resp.) in } [t_1, t_2].\end{aligned} \tag{2.2}$$

This means that *hysteresis branches* are nondecreasing, and entails that

$$\begin{aligned}&\forall\left(u, w^0\right) \in \mathrm{Dom}\,(\mathcal{F}) \text{ such that } u, w := \mathcal{F}\left(u, w^0\right) \in W^{1,1}(0,T),\\ &\tfrac{dw}{dt}\tfrac{du}{dt} \geq 0 \quad \text{a.e. in }]0,T[.\end{aligned} \tag{2.3}$$

Finally, we point out another monotonicity-type property, which holds if $\mathcal{F} := \mathcal{H}_\mu$ is a *Preisach operator* corresponding to a measure $\mu \geq 0$. (This is an especially important class of hysteresis operators, we are not going to review here; see, e.g.,

[7]; Chap. IV.) Setting $\mathrm{sign}_0(y) := -1$ if $y < 0$, $\mathrm{sign}_0(0) = 0$, $\mathrm{sign}_0(y) := 1$ if $y > 0$, in this case we have the *Hilpert inequality*

$$\left(\frac{d}{dt}[\mathcal{F}(u_1) - \mathcal{F}(u_2)]\right) \mathrm{sign}_0(u_1 - u_2) \geq \frac{d}{dt}|\mathcal{F}(u_1) - \mathcal{F}(u_2)| \quad \forall u_1, u_2 \in W^{1,1}(0,T). \tag{2.4}$$

3. P.D.E.s with Memory

In this section we deal with quasilinear parabolic P.D.E.s in which the nonlinear term is the sum of a hysteresis operator and a nonlinear time-convolution operator. Our developments can also be extended to the sum of a hysteresis operator and the time-convolution of another hysteresis operator.

At first we extend the definition of hysteresis operator to space-distributed systems. Let Ω be a domain of $\mathbf{R}^N$, and assume that

$$\mathcal{F} : C^0([0,T]) \times \mathbf{R} \to C^0([0,T]) \text{ is a hysteresis operator.} \tag{3.1}$$

We define the corresponding space-distributed operator $\tilde{\mathcal{F}}$ as follows:

$$\begin{aligned} &\forall u \in L^1\left(\Omega; C^0([0,T])\right), \forall w^0 \in L^1(\Omega), \forall t \in [0,T], \text{ for a.a. } x \in \Omega,\\ &\left[\tilde{\mathcal{F}}\left(u, w^0\right)\right](x,t) := \left[\mathcal{F}\left(u(x,\cdot), w^0(x)\right)\right](t). \end{aligned} \tag{3.2}$$

A Model Problem

Let us set $A := -\Delta \left(:= -\sum_{i=1}^N \frac{\partial^2}{\partial x_i^2}\right)$, fix three functions

$$f : Q := \Omega \times]0,T[\to \mathbf{R}, \quad \alpha : \mathbf{R} \to \mathbf{R}, \quad k :]0,T[\to \mathbf{R},$$

and couple the equation

$$\frac{\partial}{\partial t}[u + \mathcal{F}(u) + \alpha(u) * k] + Au = f \qquad \text{in } Q \tag{3.3}$$

with suitable initial- and boundary-conditions. We set $V := H_0^1(\Omega)$, and identify the space $L^2(\Omega)$ with its topological dual $L^2(\Omega)'$; in turn the latter can be identified with a subspace of V', as V is a dense subspace of $L^2(\Omega)$ with continuous injection. We denote by $\langle \cdot, \cdot \rangle$ the duality pairing between V' and V, and define the linear and continuous operator $A : V \to V'$ by

$$\langle Au, v \rangle := \int_\Omega \nabla u \cdot \nabla v \, dx \qquad \forall u, v \in V.$$

We also assume that (3.1) is fulfilled, and that

$$u^0, w^0 \in L^2(\Omega), \qquad f \in L^2(0,T;V'). \tag{3.4}$$

Problem 1. To find $u : \Omega \to C^0([0,T])$ measurable such that $u \in L^2(0,T;V)$ and, setting

$$\begin{aligned} &w(x,t) := \left[\mathcal{F}\left(u(x,\cdot), w^0(x)\right)\right](t), \qquad z(x,t) := [\alpha(u) * k](x,t) \ \ (cf.\ (1.5))\\ &\forall t \in [0,T], \text{ for a.a. } x \in \Omega, \end{aligned} \tag{3.5}$$

$w \in L^2(Q)$, $u + w + z \in H^1(0,T;V')$, and

$$\frac{\partial}{\partial t}(u + w + z) + Au = f \qquad \text{in } V', \text{ a.e. in }]0,T[, \tag{3.6}$$

$$(u + w)|_{t=0} = u^0 + w^0 \qquad \text{in } V'. \tag{3.7}$$

A more sound problem would be obtained by prescribing u in $\Omega\times]-\infty,0]$, and setting

$$z(x,t) := \int_{-\infty}^{t} \alpha(u(x,\tau))k(t-\tau)\,d\tau \qquad \forall t \in [0,T], \text{ for a.a. } x \in \Omega.$$

However, for the sake of simplicity, in this note we deal with the above setting, the extension being then fairly easy. Here u and w are not only regarded as functions of t taking values in a space of functions of x, but also as vector-valued functions of x ranging in a space of functions of t.

Theorem 3.1. *(Existence) Let Ω be a bounded domain of $\mathbf{R}^N$ of Lipschitz class, and $\mathcal{F} : C^0([0,T]) \times \mathbf{R} \to C^0([0,T])$ be causal, continuous, piecewise monotone (in the sense of (2.2)). Moreover, let (1.3) be fulfilled, $(u^0, w^0) \in \mathcal{L}$ a.e. in Ω, and assume that*

$$\begin{array}{l}\exists L, M, N \in \mathbf{R}^+ : \forall (v,\xi) \in C^0([0,T]) \times \mathbf{R}, \\ \sup_{[0,T]} |\mathcal{F}(v,\xi)| \le L \sup_{[0,T]} |v| + M|\xi| + N,\end{array} \tag{3.8}$$

$$k \in W^{1,1}(0,T), \quad \alpha : \mathbf{R} \to \mathbf{R} \text{ is of Lipschitz class, with Lipschitz constant } \hat{L}, \tag{3.9}$$

$$u^0 \in V, \quad w^0 \in L^2(\Omega), \quad f = f_1 + f_2, \quad f_1 \in L^2(Q), \quad f_2 \in W^{1,1}(0,T;V'). \tag{3.10}$$

Then Problem 1 has at least one solution such that

$$u \in H^1\left(0,T;L^2(\Omega)\right) \cap L^\infty(0,T;V), \quad w \in L^2\left(\Omega; C^0([0,T])\right). \tag{3.11}$$

Outline of the Proof. The argument follows the lines of Theorem 1.1 of [7]; Chap. IX: approximation by implicit time-discretization, derivation of a priori estimates, passage to the limit via compactness.

(i) *Approximation.* We fix any $m \in \mathbf{N}$, set $h := T/m$ and

$$\begin{cases} f_{1m}^n(x) := \frac{1}{k}\int_{(n-1)h}^{nk} f_1(x,t)dt \quad \text{a.e. in } \Omega, \quad f_{2m}^n := f_2(nh), \\ f_m^n := f_{1m}^n + f_{2m}^n, \quad u_m^0 := u^0, \quad w_m^0 := w^0, \quad \text{for } n = 1,\dots,m. \end{cases}$$

We approximate our problem by an implicit time-discretization scheme.

Problem 1_m. For $n = 1,\dots,m$, to find $u_m^n \in V$ such that, setting

$$u_m := \text{ linear time interpolate of } u_m^0,\dots,u_m^m, \text{ a.e. in } \Omega,$$

$$w_m^n(x) := \left[\mathcal{F}\left(u_m(x,\cdot), w^0(x)\right)\right](nh) \qquad z_m^n(x) := \textstyle\sum_{i=0}^n \alpha(u_m^i(x,\cdot))\, k((n-i)h)$$
$$\text{for } n = 1,\dots,m, \text{ for a.a. } x \in \Omega,$$

then

$$\frac{u_m^n - u_m^{n-1}}{h} + \frac{w_m^n - w_m^{n-1}}{h} + \frac{z_m^n - z_m^{n-1}}{h} + Au_m^n = f_m^n \qquad \text{in } V', \text{ for } n = 1,\dots,m. \tag{3.12}$$

Existence of an approximate solution can be proved step by step. For any n, at the nth step $u_m^0, \dots, u_m^{n-1}$ are known a.e. in Ω. By the causality of $\mathcal{F}$, w_m^n only depends on u_m^n; that is, $w_m^n(x) = g(u_m^n(x), x)$ for a.a. $x \in \Omega$, for a suitable Caratheodory function $g : \mathbf{R} \times \Omega \to \mathbf{R}$. (Loosely speaking, time-discretization and causality allow us to localize $\mathcal{F}$ in time.) Something similar occurs for the convoluted term.

By (2.2), for a.a. $x \in \Omega$ the function $g(\cdot, x)$ is nondecreasing. Moreover, in analogy with the continuous formula

$$\left\| \tfrac{d}{dt} \int_0^t \alpha(u(\tau)) k(t-\tau)\, d\tau \right\|_{L^2(0,T)} = \left\| \alpha(u(t)) k(0) + \int_0^t \alpha(u(\tau)) \tfrac{dk}{dt}(t-\tau)\, d\tau \right\|_{L^2(0,T)} \\ \leq \left(|u^0| + L\|u\|_{L^2(0,T)}\right) \left(k(0) + \|dk/dt\|_{L^1(0,T)}\right), \tag{3.13}$$

a direct calculation shows that

$$\left(h \sum_{j=1}^n \left| \tfrac{z_m^j - z_m^{j-1}}{h} \right|^2 \right)^{\frac{1}{2}} \leq \left(|u^0| + L \left(\sum_{j=0}^n |u_m^j|^2 \right)^{\frac{1}{2}} \right) \left(k(0) + \|dk/dt\|_{L^1(0,T)} \right) \\ \text{for } n = 1, \dots, m, \text{ for a.a. } x \in \Omega. \tag{3.14}$$

One can then prove existence of an approximate solution step by step.

(iii) *A Priori Estimates.* Let us multiply equation (3.12) by $u_m^n - u_m^{n-1}$, and sum for $n = 0, \dots, \ell$, for any $\ell \in \{1, \dots, m\}$. By (2.2),

$$\left(w_m^n - w_m^{n-1}\right) \left(u_m^n - u_m^{n-1}\right) \geq 0 \qquad \text{for } n = 1, \dots, m, \text{ a.e. in } \Omega.$$

By (3.14), a simple calculation then yields

$$\|u_m\|_{H^1(0,T;L^2(\Omega)) \cap L^\infty(0,T;V)} \leq C_2. \tag{3.15}$$

Hence u_m is uniformly bounded in $L^2(\Omega; C^0([0,T]))$, and by (3.9) we have

$$\|w_m\|_{L^2(\Omega; C^0([0,T]))} \leq C_3. \tag{3.16}$$

Setting $\bar{u}_m(\cdot, t) := u_m^n$ a.e. in Ω, if $(n-1)h < t \leq nh$, for $n = 1, \dots, m$, and defining $\bar{f}_m$ similarly, (3.12) also reads

$$\frac{\partial}{\partial t}(u_m + w_m + z_m) + A\bar{u}_m = \bar{f}_m \qquad \text{in } V', \text{ a.e. in }]0,T[. \tag{3.17}$$

(iv) *Passage to the Limit.* By the above estimates there exist u and w such that, possibly taking $m \to \infty$ along a subsequence,

$$\begin{aligned} u_m &\to u \quad \text{weakly star in } H^1\left(0,T; L^2(\Omega)\right) \cap L^\infty(0,T;V), \\ w_m &\to w \quad \text{weakly star in } L^2\left(\Omega; C^0([0,T])\right). \end{aligned} \tag{3.18}$$

As the injection $H^1\left(0,T;L^2(\Omega)\right) \cap L^\infty(0,T;V) \to L^2\left(\Omega; C^0([0,T])\right)$ is compact, by $(3.18)_1$, possibly extracting a further subsequence, we have

$$u_m(x,\cdot) \to u(x,\cdot) \qquad \text{uniformly in } [0,T], \text{ for a.a. } x \in \Omega. \tag{3.19}$$

whence

$$\alpha(u_m) \to \alpha(u) \quad \text{a.e. in } Q \text{ and strongly in } L^2(Q).$$

By taking $m \to \infty$ in (3.17), we then get (3.6). (3.7) is also easily derived.

By the continuity of $\mathcal{F}$, (3.19) entails

$$\mathcal{F}\left(u_m(x,\cdot), w^0(x)\right) \to \mathcal{F}\left(u(x,\cdot), w^0(x)\right) \qquad \text{uniformly in } [0,T], \text{for a.a. } x \in \Omega.$$

Hence, as $w_m(x,\cdot)$ is the linear time interpolate of $w_m(\cdot, nh) := \left[\mathcal{F}\left(u_m, w^0\right)\right](\cdot, nh)$ $(n = 0, \ldots, m)$ for a.e. in Ω, we have

$$w_m \to \mathcal{F}\left(u, w^0\right) \qquad \text{uniformly in } [0,T], \text{ a.e. in } \Omega.$$

By $(3.18)_2$ we then conclude that $w = \mathcal{F}\left(u, w^0\right)$ a.e. in Q. □

4. Further Results

Several regularity results can be proved for Problem 1. For instance, if $\mathcal{F}$ is linearly bounded in $H^1(0,T)$, then $w + z \in H^1\left(0,T; L^2(\Omega)\right)$, by (3.11) and (3.13). If moreover $f \in L^2(Q)$, then $\Delta u \in L^2(Q)$; hence $w \in L^2\left(0,T; H^2(\Omega)\right)$, because of the homogeneous boundary condition.

If $\mathcal{F}$ is a *discontinuous* hysteresis operator, Problem 1 becomes the weak formulation of a *free boundary problem.* Some modifications are then needed in its formulation, and the above existence result can be extended, if $\mathcal{F}$ is closed in suitable Sobolev spaces, see [7]; Chap. IX.

If either A is replaced by a nonsymmetric elliptic operator or the data are less regular, it may only be possible to derive a priori estimates by multiplying the approximate equation (3.12) by u_m^n, and then summing w.r.t. n. If $\mathcal{F}$ is a Preisach operator, existence of a solution can be proved in this case too, by extending formulation and techniques of [9].

Uniqueness of the solution of Problem 1 is an open question. However, if the convoluted term is dropped and $\mathcal{F}$ belongs to a large class of (either continuous or discontinuous) Preisach operators, the solution is unique and depends continuously on the data, see, e.g., [7]; Chap. IV. If $\mathcal{F}$ is a Preisach operator, that argument can be extended to Problem 1 (i.e., including the convoluted term), because of the following inequality. For any $u_1, u_2 \in L^1(0,T)$, a computation analogous to (3.13) yields

$$\begin{aligned} &\|[\alpha(u_1) - \alpha(u_2)] * k\|_{L^1(0,t)} \\ &\leq \left(|u_1^0 - u_2^0| + L\|u_1 - u_2\|_{L^1(0,t)}\right)\left(k(0) + \|dk/d\tau\|_{L^1(0,t)}\right) \qquad \forall t \in]0,T]. \end{aligned} \tag{4.1}$$

One can then also apply the semigroup theory to Problem 1, along the lines of [7]; Chap. VIII.

Some Other P.D.E.s with Memory Operators

(i) Let us assume that

$$\mathcal{G} : C^0([0,T]) \times \mathbf{R} \to C^0([0,T]) \text{ is also a hysteresis operator,} \tag{4.2}$$

and consider the equation

$$\frac{\partial}{\partial t}[u + \mathcal{F}(u) + \mathcal{G}(u)*k] + Au = f \qquad \text{in } Q \tag{4.3}$$

If $\mathcal{G}$ fulfils a growth condition like (3.8), the above existence result can be extended.

(ii) Existence of a solution can be proved for the semilinear hyperbolic equation

$$\frac{\partial^2 u}{\partial t^2} + \frac{\partial}{\partial t}[u + \mathcal{F}(u) + \alpha(u)*k] + Au = f \qquad \text{in } Q, \tag{4.4}$$

also if the function α is replaced by a hysteresis operator $\mathcal{G}$. However, it is the quasilinear hyperbolic equation

$$\frac{\partial^2}{\partial t^2}[u + \mathcal{H}(u)] + Au = f \qquad \text{in } Q, \tag{4.5}$$

$\mathcal{H}$ being a memory operator, that occurs more frequently in applications.

For $\mathcal{H}$ equal to a hysteresis operator, so far two special cases of the latter equation have been dealt with. If this operator can be represented by a variational inequality, well-posedness can be derived via standard techniques, cf. [7]; Chap. VII. If $\mathcal{H}$ is a *relay operator,* or more generally a *Preisach operator,* existence of a solution has recently been proved, by taking advantage of the dissipative character of these hysteresis operators, cf. [8]. The analysis of (4.5) for other classes of memory operators is an open question.

(iii) Existence results can be proved for the boundary- and initial-value problem associated with the equation

$$\frac{\partial u}{\partial t} + Au + \mathcal{H}(u) = f \qquad \text{in } Q, \tag{4.6}$$

without assuming any rate-independence or monotonicity condition on the memory operator $\mathcal{H}$.

If $\mathcal{H}$ is Lipschitz continuous in $C^0([0,T])$, uniqueness of the solution is easily derived via the contraction mapping principle, see [7]; Chap. X. If $\mathcal{H}$ is continuous in $C^0([0,T])$ but not Lipschitz continuous and not piecewise monotone, it is well-known that uniqueness may fail even if $\mathcal{H}$ is reduced to a superposition operator. Counterexamples of uniqueness can easily be constructed for $\mathcal{H}$ equal to a discontinuous and piecewise monotone hysteresis operator, see [7]; Chap. XI. If $\mathcal{H}$ is a piecewise monotone hysteresis operator, and is continuous in $C^0([0,T])$ but not Lipschitz continuous, uniqueness is nontrivial even for the corresponding O.D.E.. However, in [3] Chernorutskiĭ and Rachinskiĭ have been able to exhibit a hysteresis operator and a function g such that the Cauchy problem

$$\frac{du}{dt} + \mathcal{F}(u,0) = g(t) \quad \text{in }]0,T[, \qquad u(0) = 0 \tag{4.7}$$

has more than one solution.

(iv) In univariate systems the equation (1.6) can represent the Maxwell-Ohm equations without dispacement currents. In the corresponding vector setting, a vector extension of the Preisach model can be used, cf. [6],[7]. A suitable formulation of the vector relay operator allows then one to prove existence of a solution, cf. [9]. The same applies to the vector quasilinear hyperbolic equation representing the Maxwell equations with dispacement currents; this generalizes (4.5). Both problems are obviously relevant for applications.

Remark 4.1. *We dealt with sums of hysteresis operators and time-convolutions of either nonlinear functions or hysteresis operators. Although this is a rather wide class of memory operators, its applicative relevance is not yet clear. More specifically, this author disposes of no precise evidence that rate-dependent perturbations of hysteresis occur in that form. An exception to this is the long-time behaviour of ferromagnetism, as it is represented in the framework of the Preisach model, cf.* [1].

Acknowledgment

This work was partly supported by the project "Free boundary problems" of Italian M.U.R.S.T.

References

[1] G. Bertotti, *Hysteresis in Magnetism.* Academic Press, Boston, 1998.

[2] M. Brokate, and J. Sprekels, *Hysteresis and phase transitions.* Springer, Berlin, 1996.

[3] V. Chernorutskiĭ, and D. Rachinskiĭ, *On uniqueness of an initial-value problem for ODE with hysteresis.* NoDEA **4** (1997), 391–399.

[4] M.A. Krasnosel'skiĭ, and A.V. Pokrovskiĭ, *Systems with hysteresis.* Springer, Berlin, 1989. (Russian ed. Nauka, Moscow, 1983).

[5] P. Krejčí, *Convexity, hysteresis and dissipation in hyperbolic equations.* Gakkotosho, Tokyo, 1997.

[6] I.D. Mayergoyz, *Mathematical models of hysteresis.* Springer, New York, 1991.

[7] A. Visintin, *Differential models of hysteresis.* Springer, Berlin, 1994.

[8] A. Visintin, *Hyperbolic equations with hysteresis.* C. R. Acad. Sc. Paris **332** (2001), Série I, 315–320.

[9] A. Visintin, *Forward–backward parabolic equations and hysteresis.* Calculus of Variations, (in press).

[10] A. Visintin, *Vector Preisach model and Maxwell's equations.* Physica B, **306** (2001), 21–25.

Augusto Visintin
Dipartimento di Matematica
Università di Trento
I-38050 Povo (Trento), Italy
E-mail address: visintin@science.unitn.it